Pathways to the Origin and Evolution of Meanings in the Universe

Scrivener Publishing
100 Cummings Center, Suite 541J
Beverly, MA 01915-6106

Astrobiology Perspectives on Life in the Universe

Series Editor: Richard Gordon and Joseph Seckbach

Scope: In his 1687 book *Principia,* Isaac Newton showed how a body launched atop a tall mountain parallel to the ground would circle the Earth. We are both old enough to have witnessed the realization of this dream in the launch of Sputnik in 1957. Since then our ability to enter, view and understand the Universe has increased dramatically. A great race is on to discover real extraterrestrial life, and to understand our origins, whether on Earth or elsewhere. We take part of the title for this new series of books from the pioneering thoughts of Svante Arrhenius, who reviewed this quest in his 1909 book *The Life of the Universe as Conceived by Man from the Earliest Ages to the Present Time.*

"*Astrobiology Perspectives on Life in the Universe*" volumes will each delve into an aspect of this adventure, with chapters by those who are involved in it, and careful observers and assessors of our progress.

Publishers at Scrivener
Martin Scrivener (martin@scrivenerpublishing.com)
Phillip Carmical (pcarmical@scrivenerpublishing.com)

Pathways to the Origin and Evolution of Meanings in the Universe

Edited by
Alexei A. Sharov
Elixirgen Scientific, Baltimore, Maryland, USA
and
George E. Mikhailovsky
Global Mind Share, Norfolk, Virginia, USA

Scrivener Publishing

WILEY

This edition first published 2024 by John Wiley & Sons, Inc., 111 River Street, Hoboken, NJ 07030, USA and Scrivener Publishing LLC, 100 Cummings Center, Suite 541J, Beverly, MA 01915, USA

For more information about Scrivener publications please visit www.scrivenerpublishing.com.

Wiley Global Headquarters
111 River Street, Hoboken, NJ 07030, USA

For details of our global editorial offices, customer services, and more information about Wiley products visit us at www.wiley.com.

Library of Congress Cataloging-in-Publication Data

ISBN 978-1-119-86509-4

Cover image: The Editors
Cover design by Russell Richardson

Set in size of 11pt and Minion Pro by Manila Typesetting Company, Makati, Philippines

Printed in the USA

10 9 8 7 6 5 4 3 2 1

This book is dedicated to the memory of the founders of biosemiotics, including Jakob von Uexküll, Thomas Sebeok, Gregory Bateson, and Jesper Hoffmeyer. We gratefully remember Alexander Levich, the founder of the Theoretical Biology research group at Moscow State University, where our scientific interests have been formed.

Contents

Preface

A century ago, in the 1920s, Jakob von Uexküll discovered a new and very unusual aspect of nature, which later took shape in a new scientific discipline—biosemiotics. He demonstrated that signs and meanings exist not only in human minds and texts, but also in various living organisms that arose and have existed long before the first humans appeared. In the 1960s, Howard Pattee, Allen Newell and Herbert Simon independently rediscovered the existence of meaning in the activity of living organisms as well as in artificial intelligence agents. At the same time, the value of Uexküll's ideas had been acknowledged by the English-speaking scientific community, including by Thomas Sebeok and Gregory Bateson. Furthermore, over the last half of the century, biosemiotics has been developing intensively, and achieving interesting and significant results. However, it still has a long way to go, and this book was conceived as the next important step on this path.

The chapter authors of this book consider the origin and evolution of signs and meanings in various natural and artificial systems from different angles. Meaning is often defined as a *signified* associated in language with a signifier in the sense of Ferdinand de Saussure, or *semantics* in the sense of Charles Morris. In communication theory, meaning is the content and value of messages or information. In biosemiotics, signs and meanings are considered in a much wider sense than in linguistics, and include non-mental signification. A biosemiotic sign is a tool that thanks to its form rather than matter or energy prompts an agent (e.g., organism, cell, or protein) to perform some *meaningful* activity or change, which tends to be beneficial to the agent itself and/or associated agents (progeny, superagents). Such activity or change in response to a sign is equivalent to what biologists call *adaptation*. For example, cold temperature makes the skin of mammals grow a thicker fur, which protects an animal from cold. Besides semantics (the type of activity or change), adaptation includes a utility component, which belongs to *pragmatics* in the sense of Charles Morris. Semantics and pragmatics are two necessary and interdependent components of meanings, although in some cases we focus on either semantics or pragmatics separately. Note that syntax, one of the three components of semiotics according to Morris, is not a part of meaning, although syntax of a message (or other sign) affects its meaning.

All semiotic agents can perform meaningful actions by interpreting signs but not all of them can acquire new meanings (e.g., a ribosome). Meaning is not a material component of an agent, but rather a dynamic organization or form that supports a capacity (or affordance) of an agent to act adaptively and robustly in changing environments. Meaning can be encoded, for example, in a computer algorithm or in a text and get transferred in such a frozen state to other agents.

Biosemiotics assumes that meanings exist only in living systems, but to explain the origin of life equipped with some primordial meanings it is necessary to consider precursor conditions in non-living systems that can be interpreted as *potential meanings*. For example, Howard Pattee envisioned that life originated from "primitive geochemical matter cycles," where "it is the constraints of the primeval ecosystem which, in effect, generate the language in which the first specific messages can make evolutionary sense."[1] In fact, many causal relationships can be interpreted as potential meanings. For example, the birth of a star is caused by the ignition of a thermonuclear reaction, and thus, the appearance of a star means that a thermonuclear reaction began there. The cause of the event of a new star's appearance is simultaneously its meaning in relation to a hypothetical observer (e.g., a human astrophysicist). Such meanings exist in potential (i.e., unmanifested) form until competent agents appear who recognize their utility.

John Deely[2] discussed a similar example, where a fossilized bone underground is a sign of a dinosaur even if it is not observed. This sign, however, does not signify, and thus, to avoid confusion, we call it a "potential sign." Living organisms can get additional affordances by converting potential signs into actual form and integrating new meanings into their Umwelten (subjective worlds). Analysis of potential signs allows us to explore adjacent possible meanings that have not been actualized yet. It does not imply pansemiosis or panpsychism because potential signs do not signify unless they are sensed and interpreted by semiotic agents.

The emergence of living organisms and their spread in the Universe led to the transformation of some potential meanings into a manifested form, in other words, to their actualization. However, the actualization of potential meanings occurs only in certain kinds of organisms that can utilize these meanings. For example, the actual position of the sun in the sky does not matter to many plants and animals,[3] but it makes sense for a sunflower that turns its inflorescence to follow it. In this way, the sun plays the role of a sign for a sunflower, which interprets it as a spatial direction to follow. Expected benefits from turning to the sun include attracting pollinators (bees and flies), evaporating water on seeds which prevents fungal infection, and better photosynthesis. The more complex an organism is, the more actualized meanings it has, and the more complex connections established between these meanings become. The most highly organized animals, and especially humans, have an extremely complex organization of signs and meanings. Their appearance as a result of evolution led to the emergence of more and more complex systems of reflexes and, finally, to consciousness characterized by active learning and self-awareness.

Currently, there is no consensus on the relation between sign properties and meanings. Many molecular biologists believe that heritable features of organisms are meanings encoded in the sequence of nucleic acids in living cells. Similarly, the structuralist school in linguistics assumes that meanings are embedded in the relational structure of human language. The opposite view of the interpretational school of semiotics assumes that meanings emerge via interpretation processes, and in this respect, they do not necessarily depend on the properties of sign vehicles (i.e., material carriers of signs). The extreme version of this

---

1 Pattee, H.H. (1969), How does a molecule become a message? *Developmental Biology Supplement* 3, 1-16.

2 Deely, J. (1992), Semiotics and biosemiotics: Are sign-science and life-science coextensive? In: *Biosemiotics. The semiotic web 1991*. T.A. Sebeok and J. Umiker-Sebeok, (eds.) Mouton de Gruyter, New York: 45-75.

3 However, the presence of sun is essential for all photosynthetic plants and most animals with diurnal activity, and the length of the day (photoperiod) is an important cue that is used by many animals and plants to regulate metabolism, physiology, development, and reproduction.

view is unlimited relativism, where anything can be a sign of anything else for some agent. These two approaches are often combined, as in the theory of Peirce, who considered the meaning of symbols as conventional in contrast to the meaning of icons and indexes that depend on the features or correlations of sign vehicles.

We approach the dualism of structural vs. interpretational meaning from the position of the theory of semiotic agency. Semiotic agents are organisms, cells, autonomous cell components, families, colonies, species, ecological consortia, human organizations, nations, and autonomous human artifacts. Agents perform their functions via semiotic processes, such as sensing and interpretation of signs, as actions or representations, production and manipulation of signs, and/or communication (exchange of signs between interacting agents and subagents). The structural approach to meanings is most relevant for production and manipulation of signs by agents because it results in a construction of sign vehicles that somehow represent pre-existing meanings for certain agents, whereas the interpretational approach is most relevant for categorization and learning, where meanings are inferred from sign vehicles by agents in the context of their memory and the environment. Obviously, communication requires both approaches: structural and interpretational.

An important aspect of semiosis is semiogenesis, which is the emergence of new meanings. It can be either spontaneous or triggered by external or internal disturbance (e.g., mutations). New meanings are established as a result of the repeated interpretational activities (including misinterpretation) of agents in their adaptive evolution, development, and behavior. Mutations are not primary causes of evolutionary novelties, but some mutations are interpreted creatively by cells and/or their subagents based on their innate competence acquired in preceding evolution. Some adaptations emerge due to physiological and behavioral plasticity and are then stabilized by genetic accommodation. In cognitive animals and humans, new meanings emerge mostly by learning of new behaviors, concepts, and models of the world.

This book will be of interest not only to biosemiotcians, but also to specialists in the evolution of signs and matter, philosophers studying the dialectics of potential and actual, as well as to anyone interested in the role of the observer and agency in biology and other natural sciences (including physics) and the origin and evolution of the Universe as a whole.

Editors: Alexei A. Sharov and George E. Mikhailovsky
December 2023

Acknowledgments

We thank all the authors for their collaboration and submitting wonderful manuscripts, and the reviewers who worked hard to evaluate contributions and provide important suggestions for improvement. Also, we want to thank participants of the biosemiotics research community, the International Society for Biosemiotic Studies, and in particular, Kalevi Kull, Donald Favareau, Morten Tønnessen, Marcello Barbieri, and Terrence Deacon. Many thanks to book series editors, Richard Gordon and Joseph Seckbach, and Wiley-Scrivener for publishing the book.

Part I

THE NATURE OF MEANING AND ITS COMPONENTS

1

Introduction: Towards Integrating Studies of Meanings with Science

Alexei A. Sharov

Elixirgen Scientific, Baltimore, Maryland, USA

Online resources:
https://scholar.google.com/citations?user=RcOv5TMAAAAJ&hl=en
https://www.researchgate.net/profile/Alexei-Sharov
http://alexei.nfshost.com

1.1 Crossing the Great Paradigm Divide

Human knowledge is deeply divided into natural sciences that pursue mechanisms and their modeling, and the humanities that are focused on meaning, value, communication, and interpretation. This book is conceived as a dialogue between interdisciplinary scholars on how to cross this *great paradigm divide* and contribute to integration of natural sciences with semiotics, a theory of meaning-making and signification, also known as *semiosis*. Developments in biology clearly indicate that semiosis is not limited to humans or vertebrate animals, but exists in all living organisms [1.3] [1.38], and this fact inspired integration of biology and semiotics into biosemiotics. In order to function and survive, all organisms strive to capture the meaning of their environments and their own activities. According to the theory of meaning proposed by Jakob von Uexküll [1.80], animals develop species-specific models of their environment (umwelten) that link environmental cues with living functions. An elementary unit of such a model is a functional circle that includes sensing, perception, and action, all integrated into a self-sustaining loop [1.78] [1.79]. This introduction chapter overviews other chapters of the book in light of major concepts of biosemiotics.

Terminology is one of the challenges in studying meanings because biosemiotics has to overcome the anthropocentric nature of all terms related to signs, meanings, agency, mind, consciousness, communication, learning, and knowledge. There is no easy solution because if the meaning of terms is expanded to all living organisms, then the differences between types of semiosis in various taxonomic groups can be lost. An alternative strategy of using parallel sets of terms for each taxonomic group is also problematic because

Email: Sharov@comcast.net

such terminology would be awkward for broad evolutionary studies. Another problem with terms is related to the difference between the first-person experience (i.e., in object agent) and third-person experience (i.e., in meta-agent) [1.76]. For example, a human researcher (meta-agent), considers glucose molecules as signs that can be perceived by bacteria searching for metabolic resources. But a bacterium (object agent) does not "know" that glucose exists outside; it simply responds to incoming signals from chemoreceptors and changes its activity according to its inherited program [1.70]. Biosemiotics certainly needs a set of generic terms applicable to all organisms and its subagents, and it can be based only on meta-agent understanding. Such a set includes "sign," "interpretation," "semiosis," and "agency" [1.38]. The range of application of other terms, for example, "meaning," "mind," "consciousness," and "learning," is still debated and there is no consensus.

Donald Favareau and Kalevi Kull [1.21] present a historical overview of how signs and meanings are conceptualized in the natural sciences. Ancient studies of signs mostly considered conditions when evident facts can be used to predict the non-evident, and then used to guide human actions to change the outcome (e.g., cure a patient or succeed in other goal-directed activity). One of the early achievements was recognition of deep contextuality of any predictions by Sextus Empiricus (2-3rd century), a supporter of the school of Pyrrhonial skepticism. He assumed that objects change with time and in response to various factors; and human impressions of objects depend on position, activity, and beliefs. Hence, judgments need to be suspended until the validity of induction is carefully verified. Notably, Sextus disagreed with the notion that knowledge is not achievable, which was a typical belief (according to his writings) in the school of Academic skepticism of that time.

The great paradigm divide emerged during the age of Enlightenment (17–18th centuries), when the whole domain of science was restructured to eliminate any cognitive elements (*res cogitans*) and focus on intrinsic properties of things (*res extensa*), which were assumed fully objective and verifiable. This worldview was typical for physics and chemistry, but it also influenced the development of biology and supported mechanistic mentality and the heuristic of reductionism. However, the need for the theory of signs and meanings reemerged in the contemporary science and resulted in the development of semiotics theories that were initially associated with the humanities (i.e., philosophy, linguistics, and psychology), but recently spread to biology, cybernetics, and other sciences. Favareau and Kull [1.21] describe in detail the general semiotic theory and its further development in biosemiotics, which has advanced in formulating such terms as sign, interpretation, semiotic agency, umwelt, semiosphere, scaffolding, and habit. In contrast to the mechanistic metaphor of computation, biosemiotics considers *semiotic modeling* which is embodied, anticipatory, and supported by a hierarchy of meanings. Favareau and Kull pay special attention to the problem of *semiotic realism* presented as a middle path between radical skepticism and naïve realism. They follow Peirce's notions of pragmatic maxim and final interpretant "toward which *the actual* tends" [1.54], CP 5.312, as guiding principles. In conclusion, "it is thus through signs that we are *more fully joined* to the world" [1.21] (italics original).

Bashir Ahmad and Richard Gordon [1.1] show that the old reductionist heuristic in science, stating that higher-level phenomena can be reduced to the processes occurring at lower levels, fails as physical reality is studied at progressively smaller scales and "the burden of proof on smaller entities became ever larger." The problem is not limited to the immense combinatorial complexity of combinations at the lower level, but in addition, higher-level emergent phenomena actually change what happens at a lower level. Examples of top-down causation are most numerous in living systems, where catalytic biomolecules enable chemical reactions and quantum effects that never occur in the non-living nature. Ahmad and Gordon present numerous examples of such non-reductionist effects in physics that include the second law of thermodynamics (increasing of entropy), rounding of liquid drops, percolation, gravity, and superconductivity. Following Arthur Koestler [1.36], they argue that each hierarchical level has its own rules of operation and these rules have bi-directional effects on both lower and higher levels; thus, physics is not reductionist and the world needs to be explained by Janus-faced[1] hierarchies.

The theories of information [1.62] and cybernetics [1.83] were thought to provide a link between natural sciences and human communication, but the term "information" is interpreted differently on each side of the paradigm divide. Terrence Deacon [1.18], p. 71, describes the difference:

> At one extreme, abstraction of the formal concept of information from its referential and normative properties in everyday use and in the many areas of natural science—such as biology and cognitive neuroscience, where these nonquantifiable properties are relevant—has led to generations of efforts to formally incorporate these properties, but without widespread acceptance. At the other extreme, semiotic theories, which take issues of reference and interpretation as their focus, have largely remained confined to discussions in the humanities and social sciences, and efforts to formalize these theories have also gained limited widespread acceptance.

Deacon further explains that physicists and engineers are interested in the transmission of information no matter what it means for the end user. In contrast, biology and psychology analyze information in the context of function and reference, and linguists are interested in symbolic relations between words and their meanings. To integrate these aspects of information, Deacon turns to the semiotic theory of Charles Morris, who distinguished syntax, semantics and pragmatics as three aspects of semiosis. Deacon generalizes these aspects to make them applicable beyond human language and describes them as "*medium or signal properties* (with syntax as a special case), *referential properties* (with meaning as a special case), and *normative or functional properties* (with pragmatics as a special case)" [1.18], p. 75. In simple cases (e.g., in icons and indexes), interpretation is mediated by medium/signal properties and yields referential properties. But symbols require additional conventional and socially shared rules that integrate all three levels of semiosis.

Deacon addresses the problem of semiotic grounding, which "is the property of 'aboutness,' the non-intrinsicality of reference" [1.18], p. 77. He argues that "[s]ign vehicle properties don't therefore determine reference, instead they serve as *affordances* that aid the interpretation process." The two major classes of such affordances are: "(i) shared formal properties between sign vehicle and object (iconicity), and (ii) some physical-temporal correlation or contiguity between sign vehicle and the object of reference (indexicality)." In contrast, symbolic reference is not linked on the features of a sign vehicle but instead relies on "a kind of iconicity between different interpretive habits in a community of symbol users" (ibid., p. 79).

To develop a unified approach to the studies of physical, biological and social systems, Anton Sukhoverkhov and Arran Gare [1.73] delineate three domains of information

[1] Janus is a two-faced Roman god, associated with beginnings and entrance-exit opposition.

(or "memory") in the universe: physical, biological, and social.[2] Following Nöth and Deely, they consider physical "virtual signs" that carry information about past events no matter if they are observed or not. These signs constitute *physical memory* that can be used to reconstruct the past or make prediction of the future. At the second level, *biological memory* connects physical reality with internally represented reality, such as objects or mental associations. Finally, the third level is represented by *social memory* that stores collective experience in symbols, myths, inferences (e.g., causation models), and social activity. Social memory and language supports modeling functions such as production of resources and dwelling structures. This conceptual framework is then applied to describe the origin and evolution of human language.

A similar monistic approach to meaning is proposed by George Mikhailovsky [1.46], who considers a hierarchy of meanings that integrates the physical world, living organisms, cognition, and the noosphere of human social systems. In the physical world, meanings take the form of symptoms (meaning-in-itself) that potentially can be interpreted by organisms. In organisms, meanings represent intentions that connect perception with action; and in human language, meanings are expressed in words and other symbols. The qualitative changes of systems occur via adding new levels of hierarchy, a process called *hierarchogenesis*. Mikhailovsky describes 15 hierarchogenic steps in the evolution of the world, which are further grouped into cosmic (six steps from quark-gluon plasma to galaxies) and substantive (nine steps from heteroatomic molecules to noosphere) branches. The latter branch includes the origin of life, prokaryotic cells, eukaryotic cells, multicellular organisms, agroecosystems, and nations/states. Qualitative steps in the evolution of meanings includes actualization of potential meanings, the rise of judgments, paradigms, and worldviews. These two components of evolution (substance and semantics) show a tendency of conversion, especially since the end of the last century.

Mikhailovsky then introduces the new term "eventity"— a shorthand for "evolving entity" which can be represented by a sequence of events constituting its evolution.[3] Meanings exist only in eventities, since any meaning presupposes conditioning either by a preceding (causality) or a consequent (purposiveness) event. Abiotic eventities, such as stars, contain only potential meanings, while living eventities (agents) can actualize the meanings of other eventities as well as their own. Mikhailovsky identifies a nine-level structural semantic hierarchy of living eventities consisting of three triplets ("word" – "text" – "phrase") for several hierarchical steps of general evolution. In addition, he delineates the behavioral semantic hierarchy of living beings spanning from elementary meanings through judgments and paradigms to complex worldviews. The first two belong to biosemiotics, and the last two are unique to humans. In this way, both types of semantic hierarchy are intertwined with several later steps of general hierarchogenesis.

In contrast to Sukhoverkhov, Gare, and Mikhailovsky, who assume that meaning or information exists even in the physical world without life, Alexei Sharov [1.68] takes a biosemiotic approach to the ontology of meanings, according to which meanings exist only due to their relation with interpreters: living organisms and life-dependent semiotic agents. The latter category of agents includes ribosomes and functional proteins in living cells, and

---

[2] The last two domains are represented by life and human civilization on Earth and by hypothetical biospheres and civilizations on exoplanets.

[3] A similar term is "occurrent," which is used by Caetano-Anollés [1.13].

autonomous human artifacts (computers and robots). However, according to Sharov, the relation between meanings and interpreters can be actual or potential, and thus, *actual meanings* are those that are currently interpreted by semiotic agents, whereas *potential meanings* are those that are interpretable (i.e., knowable) by some agents, which are currently either not present or not active. In physics, potentiality is described as a field that can be measured at any point in space by an appropriate device. Sharov proposes that potential meaning can be similarly presented as a *semiotic field* measurable by some competent semiotic agent if it is placed near the meaning carrier. Natural resources are potential meanings that existed long before the origin of life, but they are recognized as "potential meanings" only retrospectively, after the emergence of organisms capable of sensing and using these resources for their own benefit. Sharov wrote:

> Such ontology of meanings is agency-dependent, similar to radical constructivism [1.77], but in addition, meanings are pragmatic and become corrected through repetition of the corresponding functional cycle. It is initially subjective but can spread via interaction and communication and become objective within the community of agents. Such collective knowledge is updated with the emergence of each new kind of agency: first, actual meanings are updated in space and time proximity of these new agents, and second, potential meanings are updated in the whole universe, as well as in the reconstructed past, and in the projected future [1.68], p. 150.

Alexander Kravchenko [1.37] develops the ontology of meanings starting from principles of radical constructivism, which "highlights the crucial role of the observer in constructing a universe as 'objectivity in parentheses.'" However, the theory of constructivism is traditionally applied to humans and therefore it is inherently anthropocentric. To overcome this limitation, Kravchenko extends this approach to non-human organisms, which, according to biosemiotics, are capable of perception and purposeful behavior. Also, in contrast to radical constructivism which treats meanings as artificial constructs, Kravchenko considers meaning as a "value-based relationship between the organism and its world of experience," following Zlatev [1.85]. In other words, meaning integrates the observer and the observed in the process of normative-controlled iterative interaction. In the case of humans, these relationships are encoded in language, and thus, meanings cannot be fully separated from the historically developed word use:

> Much, if not all, of what we think we, as observers, know about the world, is the result of our operations of distinction we make in language, or naming – specification of an entity by operationally cleaving it from a background. Once such a distinction is made, a thing with the properties that the operation of distinction specifies is added to the enlanguaged world. [...] and there are as many realities as there are observers [1.37], p. 171.

Kravchenko notes that the actual and potential are ontologically different, although in language we claim both of them as "existing." In this respect, "'past' and 'future' are mental constructs that *belong to the present* just as any other act of thinking" [1.37] (italics mine). However, in contrast to biosemiotics tradition, he rejects using the notion of "sign" from the position of a third-person observer. As a result, he proposes to apply this notion only if an organism can categorize a thing as a "sign." In particular, Kravchenko disagrees with Maritain [1.44], p. 53, that all animals "make use of signs without perceiving the relation of signification."

1.2 What is Meaning?

Meanings or ideas are generally considered as *components of mind* that comprise human understanding and knowledge [1.39], Book II, chapters I, XII. These components are brought to our attention by sensing, perception, reflection, contemplation, and by interpreting words and other language units. In linguistics, meaning appears as a *signified* associated with a *signifier*, which is a sound pattern of a word [1.17]. Gottlob Frege noticed that meanings, besides being mental components, also refer to objects in the real world. Thus, he distinguished two aspects of meaning: *reference* and *sense* [1.23], a dichotomy that is similar to opposing of *object* and *concept*. Then, Charles Peirce proposed a triadic sign relation: "I define a sign as anything which is so determined by something else, called its Object, and so determines an effect upon a person, which effect I call its interpretant, that the later is thereby mediately determined by the former" [1.55], EP2.478. One contemporary understanding of Peirce's sign is that object is akin to a signified or reference, and interpretant is a relation between the sign vehicle and its object [1.4]. In other words, the object and interpretant are two aspects of meaning communicated by the sign. Peirce assumed that sign relations are grounded in transcendental logic, and thus, they can be non-human and even non-mental. Hence, he applied it to habits of animals and plants (e.g., sunflower [1.54], CP 2.274).

But if meaning is not necessarily mental, then what is it? The authors of this book propose different answers. Favareau and Kull [1.21], p. 28, assume "that 'meaning' is always and everywhere, a function of sign relationships," understood as Peirce's triadic relations and functional circles of von Uexküll [1.80]. In other words, meaning integrates a knower (semiotic agent) with something known. Mikhailovsky considers meaning as a "component of the world that takes the form of indicator or symptom in the non-living world, intention of organism action, or something expressed in words or other symbols [in human language]" [1.46], p. 101.

According to Sharov [1.68], p. 137, "meaningful things, processes, or relations are those that have significance for organisms as indicated by either behavioral choice or regulation of innate habits and processes." Some meanings are intrinsic to non-living things (e.g., gravity, chemical valence, or catalysis); they exist in a potential form until discovered by agents, and then turn into actual meanings, as they gain agent-related normativity. Other meanings are products of evolution or learning, which are established by agents for their specific purposes. Neo-Darwinism reduces significance to a single variable—fitness (a combination of survival and reproduction rates). However, Caetano-Anollés [1.13] discusses additional dimensions of significance: economy (e.g., efficiency of metabolic reactions), mechanisms that support flexibility of function and behavior (umwelt), and robustness (e.g., preparedness for unperceived and unknown). According to Sharov [1.68], meanings outlive their carrier-organisms because due to semiosis, meanings are transferred among agents and overcome the limitations of individual matter-energy systems: "meanings once actualized keep existing and evolving."

Abir Igamberdiev [1.32], p. 266, connects meaning with final causation: "Meaning in the action of agent refers to the Aristotelian final cause, which, in his views, retrocausally directs the development of the system shaped by the material, the efficient, and the formal causes." He applies this notion to three levels of agency organization: autopoietic closure, evolution, and social systems. The normative nature of meaning is also recognized by Kobus Marais

[1.43], p. 6: "[...] significance is related to the interest of the organism and, therefore, always normative, and [...] this significance can grow in complexity, from a choice to eat or being eaten, to a large and complicated cognitive system like 'the law.'"

Alexander [1.2], p. 327, explores meaning in the context of subjective-objective duality:

> I start with the assumption that the concept of *meaning* implies a subjective perspective of a living system. Because binary logic may suggest to some that *subjective* meaning is the opposite of *objective*, I counter with the argument that subjectivity is better understood as *relative objectivity*. Subjectivity arises with the formation of internal biosemiosic relations [...] There are different levels of subjectivity; inherited or habitual and newly formed. Subjectivity is essentially selfhood, and subjective interpretations about the qualities of sign-vehicles are generally made for the sake of, or rather with the effect of, maintaining the self." (italics original)

Alexander considers three aspects of sign: *sign vehicle*, the *response*, and the *object*, which resemble but are not identical to Peirce's triadic relation. In particular, the object is defined as "the effect that the response has on the living system's ability to respond to the sign vehicle in that way. [...] the ultimate object of any semiosic process is reinforcing the conditions that allow that sign reading to recur." She further argues that biosemiosic meaning is dynamic and appears in two modes: *directionality* and *originality*. The former preserves the core aspects of the meaning despite ever-changing conditions or context, and the latter adds new components of meaning.

Marais [1.42] presents meaning as a process that includes two complementary aspects: interpretation and translation, which are inseparable as particle and wave aspects of light. He explains the difference between these two: "translation refers to a constraint-centered observation of semiosis while interpretation refers to an interpreter-centered observation." The importance of translation in biology resonates with the theory of organic codes, where meaning appears as a result of translation from one world of information units (e.g., triplets of nucleotides in mRNA) to another world (e.g., amino acids in peptides) [1.6]. Furthermore, Marais connects his understanding of translation with the theory of semiosphere developed by Yuri Lotman. He proposes that the evolution of meaning is powered by alternative translation (cf., Alexander's notion of misinterpretation [1.2]), where new interpretants provide additional degrees of freedom of functional responses of agents. Because new translations appear possible due to additional constraints, the evolution of meanings follows a paradox, where "an increase in freedom is only to be achieved through an increase in constraint" [1.42], p. 416.

In his chapter, Robert Prinz [1.56] reviews the relationship between meanings, agents, and codes in biosemiotics and in a more mechanistic theory of *organic codes* (also known as *code biology*), developed mostly by Marcello Barbieri [1.6] [1.7] [1.8] [1.9] [1.10]. The main idea of the theory of organic codes is that

> "Biological codes in this framework are arbitrary connection rules established in evolution through an adaptor that mediates between two other molecules of different type and cannot be explained based solely on physical or chemical laws. All codes in the narrow (Barbierian) sense thus consist of a set of adaptors that map a set of components from one independent world to a set of components from another independent world." [1.56], p. 245.

The notion of organic code provides a useful terminology for describing molecular-level translation processes in living systems such as protein synthesis, protein transport and degradation, mRNA splicing, DNA repair, signal transduction from membrane receptors to the nucleus, and many other molecular processes in cells [1.56]. These coding processes are mediated by adapters (e.g., aminoacyl tRNA synthetase in the genetic code) that specifically bind to coding units (e.g., triplets of nucleotides) and provide links to corresponding molecules of another type (e.g., amino acids). Translation is physically executed by a molecular agent, which Barbieri calls *codemaker,*[4] such as ribosome. Prinz provides an overview of various codes in living systems, including organic codes (at the molecular level) and neural codes (at the level of neurons and neural networks). These types of codes are further divided into structural and regulatory. Prinz argues that Barbieri erroneously ascribes meaning to material molecules (e.g., proteins) or larger units constructed in the course of the code execution. What is missing in Barbieri's theory is the role of these molecules in the life of cells and multicellular organisms, which are higher-level agents. Prinz [1.56], p. 256, proposes extending the code biology theory to the level of organism agents, where meanings appear subjective and agent-dependent:

> The outputs of coded processes are meaningful to the underlying system in terms of persistence, propagation, and survival—necessitating at the same time choices between response options. Meaning reveals two sides here: it is objective through the artifacts or outputs of coded processes, and subjective through the consequences of their readout (or interpretation) by its carrier.

To summarize these approaches: (i) meanings are always associated with agency and its goal-directedness, (ii) meanings are features and relations that are significant for organisms and other kinds of semiotic agents, (iii) by semiosis, meanings are transferred to other agents or subagents, and (iv) meaning keeps changing via evolution, development, and learning through the cycle of translation processes.

1.3 The Origin of Life

How to distinguish semiosis from non-semiotic physical and chemical processes? This question poses a serious challenge because semiotic agents do not contain any supernatural processes that are not physical and chemical, and yet, semiosis is a qualitatively different phenomenon. As the phenomenon of entropy increase in thermodynamics cannot be explained at the level of micro-states, the phenomenon of semiotic agency is associated with higher-level macro-processes that appear to be collectively meaningful. Self-organizing non-equilibrium systems, such as dissipative structures, hurricanes, or flames, somewhat resemble living organisms because they preserve their non-equilibrium state despite perturbations, which is similar to homeostasis in living cells. Some of them can even reproduce (e.g., a flame can split and continue burning in different directions). But this similarity is mostly superficial because self-organizing systems are not goal-directed and their non-equilibrium state is not encoded. Sharov and Tønnessen wrote: "Physical self-organized

[4] This term is somewhat confusing because this molecular agent executes the code rather than makes it. Prinz calls it *code user*.

systems eventually stop functioning without affecting any future self-organizing events, whereas living systems produce the next generation of living systems of the same kind" [1.69], p.127. Similarly, Sukhoverkhov and Gare wrote: "For the origin of life, living systems need, as a prerequisite, chemical autocatalytic circles (sets) with some stable memory-like structures" [1.73], p. 90. A hurricane cannot pass its experience to its progeny hurricanes to increase their stability and spread. Simple self-production (e.g., randomly breaking crystals or division of lipid-world vesicles) is also not sufficient to display goal-directedness. These systems have no means to accumulate memory of their past experience and reuse it to become more efficient in the future.

Although physicochemical self-organized systems do not qualify as living agents, they should have supported the origin of life because living systems could not emerge in a simple environment, such as uniform gas cloud. Howard Pattee [1.53], p. 3, was the first to note that molecules become messages only in a complex and meaningful environment: "[...] the first 'messages' were expressed not in the highly integrated and precise genetic code that we find today, but in a more global set of geophysical and geochemical constraints, which we could call the primeval 'ecosystem language.'" Similarly, John Deely argued that the complexity of environments is increased by "virtual semiosis," which at some point in history ignited the origin of life [1.19], p. 61. This idea is consistent with the emerging relational paradigm in physics, where complex objects are treated as *ontolons*, units of dynamic change, which play the role of primitive observers that keep traces of previous interactions [1.12]. Igamberdiev [1.32], p. 267, explained: "The observers appear as living subjects only at the highest level of organization, while at the physical level they materialize as the dynamic units of complex reality, i.e., the units of dynamic change. It has been suggested to define these units as *ontolons* to emphasize their profoundly ontological rather than epistemological nature." He also proposed that the relational framework in physics needs to be connected with contemporary developments in quantum mechanics. Mikhailovsky [1.46] called such systems *eventities*, whereas Sukhoverkhov and Gare called them *physical memory*. Sharov reformulated the idea of Deely in terms of complexity measured by the number of non-redundant potential meanings relevant for projected physicochemical systems that later gained the capacity of adaptive evolution [1.68]. With the origin of life, some of these potential meanings were converted into actual meanings by being combined into a self-supporting and evolving whole. Other potential meanings became actualized in the subsequent adaptive evolution.

Sharov [1.67] proposes that proto-life emerged in the form of self-producing simple organic molecules capable of *context-sensitive catalysis*. His scenario also assumes agency-niche coupling, where molecular agents modify their local environment (e.g., oil droplets in water) in such a way that facilitates reproduction of these agents in a niche-dependent way. In other words, the function of first primordial agents was *niche construction*.[5] Sharov rejects the hypothesis of primordial soup; he argues that even if such soup was created by some unknown forces, it would immediately degrade due to dilution, hydrolysis, photolysis, and oxidation before any life-like system would be able to emerge in it. According to Sharov, only hydrocarbons (linear, branching, and aromatic) were available as organic resources before the origin of life, whereas all other kinds of organic resources were produced by evolving living systems long after their origin via metabolic pathways that emerged in the course of simplified natural selection without nucleic acids.

[5] This term was introduced by Odling-Smee [1.52].

Similar ideas on the origin of life are discussed by Sukhoverkhov and Gare [1.73], p. 90:

> For the origin of life, living systems need, as a prerequisite, chemical autocatalytic circles (sets) with some stable memory-like structures. Such strictures can preserve and accumulate new (functional) information that is needed to maintain these reproductive circles and provide further cumulative development of such systems [...]

1.4 Semiotic Agency

The fundamental question in biosemiotics is why organisms are capable of semiosis? Sharov wrote that

> [...] organisms are not the only systems that are capable of semiosis because interpretation and production of signs also takes place in organism parts and in multi-organism systems such as colonies, populations, and ecological consortia. Thus, it makes sense to use the notion of *semiotic agency as a generalization of organism in semiotic context.* Sharov and Tønnessen [1.69], p. 153, defined semiotic agents (or simply 'agents') as systems with "a capacity for acting purposefully and using signs to make informed choices." [1.68], p. 143.

Some semiotic agents are not alive (e.g., ribosomes and robots), but they all depend by origin and/or by function on living organisms, and thus, the main assumption of biosemiotics that semiosis is coextensive with life [1.3] [1.38] still holds. Primary agents are defined as those capable of self-reproduction, whereas secondary agents are produced by other kinds of agents (e.g., robots are produced by humans and machines) [1.69]. There is no single component or single function that is uniformly present in all agents. Instead, agency is characterized by end-directed behavior[6] [1.38], p. 171, and autonomy, which means "[s]elf-sufficiency, self-government, and (relative) independence from the environment" [1.69], p. 350.

The importance of the notion of semiotic agency is reiterated in several chapters of this book. Here is a brief summary of novel ideas related to this notion. Norris and Sharov [1.50] consider bacterial cells as "composite agents whose functions are mediated by assemblies of molecules and macromolecules that are termed hyperstructures," which are capable of semiosis, and thus play the role of semiotic subagents. It appears that the network of communicating hyperstructures can support rapid change of cell function in response to stress conditions. Norris and Sharov (ibid., p. 299) wrote:

> The proposed model of *competitive coherence* describes how such a network can assemble and function through two levels of selection among components: certain molecules are selected by signalling of either preceding or co-occurring molecules to form hyperstructures, and certain hyperstructures are selected to produce a phenotype that is consistent with the environment and the needs of the cell.

The model is explained using bacterial cell cycle as an example.

Fomin [1.22] discusses the semiotic nature of memes, a term proposed by Richard Dawkins [1.15] for gene-like units of replication in human culture (e.g., jokes or opinions).

---

[6] Also known as goal-directedness, teleology, teleonomy, or normativity.

The peculiar feature of memes is that they are more than signs, and can be also seen as self-reproducing agents that "infect" people and provoke them to spread meme copies. Following the theory of semiotic agency [1.69], Fomin characterizes memes as "a particular kind of secondary subagents [i.e., not self-reproducing] that trigger the production of cultural signs" [1.22]. As signs, memes have distinct external and internal forms: a sign vehicle (e.g., spoken or written) and interpretant (mental), respectively. It is the latter one that functions as a secondary subagent. Sharov and Tønnessen wrote that although sign-role (to signify) and agent-role (to interpret signs and act) are distinct, they can be combined [1.69], as happens in memes.

1.5 Mind and Consciousness

By recognizing the semiotic nature of all organisms, biosemiotics opened a can of worms by initiating discussions on where to place the boundaries of mind, cognition, consciousness, intelligence, sentience, learning, and other notions previously restricted to human mentality.[7] There is also no consensus among biologists on how to define these notions and avoid anthropocentric connotations. The extreme position is attributing cognitive terms to all organisms, including bacteria, whereas moderate approaches suggest applying these terms to certain phylogenetic branches in the tree of life.

Arthur Reber *et al.* [1.58] in their chapter "The Sentient Cell" defend the former approach and argue that "all biological and evolutionary development is directly related to intelligent cells and their suite of sentient, problem-solving behaviors." Components of cellular consciousness include sensing and perception, memory, learning and effective problem-solving using genes and other tools. The existence of perception is supported by the presence of "semantic elements, meaning, and significance for the organism." Learning of cells is inferred from selective growth of slime mold directed by previous experience, and from anticipative response of bacteria exposed to alternating stimuli. According to Reber *et al.*, conscious properties of cells emerged early in evolution in the form of excitable plasma membrane filled with interacting biological macromolecules, which together form a bioelectric field called *senome*. However, authors assume that "non-biological entities cannot become conscious." Reber *et al.* do not deny progressive evolution from primitive cells to more robust and efficient multicellular organisms. However, they assume that "the primate mind, avian cognition, cephalopod mental representations, plant sentience are all varied tokens of an underlying cognitive system that's been an inherent component of all life forms," which is "the *ur*-sentience of prokaryote." In other words, major components of cognition are already present in bacteria, and evolutionary change is mostly quantitative rather than qualitative.

The notion of cognition in bacteria was earlier inspired by James Shapiro's book *Evolution: A View from the 21st Century* [1.63]. In response to critical reviews of his book, Shapiro [1.64], p. 211, wrote: "I described exactly what I meant by cognition (=action based on knowledge acquired through sensing)." In this citation, Shapiro wrote about bacterial

[7] Some of these opinions emerged independently from biosemiotics *per se*, but were inspired by related ideas in second-order cybernetics, systems theory, symbiogenesis, autopoiesis, and extended evolutionary synthesis.

"knowledge" from the third-person's (human scientist') point of view rather than bacterium's first-person experience, which is limited to elementary responses of molecular subagents to incoming signals [1.70].

The idea of minimal cognition in microorganisms is explored by Victoria Alexander in her chapter "Self-reinforcing Cycles and Mistakes: The Emergence of Subjective Meaning" [1.2], p. 326. She wrote that "The activity of slime mold may be considered an example of minimal cognition, displaying chemical memory and adaptable searching behaviors." To support this view she asks "Can it [slime mold] teach itself a new rule or invent a new tool for finding food?" This question draws the line between two "different levels of subjectivity; inherited or habitual and newly formed." To answer this question, Alexander cites a published experimental study, where slime mold changed its preference to temperature depending on the location of food [1.72], and interprets it as classical conditioning. However, authors of the paper do not use the term "conditioning" and describe it as "plasticity,"[8] which I think is more relevant. As discussed in the next subsection, such plasticity is inherited (i.e., prescribed), and usually short-term and low-specific.

In contrast, most biosemioticians do not employ cognition-related terms to describe signaling and regulation of molecular processes in living cells in order to avoid confusion with representational cognition and learning in conscious animals and humans. Instead, the key feature that separates life from non-living systems is *semiotic agency* with its goal-directedness, autonomy, and semiosis [1.67] [1.69]. This terminology helps to avoid over-reaching metaphors and is consistent with von Uexküll's umwelt theory [1.78] [1.79]. Functions of agency are preserved by biological memory/heredity, which is defined by Anton Sukhoverkhov [1.73] "as a set of system processes of maintenance, transmission, and reproduction of living systems by means of heritable genetic and non-genetic information." Then reproduction (and development) appears equivalent to memory retrieval: "The processes of self-reproduction could thus also be viewed as empirically observable and physically (biologically) embodied processes of memory 'retrieval'" (ibid., p. 90).

According to Sharov and Vehkavaara [1.70], semiotic processes in bacteria are much simpler than in eukaryotic cells. Thus, these authors proposed distinguishing two major types of semiosis: protosemiosis,[9] which is responding to molecular signals without perception of objects in prokaryotes, and eusemiosis, which integrates a large set of diverse and patterned signals to yield perception and categorization of objects in eukaryotes, and manifests a primitive mind. Sharov [1.66], p. 487, wrote:

> In protosemiosis, (i) individual signals are processed either directly or via simple logic gates, (ii) no reality is assumed outside the cell as a source of receptor excitation, and (iii) signals are processed in a threshold-based way. In contrast, eusemiosis, which is equivalent to minimal mind, includes (i) rich and diverse sensory input that is filtered via cascades of signaling networks, (ii) capacity to attribute sensory patterns to objects in the environment, and (iii) categorization of objects based on patterns of sensorial stimuli.

Neural version of eusemiosis is an important evolutionary transition that can be called "cognition" [1.66]. Nerves support fast signaling between organs and sensory-motor

[8] In an earlier publication, Shirakawa *et al.* [1.71] call it "associative learning" and compare it to Pavlovian conditioning.

[9] The term was introduced by Prodi [1.57].

connections. Early forms of cognition are based on innate classifiers and simple reflexes, which are further tuned by habituation and sensitization.

The dominating view in biology is that consciousness exists only in animals with a nervous system and strong integration between sensory, motor, memory, and value-generating processes [1.20]. More recently, Ginsburg and Jablonka [1.25] proposed that the core feature of consciousness is Unlimited Associative Learning (UAL), which is an "open-ended, representational, generative, and recursive form of associative learning" [1.34], p. 401. These authors wrote that consciousness is found in three lineages of animals: vertebrates, some arthropods, and some cephalopods. Similarly, Hoffmeyer and Stjernfelt [1.31] place learning, sentience, and consciousness at the highest levels of the "great chain of semiosis," well above the molecular recognition and eukaryotic cell levels (cf. [1.43]).

Despite the progress in biology, it is still not clear how the brain can support consciousness. The widely accepted connectivism paradigm, which claims that "brain's entire knowledge and thinking skills are [...] stored in the connections; and the mental operations are executed by network computations" [1.48], fails to explain the *scaling problem* of consciousness. The problem is that doubling the number of recognized objects requires as much as a ~500-fold increase in computing resources, whereas the brain handles functional complexity in a more economical way (ibid.). Nikolić [1.48] hypothesized that in addition to synaptic connectivity, consciousness is supported by the activity of neurons themselves, which can selectively activate or repress their dendritic branches via metabotropic receptors and G protein-gated ion channels. These changes can modify the activity of other neurons through synapses, and thus, specific neural subnetworks can be organized by multimodal states of individual neurons. This hypothesis is consistent with the previously discussed mechanism of dendritic computation [1.40].

Alternatively, there were attempts to explain the function of consciousness by quantum mechanics effects within microtubules composed of tubulin protein molecules [1.27] [1.28], which are present in axons and dendrites of neurons. In his chapter [1.75], Jack Tuszynski explored information processing rates in microtubules and came to the conclusion "that tubulin is not likely to operate at a quantum level if its functions include information storage and processing." However, the model indicates that "ion channels, if synchronized may be able to operate at a quantum mechanical regime." The latter conclusion is interesting because it opens a possibility that G protein-gated ion channels, identified by Nikolić as potential operators of consciousness (see above), can be synchronized and act in the quantum mode.

1.6 Semiogenesis and Learning

The term "learning" emerged in relation to the human mind, but then its meaning was extended to animals, other organisms, and even to computers and robots. Jablonka and Ginsburg [1.34] define learning as "a process leading to an experience-dependent behavioral response of a system." A similar definition of learning is found in the chapter by Caetano-Anollés [1.13], p. 229: "system's use of information obtained during one interaction with its environment to improve performance in future interactions." Although such broad definition of "learning" is relevant in systems theory, it may appear confusing in biology. For example, biologists prefer distinguishing between learning and evolution, although natural selection fits the definitions of learning mentioned above. Thus, Sharov and Tønnessen [1.69] propose using the term "semiogenesis" for all learning-like processes, including natural selection; and reserve the term "learning" for "development and storage of non-innate interpretation mechanisms, which generally require representational cognition. Learning is opposed to adaptive plasticity, which is based on an innate program." Thus, semiogenesis includes adaptive evolution, development, plasticity (e.g., habituation and sensitization), and learning.

The dominating evolutionary theory, known as "Modern Synthesis" is based on the gene-centric paradigm, which is radically mechanistic. It assumes that natural selection is a passive sieve of genetic alleles, differentiated according to their propagation rate in a population, called "fitness." According to Modern Synthesis, genes are passively copied, and their interpretation (i.e., semiotic function) is ignored [1.49]. An alternative approach was developed by Jesper Hoffmeyer and Claus Emmeche [1.30] in their "code-duality" paper, where they argue that genes (and other signs) are used by organisms in two different ways: as digital units that can be stored as memory and copied, and as analog functional forms that represent parts of a semiotic agent (cell or organism). These two representations are interdependent because the digital code can be translated into analog functional subagents, and these subagents support both copying of the digital code and its translation into functional units. Thus,

> [...] heritability of phenotypes across generations cannot be explained by the transfer of genetic information. Another and equally important channel of heredity is represented by agents capable of reading and interpreting genetic information together with necessary tools and scaffolds within cells [1.65], p. 213.

Sharov [1.68] discusses three examples of simple, intermediate, and complex biological memory: plasticity of chemoreceptor sensitivity in bacteria, hereditary function of the genome, and distributed memory in the nervous system. He emphasizes that memory may include passive potential signs that do not signify their meaning unless activated by subagents guided by syntax-signs. Sukhoverkhov and Gare [1.73], p. 91, noted that "some forms of behavior and cognition are acquired and transmitted throughout generations not via the internal, genetic (epigenetic) memory, but through the *external* behavioral, ecological and cultural 'inheritance systems' (Jablonka, Lamb), 'semiotic scaffoldings' (Hoffmeyer), or 'developmental niches' (Stotz)."

There is no doubt that bacteria have long-term genetic and epigenetic memory, as well as short-term operational memory (e.g., adjustable chemoreceptor sensitivity). But there is no consensus on whether bacterial cells are capable of learning. Reber *et al.* [1.58] and Norris [1.50] assume this capacity in bacteria is based on the system's theory definition of learning. In contrast, Sharov assumes that innate plasticity (habituation and sensitization) is responsible for adjusting of chemoreceptors sensitivity, and this effect is not true learning "since bacterial cells 'learn' only those signals for which they already have specialized devices" [1.50]. He also rejects claims that bacteria are capable of Pavlovian conditioning:

> Mitchell *et al.* [1.47] showed that *E. coli* can learn to anticipate carbon source change after exposure to another stimulus. However, in contrast to Pavlovian conditioning which develops within one and the same animal, bacterial "conditioning" requires hundreds of generations. One conclusion could be that this phenomenon is an example of associative natural selection rather than associative learning (ibid., p. 318).

Modulation of slime mold's response to temperature in the experiment of Shirakawa and Sato [1.72] also does not qualify as Pavlovian conditioning. By default, slime mold avoids cold temperature, but if the nutrient (oatmeal) is placed in the cold area far away, then after some time the slime mold overcomes the inhibition and moves into the cold area. This effect, however, is not the result of associative learning, but instead is caused by hunger and stress, which jointly repress the reflex of avoiding the cold. In behavioral studies, such effect is known as dominance [1.61], when a stronger stimulus (in this example – hunger) non-specifically overrides weaker stimuli, including temperature preference. Dominance has a short-term and low-specific effect in contrast to conditioning that is supported by long-term specific memory [1.25].

True learning (i.e., excluding plasticity) requires a strongly integrated nervous system with brain and represents another evolutionary transition towards learning-based cognition which is called consciousness [1.25] [1.66]. I agree with Jablonka and Ginsburg [1.34] that the main feature of consciousness is integration of multiple stimuli coming from diverse sense organs, or stimuli composed of multiple spatial or temporal elements. Neural integration helps sentient animals to combine elemental actions into new activity patterns and relate them to the perceived context.

According to Ginsburg and Jablonka [1.25], learning is based on the "exploration-stabilization" principle, which is the extension of the more known "variation-selection" principle in evolution. Exploration-stabilization requires agency for its both components. Exploration is represented by copying genetic mutations, sensing changes in the environment, locomotion, establishment of synapses between neurons, and free will in behavior and communication. But it can be supported also by non-agential fluctuations and processes (e.g., Brownian motion). Examples of stabilization are the system of genetic heredity supported by DNA replication, transcription, and translation. Other stabilization systems include epigenetic heredity [1.33] [1.35], developmental regulation [1.82], and neural memory [1.24]. An important addition to this principle, in relation to evolutionary and developmental growth, is proposed by Caetano-Anollés [1.13], p. 220, who developed a biphasic model of unification-diversification:

> This model predicted both the evolutionary establishment of nested arrangements (hierarchies) of modules, and the formation of tightly linked modular groups by convergence under optimization or selection. [...] The biphasic model creates novelty via two evolutionary phases that are often asynchronous. In Phase 1, parts associate in different combinations but are weakly linked to each other. [...] In Phase 2, evolutionary variants of the newly constructed modules diversify and become new parts of modules populating higher-level organization.

This model means that stabilization opens up new possibilities and is followed by a new round of diversification at the emerged level of integration. Caetano-Anollés illustrated this model by the evolution of the F_1/F_0 ATP synthase complex, and by the construction and reconstruction of the Empire State Building in New York city.

One of the fundamental problems of semiogenesis is the source of qualitatively new adaptations that cannot be reduced to quantitative adjustment or optimization of already existing functions. Interestingly, Alexander [1.2], p. 328, assumes that truly novel behaviors appear through *mistaken* choices: "When the organism takes a new direction, it forges a new path. I call this *misinterpreting*." Mistaken choice brings qualitatively new solutions in contrast to trial-and-error learning which "only helps the agent make gradual improvements on an already existing tool or procedure or just helps the agent find an optimal strategy for reaching its usual, pre-defined goal."

Organisms can improve learning by using learned models, which is one of the mechanisms of "learning to learn" [1.11], p. 173. The expression "learned models" is crucial because innate models such as simple reflexes do not provide sufficient flexibility and context. According to Sharov and Tønnessen [1.69], p. 249, "models are generalized representations [...] that remain true or bring success in a large number of similar situations." By using models, animals and humans can verify their perceptions and anticipate context-dependent outcomes of their activity. An important milestone in exploring the use of models was the development of the Modeling Systems Theory by Sebeok and Danesi [1.60].

In his chapter [1.84], Hongbing Yu explores one kind of model: modeling of danger in animals and humans. He distinguishes semiotic roles of models and signs: "models supersede, whereas signs signify." Model supersedes in the sense that it is a part of subjective existence when it is "brought to the front for salience, accessibility, and operability." Using the well-known fable of "cry wolf" he explored several aspects of the associated danger model, which, besides the danger of a wolf itself, includes "the danger of losing others' trust as a result of raising false alarms." The model uncovers the "reality-vs-illusion" dichotomy and appeals to certain social rules of conduct.

Models are routinely used for logical inference, such as deduction, when the conclusion is derived from premises (e.g., causing factors), or induction, when a general principle is uncovered in multiple observations. In both cases, the model remains in place and just applied to different objects or situations. To these kinds of reasoning Charles Peirce added *abduction*, which takes place when the model is absent and needs to be suggested to explain a given observation or set of observations [1.54], CP 5.189. Lorenzo Magnani [1.41] considers two major kinds of abduction: selecting from the set of known models – selective abduction, or constructing a new model from scratch – creative abduction. All three types of logical inference can be found in both innate models and learned models. The only difference is that abduction with innate models may require many generations and thousands of years of evolutionary change, whereas abduction with learned models can be achieved much faster, especially by humans. For example, chess players select reasonably good moves in unique positions within a minute.

The main topic of Magnani's chapter is niche construction, and specifically, cognitive niche construction. Niche construction theory has been pioneered by Odling-Smee *et al.* [1.51]. The idea is that "while organisms adapt to their environments, they also adapt to environments that they or other organisms create" [1.41]. Examples include bird nests, spider webs, and termite mounds. Magnani proposes that natural "selection favors *purposive organisms*, that is, good niche constructors."

Niche construction is especially important for conscious animals and humans because they modify their environment consciously, and thus changes in their knowledge and experience is remembered and applied systematically to their future activity. In addition, many conscious species imitate each other's behavior, which results in a socially supported hereditary habit. Cognitive niche includes artifacts that serve as "anchors of meaning" that help to acquire or store knowledge and working skills. Magnani assumes that cognition is spread across individuals, objects, artifacts, and external representations. This makes cognition distributed as it includes the following components systematized by Sutton [1.74]:

> 1) external cultural instruments and devices, artifacts, and symbol structures; 2) natural environment elements appropriately enriched with cognitive value; 3) intersubjective and social distribution or the so-called cognitive "scaffolding"; 4) embodied capacities and skills intricately entwined with our use of technological, natural, and social resources of the preceding cases; 5) internalized cognitive artifacts [1.41], p. 389.

The most remarkable component of human cognitive niche is natural language:

> The human process of constructing cognitive niches—for example, we can immediately anticipate that, in the case of human animals, natural language is a fundamental example of a cognitive niche—is a useful further integration of the intellectual framework [...] [1.41], p. 381.

Human cognition cannot be explored without considering its collective nature. Sukhoverkhov and Gare [1.73] wrote: "we need to admit that at some stage of evolution, the individual 'umwelt bubble' became collective and shared as a 'mitwelt.'" They considered social systems as higher-level distributed agents: "in addition to retrospective and historic approach to memory, this research proposes to study social memory also as 1) actual cumulative and distributed system processes that are maintained over generations and 2) semiotic processes and structures responsible for the production and re-production of the social system." Human society is integrated by communication and language, and Sukhoverkhov and Gare discuss the problem of how language could emerge without pre-existing language, a sort of chicken-and-egg dilemma. They wrote: "the language had to emerge in social reality *ex nihilo* and people were able to understand each other *naturally* without any socially imposed conventions or agreements." Such communication can start from reciprocal actions alone and then gradually transform "into more culture-specific, conventional (artificial, arbitrary) systems that require teaching and learning."

1.7 Global Dimensions of Biosemiosis

Similar to physics, which was initially focused on objects around us and then appeared relevant for distant star systems and galaxies, principles of biology and semiotics are ready to become global and applied to the extraterrestrial biospheres and habitable exoplanets. One such process is complexification of some components of the universe [1.45], which inadvertently prepared conditions for the origin of life. Deely [1.19] called it physiosemiosis, Sukhoverkhov and Gare [1.73] – physical information, Deacon [1.18] – medium or signal properties, and Sharov [1.68] – potential meaning (assigned retrospectively by humans). Mikhailovsky [1.46] considered hierarchogenesis in the combined physical and biological evolution.

The discipline of astrobiology emerged in the 1960s and 1970s together with the start of space exploration [1.14]. But initially it was mostly focused on empirical problems. Now, however, after we have realized that life has a semiotic nature and biosemiotics was born, it is time to start the analysis of meanings in the global dimension. There are many habitable exoplanets in our galaxy and beyond [1.26], and it is highly likely that a large portion of them carry microbial life because microbes are highly adaptable to extreme conditions [1.59] and hypothetically can travel in space on asteroids and rogue planets [1.5] [1.81].

In light of these discoveries and hypotheses, the call of Julian Chela-Flores [1.14] and Yogi Hendlin and Constantijn-Alexander Kusters [1.29] for the development of "astrobiosemiotics" appears quite timely. Chela-Flores discusses philosophical overlapping of humanities and astrobiology, and in particular, the relevance of the Singer's principle of equality, an ethical position that interest of non-human animals have equal weight to human interests. Chela Flores extends this principle to all forms of life because of the phenomenon of evolutionary convergence that constrains randomness, and according to which many lineages of life follow a trend that may lead to consciousness. He cites Christian de Duve [1.16], p. 291: "Life is an integral part of the universe; it is even the most complex and significant part of the known universe."

Hendlin and Kusters [1.29] extend further the ethical challenges of space exploration. They develop the line of argument that modeling of danger needs to be applied to extraterrestrial life, and doing this can be learned by caring about earthly life. They wrote: "Slow space exploration—starting with stabilizing the biotic conditions on our own planet—presents a biosemiotically-informed perspective on how organisms of various complexities can fit with (drastically) new environments with different temporal needs" [1.29], p. 453. The authors draw attention to the difference in developmental and evolutionary plasticity of various kinds of organisms, which may affect the change of biosphere on earth and exoplanets. These differences can be addressed by assuming agent-specific time scales, as proposed by Bergson in contrast to the physical notion of universal space-time or its relativistic modification [1.29]. The difference in temporal scales may cause unpredictable consequences in cases of symbiosis or mutualism between organisms from far-separated branches of the evolutionary tree, such as microbiota of prokaryotes inside various organs of animals and humans. Such interaction poses a challenge for predicting health and cognition consequences for host organisms. Besides different rates, agent-specific time is characterized by different "thickness" as proposed by Whitehead, which represents organism capacity to recall its past experiences and apply this information for anticipating the future. Hendlin and Kusters wrote:

> Taking both Bergson's and Whitehead's critique to heart, we must try and move away from the idea that any one model can span the variations which space-time might play in all possible biological relations. Moving towards a less anthropocentric and more pluralistic understanding of time experience might thus be illuminating in situations where a general theory of relativity might come up short explanatorily, such as in the cases of lifeworlds [1.29], p. 460.

1.8 Conclusions

This book shows an example of how the study of meanings crosses traditional bounds of semiotics and humanities, and can be applied to various fields of biology (molecular biology, physiology, ecology, neuroscience, evolutionary biology, and astrobiology), cognitive science, physics, cybernetics, logic, and artificial intelligence. This interdisciplinary project attempts to bridge the *great paradigm divide* and contribute to integration of natural sciences with semiotics. In particular, it explores the objective-subjective duality and offers new balanced versions of semiotic realism that acknowledge agent-specific subjectivity and the growth of objective meaning through semiotic interaction. An important achievement is the discussion of different levels of semiosis (e.g., protosemiosis, eusemiosis, cognition, and consciousness) and semiogenesis (e.g., natural selection, plasticity, and true learning). Finally, this book opens a perspective for the development of astrobiosemiotics, where meaning is systematically studied in the context of cosmic conditions and processes.

Declaration

The author declares no conflict of interest.

References

[1.1] Ahmad, B. and Gordon, R. (2023) Is it a Janus-faced world after all? Physics is not reductionist. In: *Pathways to the Origin and Evolution of Meanings in the Universe*. A.A. Sharov and G.E. Mikhailovsky, (eds.) Wiley-Scrivener, Beverly, Massachusetts, USA: 55-70.

[1.2] Alexander, V. (2023) Self-reinforcing cycles and mistakes: The emergence of subjective meaning. In: *Pathways to the Origin and Evolution of Meanings in the Universe*. A.A. Sharov and G.E. Mikhailovsky, (eds.) Wiley-Scrivener, Beverly, Massachusetts, USA: 325-344.

[1.3] Anderson, M., Deely, J., Krampen, M., Ransdell, J., Sebeok, T.A. and von Uexküll, T. (1984) A semiotic perspective on the sciences: Steps toward a new paradigm. *Semiotica* 52(1/2), 7-47.

[1.4] Atkin, A. (2022, December 2022) Peirce's theory of signs. https://plato.stanford.edu/archives/fall2022/entries/peirce-semiotics

[1.5] Balb, A. (2021) Why we should take interstellar panspermia seriously. In: *Planet formation and panspermia. New prospects for the movement of life through space*. B. Vukotić, R. Gordon and J. Seckbach, (eds.) Scrivener, Wiley, Hoboken, NJ: 7-18.

[1.6] Barbieri, M. (2003) *The organic codes: An introduction to semantic biology*. Cambridge University Press, Cambridge, New York.

[1.7] Barbieri, M. (2011) A mechanistic model of meaning. *Biosemiotics* 4(1), 1-4.

[1.8] Barbieri, M. (2012) Code biology – A new science of life. *Biosemiotics* 5(3), 411-437.

[1.9] Barbieri, M. (2014) Introduction to code biology. *Biosemiotics* 7(2), 167-179.

[1.10] Barbieri, M. (2015) *Code biology*. Springer, Heidelberg, New York.

[1.11] Bateson, G. (1987 [1972]) *Steps to an ecology of mind. Collected essays in anthropology, psychiatry, evolution, and epistemology*. Jason Aronson, Northvale, NJ.

[1.12] Brenner, J.E. and Igamberdiev, A.U. (2021) *Philosophy in reality: A new book of changes*. Springer Nature.

[1.13] Caetano-Anollés, G. (2023) Evolution of biomolecular communication. In: *Pathways to the Origin and Evolution of Meanings in the Universe*. A.A. Sharov and G.E. Mikhailovsky, (eds.) Wiley-Scrivener, Beverly, Massachusetts, USA: 217-244.

[1.14] Chela-Flores, J. (2023) Astrobiosemiotics and its frontier with astrobiology. In: *Pathways to the Origin and Evolution of Meanings in the Universe*. A.A. Sharov and G.E. Mikhailovsky, (eds.) Wiley-Scrivener, Beverly, Massachusetts, USA: 439-452.

[1.15] Dawkins, R. (1976) *The selfish gene*. Oxford University Press, New York.

[1.16] de Duve, C. (1995) *Vital dust: Life as a cosmic imperative*. Basic Books.

[1.17] de Saussure, F. (1959) *Course in general linguistics*. Philosophical Library, New York.

[1.18] Deacon, T.W. (2023) Semiotic ground and the hierarchic nature of information. In: *Pathways to the Origin and Evolution of Meanings in the Universe*. A.A. Sharov and G.E. Mikhailovsky, (eds.) Wiley-Scrivener, Beverly, Massachusetts, USA: 71-84.

[1.19] Deely, J. (1992) Semiotics and biosemiotics: Are sign-science and life-science coextensive? In: *Biosemiotics. The semiotic web 1991*. T.A. Sebeok and J. Umiker-Sebeok, (eds.) Mouton de Gruyter, New York: 45-75.

[1.20] Dehaene, S. and Changeux, J.-P. (2011) Experimental and theoretical approaches to conscious processing. *Neuron* 70, 201–227.

[1.21] Favareau, D. and Kull, K. (2023) Pathways to the understanding of signs and meanings in the biosphere: Historical and contemporary perspectives. In: *Pathways to the Origin and Evolution of Meanings in the Universe*. A.A. Sharov and G.E. Mikhailovsky, (eds.) Wiley-Scrivener, Beverly, Massachusetts, USA: 27-54.

[1.22] Fomin, I. (2023) Towards a biosemiotic account of memes as units of cultural replication and interpretation. In: *Pathways to the Origin and Evolution of Meanings in the Universe*. A.A. Sharov and G.E. Mikhailovsky, (eds.) Wiley-Scrivener, Beverly, Massachusetts, USA: 419-438.

[1.23] Frege, G. (1982) Über Sinn und Bedeutung. *Zeitschrift für Philosophie und philosophische Kritik* 100, 25-50.

[1.24] Fuster, J.M. and Bressler, S.L. (2012) Cognit activation: A mechanism enabling temporal integration in working memory. *Trends in Cognitive Sciences* 16(4), 207-218.

[1.25] Ginsburg, S. and Jablonka, E. (2019) *The evolution of the sensitive soul. Learning and the origin of consciousness*. MIT Press, Cambridge, London.

[1.26] Gowanlock, G. and Morrison, I.S. (2018) The habitability of our evolving galaxy. In: *Habitability of the Universe before Earth*. R. Gordon and A.A. Sharov, (eds.) Elsevier, Amsterdam: 149-171.

[1.27] Hameroff, S., Nip, A., Porter, M. and Tuszynski, J. (2002) Conduction pathways in microtubules,biological quantum computation, and consciousness. *BioSystems* 64(1-3), 149-168.

[1.28] Hameroff, S. and Penrose, R. (1996) Orchestrated reduction of quantum coherence in brain microtubules: A model for consciousness. *Mathematics and computers in simulation* 40(3-4), 453-480.

[1.29] Hendlin, Y.H. and Kusters, C.-A. (2023) Time horizons and biosemiotic adaptation: Taking seriously variable temporalities in the evolution of possible life forms. In: *Pathways to the Origin and Evolution of Meanings in the Universe*. A.A. Sharov and G.E. Mikhailovsky, (eds.) Wiley-Scrivener, Beverly, Massachusetts, USA: 453-470.

[1.30] Hoffmeyer, J. and Emmeche, C. (1991) Code-duality and the semiotics of nature. In: *On semiotic modeling*. M. Anderson and F. Merrell, (eds.) Mouton de Gruyter, Berlin, New York: 117-166.

[1.31] Hoffmeyer, J. and Stjernfelt, F. (2016) The great chain of semiosis. Investigating the steps in the evolution of semiotic competence. *Biosemiotics* 9(1), 7-29.

[1.32] Igamberdiev, A.U. (2023) Evolutionary growth of meanings in the relational universe of intercommunicating agents. In: *Pathways to the Origin and Evolution of Meanings in the Universe*. A.A. Sharov and G.E. Mikhailovsky, (eds.) Wiley-Scrivener, Beverly, Massachusetts, USA: 265-278.

[1.33] Jablonka, E. (2013) Epigenetic inheritance and plasticity: The responsive germline. *Prog Biophys Mol Biol* 111(2-3), 99-107.

[1.34] Jablonka, E. and Ginsburg, S. (2022) Learning and the evolution of conscious agents. *Biosemiotics* 15(3), 401-437.

[1.35] Jablonka, E. and Lamb, M.J. (2010) Transgenerational epigenetic inheritance. In: *Evolution, the expanded synthesis*. M. Pigliucci and G.B. Müller, (eds.) MIT Press, Cambridge, London: 137–174.

[1.36] Koestler, A. (1979) Janus: A summing up. *Bulletin of the Atomic Scientists* 35(3), 4.

[1.37] Kravchenko, A. (2023) A constructivist approach to "meanings in the Universe". In: *Pathways to the Origin and Evolution of Meanings in the Universe.* A.A. Sharov and G.E. Mikhailovsky, (eds.) Wiley-Scrivener, Beverly, Massachusetts, USA: 167-186.

[1.38] Kull, K., Deacon, T., Emmeche, C., Hoffmeyer, J. and Stjernfelt, F. (2009) Theses on biosemiotics: Prolegomena to a theoretical biology. *Biological Theory* 4(2), 167-173.

[1.39] Locke, J. (1959 [1689]) *An essay concerning human understanding.* Dover publications, New York.

[1.40] London, M. and Hausser, M. (2005) Dendritic computation. *Annu Rev Neurosci* 28, 503-532.

[1.41] Magnani, L. (2023) Anchors of meaning: The intertwining of signs, abduction, and cognitive niches. In: *Pathways to the Origin and Evolution of Meanings in the Universe.* A.A. Sharov and G.E. Mikhailovsky, (eds.) Wiley-Scrivener, Beverly, Massachusetts, USA: 379-400.

[1.42] Marais, K. (2023) Levels of translation, levels of freedom? In: *Pathways to the Origin and Evolution of Meanings in the Universe.* A.A. Sharov and G.E. Mikhailovsky, (eds.) Wiley-Scrivener, Beverly, Massachusetts, USA: 401-418.

[1.43] Marais, K. (2023) What does it mean to translate In: *Translation beyond translation studies.* K. Marais, (ed.) Bloomsbury Academic, London, New York: 1-14.

[1.44] Maritain, J. (1957) Language and the theory of sign. In: *Language: An Enquiry into Its Meaning and Function.* R.N. Anshen, (ed.) Harper & Bros., New York: 86–101.

[1.45] Mikhailovsky, G. (2018) From identity to uniqueness: The emergence of increasingly higher levels of hierarchy in the process of the matter evolution. *Entropy* 20(533), 1-18.

[1.46] Mikhailovsky, G. (2023) Meanings, their hierarchy, and evolution. In: *Pathways to the Origin and Evolution of Meanings in the Universe.* A.A. Sharov and G.E. Mikhailovsky, (eds.) Wiley-Scrivener, Beverly, Massachusetts, USA: 101-136.

[1.47] Mitchell, A., Romano, G.H., Groisman, B., Yona, A., Dekel, E., Kupiec, M., Dahan, O. and Pilpel, Y. (2009) Adaptive prediction of environmental changes by microorganisms. *Nature* 460(7252), 220-224.

[1.48] Nikolić, D. (2023) Where is the mind within the brain? Transient selection of subnetworks by metabotropic receptors and G protein-gated ion channels. *Comput Biol Chem* 103, #107820.

[1.49] Noble, D. (2021) The illusions of the Modern Synthesis. *Biosemiotics* 14(1), 5-24.

[1.50] Norris, V. and Sharov, A.A. (2023) A hypothesis about how bacterial cells sustain and change their lives in response to various signals. In: *Pathways to the Origin and Evolution of Meanings in the Universe.* A.A. Sharov and G.E. Mikhailovsky, (eds.) Wiley-Scrivener, Beverly, Massachusetts, USA: 299-324.

[1.51] Odling-Smee, F.J. (1988) Niche constructing phenotypes. In: *The role of behavior in evolution.* H.C. Plotkin, (ed.) MIT Press, Cambridge, MA: 31–79.

[1.52] Odling-Smee, F.J., Laland, K.N. and Feldman, M.W. (2003) *Niche construction. The neglected process in evolution.* Princeton University Press, Princeton, NJ.

[1.53] Pattee, H.H. (1969) How does a molecule become a message? *Developmental Biology Supplement* 3, 1-16.

[1.54] Peirce, C.S. (1931–1958) *Collected Papers* (eds. Hartshorne and Weiss; Burks). Harvard University Press, Cambridge, MA.

[1.55] Peirce, C.S. (1998 [1893–1913]) *The essential Peirce: Selected philosophical writings.* Indiana University Press, Bloomington.

[1.56] Prinz, R. (2023) Meaning relies on codes but depends on agents. In: *Pathways to the Origin and Evolution of Meanings in the Universe.* A.A. Sharov and G.E. Mikhailovsky, (eds.) Wiley-Scrivener, Beverly, Massachusetts, USA: 245-264.

[1.57] Prodi, G. (1988) Signs and codes in immunology. In: *The semiotics of cellular communication in the immune system*. E.E. Sercarz, F. Celada, N.A. Mitchenson and T. Tada, (eds.) Springer, BerlIn 53-64.

[1.58] Reber, A.S., Baluška, F. and Miller, W.B.J. (2023) The sentient cell. In: *Pathways to the Origin and Evolution of Meanings in the Universe*. A.A. Sharov and G.E. Mikhailovsky, (eds.) Wiley-Scrivener, Beverly, Massachusetts, USA: 281-298.

[1.59] Robinson, J.M. and Mikucki, J.A. (2018) Occupied and empty regions of the space of extremophile parameters. In: *Habitability of the Universe before Earth*. R. Gordon and A. Sharov, (eds.) Academic Press, Cambridge, MA, London: 199-230.

[1.60] Sebeok, T.A. and Danesi, M. (2000) *The forms of meaning. Modeling systems theory and semiotic analysis*. Mouton de Gruyter, New York.

[1.61] Sergeev, B.F. (2010) *Steps in the evoluttion of intelligence*. KomKniga, Moscow, 3rd.

[1.62] Shannon, C.E. (1948) A mathematical theory of communication. *Bell System Technical Journal* 27, 379–423, 623–656.

[1.63] Shapiro, J.A. (2011) *Evolution: A view from the 21st century*. FT Press Science, Upper Saddle River, NJ.

[1.64] Shapiro, J.A. (2012) Response to Pauline Hogeweg's review of my book, "Evolution: a view from the 21st century". *Evolutionary Intelligence* 5, 211.

[1.65] Sharov, A.A. (2018) Mind, agency, and biosemiotics. *Journal of Cognitive Science* 19(2), 195-228.

[1.66] Sharov, A.A. (2022) Consciousness and learning from the biosemiotic perspective. *Biosemiotics* 15(3), 483-490.

[1.67] Sharov, A.A. (2023) Chemical origins of life, agency, and meaning. In: *Pathways to the Origin and Evolution of Meanings in the Universe*. A.A. Sharov and G.E. Mikhailovsky, (eds.) Wiley-Scrivener, Beverly, Massachusetts, USA: 189-216.

[1.68] Sharov, A.A. (2023) Semiotics of potential meanings. In: *Pathways to the Origin and Evolution of Meanings in the Universe*. A.A. Sharov and G.E. Mikhailovsky, (eds.) Wiley-Scrivener, Beverly, Massachusetts, USA: 137-166.

[1.69] Sharov, A.A. and Tønnessen, M. (2021) *Semiotic agency. Science beyond mechanism*. Springer, Dordrecht.

[1.70] Sharov, A.A. and Vehkavaara, T. (2015) Protosemiosis: Agency with reduced representation capacity. *Biosemiotics* 8(1), 103-123.

[1.71] Shirakawa, T., Gunji, Y.-P. and Miyaked, Y. (2011) An associative learning experiment using the plasmodium of *Physarum polycephalum*. *Nano Communication Networks* 2, 99-105.

[1.72] Shirakawa, T. and Sato, H. (2013) Construction of a molecular learning network. *Journal of Advanced Computational Intelligence and Intelligent Informatics* 17(6), 913-918.

[1.73] Sukhoverkhov, A.V. and Gare, A.E. (2023) Ontology and semiotics of memory. In: *Pathways to the Origin and Evolution of Meanings in the Universe*. A.A. Sharov and G.E. Mikhailovsky, (eds.) Wiley-Scrivener, Beverly, Massachusetts, USA: 85-100.

[1.74] Sutton, R.S. and Barto, A.G. (2018) *Reinforcement learning: An Introduction*. The MIT Press, Cambridge, MA, 2nd.

[1.75] Tuszynski, J.A. (2023) On the energy-based limitations of the information capacity and information processing rates in the human brain. In: *Pathways to the Origin and Evolution of Meanings in the Universe*. A.A. Sharov and G.E. Mikhailovsky, (eds.) Wiley-Scrivener, Beverly, Massachusetts, USA: 345-362.

[1.76] Vehkavaara, T. (2003) Natural self-interest, interactive representation, and the emergence of objects and Umwelt. *Sign Systems Studies* 31(2), 547-587.

[1.77] von Glasersfeld, E. (1984) An Introduction to radical constructivism. In: *The invented reality*. P. Watzlawick, (ed.) Norton, New York: 17-40.

[1.78] von Uexküll, J. (1926) *Theoretical biology*. Harcourt, Brace & Co., New York.

[1.79] von Uexküll, J. (1957) A stroll through the worlds of animals and men: A picture book of invisible worlds. In: *Instinctive behaviour: The development of a modern concept*. C.H. Schiller, (ed.) International Universities Press, Inc., New York: 5–80.

[1.80] von Uexküll, J. (1982 [1940]) The theory of meaning. *Semiotica* 42(1), 25-82.

[1.81] Vukotić, B. and Gordon, R. (2021) Life in the Milky Way: The panspermia prospects. In: *Planet formation and panspermia. New prospects for the movement of life through space*. B. Vukotić, R. Gordon and J. Seckbach, (eds.) Scrivener, Wiley, Hoboken, NJ: 41-51.

[1.82] Waddington, C.H. (1968) Towards a theoretical biology. *Nature* 218(5141), 525-527.

[1.83] Wiener, N. (1948) *Cybernetics or control and communication in the animal and the machine*. Wiley & Sons Inc., New York, 2d.

[1.84] Yu, H. (2023) The peculiar case of danger modeling: Meaning-generation in three dimensions. In: *Pathways to the Origin and Evolution of Meanings in the Universe*. A.A. Sharov and G.E. Mikhailovsky, (eds.) Wiley-Scrivener, Beverly, Massachusetts, USA: 363-376.

[1.85] Zlatev, J. (2018) Meaning making from life to language: The semiotic hierarchy and phenomenology. *Cognitive Semiotics* 11(1), #20180001.

2

Pathways to the Understanding of Signs and Meanings in the Biosphere: Historical and Contemporary Perspectives

Donald Favareau[1]* and Kalevi Kull[2]

[1]*University Scholars Programme, National University of Singapore, Singapore*
[2]*Department of Semiotics, University of Tartu, Tartu, Estonia*

Online resources:
Donald Favareau:
https://nus.academia.edu/DonaldFavareau
https://scholar.google.com/citations?user=Q5jp1j0AAAAJ&hl=th
https://www.semanticscholar.org/author/Donald-Favareau/7211457
https://www.researchgate.net/profile/Donald-Favareau
Kalevi Kull:
https://scholar.google.com/citations?user=fnAqqM4AAAAJ&hl=en
https://ut-ee.academia.edu/KaleviKull
https://www.researchgate.net/profile/Kalevi-Kull

Abstract
Increasingly, the examination of the phenomena of signs and meanings in the biosphere is becoming recognized as an important area of inquiry for the natural sciences. In the current chapter, we both look back to some of the earliest history of such inquiry, as well as look ahead to how we think that these ideas are developing. The chapter is then divided into two main parts, accordingly. Part 1 presents a selectively abbreviated history of some of the more influential ways that the phenomena of signs and meanings have been conceptualized in the natural sciences from antiquity to the present, while Part 2 presents some of the key insights and concepts now being developed within the field of biosemiotics for furthering our understanding of the role of signs and meanings in the organization and interaction of living systems.

Keywords: History of sign study in science, umwelt, modeling, semiotic realism, indeterminacy, choice, semiotic scaffolding, subjective present

**Corresponding author*: favareau@gmail.com

2.1 Introduction

The undertaking of this present volume — i.e., to investigate the pathways towards the evolution of meanings in the world — is a monumental one, and one that is well advanced by the other authors and their entries in this compendium. The aims and objectives of this current chapter are much more modest, however. Instead, we wish to present a review, particularly for those with an interest in theoretical biology and the history of ideas more generally, of some of the historical and conceptual pathways towards the *thinking about* the evolution of meanings in the world — especially as such thinking has been approached in the history of the biological sciences, as well as within our own personal history in the interdiscipline of biosemiotics.

The chapter, accordingly, is divided into two main sections. In the first part, we recount some work done by ourselves and others in chronicling the circuitous history of the concept of a "sign" in the natural and biological sciences, given that "meaning" is always and everywhere a function of sign relationships. "What, then, *is* a sign?" This central question, which has been asked over more than two millennia, will be briefly surveyed in this chapter, and a thumbnail overview of this concept's history in Western science will be briefly, but hopefully profitably, recounted. In the second part of the chapter, we move on to examine *one particular history* of thinking about the concepts of signs and meanings—i.e., the interdisciplinary project of biosemiotics, which has been, arguably, the most sustained locus for such investigation in the 21st century to date.

By tracing the history of science's often conflicted thinking about the nature of signs and meanings in the natural world, and focusing on, in particular, the history of certain earlier attempts at developing a naturalistic understanding of meaning, we hope to contextualize many of the issues that the contributors to this volume are addressing, and to highlight those aspects of the undertaking that have proven most amenable, and most intractable, so far.

2.2 Part One: A Brief History of the Conceptualization of Signs and Meanings in the Natural Sciences

2.2.1 Pre-Modern Theories of Sign and Meaning

The usage of signs and meanings for navigating the world goes as far back into human pre-history as we can imagine, and almost certainly has its roots in pre-human—and most likely even pre-mammalian ways—of being [2.120] [2.121] [2.54]. For it is hard to imagine an animal of any kind that does not rely upon the signs of the world given to it by its senses as the guide by which it organizes its behavior in the world [2.106] [2.107] [2.57] [2.18] [2.89] [2.112] [2.77]. However, among all the living beings who use signs to guide their actions, apparently only human beings have developed the self-reflexive awareness that they are doing so [2.18] [2.23] [2.27].

For early humans, certainly, the efficacy of ancient hunter-gathering—and eventually agriculture, medicine, animal husbandry, migration and architecture—were all critically dependent on discovering and accurately interpreting the signs of the natural world, and in teaching their children to do likewise so that the knowledge of these hard-won "sign and meaning" relations would aid them, too. Accordingly, preserving the knowledge thus obtained through successful sign interpretation, usually through an oral or mimetic tradition, was as much of a paramount concern for pre-literate societies and individuals as it is today for us [2.101] [2.36] [2.32].

Moreover, with the rise of cultures based increasingly not only on *taking* meaning from the signs presented to them by nature, but in actively *creating* new signs to which their own arbitrary meanings could be self-consciously and conventionally assigned, the technology of spoken (and, eventually, written), language created the conditions whereby humans can explicitly reflect upon the nature of the relationship between sign and meaning *per se* [2.57] [2.18] [2.109] [2.23]. In sum, we can say that (1) all sentient beings use signs to guide their actions in the world; (2) only human beings have become aware that they are using signs, and can therefore intentionally create new signs *as* signs; and thus (3) it has only been since the beginning of semiotic study proper in antiquity, that the questions of "how do signs work" and "what is a sign *per se*" have been subject to investigation.

Indeed, Manetti [2.88] notes that perhaps the earliest recorded "scientific" treatment of the sign-meaning relationship appears in the massive *Corpus Hippocraticum*, the collection of texts attributed to the ancient physician Hippocrates of Kos (c.460–c.370 BC) and his followers, and whose principles have been said to form the basis of Western medical practice [2.52] [2.67]. For here is where we find the first sustained written treatise on the nature of the "sign," or *sēmeion* (σημεῖον), which is described by Manetti as follows:

> the inferential procedure (or *logismos*) begins with a process which is essentially 'abductive.' The individual phenomenon which the doctor observes is hypothesized as a case of some general rule. In other words, the *hekaston* ('singular phenomenon'), which is in itself insignificant, is thought of as *a semeion, a sign which refers to a system from which it receives its meaning*. The first, ascending, movement of the construction of a system of reference is followed by a second, descending, movement of verification. If the hypothesized system is valid and works, it can be proved by applying it to other cases. The sign is thus transformed in a sort of *tekmerion* ('the proof') and the method becomes deductive [2.88], p. 16 (italics added).

Often regarded as one of the first true works of scientific inquiry [2.92] [2.8] [2.55] [2.52], the texts compiled in the *Corpus Hippocraticum* have been posited to be from a medical library in Kos that Hippocrates himself may have established for his students [2.90], though treatises from both before and after that time appear to also have been included in its surviving form [2.65].

Discussions of the *Corpus Hippocraticum* such as those cited above often highlight the originality of the ancient Greeks in favoring "rational empiricism of a strictly scientific character" [2.65], p. xxi, over supernatural explanations and divination practices. Here, what Jones calls the ancient Greeks' "rational empiricism" was, for better and for worse, a far cry from the methodological rationalism and empiricism of Enlightenment science, but marks instead what the philosopher and historian of ideas John Deely characterizes as the beginning of a truly "semiotic" (i.e., self-reflexively sign-considering) understanding of nature, as opposed to a merely "semiosic" (i.e., sign-using, but without self-reflexively considering) one.

Indeed, writes Deely, "the individuals credited with the introduction of philosophical thought into human civilizations were men who speculated on what constitutes the objects of human experience so far as those objects have or involve an existence or being *independent* of what we human beings [or for that matter, any human-like gods] may think, feel, or do" [1.21], p. 3.[1] Yet this not to say that Hippocrates and his contemporaries were not at all influenced in their conceptions about signs and meaning by the conceptions of earlier cultures and traditions, from whom their own thinking was, at least in part, derived.

Of note, and as Manetti points out, the inferential "if-then" structure of Mesopotamian divination rituals offers a way of thinking about what may be our primary and more "naturally given" understanding about sign relationships (in that we share it with all sentient beings) than do such later "equivalence" models such as the Saussurean dyad of a signifier and its signified [2.17], or the one-to-one mapping of a "code" [2.88], p. 13. For in the Mesopotamian model, we see a system for choosing future-oriented (or "anticipatory") action based on one's current perception of incomplete and often anomalous perceptual events. Accordingly, notes Manetti, the many divinatory texts that we have access to

> "...consist of lengthy lists of complex propositions, each made up of a protasis and an apodosis. The protasis is introduced by the expression *summa* (the equivalent of the conjunction 'if') and has the verb in the present or past tense. This is the 'omen', that is, the ominous sign which is to be interpreted. The apodosis usually has the verb in the future tense and forms the 'oracle', that which is *indicated or revealed by the interpretation of the sign*" [2.88], p. 14.

It will do us well to spend some time thinking more deeply about this passage, because in many ways, despite divination's obviously archaic roots in superstition practices, it reveals a surprisingly modern, and even 21st century scientific understanding about what the relation of signs to meanings may be at the most foundational level (e.g., [2.39] [2.75] [2.21] [2.112] [2.99]). For in the biosemiotic understanding that we will be advancing in the second part of this chapter, signs are first and foremost resources for guiding organismic action in a non-pre-labeled world — and only on secondary (primarily human) reflection might one realize that such sign relations "point to" those (real or imputed) "hidden aspects" of things that we conventionally refer to as their "meanings."

But because we humans must, perforce, proceed first from human perspectives and understandings [2.97], this question of exactly *how* the evident can lead to veridical knowledge of the non-evident, became one of *the* most debated questions of the ancient world. Indeed, questions about inference and the nature of evidence (or what we might think of as meanings and signs, respectively) "was a key issue of debate between the Epicureans, Stoics and Sceptics" writes Allendorf [2.2], p. 321, and would return in many guises throughout the centuries in the debates between pragmatism, naive realism, and relativistic skepticism in the understanding of the relationship between signs and meanings.[2]

---

[1] Indeed, this turn away from the inescapability of subjectivity and towards what philosopher of science Thomas Nagel [2.97] refers to as an impossibly a-perspectival "view from nowhere" objectivism, would itself come to "eschew many of the abductive, intuitive and aesthetic factors needed for a proper consideration of semiosis," as astutely observed and articulated by one of the reviewers of this chapter, and as will be discussed briefly in Section 2.2.2 (and as detailed extensively in Deely [2.22] and Favareau [2.41]).

[2] Indeed, given its centrality in the mindsets and discussions of the ancient and medieval worlds, a volume devoted to explicating this debate in terms of the current debates on this matter would be most welcome. In the interim, the reader is directed to Manetti's own *Theories of the Sign in Classical Antiquity*, James Allen's *Inference from Signs: Ancient Debates about the Nature of Evidence*, Ian Hacking's *The Emergence of Probability: A Philosophical Study of Early Ideas About Probability, Induction and Statistical Analysis* [2.87] [2.1] [2.56] [2.9] and John Deely's 2001 *Four Ages of Understanding* [2.22].

The enormity and scope of such centuries-long investigations well exceeds both our present expertise, as well as the space available for its discussion here. We may point out though, that one of the most well-known and influential arguments of the classical world regarded the possibility—or impossibility—of relying on signs to provide regular and useful knowledge of the world at all. This view has come down to us through the works of Sextus Empiricus (c. late second and early third century CE) and, in particular, the *Ten Modes of Aenesidemus* arguments that form a substantial portion of his *Pyrrhoniae Hypotyposes* [2.110], and which argue for the development of an omnipresent "suspension of judgment" regarding not only the "non-evident claims" made by others,[3] but even the ability of sense perceptions to furnish any kind of verifiable knowledge beyond themselves, and thus to act as signs.

As articulated and/or paraphrased from translations by Bury [2.110] and Annas and Barnes [2.3] these second century CE arguments, as we shall see, address surprisingly modern "semiotic" concerns. In the *Ten Modes of Aenesidemus*, the following claims refer to the limitations of relying on "sense impressions" for informing us reliably about the nature of whatever may lie beyond those impressions:

1. The same impressions are not produced by the same objects, owing to the different constitutions of different animals.
2. The same impressions are not produced by the same objects, owing to the different constitutions of different human beings.
3. The same impressions are not produced by the same objects, owing to the differences among the modalities of the various senses.
4. The same object will be perceived differently depending on our condition at the moment that we perceive it.
5. The same object will be perceived differently at different times depending on its condition at the moment we perceive it.
6. That which we perceive is never direct and singular, but always plural and constituted within a medium.[4]
7. It may be impossible to recognize the same object as such throughout the different ways that it will present itself as it undergoes changes in its character, quantity, solidity, growth, decay, etc.
8. All judgements and impressions are relative, without exception.
9. Our impressions become less critical through repetition and custom. An object impresses us differently when it is familiar and when it is unfamiliar.
10. Ethics, rules of conduct, habits, laws, legendary beliefs, and dogmatic conceptions all differ drastically among individuals and cultures. There thus exists no firm ground upon which common judgement or perception of objects can be based.

[3] Sextus' other works carry the titles: *Against the Grammarians, Against the Rhetoricians, Against the Geometers, Against the Arithmeticians, Against the Astrologers, Against the Musicians, Against the Logicians, Against the Physicists* and *Against the Ethicists*, which gives one some flavor of the project of Pyrrhonic skepticism, whose aim was *ataraxia* (equanimity, tranquility) through one's non-captivation by dogma.

[4] Sextus's exact words are worth quoting in full here: "We deduce that since no object strikes us entirely by itself, but along with something else, it may perhaps be possible to say what the mixture compounded out of the external object and the thing perceived with it is like, but we would not be able to say what the external object is like by itself" [2.110], 2 CE:73, i.e., 17 centuries before Kant.

Formed as a set of detailed rhetorical arguments against the idea that veridical knowledge of the non-evident may be accessed through a consideration and interpretation of the evident — or, to put it another way, that "a sign refers to something other than itself"[5] — what the objections raised in *Pyrrhoniae Hypotyposes* are implicitly demanding for their refutation is the development of a theory of *semiotic realism* that can account both for the irrefutable subjectivity of phenomenal experience as well as for the role that non-phenomenal reality plays in ensuring such experience's sufficient veridicality—a theory, in other words, about the naturalistic evolution of signs and meanings.

Much as would happen once again at the outset of modernity some 14 centuries later, however, the non-self-reflexive study of the "signs" of nature, e.g., in the arts of medicine and agriculture, would begin its slow trajectory of developing increasingly apart from the more intensely reflexive study of "signs" *per se* that would be undertaken in considerable depth and sophistication in the Middle Ages.

Such study, often said to have begun with Boethius's commentaries on Aristotle's reflections on the nature of the propositional grammar [2.94],[6] became generalized somewhat beyond just the domain of language and human knowing by Augustine;[7] and was developed and elaborated throughout the next ten centuries by scholastics such as Thomas Aquinas and John Poinsot in their attempts to reconcile the recently re-discovered texts of Aristotle and the ancient world with the contemporary medieval theology of the Catholic Church. And only towards the very end of that time, at the outset of the 17th century, notes Deely [2.22], was an understanding of the "sign" made sufficiently general such that one might have grounds to develop, within the same one explanatory framework, a naturalistic understanding of how the relationship of a sign to its meanings can be *either* mind-dependent *or* mind-independent, or, and much more commonly, an admixture of both.

As the contingencies of history would have it, however, another 17th century development effectively sundered the study of nature from the study of sign meanings *per se*, so it behooves us at this point to survey that bifurcation point briefly before proceeding onto our discussion of the eventual re-recognition, in our own time, of the need for a science expansive enough to be able to provide for a more rigorous understanding of the nature of signs and meanings in the human, as well as in the non-human, living world.

---

[5] The formulation derives from Augustine's *De doctrina christiana*: "A sign is something which, offering itself to the senses, conveys something other than itself to the intellect" (*"Signum est res praeter speciem quam ingerit sensibus, aliud aliquid ex se faciens in cogitationem venire"*) [2.4], p. 33 (trans. Meier-Oeser [2.94] online).

[6] Semiotic scholars Deely, Eco and Meier-Oeser maintain that the ancient Greeks themselves had never really developed a general notion of a 'sign' *per se*, but dealt with only the more specific subset of 'natural signs' such as symptoms and seasonal changes, entirely apart from their consideration of 'words' as the constituents of logical thinking and communication [2.22] [2.28] [2.35] [2.34] [2.94], — and thus the true investigation into the nature of 'sign relationships' *per se* begins only with the medieval thinkers. As with everything else in this brief introductory history, the reader is directed to these more extensively researched texts for a fuller understanding of such history and its debates.

[7] We should note that Augistine's *cogitationem* (cognition based) conceptualization, as described in footnote 4, is arguably still less generally applicable for biological investigation than is the Hippocratic formulation that "a sign is something which refers to a system from which it receives its meaning" (quoted in [2.88], p. 16).

2.2.2 The Exclusion of a Theory of Meaning in Modern Science

Fuller accounts have been recounted elsewhere, detailing how the attempt to develop a naturalistic concept of "sign and meaning" that proceeded inchmeal from ancient times to the 17th century was all but abandoned with the indiscriminate rejection of scholastic thinking at the outset of the Scientific Revolution and its development of, and commitment to ideological physicalism—and how that earlier project is now being painstakingly reinitiated in our own time in the face of the failure of reductionist materialism to satisfactorily account for the complexity, cognition, and novelty-generation of living systems [2.25] [2.26] [2.41] [2.45].

For our purposes here, suffice it to say that part of the grand project of the initiators of modern science, as was advocated explicitly in the massively influential works of Bacon [2.6] Galileo [2.50] and Descartes [2.30] [2.31], was to "start anew from fresh foundations", given the cumulative failures of theological cosmology and Aristotelean physics to adequately account for the new discoveries in astronomy and the mechanics of matter and motion.

Arguably, the most radical and consequential of these well-justified attempts to put the future of scientific inquiry on a newer and more "secure foundation" were Descartes' 1637 *Discourse on Method* [2.30] and his even more broadly consequential *Meditations on First Philosophy* of 1641 [2.31]. For it is in there, wherein we find the ontological distinction made between what are claimed to be two mutually exclusive modes of being in the world—that of *res extensa* (weight, dimension, mass, measurability) and that of *res cogitans* (thought, belief, desire, subjectivity)—and the claim that it is only the latter that we can, by using the same tool of deductive rationalism that will aid us to examine the rest of nature, be most certain.[8]

Accordingly, Descartes is widely credited not only with being one of the central founders of Modern Science, but also being perhaps the central founder of Modern Philosophy, whose overriding and ultimately bio-semiotic, question then became: "How can we ever have secure knowledge of the outside world, when all our access to it comes through our limited and fallible senses?" Descartes' answer was the ontological - and thus the methodological - bifurcation of mind and body. This move, augmented by the experimentalism of Bacon and the mathematized empiricism of Newton, turned the modern investigation of nature — especially in its inanimate aspects — into a tool of understanding and discovery that is still unrivaled in all of human history. "On the scientific front," writes Deely,

> "there was almost no down-side to the modern revolution. Men and women of science went from victory to victory in the study of nature.... On the philosophical front, however, there proved to be a problem right from the start. The new method had no trouble getting the rationalist experiment of modern philosophy off the ground. The problem was getting back" (to a natural science that could include "subjectivity" itself as one of its objects of investigation) [2.22], p. 544.

[8] The similarity of the objections of the aforementioned *Pyrrhoniae Hypotyposes* to the method of hyperbolic doubt that Descartes employs in the *Meditations,* and to which he offers his Deductive Rationalism as the answer, has often been remarked upon (e.g., [2.49] [2.100] [2.105] [2.100]), lending credence to the idea that a major part of Descartes' project was the discrediting of scholasticism root and branch [2.105] [2.22] [2.26].

And indeed, so insurmountable was Descartes' challenge that the reliance on mere "signs" of the world alone could never furnish us with veridical knowledge of "how the world really is in itself", that almost all the major philosophers who came after him, including Locke, Hume, Berkeley, Kant (and later Hegel, Fichte, Schelling, Husserl and Peirce, to name just the merest handful), all in some sense made attempts to grapple with the challenge put forth by Descartes' dualism — as well as with each others' proposed "solutions" to its paradoxes.

Like Deely, historian and philosopher of science Bruno Latour sees this moment in the intellectual history of the West as an epistemological "wrong turn," the consequences of which we are only now beginning to fully appreciate. For "in asking how an isolated mind could be absolutely, as opposed to relatively, sure of anything in the outside world," observes Latour,

> "Descartes framed his question in a way that made it impossible to give the only reasonable answer ... [i.e.,] that we are *relatively* sure of the many things with which we are daily engaged... [But by Descartes' time], this sturdy relativism, based on *the number of relations established with the world,* was already in the past, a once-passable path now lost in a thicket of brambles" [2.82], p. 4; (second italics ours).[9]

Even more consequentially: so deeply entrenched became this belief that successful investigation of the world could be properly split apart into the investigation of matter removed from the consideration of subjectivity and the investigation of subjectivity removed from the consideration of matter, that over the course of the next three centuries the increasingly estranged "two cultures" of the natural sciences and the humanities [2.114] diverged to the point that each began to become both ignorant and — in far too many cases — suspicious of the validity of the methodologies and the findings of the other.

Moreover, if Deely and Latour were right that the insurmountable problem of philosophy after Descartes concerned how to get back to ground-level reality, they perhaps underestimated the complementary problem introduced by this exact same dualism into modern science. To wit: as hard as it struggles, modern science finds it hard to "get *off* the ground" level of reductionist materialist explanation in order to be able to account successfully for the higher levels of emergent relational causation and ontology that characterize the organization and interaction of living systems. To ask about the role that "signs and meanings" may play in the structuring of such other-directed and end-directed systems is to challenge an epistemological and institutional paradigm that, by this point, has become profoundly entrenched in contemporary science.

Pioneering microbiologist Carl Woese addressed this very issue in a 2004 article entitled "A New Biology for a New Century" that is worth quoting here at length:

> "The 19th was biology's defining century. There, for the first time, biology's great problems lay scientifically outlined and assembled, with all of them effectively in early stages of development ... some (like the natures of the gene and of the cell) cried out for dissection, analysis in terms of their parts; whereas others (such as evolution and morphogenesis and the significance of biological form in general) were holistic, metaphysically challenging, and not fundamentally understandable as collections of parts. ... Given

[9] See also [2.97] for an illuminating discussion of the "subjectivity *versus* objectivity" problem in science.

> the temper of the times, the entry of chemistry and physics into biology was inevitable [and] the technology that these sciences would introduce was not only welcome, but very much needed. But *the physics and chemistry that entered biology (and especially the former) was a Trojan horse, something that would ultimately conquer biology from within and remake it in its own image.* … [for the] fundamentalist reductionism of 19th century classical physics is in essence metaphysical. It is *ipso facto* a statement about the nature of the world: [i.e., that] living systems (like all else) can be completely understood in terms of the properties of their constituent parts. This is a view that flies in the face of what classically trained biologists tend to take for granted, *the notion of emergent properties.*" [2.123], pp. 184-185; 175, (italics ours).

Within such an impoverished and counterproductively restrictive metaphysics, it would take some major new discovery or technology to bring to the attention of those working within its mainstream paradigm[10] some new set of problems, needing new conceptual tools and approaches for their unraveling. And by the mid-20th century, modern science was profoundly confronted with at least one of each of these, as we shall see.

2.2.3 The Realization of the Need for a Theory of Signs and Meanings in Modern Science

According to embryologist and code-biologist Marcello Barbieri, "the discovery that life does not consist only of matter and energy, but also of *information*" was decisively forced upon scientists with the discovery of the genetic code in the 1960s, when it was established that there was no stereochemical necessity between any given sequence of nucleotides and the amino acid that it "coded" for [2.7], pp. 24, 5. "The mere presence of organic codes in every cell, starting from the genetic code," writes Barbieri, thus "requires a theoretical framework where *organic meaning* is a necessary complement of organic information," given that the coding relationship itself has a semantic aspect that joins together that which would not otherwise be joined by a physically necessary relationship of forces [2.7], p. 94.

By the time of that discovery, research undertaken in the development of mid-20th century telecommunications technology and in flexibly adaptive feedback mechanisms, such as those being developed for self-guided missile systems, had likewise made the notion of "information" an equally central one to those technologies and the industries supporting them, birthing the "information-theoretic" fields of modern computer science and cybernetics, respectively.

But the notion of "information" adopted in those latter fields is a highly technical and specific one, and one that Claude Shannon, in his pioneering development of Information Theory, made a specific point of distinguishing from the altogether semantic concept of "meaning." An electrical engineer and mathematician employed by the Bell Telephone Company's wartime research arm in 1944, Shannon's massively influential 1948 report entitled "A Mathematical Theory of Communication," begins with the assertion that

---

[10] What Kuhn [2.69] calls the paradigm of "normal science," as opposed to "revolutionary science" and what Latour [2.81] calls "ready-made-science," as opposed to "science-in-the-making" — both of which are needed, serially, for science to succeed.

> The fundamental problem of communication is that of reproducing at one point, either exactly or approximately, a message selected at another point. Frequently the messages have *meaning*; that is, they refer to or are correlated according to some system with certain physical or conceptual entities. *These semantic aspects of communication are irrelevant to the engineering problem.* [2.111], p. 379, (second italics ours).

And, indeed, for companies whose sole mission lies in the electronic transferring of messages from one point to another, this focus on the successful replication of a string of sign-*vehicles* from one point to another through a channel—leaving aside entirely, and by design, the interpretation and the "meaning" of those sign-vehicles for the agents involved in creating and interpreting them—is entirely reasonable, appropriate and sufficient for the task at hand.[11] The same cannot be said for the impoverished notion of "information" that came to be adopted within molecular biology, however [2.38].

"Information means here the precise determination of sequence, either of bases in the nucleic acid, or of amino acid residues in the protein" writes Francis Crick, in describing his self-proclaimed Central Dogma of Molecular Biology, which states that "such information cannot be transferred from protein to either protein or [back towards the originating] nucleic acid" [2.15]. For Crick, "information" here *is* the quantifiable correspondence, the specificity that a physical sequence of one kind (in the DNA or RNA) can instantiate in a physical sequence of another kind (the polypeptide chain) via the fixed physical sequencing structure of a third kind (the intermediary ribosome and its tRNA).

But so long as such determinative specificity remains fixed at the level of pre-fixed physical and deterministic correspondences, a realistic and sufficient account of biological signs and meanings cannot, to recall our earlier metaphor, "get off the ground." Indeed, "an up-to-date biology must acknowledge that the biochemical concept of information is just too impoverished to be of any explanatory use," writes molecular biologist and biosemiotician Jesper Hoffmeyer, for "information isn't just something that is transported or that streams through a channel; information is something [i.e., a set of signs] that must be understood—and which therefore can be misunderstood" [2.58], p. 61.

And yet, as Woese noted, so pervasive has been the influence of a physicalist reductionism—one wholly appropriate to the investigation of inanimate matter, yet woefully inadequate for the investigation of living beings—that the reification of the concept of "information" as "specifying cause" now permeates not only all of molecular biology, but even biology writ large — where the concepts of *signal, message, stimuli,* and *releasing mechanism* are all likewise now used promiscuously to indicate, in effect, "determinate molecular processes that we just haven't discovered yet." It is for this very reason that the late 20th century inter-discipline of biosemiotics, which we will be examining in the next section, was founded. Writes Favareau:

> "At a time when strictly materialist reductionist explanations of life and its evolution have become increasingly incompatible with what biologists are now conceding is the

[11] More fecund than its merely telecommunications application, perhaps, has been Shannon's corollary idea, derived from Boltzmann, of defining the "measure" of "information" successfully transmitted as the reduction of probabilistic "uncertainty" effected at the receiving end [2.111] — an idea with which Terrence Deacon, among others, has been working to properly augment and refine in the consideration of the origins of biological meaning [2.19] [2.20] [2.21].

> complex, adaptive and non-linear nature of organization and interaction in the natural world, the conceptual work now taking place under the aegis of the 'biosemiotic perspective' is grounded in the conviction that the living organism must be understood not only in its material organization, but also in its *relational and interactive situatedness* — and, above all, in the superordinate sign relationships that effectively join the two" [2.47], p. 227.

For despite Woese's well-justified concerns, and as in all eras always, there were and are exceptions to "the spirit of the age." In the medical faculties of many universities of 19th century Europe, for example, the medical study of signs maintained and to some extent developed the medieval tradition, in parallel with the introduction of the newer physical and chemical explanations of biology detailed by Woese. There, such "study of the signs of health and disease" was taught under its proper name of "semiotics" and in at least one case, even the term "biosemiotics" appears [2.115] [2.37] [2.33] [2.118] [2.78] — and this tradition was one of the roots from which philosophers in the last decades of the 19th century took up "the study of signs" in earnest.

This renewed investigation into the nature of signs and meanings *per se* was further developed towards the end of the 19th century perhaps most notably by Edmund Husserl's investigations into the phenomenology of experience and the various schools of thought that it inspired, and by Gottlob Frege's development of symbolic logic and the initiation of the linguistic turn in analytic philosophy. Not long after, Swiss linguist Ferdinand de Saussure's [2.17] initiation of the "synchronic turn" in linguistics ushered in whole fields of studies devoted to uncovering the structures by which signs are associated with their meanings. And from 1859–1913, American logician, scientist and philosopher Charles Sanders Peirce labored to produce a systematic "logic of sign relations" that would, in the words of the philosopher T.L. Short, provide "a naturalistic but non-reductive account of the human mind [as well as to] explain and defend the claim that the sciences are objective in their mode of inquiry and, in fact, yield knowledge of an independently existing reality" [2.113], p. ix, which was the goal of the medieval scholars before the bifurcations of modernity.

Perhaps most importantly of all for the concerns of the volume at hand, the first decades of the 20th century saw the pioneering studies of biologist Jakob von Uexküll [2.120] [2.121] investigating the centrality and establishment of sign relations in the lives and biological organization of all sentient beings, work that would later inspire the development of such fields of ethology, sensory biology, and biosemiotics [2.11] [2.95] [2.71] [2.124]. This roughly 80-year period from the 1860s to the 1940s, before the Modern Synthesis effected a sharp turn back into biological reductionism, is a period in biology that has since provided many insights and concepts that would be taken up by biosemioticians at the end of the 20th century [2.89].

And, indeed, it is the thoroughgoing influence of the ideas of Uexküll and Peirce, as well as systems theorist and anthropologist Gregory Bateson, that led one of the founding figures of the contemporary biosemiotics project, Jesper Hoffmeyer, to observe that "what biologists mean when they use the term *information* is not information at all in the sense of Information Theory, but information in the same sense as semioticians use the term *semiosis*: sign processes" (personal correspondence, 2016). It is thus to the interdisciplinary research project of biosemiotics that we now turn.

2.3 Part Two: Towards a Future Science of Signs and Meaning[12]

2.3.1 Biosemiotics: Searching for the Most General and Fundamental Aspects of the Sign Relation

Both of the present authors have written extensively elsewhere on the history of the project to "join the life sciences with the sign sciences" that was initiated by linguist Thomas A. Sebeok in the 1980s, and that continues today under the name of *biosemiotics*. We therefore have decided that it would not be an optimal use of our space in this volume for us to simply reiterate that history here again (interested readers may instead see [2.72] [2.73] [2.41] [2.45] [2.47] [2.108] [2.13] for accounts of this recent history).

What we have chosen to do instead is to present here a brief review of some of the key concepts that we believe any future "science of signs" will have to provide a satisfactory naturalistic account for. Where relevant, and in the spirit of rectifying the cleavage between the sciences and the humanities that was effected at the onset of modernity, we will also detail how such concepts have been re-conceptualized from their original articulations and disciplines, and then modified, re-defined, or naturalized in such ways that their insights would be useful for the examination of sign processes in life processes, as a result.

Below, we list a series of some of the most commonly encountered concepts that have been so incorporated and developed in biosemiotics within the last few decades. Do note, however, that since biosemiotics is an ever-developing and ultimately non-centralized project, this list reflects primarily those concepts upon which the present two authors have tended to focus in their own thinking about "the nature and evolution of signs and meanings." The list's inclusions and omissions should therefore be understood as being accordingly non-exhaustive and provisional—no one person or group has the authority to "speak for" all biosemioticians, so instead, here we two authors share some of what we have learned from being part of its interdisciplinary community for many years.

2.3.2 Major Concepts and Terms Used in Biosemiotic Analysis

Sign: A foundational understanding in biosemiotics is that signs are not independently existing relationships or entities in the world, but that the patterned regularities and entities of the world can be, and often are, used to function *as* signs by some living agent (see: Semiotic Agency). Dark clouds preceding rain may be regularities of nature, for example, but such clouds are not "signs of rain" until they *are taken to be so* by some organism, and not all organisms will take them so as such. This means that the sign-hood of any given "sign" is an agent-centric *orientation* toward the world (see: Umwelt), and that semiotic

[12] We should note here at the outset that the term "meaning" is rarely used as a *technical* concept in present-day biosemiotics. This disinclination owes its legacy, perhaps, to Charles Morris' observation that "this term, though useful enough at the level of everyday analysis, does not have the precision necessary for scientific analysis… [as it has been used to signify] any and all phases of the sign-process (the status of being a sign, the interpretant, the fact of denoting, the significatum) and frequently suggests mental and valuational processes as well; hence, it is desirable for semiotic to dispense with the term and to introduce separate special terms for the various factors which 'meaning' fails to discriminate." [2.96], p. 19. Accordingly, the move in the present text has been an emphasis on the far more technically developed corollary notion of "sign."

function inheres only in relationships, never in the individual components of those relationships *per se*.

To state the matter even more precisely: the term "sign" has use only in reference to a formal relationship that an agent participates in by using: (a) any given element in the world that it is capable of registering to (b) guide its subsequent action, in the context of (c) other elements that can be systemically associated with the registered element itself. Only with such a sufficiently broad definition can we speak sensibly of "sign" processes operative on the cellular level, as well as between organisms (see: Semiosis, Interpretation and the Gradations of Semiotic Generativity).

Sign Types: Given the above understanding that sign relations are, in the first instance, functional relations established by living systems, various ways of classifying such relationships have been proposed within biosemiotics. Kull, modifying Aristotle, distinguishes the gradations notable along the continuum of vegetative, animal, emotional, and cultural signs [2.76], while Deacon modifies logician Charles Peirce's triadic division of iconic, indexical and symbolic signs to argue for a nested evolutionary hierarchy going from perception to association to participation in systems of displaced reference [2.18], all grounded in a naturally occurring teleodynamic that instantiates a basal *ententionality* in living systems [2.20].[13] Stjernfelt [2.116], meanwhile, employs the Peircean-inspired categorizations of qualisigns, dicisigns, and legisigns in exploring the formally propositional truth-testing relations (such as cause and effect), in which all living beings are in various ways engaged.

Finally, the distinction between generative and de-generative signs is also important to biosemiotics and will be discussed in this chapter in the section entitled Interpretation and the Gradations of Semiotic Generativity. As should be clear from the forgoing discussion, however, the distinction of sign types based merely on their modalities (e.g., auditory, tactile, visual, etc.)—while an important area of study in its own right—is not an illuminating classificatory scheme within biosemiotics, given that all of the above sign types could, in principle, be instantiated using any of these modalities, as we will see below in our discussion of the animal umwelt.

Umwelt: If there is a single central organizing concept around which biosemiotics has developed in the last thirty years, it is the notion of the subjectively experienced animal *umwelt*, as introduced by pioneering 20th century physiologist Jakob von Uexküll in 1928 in his landmark volume *Theoretische Biologie* [2.120]. The biology of every organism, notes Uexküll, both enables and delimits all that organism can perceive of the surrounding world, as well as all that it can do within it—and for all organisms (including us), that fraction of the external world that we can perceive or effectively act upon is but a sliver of the existing whole. It is the organization of these delimited perceptions and actions into relationships of

[13] Deacon's coining of the neologism *ententional* to denote "all phenomena that are intrinsically incomplete in the sense of being in relationship to, constituted by, or organized to achieve something non-intrinsic" [2.20], p. 27, shares much in common with John Deely's notion of *being towards another* as an organizing principle in living nature [2.27]. Such phenomena, argues Deacon, include: "functions that have satisfaction conditions, adaptations that have environmental correlates, thoughts that have contents, purposes that have goals, subjective experiences that have a self/other perspective, and values that have a self that is benefited or harmed" [2.20], p. 27. For while "the paradoxical intrinsic quality of existing with respect to something missing, separate, and possibly non-existent is irrelevant when it comes to inanimate things," notes Deacon, "it is a defining property of life and mind" (ibid., p. 3; see also Deacon's chapter in this volume).

functional meaning for living in that largely unseen world, that constitutes the "phenomenal, or self-world" that is the organism's Umwelt [2.121].

Not to be confused with the "entirety of the outside world" that is often suggested by the word "environment," nor with those materially subsisting conditions referred to by the concept "niche," Umwelt is, rather, the meaningfully *experiential* world within which every organism (again, including us) inescapably finds itself — and, based upon which, all of its choices for action must be made. By definition, such a world is composed exclusively of the *sign-relations* between perception, action and consequence that have been established by the organism (and its lineage) in its explorations of the relations subsisting in the external world.

Accordingly (a) the umwelt of every species, and in some sense every individual, will be particular to it, and (b) for such an umwelt to be viable enough to meet the requirements for living, its constitutive sign-relations must be both sufficiently veridical *and* modifiable (see: Semiotic Realism, Habit). In the largest sense, Uexküll's umwelt concept of agents imprinting "meaning" on the objects of their perception through their actions upon those objects, and based upon the subsequent consequences experienced, underlies what semiotician John Deely means by the assertion that "a sign [relation] is what every object [of experience] presupposes" [2.24], p. 158.

Semiosis: Semiosis is the term denoting the production and transformation of such sign relations by living beings, and traditionally refers also to the acting upon such already established sign relations so as to produce meaningful interpretations and meaningfully informed actions, both upon the objects of the external world (e.g., [2.12]), as well as upon other signs and sign-users, such as in the acts of signaling, communication, translation, etc. Understanding how such processes may have arisen in the natural world prior to the appearance of human symbolic culture is the goal and purview of biosemiotics [2.57] [2.18] [2.48] [2.77] (see: Interpretation and Gradations of Sign Generativity).

Semiosphere: The concept of semiosphere, in analogy with the concept of the biosphere, was introduced by Juri Lotman in 1984, in his attempt to re-orient the locus of semiotic study from the traditionally atomistic analyses of "the sign" to the examination of the larger systems of sign relations, within which such individual signs have their meanings [2.86], p. 194. Biosemiotician Jesper Hoffmeyer uses the term similarly, writing that

> "the semiosphere is a sphere just like the atmosphere, the hydrosphere, and the biosphere. It penetrates to every corner of these other spheres, incorporating all forms of communication: sounds, smells, movements, colors, shapes, electrical fields, thermal radiation, waves of all kinds, chemical signals, touching, and so on. In short, signs of life" [2.57], p. vii.

Accordingly, the term semiosphere can refer to the total sphere of meaning relations enacted by all the living systems that currently inhabit or have at one time inhabited the Earth, or it can be applied in the analysis of a more restricted area of sign-exchange, such as that of a particular ecosystem or cultural moment in time. In either case, it is, as Lotman notes, both the result of prior sign usage and the condition for its future continuance [2.86], p. 125.

Habit and Semiotic Scaffolding: The term "habit" is an admittedly odd-sounding one that was inherited in biosemiotics through its association with the 20th century logician

Charles Sanders Peirce. It refers to those regularities found in nature (and in culture) that could have at some point in time been established otherwise, but that have since become so regular and normalized as to appear as law-like behavior, in any system "plastic" enough to take and change such system-internal norms [2.99], CP 6.621. Biosemiotics, for the most part, understands such habit-taking in ways akin to contemporary dynamical systems theory. There, the establishment of higher-order regularities within a complex adaptive (and especially a *living*) system is often the *emergent* product of multiple interacting lower-order regularities, and the systemic resolution of their conflicting effects. Such higher-order regularities may then come to exert formative "downwards" effects upon the system, sustaining the system in ongoing loops of generative self-regulation and normativity. It is by the virtue of such "emergence" in the organization and evolution of living systems, writes Hoffmeyer, that "entities and systems impose a new set of boundary conditions or constraints on their individual components [and in so doing,] establish a set of hitherto nonexistent rules" [2.58], pp. 36-37.[14]

Such "rules" (understood in the widest sense of the system-specific "norms" that such emergence and its downward effects have generated[15]) "can be seen as enabling processes of sign action unfolding at several levels of organization," writes biosemiotician Claus Emmeche, "focusing energy flow and agency of the system or subsystem upon a constrained repertoire of possibilities, thus guiding the system's behavior to follow a more definite sequence of events" in the manner of a Peircean *habit* [2.39], p. 275.

The centrality of these rules as organizing principles help govern the organism not just ontologically (e.g., in the workings of its species-specific biology), but epistemologically as well, in the form of the evolutionarily and ontogenetically acquired the "norms and habits" with which it interacts with the world and other organisms. Accordingly, writes Kull,

> "semiosis as an active meaning-seeking-making process often results in the building of some relatively static, or even quite solid, structures that somehow embed in themselves the findings of that active searching-event of semiosis. The resulting structure is a *scaffolding*. It canalizes further behavior. It is the frame for *habits*" [2.74], p. 228.

A semiosphere in this sense, can be thought of as a collection of such interacting habits. Most importantly, it is in only in the context of such dynamical systems-inspired thinking, when properly amended for living systems by the addition of a semiotic scaffolding

---

[14] Cf. Deacon, who defines "emergence" as "[the phenomena whereby] unprecedented global regularity [is] generated within a composite system by virtue of the higher-order consequences of the interactions of its composite parts" [2.19], p. 122; "here, it is the *relational properties* of the constituents (as opposed to their primary or intrinsic properties) that constitute the higher order" [2.20]. Given the ubiquity of such phenomena observed in the organization and interaction of living beings, Nicolas Rashevsky and Robert Rosen in the late 1950s attempted to develop general functional models of such systems based on their *relational* structures, under the aegis of "relational biology" [2.102]. Relationships and higher-level organizational effects and regularities — not the material analyses of structural components and their local interactions — were to underlie the 'relational biology' that Rosen and Rashevsky were hoping to inaugurate (ibid.), as captured by Rosen's oft-quoted aphorism that "if science insists on chasing particles, they will follow them right through an organism, and miss the organism entirely" (quoted in [2.53], p. 108). Rosen is thus often seen as a kindred spirit in biosemiotics for his insistence that living beings are *modeling* systems, and that their ability to *anticipate* is rooted deeply in their relational structure, which is the result of such modeling (see: Modeling).

[15] Hence the appellation "*auto-nomos*" (or "own norm" making) systems.

component, that we may most easily understand the aphorism that "organisms are their own designers" [2.70] [2.91] — as well as the biosemiotically generalized notion of "choice."

Indeterminacy, Choice, and the Subjective Present: We have seen above that signs are *relationships* that organisms establish and use, and that such establishment—both evolutionarily and ontogenetically—brings forth the organism's *umwelt;* while such use, collectively, brings forth the organisms' *semiosphere.* But the individual's *experience* of each, it should be stressed, takes place always in the "subjective now" [2.75] or "specious present" [2.64]—itself an accomplished sign relation wherein multiple sign relata are experienced *as a simultaneity* within the species-specific duration of time that constitutes each species' "individual moments" of experience. At such moments, the organism may respond to the experiencing of such sign-relational simultaneity with activity or with quiescence, and such decisions are often guided both by habit and by motivatedness—a term that Saussure introduced to designate "the limiting of arbitrariness" [2.17], p. 133, inherent in all sign relations.

In biosemiotics, such "arbitrariness" differs from "randomness" by its requirement for the energetic work of negentropic difference-making; such *arbitrarity-exploiting* sign relations "do not make themselves," nor are they "the meaningless products of random chance." Instead, they are based on the unique ability of organisms to *orient* to singular parts of the world in ways that are "informative about" its systemic relations and possibility affordances within the umwelt. Such orientation requires the ability to "assign" *relational* meanings to sensations and actions that in themselves are mere singular "events" in the subjective present.

It will often be the case, however, that the moment of the subjective present may present to the organism a simultaneity of potentially conflicting sign relata, for which it can rely on no pre-established habit nor sequential, algorithmic-like formula or heuristic for determining its interpretation or response. It is in the face of such indeterminacy that the organism must choose in a way that is not pre-determined for it. For although it may be both situationally motivated and constrained, the organism's resulting decision (or interpretation) is not the outcome of "causal necessity" but of the organism's active non-deterministic agency and *choosing* between simultaneously available possibilities.

Favareau, without wishing to stretch the analogy beyond its scientific limits, has dubbed such choice-making as "collapsing the wave function of meaning" [2.42] [2.46] while Kull concurs that "in the process of choice, options collapse, and the traces of such choosing may become a rule or habit — and these become the basis of semiotic order" [2.79], p. 87. We may also say that these are the moments when "semiosis" happens in its most fully-fledged measure and when the organism is enacting its fullest measure of semiotic agency (see Gradations of Semiotic Generativity).

Semiotic Agency: One important result of the theory of complex dynamic systems as it was developed in the second half of the 20th century has been the demonstration of conditions in which indeterminacy appears. The latter may take place in non-linear metabolic networks with autocatalytic features, these being the necessary conditions for agenthood, whereby semiotic processes become possible [2.20] [2.66]. Accordingly, every semiotic activity presupposes an agent—a metabolic system or organism that can provide physical energy for the processes of interpretation to be carried out. Hoffmeyer [2.58] [2.59] and Hoffmeyer and Stjernfelt [2.60] see the increase in biological and environmental complexity driving the increase of semiotic complexity—and thus of semiotic agency, defined by them as the measure of control that an agent can exert over their own sign processing. More recently, Sharov and Tønnessen have written that

> "A semiotic agent, which is a system capable of acting purposefully by selecting actions while making use of signs, is a distinct category of being with *an extended ontological status*, because ...[its] being is both a result of past purposeful activity [of itself and its lineage] and a source of purposeful activity in the future" [2.112], p. 343.[16]

Such purposeful activity, it has often been suggested in biosemiotics, is guided by the processes of *modeling* at play both within the organism's umwelt and its semiosphere.

Modeling: The concept of "modeling" has a long history in biosemiotics, dating back to one of its precursor feeder streams in the Moscow-Tartu school of semiotics, where it was theorized as a central organizing principle of human culture and cognition [2.45], pp. 191–196, and Sebeok and Danesi later expanded this analysis to include the semiotic activities of non-human species [2.109]. Theoretical biologist Robert Rosen, as noted earlier, made the concept of modeling central to his project of moving beyond a merely cybernetic model of the organism based on "feedback" and "reaction," no matter how complex, to articulate an explicit understanding of organisms as "anticipatory systems." These systems are organized such that "[a] present change of state depends upon [an orientation towards] future circumstances, rather than merely on the present or past" [2.103], p. v.[17] The capacity for such "anticipation" requires modeling; "organisms," writes Rosen, "comprise those systems which can make predictive models (of themselves, and of their environments) and use these models to direct their present actions" (ibid., pp. 365–366).

It is important to point out here that the notion of organismic "modeling" should not be conflated with a kind of naïve representationalism—something like a theory of "pictures in the brain that an organism carries around and must consult like a map before deciding upon its actions"—this latter idea having been rightly criticized by adherents of enactivism, which see no need for such "Cartesian theater" intermediaries [2.98] [2.51] [2.61] [2.62] [2.63]. The biosemiotic concept of modeling—which is the interactive shaping of a physically instantiated set of evolvable constraints and system biases (a process one finds operative in organisms far below the complexity at which brains or image-forming capabilities of any kind are possible)—is first and foremost a *bodily*-configured orientation for the negotiation of indeterminacy. Uexküll's functional cycle is not a map that is consulted, but—just as enactivism would have it—an embodiment that has been accomplished in interaction with a world.

The Process-Relation Explanandum: In a similar manner, previous analyses in general semiotics have tended to focus on the whole-organism (and usually, human) relations of "perception," "communication," and "semiotic experience." Such analyses, while helpful in establishing certain general principles of semiosis, yet require additional modifications and

[16] We direct the reader to the Sharov chapter in the present volume for a more detailed explication of this important new perspective on Semiotic Agency.

[17] Cf. Uexküll's dictum that: "Meaning is the guiding star that biology must follow: the rule of causality is a poor guide, [because] causal relationships deal only with antecedents and consequences, thereby completely concealing from us broad biological interrelationships and interactions" [2.121], p. 343. Likewise, although Rosen never mentions Uexküll in his work, his emphasis on holism rather than atomism, and on the centrality of relation — and most especially sign relations — in the organization and interaction of living systems, proceeds from the same foundational realizations as those evidenced in Uexküll's two earlier and programmatic quotes.

concepts before they can be applied to the workings of the cellular, or endosemiotic, realm of sign processing. There, a sign's "function" would be determined by the role that it plays in the production of what Peirce would call an "energetic" (as opposed to "emotional" or "logical") interpretant. Remembering that an interpretant is always an enacted response to the reception of a sign,[18] on the endosemiotic level, the production of a sign interpretant would be decisive in directing a semiosically organized system towards instantiating one particular next biological state or action, rather than another equally possible one (see: Interpretation and the Gradations of Semiotic Generativity).

It is because of this ability that, as the current authors have pointed out previously:

> "A principal feature of the *sign relation* is its 'aboutness'—the property of being about something other than itself. This is the feature that characterizes the phenomenal world and all knowledge...[and in *this* aspect of its being] a "sign" is an entity of a logical kind: atemporal and non-local. The *sign process* that creates the relation, however, belongs to the realm of the empirical, and thus has a duration and locus. ...[The] complex of relation-generating processes that brings the sign relation into being include the physiological processes and mechanisms needed for the relation to appear — indeed, these are what create and support the relation within and between organisms, and without which, such relations could never occur" [2.80], pp. 20-21.

Thus, we may say that the manifestation of any given sign-directed process proper consists of a specific organization of logical and atemporal relations, and of the temporal and embodied mechanisms that can instantiate a material organization in accordance with those virtual relations. "Accordingly," write the authors "every complete model of signs should describe both of these ineliminable aspects of its "sign-hood"." [2.80], p. 21.

This most potentially important area of biosemiotics is currently the most in need of development. Terrence Deacon's work on the possible origin and nature of *absential relations, ententionality,* and *teleodynamics* in living systems has been both pioneering and influential in its attempt at developing scientifically rigorous explanations of such processes.[19] The efforts of J. Augustus Bacigalupi to develop a neuromorphic dynamic systems model grounded in semiotic, information and thermodynamic theorizing, likewise seeks to realize this most urgently needed synthesis [2.5].

Interpretation and the Gradations of Semiotic Generativity: In everyday usage, an "interpretation" denotes a singular "making sense of" a particular sequence of signs (usually words, texts, artworks, or forensic evidence). While not a wholly incorrect definition, it is far too loaded with cognitive and anthropocentric connotations to be applied to the kind of sign-based processes one finds taking place within a living system. It is for this reason that biosemiotics uses Charles Peirce's deliberately anti-psychological understanding of the process of "interpretation" as: *the generation of an interpretant,* or triadically mediated response. Such response, as previously noted, can be any kind or energetic action or

[18] We thus also include here as "enacted," controversially perhaps, what are most commonly referred to as "psychological responses" and "mental states," seeing no way in which such phenomena could manifest without a bodily component, as we detail presently.

[19] As even a cursory discussion of Deacon's insights on these topics would far exceed the word limitation of the current chapter, the reader is directed instead to his chapter in the present volume, as well as his large corpus of writings, much of which is summarized in the stand-alone volumes of [2.18] and [2.20].

configurational change in a system that has been effected under the influence of a sign and its meanings (and can supply the basis for the next interpretative action).[20] Only with such a sufficiently general understanding of the unique effects brought about under the influence of semiotic organization over evolutionary and ontogenetic time can we begin to speak sensibly of cells and organs "interpreting signs" and "communicating information" with one another, without lapsing into a fruitless anthropomorphism.

Equipped with this understanding, too, we may begin to draw distinctions between the various gradations of "semiotic potential," or *interpretative generativity*, among the sign-processes we see operative on the various levels of biological organization and interaction. Towards the "lower" or "degenerativized" end of the spectrum, we find those sign-processes that have become highly regularized, canalized, "frozen" or "fixed"—such as the action of the ribosome in "translating" the codons of the mRNA into their invariantly specified amino acids. The initial correspondence "encoded" here—the well-known "genetic code"—may have begun in the limiting of randomness by systemically instantiated, arbitrarity-exploiting *joinings*, rather than by strict physical necessity. Even if initially organized by semiosis, such joinings function now almost as clockwork mechanisms, due to the immense systemic pressures of the system of which they are a part.

Accordingly, many arbitrary relations are still "sign" relations in the technical sense, but their "semiotic potential" or "interpretative degrees of freedom" have shrunk to all but zero. So, too, is the case with the majority of "second messenger molecules" (even if their resulting cascade effects are more linearly complex), and perhaps even the perceptual signs involved in the so-called "fixed-action pattern releasing mechanisms" studied by ethologists [2.85] [2.119], although there the measure of semiotic "degenerativization" may be of orders less severe than on the molecular level (e.g., [2.107] [2.89] [2.93]).

At the other extreme, we see what, at first glance at least, looks like the infinite semiotic generativity of human language and the symbol-centric culture from which it springs [2.18] [2.57]. Fictional narratives, linguistic recursivity, referential displacement, semiotic reflexivity, counterfactuality, abstract representation, poetry, art, religion and myth—all of this marks us humans as "a thoroughly symbolic species," to use Terrence Deacon's phrase [2.18], p. 436.

Yet, even here, as Bennett observes, much of our lives are marked, as well, by our continual reliance on automatized procedural memory, unthinking habit, personalized and impersonalized routinizaton, unquestioned ideology and orthodoxy, pre-given meaning, and "the resistance that already-cathected structures exert against the perception of details that would upset their [pre-determined] integrity" [2.10], p. 200.

Semiotic generativity, clearly, exists along an ever-changing continuum. Furthermore, it has recently been argued that it is only in those moments where the simultaneous incompatibility of possible interpretations is collapsed by the free choice[21] of an agent, that semiotic generativity is manifested in its purest form [2.75] [2.79] [2.80].

---

[20] Even now, however, this is a far too over-simplified articulation of Peirce's complex notion of the "interpretant" for any purpose *other* than that of this overtly abbreviated overview, written for the complete Peirce novice. Recommended introductory works on Peirce's complex semiotic logic include [2.104] [2.14] [2.83] and [2.113].

[21] The use of the term "free choice" here does not imply that such choosing is not motivated, constrained, or free from the reliance on convention or any other influencing factors. It means simply, rather, that the choice is not the product of a deterministic algorithm or its equivalent. See [2.79] [2.80] for fuller discussions on this matter.

2.3.3 Semiotic Realism

We have noted earlier how, even in antiquity, the question was asked: How could any creature that, like ourselves, is imprisoned within the limits of its own unique situatedness, history, and perceptual biology ever be able to make any truly sound judgements about the nature of objective reality? Ameliorated, for a time, by the medieval conviction that "God is no deceiver"—a claim which even Descartes [2.31], p. 46, falls back on in his project of radical skepticism and hyperbolic doubt—such reasoning was no longer deployed as a first principle even by the time of Kant and Hume, each of whom powerfully embedded some version of a "skepticism of the senses" understanding even deeper into the modern mindset, despite whatever personally lingering theistic beliefs they may have held [2.68] [2.117].

The development of semiotics over the course of the last century made clearer the understanding that a sign—of any kind—is never identical with that which it signifies, and that the entire linguistic system that humans use to think about the world is built upon the linking of arbitrary tokens to yet other arbitrary tokens. Accordingly, signs cannot ever truly *re-present* their referents *per se* but, instead, at best, *define* their "referents" through an endless chain of signifiers signifying other signifiers [2.29], p. 9. "Naïve realism," then, is hardly a tenable position for a 21st-century scientist to take, especially after what has been discovered about the biology of perception and umwelt [2.40] [2.43] [2.44] [2.70] [2.74] [2.75] [2.77]. But then so also is radical skepticism hardly a tenable position for a 21st-century scientist to take, given the ability of our models to yield accurate predictions across multiple times and places.

An effort to "justify" — if not to "save" — "the appearances," as Aristotle advocated for, is reflected in the introduction of 19th-century electro-physicist Heinrich Hertz's widely influential *Principles of Mechanics*, in which he explains that

> "We form for ourselves images or symbols of external objects; and the form that we give them is such that the *logically necessary consequents* of the images in thought are always the images of the *necessary natural consequents* of the thing pictured. ... For our purpose, it is not necessary that the images should be in conformity with the things in any other respect whatever. As a matter of fact, we do not know, nor have we any means of knowing, whether our conception of things is in conformity with them *in any other than this one fundamental respect.*[22]

Hertz's crucially important caveat in the second half of the quote recalls semiotic logician Charles Sanders Peirce's "pragmatic maxim," which states that in our search for clarity regarding the way things really are in the world, we can only "consider what effects, that might conceivably have practical bearings, we conceive the object of our conception to have. Then, our *conception of these effects* is the *whole of our conception of the object*" [2.99], CP 5.402.[23]

[22] Hertz, H. (1894 [1899]) *The Principles of Mathematics.* D.E. Jones and J.T. Walley (trans.) MacMillan, London (quoted in [2.46], p. 593). The authors are grateful to physicist Howard Pattee for bringing this important quote to their attention.

[23] It should be stressed here that Peirce's notion of the "object" of a sign — which plays a role in constraining what can serve as its sign vehicle, which in turn constrains what can emerge as its interpretant — encompasses far more than just what are conventionally thought of as "the independently existing objects of the external world" and by necessity must include the "sign objects" of pure mentation as well. But here we must re-direct the reader to the caveat offered, and the readings suggested, in footnote 20.

Taken in isolation, such scientific instrumentalism might seem weak consolation to Condillac's remonstrance that "whether we raise ourselves, to speak metaphorically, into the heavens or descend into the abyss, we do not go beyond ourselves; and we never perceive anything but our own thought" [2.16], p. 11.

But Peirce, in particular, insisted that our everyday acts of sign-use in the world are structured by ineliminably triadic relations that have one foot firmly rooted in the existing patterns of mind-independent reality, and another just as inextricably planted in the observer-specific umwelt of our inherited and ongoing biological constitution (and, for our own species, interpersonal social practices and their interactionally perpetuated conventions of knowing, as well). Semiosis—the ongoing generation of interpretants by living systems—thus incorporates, by necessity, both some formally or materially true aspect about the nature of the agent that is generating the interpretant, as well as some formally or materially true aspect about the nature of the object, even if the interpreting agent is, in fact, incorrect about that aspect in any given instance of semiosis.

In seeing a mirage of "water" in the desert, for example, one is misinterpreting, but still *capturing* some true regularity about the object being perceived—here, the tendency of light refraction caused by the difference between two layers of air temperatures to reflect the image of the sky, especially when seen through a mass of heated air. And for Peirce, it is precisely the fact that, over time, a community of inquirers have been able to successfully come to know more about the true principles and mechanisms revealed by the "appearance" of the sign that they are studying, that justifies the claim that semiosis, while fallible in any of its individual instances, is as a communal project that is self-corrective over time, such that knowledge of

> "the real, then, is that which, sooner or later, information and reasoning would finally result in, and which is therefore independent of the vagaries of me and you...[and] the one interpretative result to which every interpreter is destined to come if the sign is sufficiently considered. [In short,] the final interpretant is that toward which *the actual* tends" [2.99], CP 5.311-312.

This property of sign-hood *ensures* its underlying realism, argues Peirce; and it is through the use of signs alone (including, and perhaps especially, the use of "signs for other signs," such as those found in language and mathematics) within the transgenerational community of inquirers, that we can come to more genuinely know reality as it is—not just in its consequences, but its actual properties and qualities, continuities and discontinuities, indeterminacy and lawfulness. In this sense, we may think of the role of the "final interpretant" as "objectivity in the making" [24] with "objectivity" denoting not some state of aperspectival knowledge that is available by default, but an accomplishment of veridicality that we have to strive for collectively—and which, like the very object of our investigations, is never static or complete. In short, it is thus through signs that we are *more fully joined* to the world, and not—as the impossibly egocentric perspective would insist—epistemologically split apart from it. Likewise, it is through the

[24] We thank Alexei Sharov *(personal correspondence)* for this elegant formulation.

community of inquirers that we are joined to one another, as is well exemplified by this very volume of signs and meanings.[25]

2.4 Conclusion

The foregoing chapter was but a sketch of some of the "pathways towards the understanding of signs and meaning" in living systems that have been developed in the Western world over the course of the last 2500 years — itself but a minute fraction of the four-billon year experiment with signs and meanings that is the history of life on Earth. It was meant to set the stage for the perspectives and ideas that will follow in this volume, and to present the biosemiotic perspective on these matters, as the authors understand it today. To conclude this brief introductory chapter, then, let us say with Locke (but replacing his use of the human-centric phrase "words and ideas" to the broader category of "signs and meanings") that:

> The consideration of [signs and meanings] as the great instruments of knowledge, makes no despicable part of their contemplation to those who would take a view of human knowledge in the whole extent of it. And perhaps if [these instruments] were distinctly weighed, and duly considered, they would afford us another sort of logic and critique, other than what we have been hitherto acquainted with [2.84], p. 718

Biosemiotics, along with the rest of the contributions in the present volume, constitutes a concerted and overdue undertaking of Locke's far too long-neglected challenge.

Acknowledgments

We thank the editors of this volume, and in particular Alexei Sharov, for their confidence and persistence in inspiring us. We also thank our two reviewers, whose generous feedback has improved our text considerably. Research funding was provided by National University of Singapore AcRF Grant #A-0003116-00-00 and Grant PRG314 from the Estonian Research Council.

Declaration

The authors declare no conflict of interest

[25] In this sense, too, it should be noted that there are other arguments for semiotic realism operative in biosemiotics in addition to the one by Peirce that is here presented, focusing on "issues such as partial subjectivity (agent-dependency), uncertainty, and context" in ways that Peirce's account doesn't, as one of our reviewers has correctly pointed out. We wish to stress once again here then, that the forgoing discussion does not in any way claim to represent a "consensus" view of all biosemioticians, but instead reflects primarily those concepts upon which the present two authors have tended to focus in their own thinking about "the nature and evolution of signs and meanings." If we were able to convey even that little fragment effectively, it has been in the hopes that our readers will use the references cited here to investigate the rest.

References

[2.1] Allen, J. (2001) *Inference from signs: Ancient debates about the nature of evidence.* Clarendon Press, Oxford.

[2.2] Allendorf, T.S. (2017) Senecan signification: Troades 1055. *The Classical Quarterly* 67(1), 320-323.

[2.3] Annas, J. and Barnes. J. (1985) *The modes of scepticism: Ancient texts and modern interpretations.* Cambridge University Press, Cambridge.

[2.4] Augustine of Hippo (1963 [389 CE]) De doctrina christiana. In: *Corpus Scriptorum Ecclesiasticorum Latinorum 80: Sancti Augustini opera.* W. M. Green (ed.). Austrian Academy of Sciences, Vienna: 1-205.

[2.5] Bacigalupi, J. A. (2022) Semiogenesis: A dynamic system approach to agency and structure. *Biosemiotics* 15, 261-284.

[2.6] Bacon, F. (1863 [1620]) Novum organum. In: *Francis Bacon's works.* J. Spedding, R.L. Ellis and D. D. Heath, (eds.). Houghton, Mifflin and Company, Boston: 131-539. URL: https://onlinebooks.library.upenn.edu/webbin/metabook?id=worksfbacon

[2.7] Barbieri, M. (2003) *The organic codes: An introduction to semantic biology.* Cambridge University Press, Cambridge.

[2.8] Bauer, S.W. (2015) *The story of science: From the writings of Aristotle to the big bang theory.* W. W. Norton, New York.

[2.9] Bellucci, F. (2016) Inferences from signs: Peirce and the recovery of the σημεῖον. *Transactions of the Charles S. Peirce Society*, 52(2), 259–284.

[2.10] Bennett, T. J. (2021) *Detotalization and retroactivity: Black pyramid semiotics* (Dissertationes Semioticae Universitatis Tartuensis 35). University of Tartu Press, Tartu.

[2.11] Brentari, C. (2015) *Jakob von Uexküll: The discovery of the umwelt between biosemiotics and theoretical biology.* Springer, Dordrecht.

[2.12] Campbell, C., Olteanu, A. and Kull, K. (2019) Learning and knowing as semiosis: Extending the conceptual apparatus of semiotics. *Sign Systems Studies* 47(3/4), 352–381.

[2.13] Cobley, P., Deely, J., Kull, K. and Petrilli, S. (eds.) (2011) *Semiotics continues to astonish: Thomas A. Sebeok and the doctrine of signs.* De Gruyter Mouton, Berlin.

[2.14] Colapietro. V. (1988) *Peirce's approach to the self: A semiotic perspective on human subjectivity.* SUNY Press, New York.

[2.15] Crick, F. (1970) Central dogma of molecular biology. *Nature* 277, 561–563.

[2.16] de Condillac, E.B. (2001[1746]) *Essay on the origin of human knowledge.* H. Aarsleff, (trans). Cambridge University Press, Cambridge.

[2.17] de Saussure, F. (1959 [1916]) *Course in general linguistics.* New York: Philosophical Library.

[2.18] Deacon, T.W. (1997) *The symbolic species: The co-evolution of language and the brain.* W.W. Norton, New York.

[2.19] Deacon, T.W. (2006) Emergence: The hole at the wheel's hub. In: *The re-emergence of emergence: The emergentist hypothesis from science to religion.* P. Clayton and P. Davies, (eds.). Oxford University Press, Oxford: 111–150.

[2.20] Deacon, T.W. (2011) *Incomplete Nature: How Mind Emerged from Matter.* W.W. Norton, New York.

[2.21] Deacon, T.W. (2021) How molecules became signs. *Biosemiotics* 14(3), 537–559.

[2.22] Deely, J. (2001) *Four ages of understanding: The first postmodern survey of philosophy from ancient times to the turn of the twenty-first century.* University of Toronto Press, Toronto.

[2.23] Deely, J. (2002) *What distinguishes human understanding?* St. Augustine's Press, South Bend, Indiana.

[2.24] Deely, J. (2003) *The impact on philosophy of semiotics*. St. Augustine's Press, South Bend, Indiana.

[2.25] Deely, J. (2004) 'Σημειον' to 'sign' by way of 'signum': On the interplay of translation and interpretation in the establishment of semiotics. *Semiotica* 148(1/4), 187–227.

[2.26] Deely, J. (2008) *Descartes & Poinsot: The crossroad of signs and ideas*. Chicago University Press, Chicago.

[2.27] Deely, J. (2009) *Purely objective reality*. Mouton de Gruyter, Berlin.

[2.28] Deely, J. (2010) *Medieval philosophy redefined: The development of cenoscopic science, ad 354 to 1644 (from the birth of Augustine to the death of Poinsot*. University of Scranton Press, Scranton.

[2.29] Derrida, J. (1982) Différance. In: *Margins of philosophy*. Alan Bass, (trans). University of Chicago Press, Chicago: 3–27.

[2.30] Descartes, R. (1985 [1637]) Discourse on the method of rightly conducting the reason. In: *The Philosophical Works of Descartes, Volume I*. J. Cottingham, R. Stoothoff, and D. Murdoch (trans.). Cambridge University Press, Cambridge: 111–151.

[2.31] Descartes, R. (1985 [1641]) Meditations on first philosophy. In: *The Philosophical Works of Descartes, Volume II*. J. Cottingham, R. Stoothoff, and D. Murdoch (trans.). Cambridge University Press, Cambridge: 1–62.

[2.32] Donald, M. (1991) *Origins of the modern mind: Three stages in the evolution of culture and cognition*. Harvard University Press, Cambridge, Mass.

[2.33] Eckart, W.U. (1998) Zeichenkonzeptionen in der medizin vom 19. Jahrhundert bis zur gegenwart. In: *Semiotics: A Handbook on the Sign-Theoretic Foundations of Nature and* Culture (Vol. 2). R. Posner, K. Robering and T.A. Sebeok (eds.). Walter de Gruyter, Berlin: 1694–1712.

[2.34] Eco, U. (2014) *From the tree to the labyrinth: Historical studies on the sign and interpretation*. A. Oldcorn (trans.). Harvard University Press, Cambridge, Mass.

[2.35] Eco, U. and Marmo, C. (2000) *On the medieval theory of signs*. S. Kelly, (trans). John Benjamins, Amsterdam.

[2.36] Eddy, M. D. (ed.) (2011) Prehistoric minds: Human origins as a cultural artefact, 1780-2010. *Notes & Records of the Royal Society* 65, 1–98.

[2.37] Eich, W. (1986) *Medizinische semiotik (1750–1850): Ein beitrag zur geschichte des zeichenbegriffs in der medizin* (Freiburger Forschungen zur Medizingeschichte NF 13). Schulz, Freiburg.

[2.38] El-Hani, C.N., Queiroz, J. and Emmeche, C. (2009) *Genes, information, and semiosis*. Tartu University Press, Tartu.

[2.39] Emmeche, C. (2015) Semiotic scaffolding of the social self in reflexivity and friendship. *Biosemiotics* 8(2), 275–289.

[2.40] Favareau, D. (2002) Constructing representema: On the neurosemiotics of self and vision. *Semiotics, Evolution, Energy and Development Journal* 2(4), 3–24.

[2.41] Favareau, D. (2006) The evolutionary history of biosemiotics. In: *Introduction to biosemiotics: The new biological synthesis*. M. Barbieri (ed.). Springer, Dordrecht: 1–67.

[2.42] Favareau, D. (2008) Collapsing the wave function of meaning: The epistemological matrix of talk in interaction. In: *A Legacy of living systems: Gregory Bateson as a precursor to biosemiotics*. J. Hoffmeyer (ed.). Springer-Verlag, Dordrecht: 169–212.

[2.43] Favareau, D. (2008a) Iconic, indexical and symbolic understanding. *Journal of the American Psychoanalytic Association* 56(3), 789–801.

[2.44] Favareau, D. (2008b) Understanding natural constructivism. *Semiotica* 172(1), 489–528.

[2.45] Favareau, D. (2010) *Essential readings in biosemiotics: Anthology and commentary*. Springer, Dordrecht.

[2.46] Favareau, D. (2015) Creation of the relevant next: How living systems capture the power of the adjacent possible through sign use. *Journal of Progress in Biophysics and Molecular Biology* 119(3), 588–601.

[2.47] Favareau, D. (2015a) Why this now? The conceptual and historical rationale behind the development of biosemiotics. *Green Letters: Studies in Ecocriticism* 19(3), 22–242.

[2.48] Favareau, D. (2021) Facing up to the hard problem of biosemiotcs. *Biosemiotics* 14(3), 603–615.

[2.49] Fine, G. (2003) Sextus and external world skcpticism. *Oxford Studies in Ancient Philosophy* 24, 341–85.

[2.50] Galilei, G. (1957 [1623]) Excerpts from 'The Assayer'. In: *Discoveries and Opinions of Galileo*. S. Drake (trans). Doubleday, New York: 229-281.

[2.51] Gallagher, S. (2017) *Enactivist interventions: Rethinking the mind.* Oxford University Press, Oxford.

[2.52] Garrison, F.H. (1966) *History of medicine*. W.B. Saunders, Philadelphia.

[2.53] Gazzaniga, M. (2018) *The consciousness instinct: Unraveling the mystery of how the brain makes the mind.* Farrar, Straus and Giroux, New York.

[2.54] Godfrey-Smith, P. (2021) *Metazoa: Animal minds and the birth of consciousness.* William Collins, London.

[2.55] Goldberg, H.S. (1963) *Hippocrates, father of medicine*. Franklin Watts, New York.

[2.56] Hacking, I. (1975) *The emergence of probability: A philosophical study of early ideas about probability, induction and statistical inference.* Cambridge University Press, Cambridge.

[2.57] Hoffmeyer, J. (1996) *Signs of meaning in the universe.* Indiana University Press, Bloomington.

[2.58] Hoffmeyer, J. (2008) *Biosemiotics: An examination into the signs of life and the lives of signs.* University of Scranton Press, Scranton.

[2.59] Hoffmeyer, J. (2015) Semiotic scaffolding: a unitary principle gluing life and culture together. *Green Letters* 19(3), 1–12.

[2.60] Hoffmeyer, J. and Stjernfelt, F. (2016) The great chain of semiosis. Investigating the steps in the evolution of semiotic competence. *Biosemiotics, 9*(1), 7–29.

[2.61] Hutto, D.D. (2013) Enactivism, from a Wittgensteinian point of view. *American Philosophical* Quarterly 50(3), 281–302.

[2.62] Hutto, D.D. and Myin, E. (2013) *Radicalizing enactivism: Basic minds without content.* The MIT Press, Cambridge.

[2.63] Ingold, T. (2019) Confessions of a semiophobe. In: *Au seuil de la forêt: Hommage à Philippe Descola l'anthropologue de la nature.* G. Cometti, P. Le Roux, T. Manicone and N. Martin (eds.). Tautem, Mirebeau-Sur-Bèze: 471–483.

[2.64] James, W. (1893) *The Principles of Psychology.* Holt and Company, New York.

[2.65] Jones, W.H.S. (1923) *Hippocrates. Vol. 1.* Harvard University Press, Cambridge, Mass.

[2.66] Kauffman, S. (2012) From physics to semiotics. In: *Gatherings in Biosemiotics*. S. Rattasepp and T.J. Bennett (eds.). University of Tartu Press, Tartu: 30–46.

[2.67] Kleisiaris, C.F., Sfakianakis, C. and Papathanasiou, I.V. (2014) Health care practices in ancient Greece: The Hippocratic ideal. *Journal of Medical Ethics and History of Medicine* 7(6), 1–5.

[2.68] Kuehn, M. (2001) *Kant: A biography.* Cambridge University Press, Cambridge.

[2.69] Kuhn, T. (1962) *The structure of scientific revolutions*. University of Chicago Press, Chicago.

[2.70] Kull, K. (2000) Organisms can be proud to have been their own designers. *Cybernetics and Human Knowing* 7(1), 45–55.

[2.71] Kull, K. (2001) Jakob von Uexküll: an introduction. *Semiotica* 134(1/4), 1–59.

[2.72] Kull, K. (2003) Thomas A. Sebeok and biology: Building biosemiotics. *Cybernetics and Human Knowing* 10(1), 47–60.

[2.73] Kull, K. (2005) A brief history of biosemiotics. *Journal of Biosemiotics* 1, 1–34.

[2.74] Kull, K. (2012) Scaffolding. In: *A more developed sign: Interpreting the work of Jesper Hoffmeyer*. D. Favareau, P. Cobley and K. Kull (eds.). Tartu University Press, Tartu: 227–230.

[2.75] Kull, K. (2018) Choosing and learning: Semiosis means choice. *Sign Systems Studies* 46(4): 452–466.

[2.76] Kull, K. (2019) Steps towards the natural meronomy and taxonomy of semiosis: Emon between index and symbol? *Sign Systems Studies* 47(1/2), 88–104.

[2.77] Kull, K. (2022) The aim of extended synthesis is to include semiosis. *Theoretical Biology Forum* 115(1/2), 119–132.

[2.78] Kull, K. (2022a) The term 'Biosemiotik' in the 19th century. *Sign Systems Studies* 50(1), 173–178.

[2.79] Kull, K. (2023) Freedom in living beings: Arbitrariness and the forms of semiotic indeterminacy. In: *Open semiotics, Vol. 4: Life and its extensions*. A. Biglari (ed.). L'Harmattan, Paris: 79–95.

[2.80] Kull, K. and Favareau, D. (2023) Neurosemiotics: A brief history of its development and key concerns. *The Routledge handbook of semiosis and the brain*. A.M. García and A, Ibáñez (eds.). Routledge, New York: 13–29.

[2.81] Latour, B. (1987) *Science in action: How to follow scientists and engineers through society.* Harvard University Press, Cambridge, Mass.

[2.82] Latour, B. (1999) *Pandora's hope: Essays on the reality of science studies.* Harvard University Press, Cambridge, Mass.

[2.83] Liszka, J.J. (1996) *A general introduction to the semeiotic of Charles Sanders Peirce.* Indiana University Press, Bloomington.

[2.84] Locke, J. (1999 [1689]) *An Essay Concerning Human Understanding.* J. Manis (ed.). Pennsylvania State University Electronic Classics Series, Hazleton, PA. Available online at http://www.philotextes.info/spip/IMG/pdf/essay_concerning_human_understanding.pdf

[2.85] Lorenz, K. (1982) *The foundations of ethology.* Springer, New York.

[2.86] Lotman, J. (1990) *Universe of the mind: A semiotic theory of culture.* I. B. Tauris, London.

[2.87] Manetti, G. (1993) *Theories of the sign in classical antiquity.* Indiana University Press, Bloomington.

[2.88] Manetti, G. (2010) Ancient semiotics. In: *The Routledge companion to semiotics.* P. Cobley (ed.). Routledge, London: 13–28.

[2.89] Maran, T., Martinelli, D. and Turovski, A. (eds.) (2011) *Readings in zoosemiotics.* De Gruyter Mouton, Berlin.

[2.90] Marketos, S. G. (1997). The medical school of Kos: Hippocratic medicine. *Forum,* 7, 313–324.

[2.91] Markoš, A., Grygar, F., Hajnal, L., Kleisner, K., Kratochvíl, Z. and Neubauer, Z. (eds). (2009) *Life as its own designer: Darwin's 'Origin" and western thought.* Springer, Berlin.

[2.92] Martí-Ibáñez, F. (1961) *A prelude to medical history.* MD Publications, New York.

[2.93] Martinelli, D. (2010) *A critical companion to zoosemiotics: People, paths, ideas.* Springer, Berlin.

[2.94] Meier-Oeser, S. (2011) Medieval semiotics. In: *The Stanford encyclopedia of philosophy.* E.N. Zalta, (ed.) https://plato.stanford.edu/archives/sum2011/entries/semiotics-medieval

[2.95] Michelini, F. and Köchy, K. (eds.). (2020) *Jakob von Uexküll and philosophy: Life, environments, anthropology.* Routledge, London.

[2.96] Morris, C. (1946) *Signs, language and behavior.* Prentice-Hall, New York.

[2.97] Nagel, T. (1986) *The view from nowhere.* Oxford University Press, Oxford.

[2.98] Noë, A. (2004) *Action in perception.* The MIT Press, Boston.

[2.99] Peirce, C. S. (1931–1935; 1958) *Collected papers of Charles Sanders Peirce, Vol. 1–8*. Harvard University Press: Cambridge, Mass:

[2.100] Popkin, R. (2003 [1960]) *The history of scepticism from Savonarola to Bayle*. Oxford University Press, Oxford.

[2.101] Renfrew, C. (2008) *Prehistory: The making of the human mind*. Modern Library, New York.

[2.102] Rosen, R. (1958) A relational theory of biological systems. *Bulletin of Mathematical Biophysics* 20, 245–260.

[2.103] Rosen, R. (2010 [1985]) *Anticipatory systems: Philosophical, mathematical & methodological foundations*. Pergamon Press, Oxford.

[2.104] Savan, D. (1976) *An introduction to C. S. Peirce's full system of semeiotic*. Victoria College, Toronto.

[2.105] Schmid, S. (2019) Three varieties of early modern scepticism. In: *Sceptical paths: Enquiry and doubt from antiquity to the present*. G. Veltri, R. Haliva, S. Schmid and E. Spinelli (eds.). De Gruyter, Berlin: 181–202.

[2.106] Sebeok, T.A. (1972) *Perspectives in zoosemiotics*. Mouton, The Hague.

[2.107] Sebeok, T.A. (1990) *Essays in zoosemiotics*. Toronto Semiotic Circle, Toronto.

[2.108] Sebeok, T.A. (2001) Biosemiotics: Its roots, proliferation, and prospects. *Semiotica* 134(1), 18.

[2.109] Sebeok, T.A. and Danesi, M. (2000) *The Forms of meaning: Modeling systems theory and semiotic analysis*. Mouton de Gruyter, Berlin.

[2.110] Sextus Empiricus (1933 [c. 2 CE]) *Outlines of Pyrrhonism*. R.G. Bury (trans.). Harvard University Press, Cambridge, Mass.

[2.111] Shannon. C. (1948) A mathematical theory of communication. *Bell System Technical Journal*, 27, 379–423.

[2.112] Sharov, A. and Tønnessen, M. (2021) *Semiotic agency: Science beyond mechanism*. Springer, Dordrecht.

[2.113] Short, T.L. (2007) *Peirce's theory of signs*. Cambridge University Press, Cambridge.

[2.114] Snow, C.P. (1959) *The two cultures and the scientific revolution*. Cambridge University Press, Cambridge.

[2.115] Sprengel, K. (1815) *Handbuch der semiotik*. Alops Doll, Wien.

[2.116] Stjernfelt, F. (2014) *Natural propositions: The actuality of Peirce's doctrine of dicisigns*. Docent Press, Boston.

[2.117] Stroud, B. (1977) *Hume*. Routledge and Kegan Paul, London.

[2.118] Teichmüller, G. (1889) *Neue grundlegung der psychologie und logik*. Wilhelm Koebner, Breslau.

[2.119] Tinbergen, N. (1951) *The study of instinct*. Clarendon, Oxford.

[2.120] von Uexküll, J. (1928) *Theoretische biologie (2nd ed.)*. Julius Springer Verlag, Berlin.

[2.121] von Uexküll, J. (1992 [1934]) A stroll through the worlds of animals and men: A picture book of invisible worlds. *Semiotica* 89(4), 319–391.

[2.122] Wittgenstein, L. (1997 [1953]) *Philosophical investigations*. Oxford: Blackwell.

[2.123] Woese, C. R. (2004) A new biology for a new century. *Microbiology and Molecular Biology Reviews* 68(2), 173–186.

[2.124] Yong, E. (2022) *An immense world: How animal senses reveal the hidden realms around us*. Random House, New York.

3

Is it a Janus-Faced World After All? Physics is Not Reductionist

Bashir Ahmad[1]* and Richard Gordon[2]

[1]*Department of Physics and Astronomy, University of Manitoba, Canada*
[2]*Gulf Specimen Marine Laboratory & Aquarium, Panacea, Florida, USA*

Online resources:
Richard Gordon:
https://scholar.google.com/citations?user=LnaynB0AAAAJ&hl=en
https://prabook.com/web/richard.gordon/1943512
https://encyclopedia.pub/entry/38648

Abstract
All the higher-level phenomena in the study of life are holistic, and thereby usually Janus-faced (where higher-level phenomena affect lower levels, and *vice versa*), causing an intellectual separation of physics from biology and the conundrum of the origin of life. We examine modern physics, and the extent to which it itself is Janus-faced and/or holistic. This is done at the macroscopic, cosmological, quantum, and subatomic levels. While we do not address "meaning," except in the heads of physicists, we think it is important to note to what extent modern physics is NOT reductionist. Thus, modern physics and biology may actually have much in common.

Keywords: Top-down causation, two levels of holons, differentiation waves, emergence in physics

3.1 Introduction

In the reductionist point of view of the universe, going back at least to the ancient Greece atomists [3.122], explanation at each level is presumed to arise from phenomena at "lower" levels. As we discovered increasingly lower levels of phenomena, the burden of proof on smaller entities became ever larger. For instance, when one of us (RG) was trained in statistical mechanics, it was mostly based on the 1800's concepts of the motions of atoms and molecules. The following generation tried to take into account quantum mechanics [3.37]

**Corresponding author*: bashirahmad07@gmail.com

[3.38] [3.60] [3.146] and relativity [3.28] [3.84] [3.87] [3.88] [3.117], and even subatomic particles down to strings [3.147] [3.150] [3.154]. Furthermore, quantum entanglement entangles length scales, which might be seen as mixing levels (Table 3.1).

"Holism" is a rather overused word [3.64]. Even in scientific publications by nuclear physicists, this word might only refer to using "all existing data" [3.2]. Here we will confine our use to systems in which a higher "level" is purported to alter a lower level. This manner of thinking comes out of theories of biological evolution [3.15] [3.20] [3.24] [3.27] [3.33] [3.34] [3.35] [3.43] [3.46] [3.47] [3.55] [3.58] [3.63] [3.74] [3.78] [3.82] [3.85] [3.89] [3.90] [3.91] [3.96] [3.119] [3.120] [3.121] [3.125] [3.134] [3.135] [3.140] [3.152], where levels are recognized, with questions raised about whether or not different "laws" apply at each level:

> [...] at the level of emergence, there are subtle questions on how and why properties, which are completely new and unexpected, pop out at a certain level of complexity. [...] There are the cells, the tissues, the organs, the entire macroscopic structure, and each of these hierarchic levels is *per sé* an autopoietic system characterized by a network of processes that involves smaller components. Each of these subsystems is linked to each other [...] The term emergence describes the onset of novel properties that arise when a higher level of complexity is formed from components of lower complexity, where "novel" means that these properties are not present at the lower level of the components. This is often summarized in the popular assertion that "the whole is more than the sum of the parts," and/or with the vague term *holism* [3.104].

Note our subtle distinction: in "emergence" so-called "new" phenomena appear, which create problems for reductionism.[1] In our approach, these phenomena at the higher level actually alter what occurs at a lower level [3.54]. The presumption that lower levels affect higher levels remains, so we are concerned only with phenomena that point both ways, local ⇔ global, and are thus Janus-faced. Our approach to Janus-faced phenomena is sometimes more restrictive than Koestler's [3.93], in that we require two consecutive levels of his *holons* to actually alter each other.

Arthur Koestler made a case for Janus-faced hierarchies of explanation [3.93], where each level is a whole unto itself, with its own "rules" or "laws." Most of his examples are biological or sociological (cf. [3.104]). He points out, for instance, that:

> The cells in the growing embryo are all of identical origin, and carry the *same* set of chromosomes, i.e., the same hereditary dispositions. In spite of this, they develop into such diverse products as muscle cells, kidney cells, brain cells, toe-nails. How can they do this if they are all governed by the same set of laws, by the same hereditary canon? This is a question which, as W. H. Thorpe recently wrote, "...we are not yet able to answer" [3.141]; [3.93], p. 39–40.

[1] One reviewer suggested emergence happens with computer programs, claimed to be reductionist, even if simulating agents. Unless artificial intelligence is added, the interpretation of output as "emergent" is left to the observer of the output. For example, a simulation of H_2O condensation does not ordinarily output "ice."

3.2 Top-Down Causation in Biology

We subsequently found a possible answer in *differentiation waves* [3.64] [3.65] [3.69], with many aspects that have yet to be tested [3.73]. These are observable [3.67] substitutes for the vague concept of *morphogenetic fields* [3.16] [3.17] [3.18] [3.19] [3.39] [3.86] [3.100] [3.110] [3.112] [3.114] [3.143], and indeed raise the question of cause and effect between levels. We even concluded that differentiation waves exhibit a crude form of "purpose," insofar as they have aspects of being goal directed [3.66] [3.72] [3.73]. The interaction between levels in an embryo could be roughly summarized by saying that cells trigger differentiation waves which in turn determine what cells do next. Koestler recognized that:

> [...] self-regulating properties of holons within the growing embryo ensure that whatever accidental hazards arise during development, the end-product will be according to norm. In view of the millions and millions of cells which divide, differentiate, and move about, it must be assumed that no two embryos, not even identical twins, are formed in exactly the same way. The self-regulating mechanisms which correct deviations from the norm and guarantee, so to speak, the end-result, have been compared to the homeostatic feedback devices in the adult organism — so biologists speak of 'developmental homeostasis' [3.93], p. 42.

The ragged front of a propagating differentiation wave which smooths out or its bounded trajectory may be examples of *self-regulating properties*, analogous to *boundary smoothing* or *filling-in* in image processing [3.124].

The origin of life is an unsolved problem, that may profit by considering the causal relationships between levels [3.27] [3.68] [3.79] [3.107] [3.137] [3.138]. In this case, the "lower" level is presumed to be abiotic, and so the origin of life problem is the problem of transition from non-life to life.

An attempt has been made to suggest that top-down causation, at least in biology, can be expressed via graph theory, where the nodes of small-world graphs, while being local, are nevertheless alterable by "the state of the network as a whole" [3.12] (cf. [3.11]). We found no examples of networks resulting in top-down causation in fundamental physics, except perhaps in high energy physics [3.108].

3.3 Top-Down Causation in Physics and Chemistry

Chemistry is often regarded as emergent from physics. The basic assumption nowadays is that, if we could solve it, every molecule is describable in terms of Schrödinger's equation. But then we are confronted with situations such as the qualitative change in the shape of O_2 absorption from myoglobin to hemoglobin (Figure 3.1) [3.104].

Rather than deal with all these problems of whether or not their observed higher-level phenomena have their own rules, or if ultimately, we will discern that their rules can be explained in terms of lower levels, we will focus here on whether our current theories of physics are indeed reductionist? As physics is presumed to be the "fundamental" science, here we will examine to what extent physics partakes of holism. The more it does, the more we may end up blurring the current conceptual divide between physics and biology.

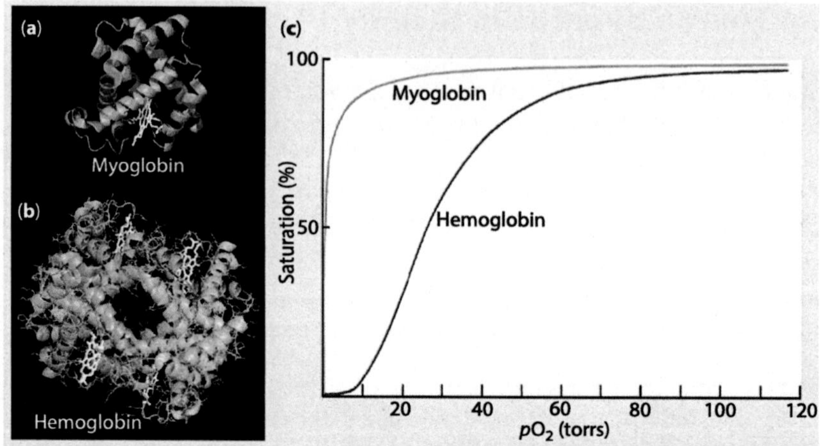

Figure 3.1 The evolutionary transition from myoglobin to hemoglobin, a "novel" emergent property, allowed efficient blood transport of O_2 because of its ranges for retaining and releasing O_2. (From [3.104] with permission of Cambridge University Press).

As holism in physics spans many phenomena, we tabulate what we have found in Table 3.1.

Quasiparticles arise out of excessively complex interactions among elementary particles [3.26] [3.101]. For example, the concept of phonons is frequently used to describe the propagating vibrational movement of atoms in a crystal lattice as well as in other solids. These quasiparticles have energies that are proportional to their frequency.

When solids form, their crystalline structure—a macro property—emerges from the underlying interactions between electrons and atoms. Yet a solid's periodic structure determines its electronic band structure which in turn determines the properties of the electrons flowing in that solid.

Electrons (and all like-charged particles) repel each other under normal circumstances. But when materials are cooled to temperatures near absolute zero, pairs of electrons may form bonds and electrons collectively form a frictionless (*superconducting*) fluid inside the material. This phenomenon is not fully understood [3.97] [3.149] [3.158]. It's a clear example of emergent phenomena, and one whose understanding carries great technological potential.

Over the past few decades, almost every work in fundamental physics, from the study of black holes to condensed matter physics, has found links to information theory. Information theory is essentially the study of correlations among systems, and how larger scale phenomena arise out of these correlations. Perhaps the most interesting concept in information theory is that of entanglement or quantum correlations, which are distinct from classical correlations. One of the most amazing conjectures in fundamental physics is that spacetime itself might arise out of entanglement [3.136] [3.144]. This idea has not been proven, but there are tantalizing clues that it may be correct, and that it may pave the way for a quantum theory of gravity.

Let's look at a simpler example, the rounding of a drop of a liquid. In learning physics, we are faced with two explanations: surface tension or molecular interactions. Which is

Table 3.1 Purported holisms spanning levels in physics.

Phenomenon	Lower level	Higher level	References
Rounding of drops	Molecules	Drop shape (surface tension)	See text
Surface shrinking of small crystals	Ions	Crystal (surface tension)	[3.113]
Band gaps in crystals	Electrons	Periodic structure of the crystal	[3.48] [3.155]
Phonons and sound waves, piezoelectricity	Atoms or molecules	Phonons or waves or mechanical compression	[3.105] [3.157]
Percolation	Channels	Conductivity	[3.9]
Quasiparticles	Elementary particle (electrons, photons, etc.) interactions	Novel properties such as fractional electric charge in Fractional Quantum Hall Effect	[3.5] [3.61] [3.76] [3.81] [3.106] [3.118] [3.126] [3.127] [3.142]
Gravity	Quantum information/ Entanglement	Space-time	[3.111] [3.156]
Superconductivity	Electrons	Electrons pairs (Cooper pairs)	[3.5]

correct? Is one more fundamental than the other? Certainly, in the concept of continuum mechanics approach one is able to ignore the fact that liquids are made of discrete molecules, at least for drop sizes that are large compared to molecules. We could take the viewpoint that surface tension is a force that "exists" and, if we must speak of molecules, alters their interactions. On the other hand, we are in some cases able to derive surface tension for simple liquids [3.14] [3.31] [3.41] [3.51], and then presume, given enough computer power, that we could do this for any liquid. A compromise is often made where both approaches are taken simultaneously, ignoring any logical conundrums, sometimes in the name of *multi-scale* systems [3.102] and *coarse graining* [3.53] [3.83]. We thus see that physicists (and engineers and biologists) are comfortable in shifting between levels of explanation. In any case, even just the rounding of a drop is an example of what we have called global/local interactions, where a global phenomenon (surface tension) alters a local phenomenon (how molecules interact), and vice versa.

Another example is crystals, which have surface tension. The interaction between the surface and bulk of a small crystal visibly shrinks it [3.113]. In shaped droplets, which form flat polygons rather than spherical drops, attempts are being made to understand the orientations of the molecules [3.70]. A holistic approach has also proven fruitful in percolation theory [3.9].

Physicists move between levels of explanation as it suits the situation at hand. Phenomena like surface tension, are "emergent" in that they happen as a result of many smaller interactions. Whether the emergence, i.e., explaining the rounding of a drop, in terms of the concept of surface tension, is in our heads or in the drop, could be debated. Gravity is like that too [3.156].

Phonons are collective excitations of the lattice points (atoms held in place following a regular pattern). So, phonons are emergent, yet they are quite "real" in that you can treat them as sound or heat particles of quantized energy. The same is in principle true about waves as well, though in practice working with waves may be somewhat more challenging. In general, when you cool a solid the behavior of phonons becomes more orderly. When you cool past some threshold temperature T_c, these phonons/lattice vibrations connect pairs of electrons, called Cooper pairs, and you get superconductivity. So, these also qualify as examples of Janus-faced phenomena.

3.4 Multilevel Probabilistic Systems

The attempt to bridge more than two levels at once is rare [3.98]. Sometimes a holistic approach is merely used as an approximation, as for solving the Schrödinger equation, for instance [3.95]. Some systems span 12 orders of magnitude [3.13]. We have found some structures that span 8 orders of magnitude in size in a single substance [3.57].

When Boltzmann and others first recognized atoms, they soon realized that trying to follow the motion (position, speed and direction) of each one and using that to calculate the overall effect is computationally impossible because there are so many. That's why statistical mechanics was invented. One important consequence was that now you didn't have just one outcome. You had many: usually an ensemble of a nearly infinite number of possible outcomes. How can one make predictions? Fortunately, only a few outcomes are very likely and all the others can usually be ignored (but cf. [3.3] [3.71] [3.156]). So, what we had and still have are "effective theories," meaning they have a cut-off threshold past which they ignore smaller details. Where one chooses the cut-off depends on the resolution by which one wants to understand a certain situation. For example, when water pours down a tap, every molecule follows a chaotic trajectory that can't be precisely predicted, but by considering gravity and the Rayleigh instability "due to" surface tension one can "predict" that the water will overall move down towards the floor and the shape it takes on the way down. That's good enough for many needs. As in toy models [3.4] [3.71], the resolution depends on the question.

Many were troubled by the fact that predictions couldn't be made with absolute certainty (as Albert Einstein's "God does not play dice" [3.45]), but only probabilistically. But as time went by it was realized that even if we have vast computational power we cannot simultaneously measure "conjugate variables" such as the position and momentum of a particle to infinite precision, hence the quantum uncertainty principle. It's becoming clearer that nature is essentially probabilistic. Why? Maybe because absolute certainty requires infinite computational power, but this power is a resource of which (every accessible section of) our universe (within our "horizon") has much but nevertheless a finite amount. For an amusing debate on the importance of quantum mechanics (and by implication the probabilistic nature of the world) to biology, see [3.1].

Let's look at another simple phenomenon: indenting a sphere. In this case, the size of the indent depends on the properties of the whole sphere, another example of global/local interaction [3.21].

Intellectually, until we actually observe a higher-level phenomenon, we do not even attempt to derive it from lower-level phenomena. The latter are usually much harder to observe and scrutinize, as they tend to be below the resolution of our senses, and require vast efforts of instrumentation and interpretation to understand. In a way, the same applies to larger scales, where we have yet to understand dark matter and dark energy, estimated as the bulk of the content in the universe, let alone whether or not our universe is alone or one of many [3.22] [3.23] [3.25] [3.40] [3.42] [3.44] [3.50] [3.56] [3.59] [3.62] [3.75] [3.80] [3.92] [3.99] [3.103] [3.123] [3.139] [3.148]. The infinite regress in both directions has not stopped.

Koestler gives an example of global/local interactions at the grandest level, i.e., in cosmology:

> An essential feature of modern physics is its increasingly holistic trend, based on the insight that the whole is as necessary for the understanding of its parts as the parts are necessary for understanding the whole. An early expression of this trend, dating from the turn of the century, was 'Mach's Principle,' endorsed by Einstein. It states that the inertial properties of terrestrial matter are determined by the total mass of the universe around us. There is no satisfactory causal explanation as to how this influence is exerted, yet Mach's Principle is an integral part of relativistic cosmology. The metaphysical implications are fundamental — for it follows from it not only that the universe as a whole influences local, terrestrial events, but also that local events have an influence, however small, on the universe as a whole. Philosophically-minded physicists are acutely aware of these implications — which remind one of the ancient Chinese proverb: "if you cut a blade of grass, you shake the Universe" [1.94], p.256.

The literature on reductionism ranges from statements that higher level causation does not exist, that "Higher levels of mechanisms are aggregated (i.e., built up from) or composed from parts that are organized into more complex spatial, temporal, and causal relations" [3.36]. Well and good, but whence that organization? Perhaps, "Thar's the rub"? Thus, we turn to endeavors such as selforganizing systems [3.6] [3.145], phase transitions [3.77] [3.153], self-organized criticality [3.8], and emergence [3.109]. While the existence of top-down causation has been called, like entanglement, "spooky" [3.8] [3.36], many cases in physics have been put forward [3.7] [3.29] [3.30] [3.32] [3.49] [3.52] [3.115] [3.116] [3.128] [3.129] [3.130] [3.131] [3.132] [3.133].

Perhaps the best example of top-down causation is one of the oldest; the classical macroscopic gas law which relates the macro properties pressure *P*, volume *V*, number of particles *N* and temperature *T* of an ideal gas:

$$PV = NK_B T,$$

where K_B is the Boltzmann constant [3.151]. This and other laws of thermodynamics are obtained (emerge) by coarse-graining the underlying atomic theory. Yet *P*, *T* and *V* in turn also impact the motion, vibrations, and reactions of constituent atoms and molecules. One key point is that large numbers of different configurations of the atoms/molecules in a gas can lead to the same macro properties. This is where the notion of entropy comes into play.

The concept of entropy has been gaining ever-increasing importance in modern physics. Classically, the entropy of a system depends on all the different ways its constituents can be arranged. Entropy is fundamental to information theory, which as stated earlier, seems to underlie most if not all physical theories. The sharp increase in the applications of the concept of entropy is perhaps yet one more indication that global phenomena (i.e., the overall properties of the system) emerging from local interactions (i.e., how parts in a system are arranged) is a fundamental feature of our universe.

Thus, for now, we are stuck in the middle: we can look down to lower levels and struggle to create reductionist explanations for higher levels, or we can assign causality to higher levels, and try to tease out whether or not they affect lower levels. Hypothesizing both, we find ourselves practicing Janus-faced logic. Is it real? Hard to say. Certainly the impact on how we perceive ourselves is great [3.94] [3.122], but our self-perception is usually not considered relevant to scientific explanation, though it is certainly relevant to the politics of our time [3.10] [3.94]. Koestler flirted with Lamarckism and parapsychology, and perhaps for this reason is ignored in philosophy of biology [3.122], but Janus-faced logic even in physics may deserve serious consideration.

References

[3.1] Abbott, D., Gea-Banacloche, J., Davies, P.C.W., Hameroff, S., Zeilinger, A., Eisert, J., Wiseman, H., Bezrukov, S.M. and Frauenfelder, H. (2008) Plenary debate: Quantum effects in biology: trivial or not? *Fluctuation and Noise Letters* 8(1), C5-C26.

[3.2] Aksyutina, Y., Aumann, T., Boretzky, K., Borge, M.J.G., Caesar, C., Chatillon, A., Chulkov, L.V., Cortina-Gil, D., Pramanik, U.D., Emling, H., Fynbo, H.O.U., Geissel, H., Ickert, G., Johansson, H.T., Jonson, B., Kulessa, R., Langer, C., LeBleis, T., Mahata, K., Munzenberg, G., Nilsson, T., Nyman, G., Palit, R., Paschalis, S., Prokopowicz, W., Reifarth, R., Rossi, D., Richter, A., Riisager, K., Schrieder, G., Simon, H., Summerer, K., Tengblad, O., Weick, H. and Zhukov, M.V. (2013) Structure of the unbound nucleus ^{13}Be: One-neutron knockout reaction data from ^{14}Be analyzed in a holistic approach. *Physical Review C* 87(6), 12, #064316.

[3.3] Alicea, B., Ahmad, B. and Gordon, R. (2023) An ensemble approach to the origin of life [invited]. *Life*(Special Issue Computer Simulation of the Origin of Life, editors Wentao Ma & Paul Higgs), Submitted Y2023m2003d2015 Manuscript ID: life-2314399.

[3.4] Alicea, B. and Gordon, R. (2014) Toy models for macroevolutionary patterns and trends. *BioSystems* 122, 25-37.

[3.5] Anonymous (2020) Emergent superconductivity. *Nature Physics* 16(7), 705-705.

[3.6] Ashby, W.R. (1962) Principles of the self-organizing system. In: *Principles of Self-organization*. H. Von Foerster and G.W. Zopf Jr., (eds.) Pergamon Press, London: 255-278.

[3.7] Babak, P. (2007) Nonlocal problems involving spatial structure for coupled reaction-diffusion systems. *Applied Mathematics and Computation* 185(1), 449-463.

[3.8] Bak, P. (1996) *How Nature Works: The Science of Self-Organized Criticality*. Springer-Verlag, New York.

[3.9] Balberg, I. (2021) Simple holistic solution to Archie's-law puzzle in porous media. *Physical Review E* 103(6), #063005.

[3.10] Becht, G. (1974) Systems theory, the key to holism and reductionism. *BioScience* 24(10), 569-579.

[3.11] Bechtel, W. (2017) Explicating top-down causation using networks and dynamics. *Philosophy of Science* 84(2), 253-274.

[3.12] Bechtel, W. (2017) Systems biology: Negotiating between holism and reductionism. In: *Philosophy of Systems Biology: Perspectives from Scientists and Philosophers*. S. Green, (ed.) Springer International Publishing Ag, Cham. 20: 25-36.

[3.13] Bell, A.T. (2007) Special issue - Frontier of chemical engineering - Multi-scale bridge between reductionism and holism - Foreword. *Chemical Engineering Science* 62(13), 3283-3283.

[3.14] Bellemans, A. (1963) Statistical mechanics of surface phenomena. III. Relation between surface tension and curvature. *Physica* 29(5), 548-554.

[3.15] Beloussov, L.V. (1986) Interlevel relations in ontogenesis and their evolutionary significance [Russian] [Межуровневые отношения в онтогенезе и их эволюционное значение]. In: *Animal Morphology and Evolution [Russian] [В кн. «Морфология и эволюция животных» (Э.И. Воробьева, Н.С. Лебедкина ред.)]*. E.I. Vorobieva and N.S. Lebedkina, (eds.) Nauka Publishers, Moscow, USSR: 75-86.

[3.16] Beloussov, L.V. (2001) Morphogenetic fields: Outlining the alternatives and enlarging the context [English]. *Rivista di Biologia - Biology Forum* 94(2), 219-235.

[3.17] Beloussov, L.V. (2015) Morphogenetic fields: History and relations to other concepts [English]. In: *Fields of the Cell*. D. Fels, M. Cifra and F. Scholkmann, (eds.) Research Signpost, Kerala, India: 271-282.

[3.18] Beloussov, L.V. and Gordon, R. (2018) Two ways for interpreting Driesch's law: "Positional information" and morphogenetic fields. *Biosystems* 173, 7-9.

[3.19] Beloussov, L.V., Opitz, J.M. and Gilbert, S.F. (1997) Life of Alexander G. Gurwitsch and his relevant contribution to the theory of morphogenetic fields. *Int J Dev Biol* 41(6), 771-779.

[3.20] Blackstone, N.W. and Ellison, A.M. (2000) Maximal indirect development, set-aside cells, and levels of selection. *Journal of Experimental Zoology* 288, 99-104.

[3.21] Boulbitch, A. (2000) Deformation of the envelope of a spherical Gram-negative bacterium during the atomic force measurements. *Journal of Electron Microscopy* 49(3), 459-462.

[3.22] Bousso, R., Hall, L.J. and Nomura, Y. (2009) Multiverse understanding of cosmological coincidences. *Physical Review D* 80(6), 19, #063510.

[3.23] Bousso, R. and Susskind, L. (2012) Multiverse interpretation of quantum mechanics. *Physical Review D* 85(4), 22, #045007.

[3.24] Brunet, T.D. and Doolittle, W.F. (2015) Multilevel Selection Theory and the Evolutionary Functions of Transposable Elements. *Genome Biol Evol* 7(8), 2445-2457.

[3.25] Carr, B. and Ellis, G. (2008) Universe or multiverse? *Astronomy & Geophysics* 49(2), 29-33.

[3.26] Chandler, D.L. (2010) Explained: Phonons. https://news.mit.edu/2010/explained-phonons-0706

[3.27] Chao, L. (1991) Levels of selection, evolution of sex in RNA viruses, and the origin of life. *J Theor Biol* 153(2), 229-246.

[3.28] Chavanis, P.H. (2020) Statistical mechanics of self-gravitating systems in general relativity: II. The classical Boltzmann gas. *European Physical Journal Plus* 135(3), #310.

[3.29] Chen, K. and Bak, P. (2000) Scale-dependent dimension in the forest fire model. *Phys Rev E Stat Phys Plasmas Fluids Relat Interdiscip Topics* 62(2 Pt A), 1613-1616.

[3.30] Chialvo, D.R. and Bak, P. (1999) Learning from mistakes. *Neuroscience* 90(4), 1137-1148.

[3.31] Cho, H.J., Sresht, V. and Wang, E.N. (2018) Predicting surface tensions of surfactant solutions from statistical mechanics. *Langmuir* 34(6), 2386-2395.

[3.32] Christensen, K., Danon, L., Scanlon, T. and Bak, P. (2002) Unified scaling law for earthquakes. *Proc Natl Acad Sci USA* 99 Suppl 1, 2509-2513.

[3.33] Clarke, E. (2011) Plant individuality and multilevel selection theory. In: *The Major Transitions in Evolution Revisited.* B. Calcott and K. Sterelny, (eds.) MIT Press, Cambridge, MA USA: 227-250.

[3.34] Corel, E., Pathmanathan, J.S., Watson, A.K., Karkar, S., Lopez, P. and Bapteste, E. (2018) MultiTwin: A Software Suite to Analyze Evolution at Multiple Levels of Organization Using Multipartite Graphs. *Genome Biology and Evolution* 10(10), 2777-2784.

[3.35] Cracraft, J. (1990) *The Origin of Evolutionary Novelties: Pattern and Process at Different Hierarchical Levels.* University of Chicago Press, Chicago, Illinois, USA.

[3.36] Craver, C.F. and Bechtel, W. (2007) Top-down causation without top-down causes. *Biology & Philosophy* 22(4), 547-563.

[3.37] Cuccoli, A., Giachetti, R., Tognetti, V., Vaia, R. and Verrucchi, P. (1995) The effective potential and effective Hamiltonian in quantum statistical mechanics. *Journal of Physics-Condensed Matter* 7(41), 7891-7938.

[3.38] D'Alessio, L., Kafri, Y., Polkovnikov, A. and Rigol, M. (2016) From quantum chaos and eigenstate thermalization to statistical mechanics and thermodynamics. *Advances in Physics* 65(3), 239-362.

[3.39] Davidson, E.H. (1993) Later embryogenesis: regulatory circuitry in morphogenetic fields. *Development* 118(3), 665-690.

[3.40] Davies, P.C.W. (2004) Multiverse cosmological models. *Modern Physics Letters a* 19(10), 727-743.

[3.41] De Coninck, J., Sole, S. and Ruiz, J. (2005) On the statistical mechanics and surface tensions of binary mixtures. *Journal of Statistical Physics* 119, 597-642.

[3.42] De Simone, A., Guth, A.H., Linde, A., Noorbala, M., Salem, M.P. and Vilenkin, A. (2010) Boltzmann brains and the scale-factor cutoff measure of the multiverse. *Physical Review D* 82(6), #063520, #063520.

[3.43] Deacon, T.W. (2003) The hierarchic logic of emergence: Untangling the interdependence of evolution and self-organization. In: *Evolution and Learning: The Baldwin Effect Reconsidered.* B.H. Weber and D.J. Depew, (eds.) MIT Press, Cambridge, Massachusetts, USA: 273-308.

[3.44] Deutsch, D. (2001) The Structure of the Multiverse. http://arxiv.org/abs/quant-ph/0104033

[3.45] Dickerson, K. (2015) One of Einstein's most famous quotes is often completely misinterpreted. https://www.businessinsider.com/god-does-not-play-dice-quote-meaning-2015-11

[3.46] Dugatkin, L.A. and Reeve, H.K. (1994) Behavioral ecology and levels of selection: Dissolving the group selection controversy. *Advances in the Study of Behavior* 23, 101-133.

[3.47] Eldredge, N. (1982) Phenomenological levels and evolutionary rates. *Systematic Biology* 31(4), 338-347.

[3.48] Electrical4U. (2020) Energy Bands in Crystals. https://www.electrical4u.com/energy-bands-in-crystals/

[3.49] Ellis, G. (2016) The foundations: Physics and top-down causation. In: *How Can Physics Underlie the Mind?: Top-down Causation in the Human Context.* Springer-Verlag Berlin, Berlin: 243-290.

[3.50] Ellis, G.F.R., Kirchner, U. and Stoeger, W.R. (2004) Multiverses and physical cosmology. *Monthly Notices of the Royal Astronomical Society* 347(3), 921-936.

[3.51] Errington, J.R. (2003) Evaluating surface tension using grand-canonical transition-matrix Monte Carlo simulation and finite-size scaling. *Physical Review E* 67(1), #012102.

[3.52] Flyvbjerg, H., Sneppen, K. and Bak, P. (1993) Mean field theory for a simple model of evolution. *Physical Rev. Letters* 71(24), 4087-4090.

[3.53] Frieden, B.R. and Hawkins, R.J. (2010) Quantifying system order for full and partial coarse graining. *Phys Rev E Stat Nonlin Soft Matter Phys* 82(6 Pt 2), 066117.

[3.54] Friesen, M.R.P., Gordon, R. and McLeod, R.D. (2014) Exploring emergence within social systems with agent based models [invited]. In: *Interdisciplinary Applications of Agent-Based Social Simulation and Modeling*. D.F. Adamatti, G.P. Dimuro and H. Coelho, (eds.) IGI Global, Hershey, Pennsylvania, USA 52-71.

[3.55] Gardner, A. (2015) The genetical theory of multilevel selection. *Journal of evolutionary biology* 28(2), 305-319.

[3.56] Gedalia, O., Jenkins, A. and Perez, G. (2011) Why do we observe a weak force? The hierarchy problem in the multiverse. *Physical Review D* 83(11), #115020.

[3.57] Ghobara, M.M., Mazumder, N., Vinayak, V., Reissig, L., Gebeshuber, I.C., Tiffany, M.A. and Gordon, R. (2019) On light and diatoms: A photonics and photobiology review [Chapter 7]. In: *Diatoms: Fundamentals & Applications [DIFA, Volume 1 in the series: Diatoms: Biology & Applications, series editors: Richard Gordon & Joseph Seckbach]*. J. Seckbach and R. Gordon, (eds.) Wiley-Scrivener, Beverly, MA, USA: 129-190.

[3.58] Gibson, A.H., Kwapich, C.L. and Lang, M. (2013) The roots of multilevel selection: Concepts of biological individuality in the early Twentieth Century. *History and Philosophy of the Life Sciences* 35(4), 505-532.

[3.59] Gleiser, M. (2010) Drake equation for the multiverse: From the string landscape to complex life. *International Journal of Modern Physics D* 19(8-10), 1299-1308.

[3.60] Gogolin, C. and Eisert, J. (2016) Equilibration, thermalisation, and the emergence of statistical mechanics in closed quantum systems. *Reports on Progress in Physics* 79(5), 51, #056001.

[3.61] Goldhaber, A., Parwani, R. and Singh, H. (1996) On the fractional electric charge of a magnetic monopole at nonzero temperature. *Physics Letters B* 386(1-4), 207-210.

[3.62] González-Diaz, P.F. (2011) Life originated during accelerating expansion in the multiverse. *Physics Essays* 24(3), 445-453.

[3.63] Goodman, M., Weiss, M.L. and Czelusniak, J. (1982) Molecular evolution above the species level: Branching pattern, rates, and mechanisms. *Systematic Biology* 31(4), 376-399.

[3.64] Gordon, N.K. and Gordon, R. (2016) *Embryogenesis Explained*. World Scientific Publishing, Singapore.

[3.65] Gordon, R. (1999) *The Hierarchical Genome and Differentiation Waves: Novel Unification of Development, Genetics and Evolution*. World Scientific & Imperial College Press, Singapore & London. 2 vols.

[3.66] Gordon, R. (2021) Are we on the cusp of a new paradigm for biology? The illogic of molecular developmental biology versus Janus-faced control of embryogenesis via differentiation waves. *BioSystems*(Waves in Fertilization, Cell Division and Embryogenesis, Guest Editors: Jack Tuszynski, Luigia Santella & Richard Gordon), #104367.

[3.67] Gordon, R. and Björklund, N.K. (1996) How to observe surface contraction waves on axolotl embryos. *International Journal of Developmental Biology* 40(4), 913-914.

[3.68] Gordon, R. (2023) *Origin of Life via Archaea: Shaped Droplets to Archaea First, With a Compendium of Archaea Micrographs [OOLA, Volume in the series Astrobiology Perspectives on Life of the Universe, Eds. Richard Gordon & Joseph Seckbach, in preparation]*. Wiley-Scrivener, Beverly, Massachusetts, USA.

[3.69] Gordon, R. and Gordon, N.K. (2019) The differentiation code [invited] [DFCD]. *BioSystems* 184(Second Special Issue in Code Biology: the study of all Codes of Life, Guest Editors Marcello Barbieri and Jan-Hendrik Hofmeyr), #104013.

[3.70] Gordon, R., Hanczyc, M.M., Denkov, N.D., Tiffany, M.A. and Smoukov, S.K. (2018) Chapter 18: Emergence of polygonal shapes in oil droplets and living cells: The potential role of tensegrity in the origin of life. In: *Habitability of the Universe Before Earth [Volume 1 in series: Astrobiology: Exploring Life on Earth and Beyond, eds. Pabulo Henrique Rampelotto,*

Joseph Seckbach & Richard Gordon]. R. Gordon and A.A. Sharov, (eds.) Elsevier B.V., Amsterdam, Netherlands: 427-490.

[3.71] Gordon, R., Raj Vansh Singh, S., Katyal, K. and Gordon, N.K. (2023) The fish ladder toy model for a thermodynamically at equilibrium origin of life in a lipid world in an endoreic lake (Chapter 17). In: *Conflicting Models for the Origin of Life [COLF, Volume in the series Astrobiology Perspectives on Life of the Universe, Eds. Richard Gordon & Joseph Seckbach]*. S.K. Smoukov, J. Seckbach and R. Gordon, (eds.) Wiley-Scrivener, Beverly, Massachusetts, USA: 425-458.

[3.72] Gordon, R. and Stone, R. (2016) Cybernetic embryo. In: *Biocommunication*. World Scientific Press, Singapore: 111-164.

[3.73] Gordon, R. and Stone, R. (2021) A short tutorial on the Janus-faced logic of differentiation waves and differentiation trees and their evolution. *BioSystems* 205(Waves in Fertilization, Cell Division and Embryogenesis, Guest Editors: Jack Tuszynski, Luigia Santella & Richard Gordon), #104414.

[3.74] Gould, S.J. and Lloyd, E.A. (1999) Individuality and adaptation across levels of selection: How shall we name and generalize the unit of Darwinism? *Proceedings of the National Academy of Sciences* 96(21), 11904-11909.

[3.75] Hall, L.J. and Nomura, Y. (2008) Evidence for the multiverse in the standard model and beyond. *Physical Review D* 78(3), #035001.

[3.76] Halyo, V., Kim, P., Lee, E.R., Lee, I.T., Loomba, D. and Perl, M.L. (2000) Search for free fractional electric charge elementary particles using an automated Millikan oil drop technique. *Physical Review Letters* 84(12), 2576-2579.

[3.77] Hill, T.L. and Stein, L. (1979) Approximate steady-state properties of lattices of interacting three-state enzyme molecules: A novel phase transition. *Proceedings of the National Academy of Sciences of the United States of America* 76(3), 1014-1017.

[3.78] Hogeweg, P. (2012) Toward a theory of multilevel evolution: long-term information integration shapes the mutational landscape and enhances evolvability. *Adv Exp Med Biol* 751, 195-224.

[3.79] Hogeweg, P. and Takeuchi, N. (2003) Multilevel selection in models of prebiotic evolution: Compartments and spatial self-organization. *Origins of Life and Evolution of Biospheres* 33(4-5), 375-403.

[3.80] Hooper, R. (2014) How to live in the multiverse Does anything matter when everything happens? How to live in the multiverse. *New Scientist* 223(2988), 32-37.

[3.81] Hou, C.Y., Chamon, C. and Mudry, C. (2010) Deconfined fractional electric charges in graphene at high magnetic fields. *Physical Review B* 81(7), #075427.

[3.82] Humphreys, A.M., Rydin, C., Jonsson, K.A., Alsop, D., Callender-Crowe, L.M. and Barraclough, T.G. (2016) Detecting evolutionarily significant units above the species level using the generalised mixed Yule coalescent method. *Methods in Ecology and Evolution* 7(11), 1366-1375.

[3.83] Ingolfsson, H.I., Lopez, C.A., Uusitalo, J.J., de Jong, D.H., Gopal, S.M., Periole, X. and Marrink, S.J. (2014) The power of coarse graining in biomolecular simulations. *Wiley Interdisciplinary Reviews-Computational Molecular Science* 4(3), 225-248.

[3.84] Israel, W. and Kandrup, H.E. (1984) Nonequilibrium statistical mechanics in the general theory of relativity .1. A general formalism. *Annals of Physics* 152(1), 30-84.

[3.85] Jablonka, E. (1994) Inheritance systems and the evolution of new levels of individuality. *J Theor Biol* 170(3), 301-309.

[3.86] Jaffe, L.F. (1986) Calcium and morphogenetic fields. *CIBA Found. Symp.* 122, 271-288.

[3.87] Kaniadakis, G. (2002) Statistical mechanics in the context of special relativity. *Physical Review E* 66(5), 17, #056125.

[3.88] Kaniadakis, G. (2005) Statistical mechanics in the context of special relativity. II. *Physical Review E* 72(3), 14, #036108.

[3.89] Keller, L. (2019) *Levels of selection in evolution*. Princeton University Press, Princeton. New Jersey, USA.

[3.90] King, M.C. and Wilson, A.C. (1975) Evolution at two levels in humans and chimpanzees. *Science* 188(4184), 107-116.

[3.91] Klingenberg, C.P., Duttke, S., Whelan, S. and Kim, M. (2012) Developmental plasticity, morphological variation and evolvability: A multilevel analysis of morphometric integration in the shape of compound leaves. *J Evol Biol* 25(1), 115-129.

[3.92] Kockelman, P. (2011) Biosemiosis, technocognition, and sociogenesis selection and significance in a multiverse of sieving and serendipity. *Current Anthropology* 52(5), 711-739.

[3.93] Koestler, A. (1978) *Janus: A Summing Up*. Random House, New York, NY, USA.

[3.94] Koestler, A. (1979) Janus: A summing up. *Bulletin of the Atomic Scientists* 35(3), 4.

[3.95] Kolesik, M., Wright, E.M., Andreasen, J., Brown, J.M., Carlson, D.R. and Jones, R.J. (2012) Space-time resolved simulation of femtosecond nonlinear light-matter interactions using a holistic quantum atomic model: Application to near-threshold harmonics. *Optics Express* 20(14), 16113-16128.

[3.96] Korb, J. (2010) Social insects, major evolutionary transitions and multilevel selection. In: *Animal Behaviour: Evolution and Mechanisms*. P. Kappeler, (ed.) Springer-Verlag Berlin, Berlin, Germany: 179-211.

[3.97] Kulic, M.L. (2006) Importance of the electron-phonon interaction with the forward scattering peak for superconducting pairing in cuprates. *Journal of Superconductivity and Novel Magnetism* 19(3-5), 213-249.

[3.98] Kwauk, M. (2007) Special issue - Frontier of chemical engineering - Multi-scale bridge between reductionism and holism - Summary. *Chemical Engineering Science* 62(13), 3287-3294.

[3.99] Larsen, G., Nomura, Y. and Roberts, H.L.L. (2011) Cosmological constant in the quantum multiverse. *Physical Review D* 84(12), #123512.

[3.100] Levin, M. (2012) Morphogenetic fields in embryogenesis, regeneration, and cancer: non-local control of complex patterning. *BioSystems* 109(3, SI), 243-261.

[3.101] Lewton, T. (2021) The Near-Magical Mystery of Quasiparticles. https://www.quantamagazine.org/the-near-magical-mystery-of-quasiparticles-20210324/

[3.102] Li, J.H. and Ge, W. (2007) Special issue - Frontier of chemical engineering - Multi-scale bridge between reductionism and holism - Preface. *Chemical Engineering Science* 62(13), 3285-3286.

[3.103] Liu, D., Ramya, R.C.S. and Mueller-Cajar, O. (2017) Surveying the expanding prokaryotic Rubisco multiverse. *FEMS Microbiology Letters* 364(16), #fnx156.

[3.104] Luisi, P.L. (2016) *The Emergence of Life: From Chemical Origins to Synthetic Biology*. Cambridge University Press, Cambridge, UK, 2nd.

[3.105] Ma, J.H. (2022) Phonon engineering of micro- and nanophononic crystals and acoustic metamaterials: A review. *Small Science*.

[3.106] Mar, N.M., Lee, E.R., Fleming, G.R., Casey, B.C.K., Perl, M.L., Garwin, E.L., Hendricks, C.D., Lackner, K.S. and Shaw, G.L. (1996) Improved search for elementary particles with fractional electric charge. *Physical Review D* 53(11), 6017-6032.

[3.107] McMullin, B., Kelly, C. and O'Brien, D. (2007) Multi-level selectional stalemate in a simple artificial chemistry. In: *Proceedings of the 9th European conference on Advances in artificial life*. Springer-Verlag, Lisbon, Portugal: 22–31.

[3.108] Montangero, S., Rico, E. and Silvi, P. (2022) Loop-free tensor networks for high-energy physics. *Philosophical Transactions of the Royal Society A-Mathematical Physical and Engineering Sciences* 380(2216), 19, #20210065.

[3.109] Morowitz, H.J. (2002) *The Emergence of Everything, How the World Became Complex.* Oxford University Press, Oxford, UK.

[3.110] Morozova, N. and Shubin, M. (2012) The Geometry of Morphogenesis and the Morphogenetic Field Concept. http://arxiv.org/abs/1205.1158v1

[3.111] Musser, G. (2018) Quantum gravity: What is spacetime? *Nature* 557, S4-S6.

[3.112] Needham, J. (1989) Chemical aspects of morphogenetic fields. In: *Perspectives in Biochemistry.* H. Neurath, (ed.) American Chemical Society, Washington, DC, USA. 1: 66-80.

[3.113] Nicholson, M.M. (1955) Surface tension in ionic crystals. *Proceedings of the Royal Society of London. Series A. Mathematical and Physical Sciences* 228(1175), 490-510.

[3.114] Nieuwkoop, P.D. (1967) The "Organization centre" III. Segregation and pattern formation in morphogenetic fields. *Acta Biotheor.* 17, 178-194.

[3.115] Norrelykke, S.F. and Bak, P. (2002) Self-organized critical system with no stationary attractor state. *Phys Rev E Stat Nonlin Soft Matter Phys* 65(3 Pt 2A), #036147.

[3.116] Paczuski, M., Maslov, S. and Bak, P. (1994) Field theory for a model of self-organized criticality. *Europhys. Lett.* 27, 97-100.

[3.117] Peres, A. (1981) Relativity, quantum-theory, and statistical mechanics are compatible. *Physical Review D* 23(6), 1458-1459.

[3.118] Perl, M.L. and Lee, E.R. (1997) The search for elementary particles with fractional electric charge and the philosophy of speculative experiments. *American Journal of Physics* 65(8), 698-706.

[3.119] Rainey, P. and Kerr, B. (2011) Conflicts among Levels of Selection as Fuel for the Evolution of Individuality. In: *The Major Transitions in Evolution Revisited.* B. Calcott and K. Sterelny, (eds.) MIT Press, Cambridge, MA USA: 141-162.

[3.120] Rensch, B. and Altevogt, R. (1959) *Evolution Above the Species Level.* Methuen.

[3.121] Roze, D. and Michod, R.E. (2001) Mutation, multilevel selection, and the evolution of propagule size during the origin of multicellularity. *Am Nat* 158(6), 638-654.

[3.122] Ruse, M. (2003) Darwin and Design: Does Evolution Have A Purpose?

[3.123] Sandora, M. (2019) Multiverse predictions for habitability: Number of potentially habitable planets. *Universe* 5(6), #157.

[3.124] Satoh, S. (2012) Computational identity between digital image inpainting and filling-in process at the blind spot. *Neural Computing & Applications* 21(4), 613-621.

[3.125] Simpson, C. (2011) How many levels are there? how insights from evolutionary transitions in individuality help measure the hierarchical complexity of life. In: *The Major Transitions in Evolution Revisited.* B. Calcott and K. Sterelny, (eds.) MIT Press, Cambridge, MA USA: 199-225.

[3.126] Sitenko, Y.A. and Gorkavenko, V.M. (2004) Fractional electric charge of a magnetic vortex at nonzero temperature. *Nuclear Physics B* 679(3), 597-620.

[3.127] Slansky, R., Goldman, T. and Shaw, G.L. (1981) Observable fractional electric charge in broken quantum chromodynamics. *Physical Review Letters* 47(13), 887-891.

[3.128] Sneppen, K., Bak, P., Flyvbjerg, H. and Jensen, M.H. (1995) Evolution as a self-organized critical phenomenon. *Proc. Natl. Acad. Sci. USA* 92(11), 5209-5213.

[3.129] Sole, R.V., Manrubia, S.C., Benton, M., Kauffman, S. and Bak, P. (1999) Criticality and scaling in evolutionary ecology. *Trends in Ecology & Evolution* 14(4), 156-160.

[3.130] Sole, R.V., Manrubia, S.C., Benton, M.J. and Bak, P. (1997) Self-similarity of extinction statistics in the fossil record. *Nature* 388(6644), 764-767.

[3.131] Sole, R.V., Manrubia, S.C., Benton, M.J., Kauffman, S. and Bak, P. (1999) Criticality and scaling in evolutionary ecology. *Trends in Ecology & Evolution* 14(4), 156-160.

[3.132] Sole, R.V., Manrubia, S.C., Benton, M.J., Kauffman, S. and Bak, P. (1999) Reply from R.V. Sole, S.C. Manrubia, M.J. Benton, S. Kauffman and P. Bak. *Trends in Ecology and Evolution* 14(8), 321-322.

[3.133] Sole, R.V., Manrubia, S.C., Benton, M.J., Kauffman, S. and Bak, P. (1999) Self-organized criticality in ecology and evolution - Reply. *Trends in Ecology & Evolution* 14(8), 321-322.

[3.134] Stanley, S.M. (1975) A theory of evolution above the species level. *Proceedings of the National Academy of Sciences of the United States of America* 72(2), 646-650.

[3.135] Stebbins Jr., G.L. (1974) *Flowering Plants, Evolution above the Species Level.* Harvard University Press, Cambridge, Massachusetts, USA.

[3.136] Swingle, B. (2018) Spacetime from entanglement. In: *Annual Review of Condensed Matter Physics, Vol 9.* S. Sachdev and M.C. Marchetti, (eds.) Annual Reviews, Palo Alto. 9: 345-358.

[3.137] Szathmáry, E. (1992) Viral sex, levels of selection, and the origin of life. *J Theor Biol* 159(1), 99-109.

[3.138] Takeuchi, N. and Hogeweg, P. (2009) Multilevel selection in models of prebiotic evolution II: A direct comparison of compartmentalization and spatial self-organization. *PLoS Computational Biology* 5(10), #e1000542.

[3.139] Tegmark, M. (2014) Multiverse musings. *New Scientist* 221(2956), 34.

[3.140] Tessera, M. and Hoelzer, G.A. (2013) On the thermodynamics of multilevel evolution. *Biosystems* 113(3), 140-143.

[3.141] Thorpe, W.H. (1974) *Animal Nature and Human Nature.* Methuen.

[3.142] Trower, W.P. (1985) Detection of free fractional electric and magnetic charge. *Nuclear Physics B* 252(1-2), 285-294.

[3.143] Tyler, S.E.B. (2014) The work surfaces of morphogenesis: The role of the morphogenetic field. *Biological Theory* 9(2), 194-208.

[3.144] Ungar, A.A. (2020) A spacetime symmetry approach to relativistic quantum multi-particle entanglement. *Symmetry-Basel* 12(8), 40, #1259.

[3.145] Von Foerster, H. and Zopf, G.W. (1962) *Principles of Self-Organization: Transactions of the University of Illinois Symposium on Self-Organization, Robert Allerton Park, 8 and 9 June, 1961.* Symposium Publications Division, Pergamon Press, New York, NY, USA.

[3.146] Voth, G.A. (1996) Path-integral centroid methods in quantum statistical mechanics and dynamics. *Advances in Chemical Physics, Vol Xciii* 93, 135-218.

[3.147] Wadia, S.R. (2001) A review of the microscopic modeling of the 5-dim. black hole of IIB string theory. *Pramana-Journal of Physics* 56(1), 1-46.

[3.148] Wallace, D. (2012) *The Emergent Multiverse: Quantum Theory According to the Everett Interpretation.* OUP Oxford.

[3.149] Wang, F. and Lee, D.H. (2011) The electron-pairing mechanism of iron-based superconductors. *Science* 332(6026), 200-204.

[3.150] Wheater, J.F. (1994) Random surfaces: From polymer membranes to strings. *Journal of Physics A-Mathematical and General* 27(10), 3323-3353.

[3.151] Wikipedia. (2021) Ideal gas law. https://en.wikipedia.org/wiki/Ideal_gas_law#Common_forms

[3.152] Williams, G.C. (1992) *Natural Selection: Domains, Levels, and Challenges.* Oxford University Press, New York, NY, USA.

[3.153] Wittebrood, M., Luijendijk, D., Stallinga, S., Rasing, T. and Muševič, I. (1996) Thickness-dependent phase transition in thin nematic films. *Physical Review E* 54(5), 5232-5234.

[3.154] Wu, F.Y. (1992) Knot theory and statistical mechanics. *Reviews of Modern Physics* 64(4), 1099-1131.

[3.155] Yablonovitch, E. (1993) Photonic band-gap crystals. *Journal of Physics-Condensed Matter* 5(16), 2443-2460.

[3.156] Yang, H.S. (2009) Emergent spacetime and the origin of gravity. *Journal of High Energy Physics*(5), 77, #012.
[3.157] Yang, J. (2006) A review of a few topics in piezoelectricity. *Applied Mechanics Reviews* 59(1-6), 335-345.
[3.158] Yu, R., Hu, H.Y., Nica, E.M., Zhu, J.X. and Si, Q.M. (2021) Orbital selectivity in electron correlations and superconducting pairing of iron-based superconductors. *Frontiers in Physics* 9, 16, #578347.

4

Semiotic Ground and the Hierarchic Nature of Information

Terrence W. Deacon

Department of Anthropology & Cognitive and Brain Sciences Institute, University of California, Berkeley, California, USA

Online resources:
https://anthropology.berkeley.edu/terrence-w-deacon
https://www.researchgate.net/profile/Terrence-Deacon
https://philpeople.org/profiles/terrence-w-deacon

Abstract
At one extreme, the formal concept of information ignores the referential and normative properties that characterize both the colloquial use of the term and its use in such scientific fields as biology and cognitive neuroscience. At the other extreme, semiotic theories, which take issues of reference and interpretation as their focus, have largely remained confined to discussions in the humanities and social sciences. Efforts to formalize semiotic theories have also gained limited widespread acceptance. This has left the biological sciences, which must make use of insights from both domains, without critical analytic tools. In this chapter, I argue that there is a non-obvious and indirect isomorphism between three nested levels of information, framed in linguistic terms as syntactic, semantic, and pragmatic information, and the semiotically defined modes of referential relations, framed in terms of iconic, indexical, and symbolic interpretive processes. Recognizing this cryptic isomorphism has been complicated by the very different definitions of terms (such as "symbol") used in these fields. By disambiguating these uses and identifying some previously unrecognized isomorphisms I offer some first steps to provide a bridge theory linking these paradigms.

Keywords: Peirce, Shannon, referential relation, normativity, icon, index, symbol, syntactic, semantic and pragmatic information

Email: deacon@berkeley.edu

4.1 From Information to Semiosis: Why We Can't Seem to Even Start

Reference and significance are not properties that are attributable to any specific physical substrate or signal medium (e.g., in the way that mass and charge are). These defining features of information in the nontechnical sense are neither physical nor mathematical properties of a physical signal. They are *relational properties* and are "assigned" by an extrinsic interpretive process: *semiosis.* So, it is not possible to explain reference and significance using a conception of information that focuses on signal medium properties alone and ignores its relational properties. This limits the application of an information concept that is defined with respect to only the physical-statistical properties of the medium to uses where referential and normative properties can be ignored (e.g., physics and engineering contexts). Claude Shannon's 1948 demonstration of a quantitative method for analyzing and measuring the information-bearing properties of communication media [4.15] revolutionized nearly every scientific field that dealt with information but as the philosopher John Collier put it ... "The great tragedy of formal information theory is that its very expressive power is gained through abstraction away from the very thing that it has been designed to describe" [4.1]. Although Shannon's approach was generic to the *form* of information (including continuous signals) its focus reflects the influence of his background in encryption and Boolean algebra. This background is exemplified by his 1938 Master's thesis ("A Symbolic Analysis of Relay and Switching Circuits") [4.13] and a classified memorandum written at the end of the war in 1945 ("A Mathematical Theory of Cryptography") [4.14]. This formal re-conception of the concept of information was recognized as an abstraction from the colloquial use of the term. Shannon makes this clear in his opening paragraph when he says: "The fundamental problem of communication is that of reproducing at one point either exactly or approximately a message selected at another point." ...and although communications often have meaning, "These semantic aspects of communication are irrelevant to the engineering problem," which is to assess the demands of the transmission, storage, and transformations of signals in all forms. In recognition of this special reduced sense of "information" [4.15], Warren Weaver (in his short introduction published along with Shannon's theory) made clear that this was a highly restrictive technical use. Thus, "Information has 'nothing to do with meaning' although it does describe a 'pattern'" [4.21]. This accomplished something critical, however: 'information' in this new sense became precisely measurable (i.e., in bits), and thus, became a useful tool for a wide range of scientific applications.

Because of this widespread use of information theory in many fields of study during the decades following this ground-breaking publication, the distinction between this technical concept and the colloquial concept has often been ignored. So, in a 2001 interview with another pioneer of mathematical information theory, Robert Fano felt the need to correct this misunderstanding this way: "I didn't like the term Information Theory. Claude didn't like it either. You see, the term 'information theory' suggests that it is a theory about information—but it's not. It's the transmission of information, not information. Lots of people just didn't understand this... I coined the term 'mutual information' to avoid such nonsense: making the point that information is always about something. *It is information provided by something, about something*" (italics added) [4.5]. Shannon had also warned of such an overextension much earlier in a short comment titled "The bandwagon" [4.16]. I would add to Fano's characterization that information is provided *by* something, *about* something, and *for* something, but I would disagree with Fano's claim that the concept of mutual information resolves the problem of aboutness (see below).

The great attraction of this formal reduction of the concept of "information" to the statistical properties of a communication medium was that the philosophically troublesome teleological aspects of the concept could be set aside, making the concept of information "safe" for reductionistic science. By bracketing any consideration of informational content/reference, functional significance/usefulness, and assessing "information" only in terms of the statistical properties of a physical medium, it became a precisely measurable phenomenon that was consistent with the strictly materialistic metaphysical assumptions that were dominant in the sciences of the 20th Century. By additionally relegating semiotic issues to a "subjective" context, they could effectively be treated as epiphenomenal and ignored.

This was therefore a useful methodological compromise. The technical conception of information was applicable across the sciences in part because it was consistent with a tacit metaphysical principle assumed ubiquitously in the contemporary physical sciences: i.e., that any and all mentalistic or semiotic properties (e.g., representation, intentionality, agency) should be excluded from playing explanatory roles in scientific theories. But the explanatory power thus gained has complicated the application of this concept of information in fields distinguished by the need to explain function and reference, such as evolutionary biology, cognitive neuroscience, psychology, and the social sciences.

Less obvious is the way that this conception of information is limited by virtue of its relationship to a similarly reduced conception of language. By this I mean a construal of linguistic reference and meaning as a code-like mapping relationships between what Saussure [4.12] called "signifiers" and their meaningful "content" which he called "signifieds." This simplifying assumption allows the analysis of media (e.g., texts and word sounds; "sign vehicles" in general) properties to be conducted independent of any consideration of how they acquire significance. Although this enabled linguists to develop highly sophisticated tools for formal and quantitative analysis of language structures, it is generally acknowledged that these properties are abstracted from a far richer semiotic and pragmatic reality. Unfortunately, this idealization inverts the underlying genealogy of language. Language is a special derived case of semiotic communication, not the generic case. It is the most recently evolved, and the most divergent referential form of communication to ever have evolved. It is regularly used by only one species (*Homo sapiens*), is only minimally trainable in other species, and it evolved from and depends on other nonlinguistic forms of communication [4.2]. To treat all forms of information relations as though related to referential content by a code-like mapping relationship obscures the complex semiotic infrastructure from which linguistic properties emerge.

4.2 Three Embedded Information Theories

Many controversies arise because the term "information" actually refers to different phenomena depending on the context of use that is presupposed. Following a philosophical tradition that dates to the work of Charles Morris and from his use of distinctions originally developed by Charles Sanders Pierce, I will distinguish three nested attributes that are commonly invoked by the term "information" used in different contexts. Morris distinguished three methodological distinctions that have since come to characterize the standard subdivisions of linguistic research: syntax, semantics, and pragmatics [4.18]. In syntactic analysis one abstracts the structure of a sentence from its meaning or use in order to study only the relations between sign tokens, including "... [the] relations of signs to one another in abstraction from the relations of signs to objects or to interpreters" [4.8], p.13. In semantic analysis one abstracts from the context of use and analyzes only the relations of signs to their meanings; "... the relation of signs to their designate and so to the objects which they may or do denote" [4.8], p. 21. Pragmatic analysis takes into account the relevance of the meaning to the user, a particular use, or the influence and relevance of the context, i.e., "... the relation of signs to their interpreters" [4.8], p.30.

Although, Morris was writing in the context of behavioristic psychology and at a time when the modern concept of information had not been developed, these distinctions have now become widely recognized as also applying to the concept of information in general [4.4]. Paralleling this tripartite division, information philosophers have distinguished corresponding framings of the concept of information: syntactic information, semantic information, and pragmatic information. Thus, syntactic information refers to what is commonly associated with mathematical information theory. Semantic information refers to what is commonly described as the content of information; i.e., what a given source of information is "about." And pragmatic information refers to the functional attributes of a given signal that are responsible for controlling or modifying some process.

Unfortunately, these distinctions import a significant linguistic bias into the analysis. As noted above, this is particularly troublesome for any use of the concept of information in domains where symbolic reference is not applicable. Consider, for example, a fever. When a person's body temperature is in excess of 100° Fahrenheit it is presumed to be information about that person's state of ill health. Although one could describe the increase in temperature as syntactic information, and the inferred cause as semantic information, these uses overstretch the linguistic analogy. The combinatorial implications of the term "syntax" and the connotations of the term "semantic" only apply in a metaphoric sense in nonlinguistic contexts.

As a result, the linguistic framing of the concept of information has been problematic. When considered from a biological and evolutionary perspective, language is the extreme exception as a medium for the transmission and preservation of information. As noted above, linguistic communication evolved only once, in only one species—*Homo sapiens*—and exists alongside many other media for social communication and adaptation. In addition, the acquisition and use of language depends extensively on this vast and ancient nonlinguistic semiotic infrastructure that evolved as the primary modes of interpreting and communicating in the vast diversity of other animal species with brains. So, although language is for us a familiar information-bearing medium, its atypically specialized features can mislead us into ignoring features of the information relation that diverge from its form.

For example, language components can be analyzed into discrete units at many levels: such as sound units (phonemes) and meaningful units (morphemes and words), that convey information by being combinable into a large number of possible higher-order units like phrases and sentences. But this digital organizational principle is not relevant to many other forms of information. Recognizing the approach of a friend because of a familiar tone of voice or sensing the presence of a skunk due to an unpleasant odor are sources of information, but it seems odd to think of their form as a distinctive combination of units or the content being conveyed as expressible by a definition. Language is the special case, not the general case.

So, although this tripartite distinction reflects a useful categorical distinction, I suggest that it would be more accurate to reformulate it in more general terms, that differ both from the linguistic distinction that derives from Morris and from that used by information philosophers [4.4]. Indeed, this analytic-methodological distinction almost certainly reflects a more general semiotic logic. In this respect, I propose that the underlying tripartite distinction can be more generally framed in terms of *medium or signal properties* (with syntax as a special case), *referential properties* (with meaning as a special case), and *normative or functional properties* (with pragmatics as a special case). We can abbreviate these as structural information, referential information, and normative information, while recognizing that a full concept of information involves all three. I will use this terminology throughout the remainder of this discussion.

Both the linguistically relevant categorization and my more generic reformulation have been influenced by the semiotic taxonomy first systematically articulated by the philosopher Charles Sanders Peirce. In his efforts to encompass all forms of representational relationships, Peirce distinguished sign vehicle properties from the properties that link sign vehicles to the objects they refer to, and from the properties of the interpretive process that establish this link (CP 2.92).

Peirce's use of the categories of First, Second, and Third to construct his taxonomy of sign relation suggests that these three aspects of any representational relation are hierarchically nested and interdependent. In his 1903 *Syllabus* [4.7] (CP 2.243) Peirce describes a semiotic taxonomy in which a sign-vehicle (representamen) is a First, the sign-object relation is a Second, and the relation of the sign to an interpretant sign is a Third. Using the analytic operation that he termed prescinding, I would re-interpret this to mean that sign vehicle properties can be analyzed irrespective of referential relationships, which can be analyzed irrespective of interpretive processes that make use of them. In other words, one can dissect away higher-order semiotic properties to analyze lower-order semiotic properties, but not vice versa [4.9] (CP 1.549).

When re-analyzed in these terms, this semiotic hierarchic asymmetry has a clear analytic counterpart in the hierarchy of information relationships.

Thus, formal information theories analyze medium properties irrespective of any consideration of what they refer to or are used for. This ability to set these aspects aside is what made Shannon's analysis of information possible. And although Shannon goes on to use the concepts of "message" and "noise" in his analysis, these are in effect "normative" distinctions that ultimately need to be established by an interpretive process, and are not distinguishable by analyzing signal properties alone. Thus, "noise" can be a "message" to someone trying to repair a communication device. A message in this sense has a specific object of reference, and "noise" is any signal properties that make it difficult to determine this reference. Additionally, although one can specify the possible "messages" that a given signal can convey irrespective of how this information might be used, one can't consider how information could be used unless a particular reference is specified.

This analytic separability does not mean, however, that information is thereby reducible to its most analytically minimalistic attributes. The referential and normative aspects of any source of information cannot be determined given only the statistical attributes implicit in Shannon's analysis. Ultimately, the concept of the use or function of an information source is merely assumed in the formal analysis, if only implicitly. Otherwise "information" becomes merely another way of describing some order parameter of the physical medium. Indeed, one of the great contributions of this analytic reduction is that it led to more sophisticated ways to analyze the concepts of order and disorder. Because this most elementary aspect of information can be analyzed without explicitly specifying reference or use, it is a useful shorthand to call these physical properties themselves information. As an operational definition employed for engineering purposes or for physical analysis this is completely legitimate, but it can mistakenly suggest that the referential and normative properties that are implicitly assumed are instead merely heuristic subjective fictions.

To summarize: medium properties can be analyzed irrespective of reference or use, referential properties can be analyzed irrespective of use but not irrespective of medium properties, and functional properties cannot be analyzed without considering both referential and medium properties.

4.3 Computation and Structural Information

In classic computational theories reference is assumed to be assigned extrinsically by an arbitrary one-to-one mapping function. Computation thus requires finding or constructing an isomorphism between a formal operation and a specific mechanistic process. This could, for example, be accomplished by assigning a symbolic value to each different distinct machine state so that the operation of the machine produces a change in state that corresponds to a specific manipulation of symbol tokens. Thus, even devising a mapping between the signs and transformations of some formal operation and the states and structural changes of the operation of an internal combustion engine could enable the engine to "compute" that formal operation (though it would be trivially simple and repetitive). Turing's genius was to show how any finite series of manipulations of sign tokens could be mapped to a machine operation.

The machine metaphor has been the paradigm model of scientific explanation since the enlightenment. But it is important to recognize that a machine is an idealization in which only some subset of what it does is relevant to serving its designed function. And the function that informs its design is imposed from without. A device such as a mechanical clock can thus be analytically described merely in terms of how changes in the states of its parts at one moment produce each subsequent physical state. From a physicist's perspective, its designed function—marking the incremental passage of time—is completely ignorable, and might not be even guessed by an extraterrestrial visitor with no prior knowledge of human cultures.

So, although the design of a clock is determined to serve some informational function, once built and set running, this "function" is not any of its material or physical properties. It is merely a description of one possible way of mapping its physical pattern of change onto some other feature of the world; in this case marking the passage of time. Notice also that if somehow the gears get linked up in an inappropriate way (e.g., because of a loose screw) it may still physically change from state to state in a precisely determinate way but will no longer provide the designed informational function because these changes no longer conform to the "preferred" mapping. In this respect, a clock's function is entirely parasitic on the use that its human users can make of its operation.

So, the information that a clock provides to its users is only "afforded by" its mechanical operation, not determined by it. Similarly, the formal details of a computation or physical information transmission process don't themselves determine what it will be interpreted to be about or how this information will be used. And so, although information is often distinguished from noise, this distinction is not due to any intrinsic feature of the signal itself. This distinction is normative. This is why assessing the deviation from the designed correspondence between the rate of mechanical change of the clock and the passage of a second is useful information for someone trying to repair it. In this case noise becomes signal; indicating what's wrong.

Reflecting on the clock example also helps to distinguish two correlates of information related to its normative consequences. First, notice that only a very minimal set of the physical properties of the clock are used to provide information about time. This is also a characteristic feature of all physical media used to convey information. As in the case of the clock, many "incidental" consequences of its operation, such as the sound it makes, are irrelevant to its normal informative function. And yet, deviations of physical properties that are incidental to this functioning may be informative to a repairman. Second, the form of the dynamical constraints that enable a physical process to be used to "tell time" can be realized by a vast number of mechanisms. This has been described as *multiple realizability* in the philosophical paradigm known as functionalism. This follows from the fact that neither reference nor function are attributes of a physical signal. Both the fact that only some physical attributes of an information-bearing medium are relevant and that many physical configurations can convey the same information make it clear that normativity and reference are properties imposed externally, i.e., by interpretation.

In this respect. formal information theory and semiotic theory are methodological duals of one another in that they each focus on aspects of information that the other ignores. Because referential and normative properties are extrinsically imposed and interpretation-dependent they are typically excluded from formal information analyses. Semiotic theories instead explicitly address the problem of reference and normativity, but ignore physical and statistical features of the communication medium.

4.4 Semiotic Grounding

In order to unify information theories and semiotic theories into a single synthetic theory of semiotic information it is necessary to deal with the problem of referential displacement. This is the property of "aboutness," the non-intrinsicality of reference. This is an ancient and fundamental problem of epistemology: explaining how sign vehicles, such as thoughts and words, can provide a causally useful connection to other objects and processes in the physical world, even though they are not in any way intrinsically involved in or part of what they refer to.

A computationally framed variant of this classic problem was described as the "symbol grounding problem" by Steven Harnad in a celebrated paper in 1990 [4.6]. In this usage of the term "symbol" Harnad is treating it as a generic term for any sign vehicle, with a sort of linguistic analogy in mind. This poses a common terminological problem for any effort to unify information theory with semiotic theory, which needs to be addressed before going further. In a semiotic context, a symbolic relationship is a specialized subcategory of sign relation, not the generic form. From this perspective, linguistic reference is understood as a highly derived form of referential relationship, not its base form. So, failure to distinguish these uses can lead to serious confusion. To avoid this, I will assume that Harnad's analysis involves the generic case, and not merely linguistic reference, though his definition of "symbol" as an arbitrary token suggests that he considers the linguistic model as generic. To try and avoid confusion, then, (as noted above) I will continue to use the term "sign vehicle" generically, to refer to the physical token that mediates between the object of reference and the process of interpretation.

In general, the term "ground" refers to the real-world connection that gives the representation relation its potential physical relevance. But "grounding" is not "given" as an intrinsic sign vehicle property, it must be generated by an interpretive process. Sign vehicle properties don't therefore determine reference, instead they serve as *affordances* that aid the interpretation process.[1] From a semiotic perspective, then, we can identify two major classes of affordances for establishing referential grounding. They are (i) shared formal properties between sign vehicle and object (iconicity), and (ii) some physical-temporal correlation or contiguity between sign vehicle and the object of reference (indexicality). Indeed, Harnad refers to these relations as providing possible contributions to the "symbol grounding problem," but does not consider them to be semiotic in the same sense. This can be analytically disambiguated by distinguishing between sign vehicles and their formal and physical properties taken as referential affordances.

In this respect, let's briefly revisit the concepts of iconic and indexical reference reconceived with respect to the affordances that are being utilized.

Iconic reference depends on an interpretive process taking advantage of shared formal properties between the sign vehicle and what it refers to. For example, with a bit of imagination a face can be discerned on the full moon, or in a cloud formation, and it might even remind you of someone you know. But iconism can also be highly abstract. A mathematical equation refers iconically, once you know how to discern its symbol-mediated isometry (e.g., between the structure of the interpreted equation and a corresponding geometric or dynamical relationship). But an equation can be interpreted to be iconic (e.g., of a parabolic trajectory) only if you know how to discern the way that differences in the values or operations directly correspond to differences in the geometric object of reference.

Indexical reference depends on an interpretive process taking advantage of some physical correlation between a sign vehicle and what it is taken to refer to. Thus, contiguity in space or time, or having been derived from a common physical source or substrate can provide such an affordance. A simple correlation can therefore be the ground for indexical reference. A lipstick smear on a man's shirt collar can be a troublesome indication to his wife, a urine scent on a branch can be a sexual index to a female lemur, and the mobbing call of a small bird can indicate the present of a raptor. What gets correlated and how (accidental, cultural, evolutionary) is arbitrary, but the fact of correlation is not. So, a rat in a Skinner box pressing a bar in response to a bell in order to get a water reward has learned that the bell is an index of the state of the apparatus, even though this pairing was the whim

[1] Although the physical properties of the sign vehicle medium constrains the possible semiotic affordances that are available, so do the competence and "interests" of the interpreting agent.

of an experimenter. The arbitrary pairing doesn't make the one a symbol of the other. For the rat, what matters is discovering this correlation so that it can use the bell to indicate the availability of water. Though the particular physical properties of the bell sound have no intrinsic relationship to water, learning the reliable temporal correlation with the water reward provides the correlated indexical affordance.

Analogously, repeated pairing over the course of generations in the evolution of a species can be the basis for an innate tendency to interpret something indexically. This is the case for the indexicality of a vervet monkey alarm call sound and its correlation with the presence of a type of predator and the urge to engage in an appropriate defense activity or escape behavior. Conventionalized indices are also quite common. For example, consider the white line painted down the middle of two-lane roads or the salute given by a soldier to a superior officer. Though these are conventionalized sign vehicles, they refer indexically, not symbolically.

Symbolic reference differs from iconic and indexical modes of referring because referential affordance is not provided by features intrinsic to the sign vehicles. Symbolic competence is therefore afforded by a kind of iconicity between different interpretive habits in a community of symbol users. In other words, the supportive affordances for symbolic reference are distributed, not located in sign vehicle properties or medium structure. They have been shifted to between-interpreter and between-sign relations. As a result, because sign vehicle properties can be chosen irrespective of intrinsic referential affordances, their intrinsic properties are free to be used for other purposes and their iconic and indexical relationships to one another can become separately informative.

The way Harnad articulated the symbol grounding problem followed generations of scholars using the term "symbol" to refer to any conventional sign vehicle. Alphanumeric characters are conventional sign vehicles that have been explicitly developed over millennia to express symbolic relations, so this terminology is generally unproblematic in these contexts. However, when your computer begins randomly spewing alphanumeric characters onto your screen these events are interpreted as indices of a malfunction, not as symbolizing anything. So, while this terminological distinction is irrelevant in information and computation theories, it cannot be ignored when considering the problem of referential grounding in both biological and cognitive contexts.

Equating symbolic reference with mere conventional correspondence also ignores a complex semiotic infrastructure, since convention itself is a consequence of prior semiotic processes. Conventions don't just exist by fiat. They are shared interpretive habits that must be created or discovered. Social conventions in particular need to be created semiotically, i.e., by communicating. To avoid circularity, then, we need to understand how conventional forms of communication are created using nonconventional semiotic processes.

As the father of semiotic theory Charles Peirce pointed out over a century ago, we must distinguish properties of a sign vehicle (e.g., a drawing or spoken word), from properties taken to link it to its object of reference. Symbolic reference involves a conventional type of sign vehicle that additionally represents its object of reference in a conventionally-mediated way. Something can be considered symbolic, then, only if the property determining its relationship to what it refers to is also due to a shared rule or convention. Symbols are in this sense doubly conventional. They are conventionally derived sign vehicles that refer to things via conventionally determined means.

This hierarchy of semiotic affordances also has implications for the relationship between referential and normative information. Use and function are doubly displaced from sign vehicle properties, analogous to the way symbolic reference is displaced from its iconic and indexical referential grounding. Thus, even though we know nothing of the beliefs and rituals of the Stone Age people who painted on cave walls in Southern France, we can easily recognize the idealized forms of animals and humans painted by these Stone Age "artists." Interpreting their iconic content is almost trivial, while interpreting their symbolic significance (if there is any) or their normative function in the lives of these people is beyond our reach, because we can't reconstruct the distributed social habits they were created within. In order to be able to interpret the significance of cave paintings or the meanings of words in an ancient lost language one would need some prior experience with the patterns and habits of use of these artifacts within a cultural network of other semiotic activities.

4.5 A Possible Bridge?

How can formal information theory be expanded to incorporate these semiotic referential and normative distinctions? Although formal information theory analyzes only the statistical properties of the sign vehicle medium, by also considering the relationships between these formal properties *and* the physical properties of that medium, information theory can be situated in a physical context where external relational features become relevant. The material-energetic substrates that embody the formal properties of information provide the critical affordances necessary for establishing these semiotic properties.

For example, consider the concept of *mutual information*. In nontechnical terms mutual information is the amount of uncertainty about one information source that is reduced by knowing the information in another. In this sense it is effectively a measure of similarity or shared form. It provides a way to formalize the concept of iconic affordance in information theoretic terms. But iconicity cannot provide specific reference, only the possibility of reference, since merely sharing a common form leaves open an indefinite class of relations. To specify which instance of shared form is referred to, the individuating role of physical indexicality is additionally required.

The information theory analogue of indexicality is a bit more complex to identify, however. This is because it involves the physicality of the sign vehicle. One way to analyze this physicality is to consider the concept of entropy in both its informational and thermodynamic senses. Calling these both "entropy" is confusing, however, because their relationship is not one of identity, as many have argued (e.g., [4.7] [4.19] [4.20]). Shannon famously named his measure of information "entropy" because of this mathematical similarity to the measure of thermodynamic entropy as first formulated by Ludwig Boltzmann. By paying attention to both the similarities *and* differences between these two "entropies" it is possible to recognize a relevant connection (beyond the superficial similarity of their mathematical representations) [4.3].

This "entropy connection" relies on the fact that any sign medium is a physical medium. The information entropy of the medium is therefore necessarily a function of its physical entropy. But each particular state of the physical medium is also potentially subject to modification by extrinsic physical conditions (whether due to human or other physical influences). This means that a change in information entropy necessarily also corresponds to some correlated change in its physical entropy, although typically only involving a fraction of its physical entropy. As a result, a change in information entropy *correlates with* a change in the physical entropy of the medium produced by some extrinsic influence that modified it and which it thereby *indicates*.

So, the gap between information theory and semiotic theory can be bridged, at least with respect to iconic and indexical relations. Extending this analysis to account for the semiotic notion of symbolic reference, where there are no formal or physical affordances available to ground reference, requires understanding how conventional relations grow from iconic and indexical communications. In this way, we can recast the problem as a sign vehicle *un*-grounding process, while paradoxically maintaining referential grounding.

An illustrative biological example of informational-referential continuity despite a lack of direct grounding is provided by the familiar genetic "code." The code analogy is apt because the chemical composition of nucleotide bases is unrelated to the chemical composition of amino acids. The grounding of genetic information on protein function is mediated by the codon-anticodon iconism between DNA, mRNA, and tRNA. The indirect iconism between a DNA codon sequence and the amino-acid sequence of a resulting protein is additionally mediated by the physical correlation between individual tRNAs and their contiguous amino acids. Finally, the amino acid sequence influences the eventual 3D structure of the folded protein and thereby its cellular function. Thus, protein functional information is indirectly recorded as DNA sequence information by virtue of the mediation of linked iconic and indexical affordances. This mediation enables preservation of informational continuity (grounding) despite the lack of structural similarity or direct chemical interaction. In this respect, the relationship between DNA structures and cellular functions is roughly analogous to the conventionally mediated grounding of words in a language by virtue of supportive iconic and indexical forms of pragmatic communication.

So, an expansion of the concept of information to include its referential and normative ground requires both an analysis of the form and physical characteristics of the medium. And to align the information theoretic concept of "symbol" with the corresponding semiotic concept requires demonstrating the way that iconic (mutual informational) and indexical (entropic) affordances can combine to enable referential grounding to be displaced from one sort of sign vehicle to another quite different one. This is only a first step, however. A full explanation of how these are leveraged by an interpretive process still requires an account of what sort of process is able to utilize these affordances. This is a task that should be a core challenge for future biosemiotics.

4.6 Preliminary Conclusions

The possibility of formalizing the relationship between information theories and semiotic theories has been impeded by four erroneous assumptions: 1) that information is a single level concept when in fact it is a nested hierarchy of concepts; 2) that language is the appropriate exemplar of base level generic information relationships in which reference can be understood as arbitrary mapping; 3) that only intrinsic structural information is physical and that referential and normative information relationships are "subjective" and epiphenomenal; and 4) that it is unnecessary to consider the material-energetic features of signal (sign vehicle) media to explain referential and normative properties.

These assumptions need to be abandoned in order to formulate a unified semiotic information theory. The dilemmas created by these assumptions can only be resolved by recognizing how information theory is embedded within semiotic theory and how both are embedded within physical and biological contexts. The critical bridge between these paradigms requires identifying how sign vehicle properties can be analyzed so that they corresponds to the semiotic affordances described as iconic, indexical, and symbolic referential relations. Identifying this correspondence requires distinguishing two structural features of any information-bearing medium: its formal relationships with another medium and its physical relations with another medium. The information theoretic concept of *mutual information* captures the formal correspondence essential to account for iconic affordance, and the relationship between the information entropy and thermodynamic entropy captures the physical relation essential to account for indexical affordance. These two non-equivalent concepts of entropy are linked by the physicality of the informing medium and its embedding in a larger physical context. The indexical link is thus mediated by virtue of the fact that the physical context can produce a change in the medium's physical configuration (which results in a change in its thermodynamic entropy) and this can potentially be reflected in a change in its information entropy. Finally, symbolic referential affordance is supported by the way linked iconic and indexical signs can maintain referential grounding between otherwise unrelated sign vehicles and referents.

This analysis has provided only a preliminary first step in the process of articulating a semiotic information theory; i.e., a theory that can account for both the referential and the normative properties that characterize the full meaning of information. What remains is to develop a theory of what constitutes an interpretive process that can take advantage of these semiotic affordances to provide the mediating link between sign vehicles and objects of reference. Our previous inability to identify the common ground between basic information theoretic properties and the primary semiotic units of analysis has up to now blocked any effort to bring these fields together. In this chapter, I have argued that recognizing how both theoretical paradigms may need to reconsider their formal structure and their material-energetic instantiation before it is possible to establish common ground. This can at least provide a critical first step towards development of a theory able to bridge between the many sciences that must deal with information in all its many aspects.

References

[4.1] Collier, John (2003) Hierarchical dynamical information systems with a focus on biology. *Entropy* 5, #102.

[4.2] Deacon, T. (1997) *The Symbolic Species: The Coevolution of Language and the Brain.* New York: W.W. Norton & Co.

[4.3] Deacon, T. (2017) Information and reference. In: *Representation and reality: humans, animals and machines.* Gordana Dodig-Crnkovic & Raffaela Giovagnoli, eds. Springer, Heidelberg: 3-15.

[4.4] Floridi, L. (2010) *Information: A very short introduction.* Oxford University Press, Oxford, New York.

[4.5] Fano, Robert (2001) Interview in O. Aftab, P. Cheung, A. Kim, S. Thakkar, N. Yeddanapudi (2001) *Information theory and the digital age,* Massachusetts Institute of Technology, 6.933, 24. URL: https://web.mit.edu/6.933/www/Fall2001/Shannon2.pdf

[4.6] Harnad, Steven (1990) The Symbol Grounding Problem. *Physica D*, 42, 335-346. URL: ftp://ftp.princeton.edu/pub/harnad/Harnad/HTML/harnad90.sgproblem.html

[4.7] Mirowski, Philip (2002) Machine Dreams: Economics Becomes a Cyborg Science. Cambridge University Press, Cambridge.

[4.8] Morris, Charles W. (1938) Foundations of the theory of signs. In: Otto Neurath *et al.* (eds.). *International encyclopedia of unified science,* Vol. I, No.2. The University of Chicago Press, Chicago, Illinois: 1-59.

[4.9] Peirce, Charles Sanders (1958-1966) *Collected Papers of Charles Sander Pierce.* Vols. I-VI edited by Charles Hartshorne and Paul Weiss; Vols. VII-VIII edited by A. W. Burks, published by the Belknap Press of Harvard University Press.

[4.10] Peirce, Charles Sanders (1904) *Collected Papers of Charles Sander Pierce* Vol. IX Letters to Lady Welby, SS 33.

[4.11] Peirce, Charles Sanders (n.d.) MS 1345 in Robin, Richard (1967) *Annotated Catalogue of the Papers of Charles S. Peirce.* University of Amherst Press, Amherst.

[4.12] de Saussure, Ferdinand (1916; English translation 1959). *Course in General Linguistics.* Translated by Wade Baskin. Columbia University Press, New York.

[4.13] Shannon, Claude (1938) *A Symbolic Analysis of Relay and Switching Circuits.* MIT Department of Electrical Engineering M.S. 1938. MIT Institute Archives.

[4.14] Shannon, Claude (1945) *A Mathematical Theory of Cryptography.* URL: https://www.iacr.org/museum/shannon/shannon45.pdf

[4.15] Shannon, Claude (1948) The Mathematical Theory of Communication. *The Bell System Technical Journal* 27, 379–423, 623–656.

[4.16] Shannon, Claude (1956) The bandwagon, *IEEE Transactions on Information Theory* 2(3), 3.

[4.17] Shannon, Claude & Weaver, Warren (1949) *The Mathematical Theory of Communication,* University of Illinois Press, Urbana.

[4.18] Silverstein, Michael (1972) Linguistic Theory: Syntax, Semantics, Pragmatics. *Annual Review of Anthropology* 1, 349-382. URL: doi.org/10.1146/annurev.an.01.100172.002025

[4.19] Ter Haar, Dirk. (1954). *Elements of Statistical Mechanics.* Rinehart & Co., New York.

[4.20] Ueltschi, Daniel (2011) Shannon Entropy, in: Publication (§6). URL: ueltschi.org/teaching/chapShannon.pdf

[4.21] Weaver, Warren (1949) Recent contributions to the mathematical theory of communication. In: C.E. Shannon and W. Weaver (eds.) *The Mathematical Theory of Communication,* University of Illinois Press, Urbana.

5

Ontology and Semiotics of Memory

Anton V. Sukhoverkhov[1]* and Arran E. Gare[2]

[1]*Department of Philosophy, Kuban State Agraria n University, Krasnodar, Russia*
[2]*Philosophy and Cultural Inquiry, Swinburne University of Technology, Hawthorn, Victoria, Australia*

Online resources:
Anton V. Sukhoverkhov:
https://kubsau.academia.edu/AntonSukhoverkhov
https://philpeople.org/profiles/anton-sukhoverkhov
https://www.researchgate.net/scientific-contributions/Anton-Sukhoverkhov-2072757047
Arran E. Gare:
https://www.researchgate.net/profile/Arran-Gare
https://swinburne.academia.edu/ArranGare
https://philpeople.org/profiles/arran-gare

Abstract
This article presents an ontological approach to memory by focusing on the fundamental principles responsible for the maintenance, transmission, reproduction and evolution of physical, biological and social systems. We defend semiotic realism wherein information is considered an essential property of the Universe. Such a view grounds cognition in reality and explains why living organisms can interact effectively with their environment, why different species can to some degree "understand" the intentions of each other, and how the origin of communication and language became possible in spite of a multiverse of "umwelts." The work analyzes the ontology of memory from an evolutionary perspective, describing its three major types: physical, biological and social. *Physical memory* is a natural retention of physical states (that can be used as a source of information) in an organism or the environment. It includes detectable differences, "recorded" in physical structures or processes (e.g., relic radiation, acoustic signals, fossil records, etc.), carrying potential information for their interpreters as "virtual signs" (Deely). The ability of matter to "preserve" information in its structures is considered as the objective ground/precondition for the origin of life, heredity, cognition and communication. We define *biological memory* broadly, including homeostasis, cellular differentiation, the honeybee waggle dance, seasonal migrations of butterflies and other biological processes built upon genetically and epigenetically inherited information. *Social memory* is presented as a distributed process in nature and society that is scaffolded by different external semiotic systems and

**Corresponding author*: sukhoverkhov.ksau@gmail.com

maintained throughout generations via behavioral, ecological and cultural inheritance. It is shown that the modern approaches often follow heuristics of extreme theoretical dichotomies (antinomies) in memory studies. Researchers focus either on *material structures*: traces, sign vehicles, artifacts that represent memory, or on cognitive *processes* that interpret or implement it. We argue that a comprehensive approach requires integration of these complementary approaches.

Keywords: Ontology of memory, non-genetic inheritance, semiotic scaffolding, semiotic constructivism, semiotic realism, origin of language

> *Everything in reality is composed of information.* [5.69]: p. 173

5.1 Semiotic Realism versus Semiotic Constructivism

Let us start with a simple question: Where is the content (meaning) of a book? Is it in the book or in the mind of a reader? The usual answer would be that it is in the book because we have many books on our bookshelves that in different ways influenced us. Some entertained us, some taught (e.g., textbooks), and others instructed us on how to do something (e.g., recipe books). The bookshelf is like an external memory system that keeps potential information on various subjects and transmits it throughout time to us, our children or grandchildren.

However, if a reader does not have knowledge (memory) of a particular language he/she is unable to "extract" information from these books. For example, there are many historic texts written in unknown languages that are still undeciphered, such as the Phaistos Disc or the the rongorongo script of Easter Island [5.3] [5.21]. Thus, the more literate and educated the reader is, the more information he or she can get from the text. From this perspective, the information or content of a book belongs to the mind of the reader who assigns meanings to "meaningless" letters/words (sign vehicles) in the book [5.47].

If we stick to the second approach, it would mean that all informational content belongs solely to the mind of the interpreter (reader) and books (symbols, letters) do not carry any information intrinsically, by themselves. Similarly, photons of light that "transmit" information about reality also would not have any information in themselves, because any "meanings" of particles and their patterns belong to an observer who interprets ("reads") them. Then, doctors would skeptically observe "symptoms" of diseases and archeologists would look with doubt at "prehistoric" fossils and results of radiocarbon dating. If we take such a radical, constructivist approach in epistemology and semiotics, then it would lead us to fruitless relativism and agnosticism. Of course, we have to admit that, for instance, the *Tao Te Ching* by Lao Tzu has multiple translations and interpretations [5.39]. Charles Peirce's works also present differing ideas that have led to various interpretations and conceptualizations of his thoughts [5.52]. Nevertheless, researchers also acknowledge that in these books unique ideas are presented that have been discussed for centuries. In everyday life, we also follow common-sense realism because we acknowledge that a check for dinner in a restaurant or a utility bill for our house have "meanings" in themselves, not in the mind of a "reader."

If we consider DNA as a "book of life," we have a similar problem: do the "meanings" of the genomic sequence belong intrinsically to the structure of DNA or to living systems that interpret it in the process of semiosis [5.4] [5.25] [5.27] [5.47]? As in the debates on the nature of light (a particle versus wave), this research proposes to accept the "complementarity principle" and admit that life is based on both material "sign vehicles" and processes of "semiosis" (i.e., "writing" and "reading"). Living systems are not only internal sign vehicles ("texts") but also interpreters of these. To find food or a partner, avoid predators or adjust to ever-changing environments, organisms should detect, recognize and read "external" environmental clues. At the same time, living systems are autopoietic and self-referential, as argued by radical constructivists [5.9] [5.20] [5.42]. Indeed, different species have their own "umwelt bubble" [5.70]; for instance, the search for food is specific for each species and involves the pursuit for different clues and "meanings" in their environments. However, we have to acknowledge that clues do exist in the external environment; they enable the "affordances" of organisms for actions and cognition. For example, physical movement from point A to point B must be "grounded" in reality (e.g., as in the case of bats' echolocation). Seawater is "salty" because it contains salt, which, in turn, determines the salty *taste* of water for organisms that have a sensory system to detect salt (or other chemicals) in the environment.[1] Thus, the environment somehow contains information in itself (in the form of detectable differences), otherwise any theory of perception, interpretation and cognition will fall into the cul-de-sac of subjective idealism and solipsism.

Confirmation of such an "essentialist" statement in semiotics and information theory can be found in modern physics, which tends to consider the physical world more as an informational system than just plain matter [5.14] [5.41]. In other sciences, information is also considered a fundamental property of the Universe, along with matter and energy, that defines the very essence of biological life and human society [5.10] [5.69]. This does not mean that "information" exists independently of relations, as "atomic substances" floating around [5.24]. For instance, in arguing on the basis of quantum theory that "It" comes from "Bit," that is, what exists comes from information, the physicist John Wheeler claims that

> [...] every particle, every field of force, even the spacetime continuum itself — derives its function, its meaning, its very existence entirely — even if in some contexts indirectly — from the apparatus-elicited answers to yes-or-no questions. [...] [T]hat which we call reality arises in the last analysis from the posing of yes-no questions and the registering of equipment-evoked responses; in short, that all things physical are information-theoretic in origin and this is a participatory universe [5.73]: p. 5.

Wheeler could have mentioned that such information can be and usually is recorded, and is acted upon. So, information is basic to what exists, but only in the context of and in relation to observing systems and interpretative schemes.

For now, it is hard to define theoretically such physically understood information as well as to make a clear-cut distinction between information and meanings (that we have in the case of *interpreted* information). Maybe new discoveries in physics or the development of transdisciplinary integration of physics, semiotics, and cybernetics will give us a better and more solid understanding of the nature of information existing before the emergence of

---

[1] The origin of the representation (mediation) of reality by sense organs (receptors) is supposedly paradoxical. How can the organism produce a sensory system for detecting novel aspects of reality without a basic cognitive system that has already detected that reality? For the organism, the revelation of a new aspect of reality requires a sensory system, but to produce such a system the organism needs to know (recognize) aspects of reality for which the sensory system will be developed.

living systems [5.8] [5.9]. However, in the instance of physical information, we can resort to the definition of information as detectable differences that could potentially carry or convey further information. It can also be presupposed that the nature of information can evolve and have more complicated and "purified" ideal forms. For instance, more complex forms of life generate more complex semiosis, from "functional information" [5.54], associated with the first steps of the most basic forms of life, to the object-oriented information associated with animal semiosis, and to the highly developed information processing identified as such through cultural semiosis [5.37].

If we acknowledge that information is an intrinsic component of the Universe, then cognition could be grounded in reality and epistemology could explain why any living organism (including scientists) is able to learn its environment, understand the present and the past, and predict possible future events. Furthermore, we can explain why not only members of the same species, but members of different species, to some degree can "understand" each other, including their intentions, and how this explains the origin of communication and language in spite of a multiverse of "umwelts" (meaningful surrounding worlds of perceivers). Finally, according to Thomas Sebeok and Winfried Nöth, this approach could help to understand the presemiotic stages between the physical and the biological worlds, and the necessary intermediate steps from physiosemiosis to biosemiosis [5.45]. Let us consider the problem of protosemiosis in detail.

5.2 Protosemiosis: Recording Without Reading or 'Heritability' of the Universe Before Earth

Does a book exist as a book if nobody reads it, or will signs remain signs if nobody interprets them? According to Nöth, a radical group of semioticians state that no reading means no signs. However, in his view, information is real; furthermore, the idea of information can explain a gradual transition between the physical and semiotic worlds. Nöth agrees with John Deely that semiosis or protosemiosis could exist before the origin of life [5.55] and could be found in a natural phenomenon before it is interpreted as a natural sign, and such type of sign could be called "a virtual sign" [5.16] [5.45]. Deely gives as an example of such a "virtual sign" being a dinosaur bone in his garden that carries information about events that happened seventy million years ago [5.17]. In his view, virtual signs and the information in them do exist, no matter if anyone reads them or not. However, can we call these historic recordings "semiosis" or "protosemiosis" ("physicosemiosis" or "physiosemiosis") and treat other physical processes as semiotic? Are relic radiation, sedimentary rocks or rings of a tree examples of physical semiosis? Does the universe naturally "write" or record "books" (or "proto-books" or "virtual books")?

Higuera shows that it is not enough to accept that objects have *innate* semiotic properties [5.49], the question also is whether things do have a specific type of action between themselves, which is replicated later by biological forms, so that "we could interpret from the mere notion of physiosemiosis as a type of semiosic action that occurs in the physical world" [1.34]: p. 6. If we take *any fixed changes* in the physical environment as metaphorical "writing" ("recording") then, indeed, these processes could be interpreted as physiosemiosis or protosemiosis that carries (or produces) information. We accept this view, but to avoid terminological misunderstanding, we call this *reduced semiosis*, which presupposes just "recording" and not "reading," as "physiosemiosis" or "virtual semiosis." Whereas the basic informational (meaningful) processes in living systems that involve "reading" (reading-as-action) of virtual signs is referred to as "protosemiosis," as coined by Prodi (cited from [5.58]), and elaborated by Sharov and Vehkavaara [5.55].

In practice, such hypothesis of natural physiosemiosis can help to explain other types of classical (i.e., *intended and addressed*) semiosis, because in everyday communication we also resort to changes in the physical environment to transmit information (e.g., by making sounds or carving letters in stone). This phenomenon, where one physical object or process naturally keeps information about the impact made by other physical processes, is an example of physical memory. Examples of this type of memory are the relic radiation that "recorded" the dawn of the Universe, sedimentary rocks that preserve steps in the origin of life, radio signals, or paper books that keep information in their physical structure for possible receivers. Carriers of such information could be physical, natural (chemical components of DNA or tree rings) or artificial (electronic books or road signs). Sometimes stable, "directed" processes in nature can emerge in the form of "dynamic attractors" that determine trajectories in the development of biological/social systems, or "ecological memory" that influence the dynamics of ecosystems [5.36], [5.58]: p. 13. If we imagined the Universe, where all is in flux, we could never have memory, information or living systems. Therefore, these phenomena presuppose the ontological dialectics of dynamic and stable states of matter. From the physical point of view, life can be seen as a cumulative inertial system. Later in evolution it acquires various biological and social constraints [5.25] but all of them have to be grounded on physical memory.

Organisms in nature and scientists in a society use these "recordings" to recover information about past events, make predictions or transmit data to the future generations. Nöth, Weber and Deacon also agree that memory in physical systems could be considered as protosemiotic, because memory is "an eminently semiotic process" [5.45]. In this regard, we propose that not just the idea of information, but also the concept of physical memory can be used for analysis of the intermediate stages between matter and information, nonsemiotic and protosemiotic.

5.3 Biological Memory: Between a Particle and a Wave

Genuine semiosis requires "reading" or transition from direct (physical) reality to represented reality (i.e., from sign vehicles to representations). As in the case of a book, information is not given to us directly and requires recognition, interpretation and learning of various "natural signs" or "indexical signs" in the environment. Perceivers do not need photons as such; they need them as tokens (clues) revealing objects in the world that have reflected and transformed that light. Readers do not need paper and letters in the book *per se*; they are searching for reflections of the author that stands "behind" the letters. However, to "unfold" the content of physical memory or virtual signs, the "reader" needs memory. In classical semiosis (with interpretant and interpreter), memory comes from memory (compare *omne vivum ex vivo*). Ancient cavemen, to match the sound of a song with a specific bird, needed to experience them together, and the more experience they had, the more "signs" in the forest they could "read" and anticipate more events. To turn a bone into a sign of past events or a sign of a dinosaur, a researcher, who had discovered this bone, needs to have historic apperception (historical *a priori*) [5.11], knowledge or category of the past as a prerequisite. To read the Phaistos Disc, the reader needs to know the language and culture of the people who made it. Such knowledge could enable the transition from the present to the virtually reconstructed past or the future. The more knowledge the organism, person or society has (or inherits), the bigger picture of the past they can reconstruct; and the bigger the picture is, the more facts in reality are involved in the process of historic reconstruction, which is a kind of "hermeneutic circle."

For the origin of life, living systems need, as a prerequisite, chemical autocatalytic circles (sets) with some stable memory-like structures. Such strictures can preserve and accumulate new (functional) information that is needed to maintain these reproductive circles and provide further cumulative development of such systems [5.56] [5.57] [5.58]: pp. 124-126. In particular, the self-reproduction of organisms requires two aspects of "memory" processes: structural semiotic aspects (e.g., RNA or DNA) and processual aspects ("semiosis," "reading," "transcription," "interpretation," etc.). In the process of evolution, hereditary information is preserved via the interplay of copying and construction processes [5.56].

Some researchers hypothesize that self-reproduction could emerge without the material structures that "record" memory.

> Ensembles of processes can take on very stable configurations that can serve as memory until such time as a material structure might appear to record memory. Terrance Deacon, for example, believes that the precursors of RNA originally performed some function like energy storage and/or transfer and only later were exapted to serve as a memory repository [1.68]: p. 24.

However, even if RNA did not work originally as fully informational four-letters genetic "language," it could work as energy "storage" that maintained metabolic processes through its physical/chemical structure, and thus functioned as a physical memory.

Methodologically, it is important to emphasize that memory processes in living systems emerged not in the form of digital "memory," but as regular biological processes. In a narrow (reductionist) sense, biological memory can be viewed as a *specific* process or structure that accumulates, transmits and reproduces some data through time and space by means of "footprints" (or "blueprints"). However, in the broad sense, many *actual* system processes, such as homeostasis, meiosis, cellular differentiation, the waggle dance of bees, seasonal migrations of butterflies, etc., can be considered as embodied, distributed and empirically observable memory processes.

It can be hypothesized that the very essence of life is a fight against oblivion (obliteration), and this self-preservation instinct could be revisited as a "memory instinct." The processes of self-reproduction could thus also be seen as empirically observable and physically (and biologically) embodied processes of memory "retrieval." In this regard, we propose to consider *biological memory*, generally, as a set of system processes of maintenance, transmission, and reproduction of living systems by means of heritable genetic and non-genetic information. It must be emphasized that the studies of biological memory, as well as social memory (see below), often result in theoretical dichotomies (antinomies) of structure/process, code/semiosis, genes/environment, nature/nurture [5.2] [5.5] [5.30] [5.74]. However, a comprehensive approach in memory studies should be based on the complementarity principle.

As it has been shown in many studies, the reproduction of living systems cannot be reduced to a sequence of DNA codes or epigenetic marks [5.13]. It is a decentralized, multi-leveled, distributed process that very often goes beyond individual organisms [5.38] [5.65]. Composite biological/ecological systems, such as bee or ant colonies, can be seen as such distributed, multi-agent memory systems (with external and internal systems of determination and regulation) that maintain, transmit, and reproduce the whole system and its elements. An effective methodological toolkit for the analysis and description of such type of evolving systems was elaborated in the inclusive or extended evolutionary synthesis [5.13] and in developmental systems theory that revised and extended gene-centered approaches to inheritance and evolution [5.46].

5.4 Social Memory: From Nature to Nurture

As has been shown by many researchers, some forms of behavior and cognition are acquired and transmitted throughout generations not via the internal, genetic (epigenetic) memory, but through *external* behavioral, ecological and cultural "inheritance systems" (Jablonka, Lamb), "semiotic scaffoldings" (Hoffmeyer), or "developmental niches" (Stotz) [5.33] [5.35] [5.60]. For instance, the study of feral children has taught scientists that to be born to human parents does not necessarily mean to be human. A newborn child needs social interactions through the physical and symbolic tools with other people to acquire those habits that are normally identified with "human nature" [5.60] [5.72]. Therefore, society is an important factor that supports the development and identity of individuals, who form and reproduce that society. This means that individual memory (mind, consciousness) originates from and is mediated by transgenerational and cumulative social memory (e.g., language and traditions).

Thus, we need to admit that at some stage of evolution, the individual "umwelt bubble" became collective and shared as a "mitwelt." For instance, in the mythological worldview people shared and socially transmitted models of reality based on symbolic and anthropomorphic representations of the physical world and natural processes (e.g., the myth about Phaethon). In late Medieval and Renaissance societies, myths were replaced by Aristotelian models of reality based on explaining everything through material, formal, final and efficient causes. In modern times, scientists replaced this with models based on mathematical representations. Other practical and observable processes such as fire-making, building houses, reading texts or speaking a language also have both an individual and a social nature. They are formed and maintained by individual memory and, at the same time, were formed by social traditions and inherited through active teaching, learning and practice.

Usually, social memory is conceptualized in the studies as 1) information about the past, 2) traditions that came from the past, or as 3) in the present, culturally-inherited historic artifacts. Nevertheless, many actual biological and social processes (seasonal migration, song "dialects" of birds, usage of days of the week, sport activities, educational organizations, etc.) are tangible products of social memory that have a future-directed nature and constantly incorporate novel elements. In this context, an interesting example of "narratives in building" can be borrowed from Arran Gare. He states that "... signs can be more than signs of objects in an external world. They can play a creative role, creating objects or processes in this world" [1.26]: p. 229, that is, stand for something that only will appear in the future. However, for that, it should be in the form of a project (blueprint, "virtual sign") in the present and, for the accomplishment of the project, builders need skills (narration, semiosis, traditions) borrowed from the past.

Gare also writes that before the twentieth century, architects did not design every aspect of construction plans and left the details to artisans. Furthermore, construction of cathedrals and cities in the Middle Ages could be stretched over many generations of builders [5.26]: p. 229. For those reasons, the constructive "narration" unfolded itself through many generations, adopting novel elements and new technologies into the project. In terms of the presented research, building narration worked as a "memory process" open to "updates." It should be noted, that the realization of such immense, long-term projects can be described from different theoretical perspectives. For instance, it can be described in terms of the theory of distributed cognition, labor division theory, educational theory (the process involves new people who are taught to be builders and artisans), the theory of semiosis/narration and, also, through the idea of social memory and social inheritance.

Narratives in building are not limited to verbal communication or "communicative actions" (J. Habermas). Gare shows that:

> Builders also orient themselves by what has already been constructed and to the identities and actions of other participants in building. Incomplete buildings are a field of signs of what had been done... Interpretants of these signs include productive actions and further building. What is built in turn become signs. Most importantly, all these signs are components of the narrative of the unfinished project [5.26]: pp. 229-230.

This example embodies the nitty-gritties of an externalist, distributed approach to semiosis together with the internalist, interpretative approach because it shows that shared memory and narrative processes become "sedimented" in external artifacts and thus semiotically scaffold further actions, enabling the inheritance of the process and its relevant content ("meanings") to be further built upon.

It is important to emphasize, that no matter what type of symbolic "narration" we take as our example in society (agricultural, architectural or scientific), it is always distributed and cumulative [5.40]. As Isaac Newton observed, "if I have seen further [than others], it is by standing on the shoulders of giants." Étienne Bonnot de Condillac maintained similar ideas in his treatise "Essay on the Origin of Human Knowledge." He stated that people discovered their greatness in all kinds of art and science almost simultaneously not just because of geographic factors, but also due to the development of sign systems such as algebraic and geometric signs and the language system. They contribute to this by developing the memory, imagination and thinking of people who have studied them [1.15]: pp. 286-292. Condillac also noted that the systems of signs of language are similar to the signs of geometry, allowing new knowledge to be born and expanding the cognitive capabilities of researchers. Newton, according to the author, also owed his discoveries in mathematics and physics to the developed language and the contrivance of methods of calculating, which allowed him to create outstanding discoveries [5.15]: p. 288.

This characterization supports the proponents of a soft version of "linguistic determinism" or, in our case, "semiotic determinism," because different types of sign systems used in society can accumulate practical achievements, meanings, concepts and theoretical knowledge that scaffold further development of human actions, mind and society.[2] It also means that comprehensive semiotic theory should integrate into its structure the externalists' (e.g., the ecological approach to language and language acquisition) [1.32] [1.63] and the internalists' perspectives. In this regard, in our previous work, the idea of the "dual inference" of language was proposed [5.58]: pp. 157-158. Language emerges both from outside and inside, not just from inside (Noble, Tønnessen) [5.67]: p. 123, because in society cognition and actions are situated, extended, distributed and embodied [5.32] [5.40]. System (society) and its elements (people) are integrated.

As mentioned before, in the studies of memory, theoretical dichotomies can be found (structure/process, form/movement, containment/flow) similar to those found in biological memory studies [5.53] [5.71]. Daniele Salerno writes "...when analyzing memory, we are asked to acknowledge its puzzling nature: memory is always localized and always displacing, living in its transmission from one context to another, between local *mise en forme* and movement" [5.53]: pp. 87-88. Social memory could be viewed as a form of semiotic materializations (texts, monuments, images, music, performances) and as a process involving interpretation. According to Salerno the modern trend in the studies of memory is to "transcend the 'container'" and focus on the idea of 'multidirectional' movement and change [5.53]: pp. 89-90. However, as in the case of biological memory, we have to resort to the complementarity principle in memory studies and see both the *process* of semiosis (narration) and its *symbolic embodiments* that scaffold and accumulate it.

Summing up, in addition to the retrospective and historic approach to memory, this research proposes to study *social memory* also as 1) actual *cumulative* and *distributed* system processes that are maintained over generations and 2) semiotic processes and structures responsible for the production and reproduction of the social system (e.g., education and educational systems).

The maintenance and development of biosocial and social systems mediate and are mediated by inherited information and social learning, and can be implemented at different time scales [5.28] [5.71]. Furthermore, according to the "dual inheritance theory" (Cavalli-Sforza, Feldman), the theory of evolution in four dimensions (Jablonka, Lamb) and the "reticulate model" of evolution (Gontier), there is also a constant horizontal, vertical and reticulate transmission of information (traits) within different levels and systems of inheritance (genetic, epigenetic, behavioral, ecological and cultural) [5.29] [5.38] [5.43].

5.5 Memory and the Origin of Language

There are two extreme perspectives on the nature of language. The first one examines it as *innate* competence or a special module located in the brain. This view considers language as relatively independent of the external, social symbolic systems and processes of communication (e.g., nativism of Chomsky and Pinker) [5.48]. The opposite theory (e.g., the sociocultural approach of Vygotsky or Everett) views language as cultural by its nature [5.18] [5.72]. It is predominately an external system (tool) that should be *internalized* during the process of socialization (enculturation), and "linguistic competence" is a skill that is acquired from society (as compared with "the critical period hypothesis"). Chomsky called

[2] However, some human achievements, such as the pyramids at Giza, still make us ponder how they were made.

these two types of languages *I-Language* and *E-Language* to refer to the internal/internalized and external/externalized language of the speaker [5.1].

However, both theories lead to paradoxes in the explanation of the origins of language. If the I-language view is right, then *linguistic competence* should emerge through some *random* genetic mutation in *a particular* individual (in a particular brain). However, if we admit that this mutation is random, and the linguistic (semiotic) output is unpredictable, then how would these unique individuals understand each other? They will produce unknown signs and signal to each other through dissimilar channels of communication. A good example of this paradox is the 52-hertz whale who was called "the loneliest whale in the world" because of the strangeness of its song [5.44]. In this regard, this problem of the I-language origin can be named after it as the "lonely whale paradox."

It should also be mentioned that the problem could be solved if the aforementioned mutation had happened to everyone and symbolic/semiotic output was similar for everyone. Yet, this hypothesis presupposes abandoning random mutation theory as the main driving force of evolution and joining the club of teleonomic and system models in evolution (e.g., Berg's nomogenesis theory) [5.7] [5.12] [5.50].

If we admit that language is an *external symbolic system* and a product of *social agreement or convention,* then linguistic competence can be acquired only from already existing language. However, how could these means and rules of communication come into existence in the first place? The first "language-makers" could not introduce language without prior language or without having internalized skills (means) of communication already available in external language. The only solution to this problem is the theory that language had to emerge in social reality *ex nihilo* and people were able to understand each other *naturally* without any socially imposed conventions or agreements. This problem, which is common in sociocultural approaches, can be called "the linguistic convention paradox" or "the paradox of linguistic agreement." In short, it can be expressed in the form of the question: "How can language be introduced without language?"

As a solution to these paradoxes, this research proposes a semiotic approach. In the first case, we have the problem: if language (linguistic competence) is a product of an individual mind (brain), how to introduce it to someone who does not have it (*solipsism bottleneck*)? In the second case, we cannot explain by what (linguistic) means a language could be introduced in a society for individuals if it is a unique construct of that society (*chicken-and-egg dilemma*). The solution of these paradoxes requires the hypothesis of *natural* (non-conventional) and *universal* means of communication and the assumption that living organisms have common cognitive modules for their recognition (perception, interpretation).

We propose that *actions* can play the role of natural and more or less universal "language" for different species. From our everyday experience, we know that domestic animals can communicate with us and with each other via actions ("expressive behavior"), and we can do the same with them [5.6]. Human toddlers also learn to communicate with their parents by actions. For instance, when a mother sees how a baby is trying to grasp a toy, she gives it to him or her. By detecting this correlation (indexical signs, natural signs), the baby next time just *points* at a toy, expecting via this act of communication, that the mother will understand and supply what is pointed at. That is, semiotically, the first sparkles of communication in nature could come from usage (sometime deceptive usage) of indexical and iconic signs [5.59] [5.62].

The specific theories of language origin, such as gestural, pantomimic, and storytelling (narrative), also incorporate action theory and iconic signs in their explanatory models [5.31] [5.19]. However, in the case of gestural theory, researchers often oppose it to the acoustic explanation, because, supposedly, it was replaced by *conventional* acoustic sounds during language evolution. The current research argues that, for the broad semiotic and action approach, the opposition of gestures and acoustic symbols is inappropriate. As shown by *the motor theory of speech perception*, individuals perceive not sounds *per se*, but *vocal tract gestures* by which they were produced [5.22] [5.23] [5.61]. Thus, from the point of action theory, it does not matter what means of communication (visual gestures or acoustic sounds) we use, the matter is what we recognize. We recognize actions and intentions that have produced these signals. However, the transition from gestures to acoustic sounds (and later to various writing systems), indeed, led to the gradual evolution (transformation) of language from the universal and natural communicative medium into more culture-specific, conventional (artificial, arbitrary) systems that require teaching and learning.

At that stage of language evolution, we need to incorporate into the model of its origin the idea of social memory and behavioral/cultural inheritance. The emergence of various local "traditions" of communication led to the externalization and social inheritance of language (means of communication) and skills of communication ("languaging"). At that stage, language and "linguistic rules" became distributed and relatively independent in particular individuals that produced them; it is so-called "second-order language" [5.63] [5.64] [5.66]. Furthermore, as shown above, in such distributed and culturally embodied form, language started to scaffold individual and social (linguistic, cognitive, and scientific) development.

It is important to note that the evolution of mind should not be reduced just to the analysis of the origin of language. For instance, Chomsky, Reboul and some other researchers believe that language originally was engendered not for communication but for cognition, for inner speech, for usage in intrapersonal cogitation (the "language of thought" hypothesis) [5.51]. However, later the development of language and cognition was affected by the origin and use of mathematical notation, logic and algebra, which resulted in the rise of science, architecture, engineering and economy. This means that, evolutionarily, we were endowed not just with cognitive modules for language but with a broad ability for the production of *various symbolic systems* that relate to different cognitive skills or activities (other examples include musical notation, road signs, etc.). In addition, human brain plasticity enabled the acquisition of numerous sign systems invented in society. Such progressive acquisition of sign systems supported the accumulation and transmission of human achievements throughout generations, which constitutes social memory and its inheritance.

5.6 Conclusions

It is common to describe the nature of memory as either 1) an ability to retain "footprints" of past events (similar to Plato's wax tablet holding stamped images), 2) knowledge about events that happened in the past, or 3) an individual epistemological ability of a particular organism to retain and retrieve this knowledge. This research proposal suggests that there is a possibility and necessity to elaborate the ontology of memory embracing systemic, biologically/socially distributed and future-directed memory processes in the Universe. The

Universe, life and society themselves could be considered as evolving memory systems standing for their preservation and development.

Physical, biological and social types of memory could be considered as three major steps in the evolution of memory and could be used as the main objects of investigation and comparison in the ontological studies of memory. Such studies can be based on empirical research and on data from natural sciences. Available methodological means for the study of the origin and evolution of biological and social memory can be found in extended evolutionary synthesis, theories of non-genetic inheritance ("extended inheritance," "soft inheritance," "inclusive heritability") and semiotic theory. Along with them, developmental systems theory, the theory of distributed actions and distributed cognition can be incorporated into this research methodology.

Declaration

The authors declare no conflict of interest.

References

[5.1] Araki, N. (2017) Chomsky's I-language and E-language. *Hiroshima Institute of Technology Research* 51, 17-24.

[5.2] Arshavsky, Y. I. (2003) Long-term memory: does it have a structural or chemical basis?. *Trends in neurosciences*, 26(9), 465-468.

[5.3] Balistier, T. (2000) *The Phaistos Disk: An Account of Its Unsolved Mystery*. Verlag Dr. Thomas Balistier.

[5.4] Barbieri, M. (2006) Life and semiosis: The real nature of information and meaning. *Semiotica* 158, 233-254.

[5.5] Barbieri, M. (2018) What is code biology?. *Biosystems* 164, 1-10.

[5.6] Bar-On, D., & Priselac, M. (2013) Triangulation and the Beasts. In: *Triangulation: from an epistemological point of view*. M. C. Amoretti & G.Preyer (Eds.), Walter de Gruyter: 40, 121-152.

[5.7] Berg, L.S. (1922) *Nomogenesis: Or, Evolution Determined by Law*. M.I.T. Press, Cambridge, UK.

[5.8] Brier, S. (1998) Cybersemiotics: a transdisciplinary framework for information studies. *BioSystems, 46*(1-2), 185-191.

[5.9] Brier, S. (2003) The cybersemiotic model of communication: An evolutionary view on the threshold between semiosis and informational exchange. *tripleC: Communication, Capitalism & Critique,* 1(1), 71-94.

[5.10] Chalmers, D.J. (2022) *Reality+: Virtual worlds and the problems of philosophy*. Penguin UK.

[5.11] Collingwood, R.G. (1994) The idea of history. Oxford University Press.

[5.12] Corning, P.A. (2014) Evolution 'on purpose': how behaviour has shaped the evolutionary process. *Biological Journal of the Linnean Society* 112(2), 242-260.

[5.13] Danchin, E., Charmantier, A., Champagne, F. A., Mesoudi, A., Pujol, B., & Blanchet, S. (2011) Beyond DNA: integrating inclusive inheritance into an extended theory of evolution. *Nature reviews genetics* 12(7), 475-486.

[5.14] Davies, P., & Gregersen, N. H. (Eds.) (2014) Information and the nature of reality: from physics to metaphysics. Cambridge University Press.

[5.15] de Condillac, E. B. (1756) *An essay on the origin of human knowledge: Being a supplement to Mr. Locke's essay on the human understanding.* Translated and edited by Hans Aarsleff. London: J. Nourse.

[5.16] Deely, J. (2001) Physiosemiosis in the semiotic spiral: A play of musement. *Σημειωτκή-Sign Systems Studies* 29(1), 27-48.

[5.17] Deely, J. N. (2008) How does semiosis effect renvoi?. *The American Journal of Semiotics* 11(1/2), 11-61.

[5.18] Everett, D. L. (2012) *Language: The cultural tool.* Vintage.

[5.19] Ferretti, F., Adornetti, I., Chiera, A., Nicchiarelli, S., Magni, R., Valeri, G., & Marini, A. (2017) Mental Time Travel and language evolution: a narrative account of the origins of human communication. *Language Sciences* 63, 105-118.

[5.20] Fischer, B. (1997) Toward a constructivist epistemology: Johann Gottfried Herder and Humberto Maturana. *The European Legacy* 2(2), 304-308.

[5.21] Fischer, S. R. (1997) *Rongorongo: the Easter Island script: history, traditions, texts* (Vol. 14). Oxford University Press.

[5.22] Fowler, C. A., & Rosenblum, L. D. (1991) The perception of phonetic gestures. In: *Modularity and the motor theory of speech perception.* I. G. Mattingly & M. Studdert-Kennedy (Eds.), Lawrence Erlbaum Associates, Inc.: 33–59.

[5.23] Galantucci, B., Fowler, C. A., & Turvey, M. T. (2006) The motor theory of speech perception reviewed. *Psychonomic bulletin & review* 13(3), 361-377.

[5.24] Gare, A. (2020) Semiosis and information: meeting the challenge of information science to post-reductionist biosemiotics. *Biosemiotics* 13(3), 327-346.

[5.25] Gare, A. (2021) Code biology and the problem of emergence. *Biosystems* 208, 104487.

[5.26] Gare, A. (2022) Life Processes as Proto-Narratives: Integrating Theoretical Biology and Biosemiotics through Biohermeneutics. *Cosmos and History: The Journal of Natural and Social Philosophy* 18(1), 229.

[5.27] Godfrey-Smith P. (2007) Information in biology. In: *The Cambridge Companion to the Philosophy of Biology.* D. Hull, and M. Ruse (Eds), Cambridge, UK: Cambridge University Press: 103–119.

[5.28] Gontier, N. (2021) Hierarchies, networks, and causality: The applied evolutionary epistemological approach. *Journal for General Philosophy of Science* 52(2), 313-334.

[5.29] Gontier, N., & Sukhoverkhov, A. (2022) Reticulate evolution underlies synergistic trait formation in human communities. *Evolutionary Anthropology: Issues, News, and Reviews,* 32(1), 26-38. https://doi.org/10.1002/evan.21962

[5.30] Griffiths, P. E, Stotz K. (2017) Developmental systems theory as a process theory. In: *Everything flows: towards a processual philosophy of biology.* D.J. Nicholson, J. Dupre (Eds), Oxford, NY: Oxford University Press: 225-246.

[5.31] Hawhee, D. (2006) Language as sensuous action: Sir Richard Paget, Kenneth Burke, and gesture-speech theory. *Quarterly Journal of Speech* 92(4), 331-354.

[5.32] Hodges, B. H., & Fowler, C. A. (2010) New affordances for language: Distributed, dynamical, and dialogical resources. *Ecological Psychology* 22(4), 239-253.

[5.33] Hoffmeyer, J. (2008) Semiotic scaffolding of living systems. In: *Introduction to biosemiotics.* M. Barbieri (Ed.), Springer, Dordrecht: 149-166.

[5.34] Higuera, C. J. R. (2012) *A Typology of Arguments for the Existence of Physiosemiosis.* MA diss., University of Tartu, Estonia.

[5.35] Jablonka, E., & Lamb, M. J. (2020) *Inheritance systems and the extended evolutionary synthesis.* Cambridge University Press.

[5.36] Khalighi, M., Sommeria-Klein, G., Gonze, D., Faust, K., & Lahti, L. (2022) Quantifying the impact of ecological memory on the dynamics of interacting communities. *PLOS Computational Biology* 18(6), #e1009396.

[5.37] Kull, K. (2009) Vegetative, animal, and cultural semiosis: The semiotic threshold zones. *Cognitive semiotics* 4, 8-27.

[5.38] Kull, K. (2010) Ecosystems are made of semiosic bonds: Consortia, umwelten, biophony and ecological codes. *Biosemiotics* 3(3), 347-357.

[5.39] LaFargue, M., & Pas, J. (1998) On Translating the Tao-te-ching. In: *Lao-tzu and the Tao-te-ching*. L. Kohn and M. LaFargue (Eds), University of New York Press: 277-301.

[5.40] Linell, P. (2013) Distributed language theory, with or without dialogue. *Language Sciences* 40, 168-173.

[5.41] Litchfield, G. (2022) Matter-Information Equivalence. *Cosmos and History: The Journal of Natural and Social Philosophy* 18(2), 457-466.

[5.42] Maturana, H. R., & Varela, F. J. (1991) *Autopoiesis and cognition: The realization of the living (Vol. 42)*. Springer Science & Business Media.

[5.43] McElreath, R. (2010) The coevolution of genes, innovation and culture in human evolution. In: *Mind the gap: tracing the origins of human universals*. P.M.Kappeler, J.B. Silk (Eds.), Berlin, Heidelberg: Springer-Verlag: 451-474.

[5.44] Mulvaney, K. (2021, July 13) The search for the loneliest whale in the world. *The Gardian*. https://www.theguardian.com/environment/2021/jul/13/loneliest-whale-in-the-world-search

[5.45] Nöth, W. (2001) Protosemiotics and physicosemiosis. *Σημειωτκή-Sign Systems Studies* 29(1), 13-26.

[5.46] Oyama, S., Gray, R. D., & Griffiths, P. E. (Eds.). (2003) *Cycles of contingency: Developmental systems and evolution*. MIT Press.

[5.47] Pharoah, M. (2020) Causation and Information: where is biological meaning to be found?. *Biosemiotics* 13(3), 309-326.

[5.48] Pinker, S. (2013) *Language, cognition, and human nature: selected articles*. Oxford University Press.

[5.49] Piatigorsky, A. M. (1976) Quelques prémises theoriques de la semiotique. *Travaux sur les systemes de signes: école de Tartu*. Bruxelles: Editions Complexe, 243-246.

[5.50] Popov, I. (2018) *Orthogenesis versus darwinism*. Springer International Publishing.

[5.51] Reboul, A. C. (2015) Why language really is not a communication system: a cognitive view of language evolution. *Frontiers in Psychology* 6, 1434.

[5.52] Rosenthal, S. B. (1994) *Charles Peirce's pragmatic pluralism*. SUNY Press.

[5.53] Salerno, D. (2021) A semiotic theory of memory: between movement and form. *Semiotica* 241, 89-90.

[5.54] Sharov, A. A. (2009) Role of utility and inference in the evolution of functional information. *Biosemiotics* 2(1), 101-115.

[5.55] Sharov, A. A., & Vehkavaara, T. (2015) Protosemiosis: Agency with reduced representation capacity. *Biosemiotics*, 8, 103-123.

[5.56] Sharov, A. A. (2016) Evolution of natural agents: Preservation, advance, and emergence of functional information. *Biosemiotics* 9(1), 103-120.

[5.57] Sharov, A. A., & Gordon, R. (2018) Life before earth. In: *Habitability of the Universe before Earth*. R .Gordon, & A. Sharov (Eds). Astrobiology: Exploring Life on Earth and Beyond, Academic Press: 265-296.

[5.58] Sharov, A., & Tønnessen, M. (2021) *Semiotic Agency: Science Beyond Mechanism*. Springer Nature.

[5.59] Sonesson, G. (2019) On mimicry, signs and other meaning-making acts. Further studies in iconicity. *Biosemiotics* 12(1), 99-114.

[5.60] Stotz, K. (2010) Human nature and cognitive–developmental niche construction. *Phenomenology and the Cognitive Sciences* 9(4), 483-501.

[5.61] Studdert-Kennedy, M., & Goldstein, L. (2003) Launching language: The gestural origin of discrete infinity. *Studies in the Evolution of Language*, 3, 235-254.

[5.62] Sukhoverkhov, A. (2012) Natural signs and the origin of language. *Biosemiotics* 5(2), 153-159.

[5.63] Sukhoverkhov, A. (2014) Current trends and developments in ecolinguistics. *Yazik i kultura–Languagc and culturc* (27), 166 175.

[5.64] Sukhoverkhov, A. V., & Fowler, C. A. (2015) Why language evolution needs memory: systems and ecological approaches. *Biosemiotics* 8(1), 47-65.

[5.65] Sukhoverkhov, A. V., & Gontier, N. (2021) Non-genetic inheritance: Evolution above the organismal level. *Biosystems* 200, 104325.

[5.66] Thibault, P. J. (2011) First-order languaging dynamics and second-order language: The distributed language view. *Ecological Psychology* 23(3), 210-245.

[5.67] Tønnessen, M. (2021) Making the umwelt bubble of the modern synthesis burst. *Biosemiotics* 14(1), 121-125.

[5.68] Ulanowicz, R. E. (2018) Dimensions missing from ecology. *Philosophies* 3(3), 24.

[5.69] Vedral, V. (2018) *Decoding reality: the universe as quantum information*. Oxford University Press.

[5.70] von Uexküll, J. (2010) *A foray into the worlds of animals and humans: with a theory of meaning*. Trans. J. D. O'Neil. Minneapolis: The University of Minnesota Press.

[5.71] Võsu, E., Kõresaar, E., & Kuutma, K. (2008) Mediation of Memory: towards transdisciplinary perspectives in current Memory Studies. *Trames* 12(3), 343-363.

[5.72] Vygotsky, L. S., & Luria, A. (1994) Tool and Symbol in Child Development. In: *The Vygotsky Reader*. R. Van Der Veer, & J. Valsiner (Eds.), Oxford: Blackwell Publishers: 99-175.

[5.73] Wheeler, J. A. (2018) Information, physics, quantum: The search for links. In: *Complexity, Entropy, and the Physics of Information*. W.H. Zurek, (Ed.), Addison-Wesley, Redwood City: 309-336.

[5.74] Zlotnik, G., & Vansintjan, A. (2019) Memory: An extended definition. *Frontiers in Psychology* 10, 2523.

6

Meanings, Their Hierarchy, and Evolution

George E. Mikhailovsky

Global Mind Share, Norfolk, Virginia, USA

Dedicated to the glowing memory of my closest friend and colleague Professor Sergey Titov.

Online resources:
https://www.researchgate.net/profile/George-Mikhailovsky
https://www.linkedin.com/in/george-mikhailovsky-b3a5a885

Abstract

Meaning is a component of the world that takes the form of indicator or symptom in the non-living world, intention of organism action, or something expressed in words or other symbols. I distinguish between meanings-in-itself or potential meanings and manifested meanings that are perceived or generated by agents. Inanimate objects can manifest only meanings-in-itself, primarily in the form of laws of nature and their components. An evolving macroscopic object can be called "*eventity*," which is short for "evolving entity" that exists due to events constituting its evolution. The earliest and most important *eventities* are stars. They were not only the first macro-objects in the universe, but also (in combination with planets) the most important conditions for the origin of living systems—the most primitive agents. Evolution of life and meanings have gone hand in hand, though they have not directly conditioned each other. The main milestones along the way were the emergence of new levels of hierarchy, i.e., hierarchogenesis. For life, and later for human history, such stages of hierarchogenesis were the emergence of prokaryotic cells, eukaryotic cells, multicellular organisms, agro-ecosystems, and nations/states. For meanings, hierarchogenetic stages include the actualization of potential meanings, the rise of judgments, paradigms, and worldviews. Although these two strands of evolution followed their own paths and had completely different dynamics, in the last few millennia, and especially in the last century, these paths have begun to converge. The possible prospects and consequences of this convergence are discussed.

Keywords: Meaning; evolution, hierarchy, hierarchogenesis, agency, semantic prism

---

Email: gmikhai@yahoo.com

6.1 Introduction

The word "meaning" is repeated many times throughout this book, and this chapter is no exception. This alone makes the book literally "meaningful." This obviously overly optimistic conclusion shows the ambiguity of the word and the concept it expresses.

In general, the word "meaning" can be used primarily in three situations:

1. It is an indicator or symptom of a phenomenon.
2. It denotes the intention of some action.
3. It is something that is indicated or expressed in words or other symbols.

The 1st situation can be illustrated by the following examples: "low dark clouds mean rain," "the discovery of trilobite fossils means that there was an ocean here," "the great similarity of coastlines between the east coast of South America and the west coast of Africa means that it was once one continent," etc. Examples of the 2nd situation might be "choosing quantum physics as a major at university means that the student is planning to become a quantum physicist" or "an intense search for prey by a predator means that it is hungry." Finally, the 3rd type could include situations like "this word means this and that in the Oxford Dictionary," "the '&' sign means the word 'and,'" or "the '!' sign means the factorial in mathematics."

Semantics in the narrow sense, dealing with meanings in human and social consciousness, deals mainly with the 3rd situation. Semantics in biology or biosemantics treats the concept of meaning more broadly and actively includes consideration of the 2nd and partially 1st situations. This creates a charming ambiguity in the concept discussed above. Motivated by this charm, I am going to use "meaning" in the broadest sense in this chapter, including all three types, but above all the first two types, since the third one has already been sufficiently developed in linguistics and semiotics.

What all three of these situation types have in common is that meaning emerges as a result of comprehending a particular situation. But who is the one who comprehends it? Who reveals the meaning of the "low dark clouds," the "intense hunt for prey," or the very word "meaning" that we are now trying to parse?

The answer to these questions is an "agent" of one kind or another.

The notion of the agent as a self-organizing entity was introduced by Stuart Kauffman [6.42]. After a few years, he placed the concept in a semiotic context, treating autonomous agents as "semiotics creatures" [6.43]. Later, Jesper Hoffmeyer [6.36] and then Alexei Sharov [6.83] [6.84] and Morten Tønnessen [6.98] elaborated on the concept of a semiotic agent in more detail.

Such agents can be people and their groups, animals and plants, their populations and ecosystems, and even single-celled organisms, including prokaryotes. We will examine the validity of this statement later, but for now I will simply take it as it is and analyze all three types of situations one by one, starting with the third, most traditional in semantics, using Frege's triangle: "Referent - Symbol - Concept or Meaning" [6.66].

In this 3rd situation, a word in a dictionary or encyclopedia can be seen as a *symbol* of this or that object, phenomenon, concept, etc., which is the *referent* of the symbol, and the definition or explanation of the word is its *meaning*. This meaning reflects the sense held in a language rather than a person's personal understanding. Dictionaries and encyclopedias are created precisely to express this sense and to fix the norm of word usage. Moreover, words in a dictionary can have one or more meanings. For example, the word "fern" in *Webster's College Dictionary* [6.109] has only one meaning: "a nonflowering vascular plant of the class *Filicianae*, having fronds and reproduces by spores." The word "meaning" itself, on the other hand, has (as a noun) two different meanings: "1. what is intended to be or actually is expressed or indicated" and "2. the end, purpose, or significance of something," according to the same dictionary. As you can see, the 2nd meaning above corresponds to two situations at once (1st and 2nd).

Some words have many meanings. For example, the word "set" has as many as 430 meanings [6.29]. On the other hand, some different words (synonyms) have the same or at least very similar meanings. Thus, the relationship between words (symbols) and meanings is not necessarily "one-to-one" or even "one-to-many," but "many-to-many." This is also true for relations between referents and words. There are about 8 billion referents for the word *human* (the current population of the Earth). At the same time, the words describing these referents are also very numerous: *man, woman, engineer, astronaut, scientist,* and so on. Therefore, the semantic triangle is rather a prism, each edge of which represents a set of referents, symbols, or meanings (Figure 6.1).

The number of referents corresponding to a given symbol is a quantity that can be called the "width" of meaning because it shows how many real objects fits to it. Another quantitative parameter of meaning — its "depth" — could also be defined as the number of meanings for a given symbol. For example, the depth of the meaning "electron" is 1, since it probably contains no other meanings except itself. The meaning of "mass" (as a parameter of material objects, leaving aside "mass" as a tumor, a pile of stuff, or the Catholic Mass), on the other hand, has a depth of at least 2, since it contains such values as "gravitational mass" and "inertial mass." The depth of the meaning "human" is very large, because it contains a huge number of meanings, including all the meanings in this book, in all the other books, in all the movies, on the Internet, etc.

In the 2nd situation, the action serves as a *referent*, the *symbol* of which is the goal or task of action, and the motivation of action can be seen as *meaning*. In contrast to the previous situation, this meaning exists mainly at the individual level, although the individual can pass it on to other agents. A goal becomes meaningful only if someone is motivated to achieve it. If, for example, parents force their son to choose marrying an upper-class girl

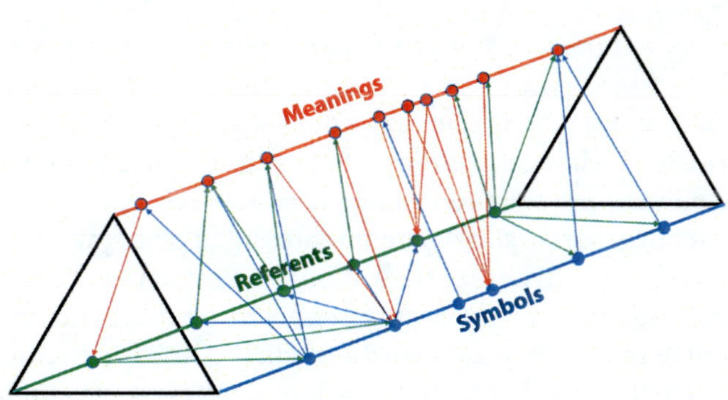

Figure 6.1 Semantic prism.

(as a goal), he may find it either meaningful or meaningless. In the first case, he will be motivated to achieve it, in the second case he, as an obedient son, may accept and perhaps even achieve such a goal, but without motivation. Zero motivation can even be associated with goals that someone chooses for himself without any pressure from others. A daughter decides to become a doctor because both her parents are doctors. At the same time, she may be not motivated to do so and may not feel pressured by her family, but simply follows in her parents' footsteps. Of course, any agent observing another agent's actions can speculate about its goals and motivations, but the agent's internal goals always precede its actions. Thus, an agent, as a source of information, may conceal his/her real goals and transmit false ones, to misinform real or potential enemies and competitors. On the other hand, an observing agent may misinterpret another agent's real goals, even if it is not trying to hide them.

Here again we encounter the "many-to-many" relationships between sides of the semantic prism. Obviously, an action can have several goals, and the same goal can be achieved by several actions. At the same time, the same goal or action can have different motivation, and the same motivation can lead to different goal setting and different types of action. However, the number of all such choices and types is much smaller than in the 3rd situation. Accordingly, the "length" of the semantic prism here is significantly shorter.

The 1st situation is perhaps the most interesting for us as scientists, since it was these meanings that led in their time to the emergence of science itself. In this situation, a phenomenon or event acts as a *symbol*, its cause acts as a *referent*, and the awareness of the connection between them can be interpreted as *meaning*.

Generally speaking, a natural phenomenon can have several different causes, one cause can produce different phenomena, and the same cause-effect relationship can have different explanations and, accordingly, meanings. However, the "thickness" of all these "many-to-many" relationships is usually small, and each of them often includes no more than a few variants. As a result, the "length" of the semantic prism in this 1st situation is the smallest. Moreover, scientists studying a natural phenomenon always try to reduce this "many-to-many" relation to the "one-to-one" relation, deliberately simplifying the situation and considering the main cause of the phenomenon, trying to give a unified explanation (to reveal the unified meaning) of the relationship between cause and effect. In other words, typical scientific research considers only the semantic triangle as a certain slice of the semantic prism at the cost of discarding a great deal of information. In recent years, however, exceptions have appeared, such as machine learning and multiphysics modeling, but these are still rare exceptions.

It is also interesting to note that if in the material world the killer of information is free energy, the impact of which destroys established structures [6.62], in the world of meanings this role is played, oddly enough, by scientific knowledge. It is such knowledge that formulates concepts and discovers laws, discarding all unexplored information as irrelevant in the chosen aspect, simplifying the situation as much as possible. And the simpler the system (for example, a rather degenerate solar system represented by a system of material points), the more effective is the scientific approach. On the contrary, science is less effective in describing complex systems carrying large amounts of essential information that cannot be reduced to a simple model. In other words, predicting the weather for a week is much harder and less reliable than predicting the position of the planet for millions of years.

Another significant difference between the 1st situation and the other two is the different involvement of agents in the entities and processes behind the semantic triangle components. In situations 2 and 3, each corner and side of the corresponding triangles and prisms need agents as obligatory actors in their respective processes. In situation 1, however, many causes and their effects exist without agents in any form.

For example, the rotation of Earth and other planets around the Sun and their causes follow Kepler's laws and the law of universal gravitation that obviously existed before life and, respectively, any agencies in the Solar system. The meanings of these phenomena, as comprehended by Kepler and Newton, existed long before humans realized them. These causal relations existed before any agents realized their meanings, and these meanings discovered much later by humans were unambiguously predetermined by these relations. Such future meanings can be called "potential meanings" [6.85] or "meanings-in-themselves."

In general, all meanings can be divided into "meanings-in-themselves" and "manifested meanings." An example of "manifested meanings" in its purest form is mimicry, which has no direct meaning for the organism itself, but gives false meaning to predators. It certainly helps the organism not to be eaten, but this meaning is not "meaning-in-itself," because it immediately disappears in the absence of the predator. Most meanings are both "meanings-in-themselves" and "manifested meanings" at the same time.

There is another fundamental difference between the 1st situation and the other two. The causal meanings relating to this situation are never atomic but link atomic meanings of the corresponding cause and effect. In other words, such meanings are judgments constructed from atomic meanings and represent the next level of the semantic hierarchy as compared to atomic meanings, which will be discussed in Section 6.5.

6.2 Hierarchogenesis and Its Stages During the General Evolution of the Universe

The concept of "hierarchy" has several different meanings. Originally, it denoted relations of status or authority between individuals or groups of people within the church [6.102]. There are many definitions of hierarchy in science, sometimes significantly different from one another. Here I describe hierarchy and hierarchogenesis (with its stages, i.e., hierarchogenetic events) in the context of the evolution of the universe. Let us consider an event as hierarchogenetic if it results in the appearance of a system with the features listed in Table 6.1.

The first condition in Table 6.1 excludes systems such as free radicals, cell organelles, or organs (and organ systems) of multicellular organisms. The second condition excludes hierarchical systems in their original, ecclesiastical sense. For example, the alpha male in a pack of monkeys is the highest level of hierarchy, but this level does not consist of beta males, females, juveniles, etc. The third condition excludes systems composed of monotypic or near-monotypic subsystems, such as homopolymers, colonies (with rare exceptions like insect societies), populations, or multicellular prokaryotes [6.30] [6.53]. Finally, the fourth condition in Table 6.1 does not, for example, allow us to consider each of the numerous appearances of multicellular organisms in different clades [6.53] [6.16] [6.44] as separate hierarchogenetic events, unlike the appearance of eukaryotes, which occurred probably

Table 6.1 Defining a hierarchical level based on its four conditions (prerequisites), which we will use in this chapter.

Prerequisite	Obligatory attributes of hierarchical level
1	System can exist by itself, not just as part of a super-system at a higher hierarchical level(s).
2	System consists of subsystems belonging to one or more lower hierarchical levels.
3	Its subsystems belong to several types, radically different from each other.
4	The interconnections between these subsystems result in an entity that did not exist before, i.e., novelty.

only once in the history of life [6.125] [6.11]. Thus, according to this condition, the emergence of both eukaryotes and multicellular organisms should be considered as two single hierarchogenetic events.

These four conditions do not contradict the direct relationship between the number of hierarchical levels and system complexity discussed by Bradley Alicea [6.1]. However, they impose much more serious limitations on the concept of hierarchy than Alicea suggested. The work of Stanley Salthe [6.79] [6.80] demonstrates an approach to understanding hierarchy similar to mine. He calls it the compositional hierarchy and contrasts it with the classification hierarchy used in biology.

According to the definition of the hierarchy formulated in Table 6.1, there were only 15 such hierarchogenetic events in the history of the universe with two branches of the evolutionary tree: Cosmic and Substantive (Table 6.2).

The numbers in the 3rd column of Table 6.2 are approximated or averaged dates of each hierarchical step found in various sources. On their basis, the values of the 4th column were calculated. The Big Bang (zero point for dates) time was assumed to be 13.8 ± 0.02 billion years ago [6.19]. The appearance of quark-gluon plasma ("quark soup") and heavy particles is estimated 10^{-12} and 10^{-6} seconds after the Big Bang, respectively [6.2]. The first nuclei appeared from the first second to several minutes of the existence of the universe [6.111]. Thus, the time from the Big Bang through these first three hierarchogenetic steps is practically zero (in gigayears' time scale).

Then, based on the current Standard Model of Cosmology (SMC), after about 380 kiloyears, when the universe cooled down enough, the first atoms appeared [6.81]. After about 100 megayears (My), the first stars were lit [6.121]. In my previous works, I used the estimate of 180 My made by Bowman *et al.* [6.14], but recent data allowed me to adjust this number. I previously estimated the period between the appearance of the first stars and the first galaxies as 310 My, based on the age of the oldest galaxy (490 My after the Big Bang) discovered by NASA in 2018 [6.74]. Now, however, the James Webb Space Telescope has discovered a galaxy with an age of 350 My after the Big Bang [6.117], which led me to adjust that number as well. It is worth noting that both of these adjustments made the curve in Figure 6.4 even smoother and more regular compared to its previous version [6.59]. This may indicate that this curve, as well as the Table 6.2 behind it, in general correctly reflects reality.

Table 6.2 Hierarchogenetic branches in the main material evolution of the universe, their stages, and types of associated meanings. Question marks denote interpolated (macromolecules and living droplets) and extrapolated (noosphere) values. The last column (Meanings) will be discussed in Sections 6.3 and 6.4.

Hierarchogenetic branch	Hierarchogenetic step	Time after Big Bang (in years)	Time for the next step (in years)	Related field(s) of science	Meanings
Cosmic	1-Appearance of the rest mass and light particles (quark-gluon plasma)	3.17×10^{-20}	3.17×10^{-20}	Elementary particle physics	No clear meanings
	2-Appearance of hadrons (heavy particles)	3.17×10^{-14}	3.17×10^{-14}	Physics of strong forces	
	3-Appearance of nuclei	3.17×10^{-7}	3.17×10^{-7}	Nuclear physics	
	4-Appearance of atoms	3.80×10^{5}	3.80×10^{5}	Quantum mechanics, Spectrometry	
	5-Appearance of stars and planetary systems	1.00×10^{8}	1.00×10^{8}	Astrophysics	Potential meanings without agents
	6-Appearance of galaxies	3.50×10^{8}	2.50×10^{8}	Astrophysics	
Substantive	7-Appearance of heteroatomic molecules (monomers)	1.10×10^{9}	6.10×10^{8}	Chemistry	
	8-Appearance of macromolecules (heteropolymers)	2.40×10^{9} (?)	1.30×10^{9} (?)	Biochemistry	
	9-Appearance of life	4.10×10^{9} (?)	1.70×10^{9} (?)	Biochemistry of RNA/protein/ coenzyme worlds	
	10-Appearance of prokaryotic cells	9.75×10^{9}	5.65×10^{9} (?)	Microbiology	Actual meanings accepted or generated by agents
	11-Appearance of eukaryotic cells with mitotic cycles	1.17×10^{10}	1.95×10^{9}	Protistology	
	12-Appearance of eukaryotic multicellular organisms with continuing differentiation [6.31] (Gordon,1999), and thus embryogenesis	1.30×10^{10}	1.34×10^{9}	Embryology	
	13-Appearance of artificial environment (agroecosystems), i.e., Neolithic revolution	1.38×10^{10}	7.65×10^{8}	Anthropology, Agronomy, Veterinary	

(*Continued*)

Table 6.2 Hierarchogenetic branches in the main material evolution of the universe, their stages, and types of associated meanings. Question marks denote interpolated (macromolecules and living droplets) and extrapolated (noosphere) values. The last column (Meanings) will be discussed in Sections 6.3 and 6.4. (*Continued*)

Hierarchogenetic branch	Hierarchogenetic step	Time after Big Bang (in years)	Time for the next step (inyears)	Related field(s) of science	Meanings
	14-Appearance of nations and states with armies and governments, i.e., history	1.38×10^{10}	$\approx 5.0 \times 10^3$	History, Economics, Politics	
	15-Appearance of noosphere	1.38×10^{10}	5.0×10^3 (?)	Crowd Thinking, Social Networking, Politics	

The formation of galaxies completed the cosmic branch of general hierarchogenesis, although several hundred million years later galaxies began to unite into groups and clusters. This dating follows from the discovery of a group of galaxies that began to clump together about 13 billion years ago, forming a protocluster [6.17]. However, these formations do not satisfy our 3rd necessary condition for a hierarchogenesis event. In fact, these groups and clusters have been, as is currently known, only a kind of "colonies" of galaxies, not true novelties. Thus, neither groups nor clusters of galaxies were included in my basic material hierarchy.

The next stages of hierarchogenesis belong to its second (Substantive) branch, which separated from the first branch (or trunk) after the 2nd billion years of the existence of the universe. This branch began with the appearance of substance in the form of interstellar dust consisting of heteroatomic molecules. It is almost impossible to directly determine the time of the appearance of the first relatively complex heteroatomic molecules. However, it can be approximated by the time of the appearance of interstellar dust, i.e., as 1.1 ± 0.3 gigayears (Gy) after the Big Bang [6.57] or 600–800 My after the appearance of galaxies.

Unfortunately, there are no data on the time of the appearance of heteropolymers. As for the appearance of the first life (the first extraterrestrial living systems), this occurred, as shown by Sharov and Gordon [6.86] [6.87], 9.7 ± 0.3 Gy ago or 4.1 ± 0 .3 Gy after the Big Bang. This means, as the authors point out, that life first appeared not on Earth or even in the Solar system, which is only 4.571 Gy old, but on some other planets orbiting other stars. And only then, about 6 Gy after its appearance in the universe and a few million years after the formation of the Earth, life was brought to Earth by comets and meteorites.

It is probably worth mentioning that a few years later, Richard Gordon [6.32] suggested that the origin of life on Earth could have been greatly accelerated by the abundance of day/night cycles as a result of the short duration of such a cycle (up to 2 hours) during this period. I helped him with calculations confirming the meaningfulness of this hypothesis. However, this is only a hypothesis, and one of its consequences is the slowing down of the rate of biological evolution on the Earth with the increase in the day/night cycle length from the Moon formation to the present, which contradicts its obvious acceleration, especially after the so-called "Cambrian explosion" [6.61]. In addition, the slowing down of the Earth's rotation during this period, in turn, is the result of friction due to lunar tides. Thus, such an assumption can be applied only to planets with a relatively giant satellite, such as the Moon. Obviously, this is a very rare case and cannot change the overall estimate of 9.7 ± 0.3 Gy used in Table 6.2.

Richard Gordon and I [6.60] analyzed a lot of data on the time of appearance of the first prokaryotic and first eukaryotic cells on Earth. As a result, we obtained the following estimates for these hierarchogenetic events: 4.05 ± 0.25 Gy and 2.1 ± 0.6 Gy ago, or 9.75 ± 0.25 and 11.7 ± 0.6 Gy of universe history, respectively.

The next hierarchogenetic step was the emergence of multicellular eukaryotic organisms. This occurred 765 ± 25 My ago [6.28] [6.31] or 13.035 ± 0.025 Gy after the Big Bang. This was followed by the emergence of artificial ecosystems (agroecosystems) as systems consisting of groups of people, populations of the animals they bred, and populations of the plants they grew. In other words, one species, *Homo sapiens*, created the next, higher hierarchical level system consisting of a set of domesticated species of plants and animals, ensuring itself a much more comfortable existence.

Finally, the last known step of hierarchogenesis was the formation of states and nations as systems of agroecosystems and people united by territory, economic ties, language, mentality, etc. These last two steps occurred about 10,000 [6.3] [6.37] and ≈ 5,000 [6.25] [6.114] years ago, respectively. However, these events, as well as the very first steps, are virtually indistinguishable (in our gigayears time scale) from the present time in which we live.

All these steps and the two branches to which they relate are shown in Figure 6.2.

However, this main, Substantive branch:

> *Monomers → Heteropolymers/Macromolecules → Living Droplets → Prokaryotic Cells → Eukaryotic unicellular organisms → Multicellular organisms → Agroecosystems → Nations/States → Noosphere (?)*

is not the only possibility.

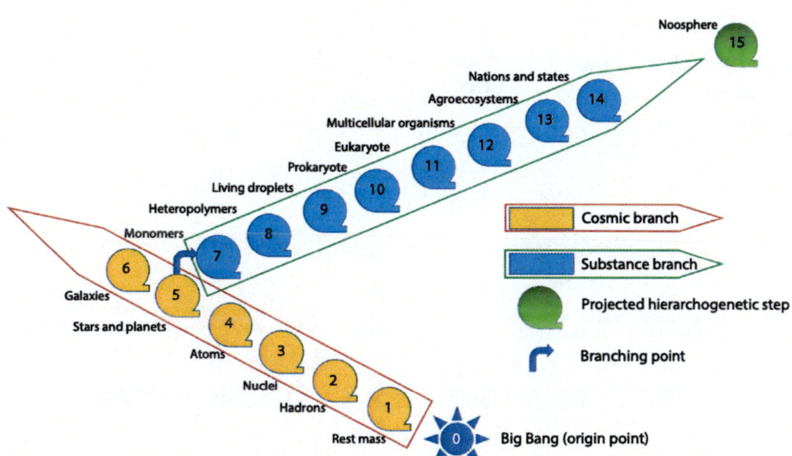

Figure 6.2 Hierarchogenetic steps and branches in the main material evolution of the universe. The step numbers are the same as in Table 6.2.

It is easy to discover other substantive hierarchies as well. For example:

Interstellar dust → Planetesimals → Planets

or

Living organisms → Ecosystems → Biosphere

which developed in parallel with the main direction of hierarchogenesis.

Although they do not belong to the mainstream evolution of the universe, at least from the human point of view, the evolution corresponding to the alternative biological hierarchy (*Organisms → Ecosystems → Biosphere*) is not without interest. Graphically, it is shown in Figure 6.3 (left).

To begin with, we do not know when the biosphere appeared. It is obvious that life cannot exist outside of ecosystems, which ensure the closure of biogeochemical cycles. Accordingly, the evolutionary time between an organism and an ecosystem is practically zero. However, it is not entirely clear when the biosphere arose as a global conglomerate of ecosystems. It is only certain that by the Great Oxidation Event (2.4–2.3 gigayears ago), when cyanobacteria changed the composition of the Earth's atmosphere by making it oxygenic, the biosphere was already playing its planetary role.

However, there is a point of view based on the idea of Kamshilov [6.41], developed and supported on the basis of many facts by Smith and Morowitz [6.89], that evolution corresponding to this hierarchy went in the opposite direction, i.e., it began with the emergence of the biosphere as a system of global biogeochemistry cycles, which then broke up into many ecosystems specific to different habitats and, eventually, into individual organisms (Figure 6.3, right). We will return to this concept in connection with the hierarchies of meanings in Section 6.5.

The fifteenth step in Table 6.2 is the noosphere, i.e., the sphere of human thinking. It can be regarded as a possible candidate for the next hierarchogenetic event. The concept of the noosphere was formulated by Teilhard de Chardin in 1922 [6.94] and then developed by him in his main book *The Phenomenon of Man* [6.95] and by Vladimir Vernadsky [6.103] [6.104] as a possible next stage in human evolution. However, it is difficult to predict exactly how the noosphere will be realized as the next hierarchical step, because it is impossible to foresee the further evolution of the system at its bifurcation point or just before it, where

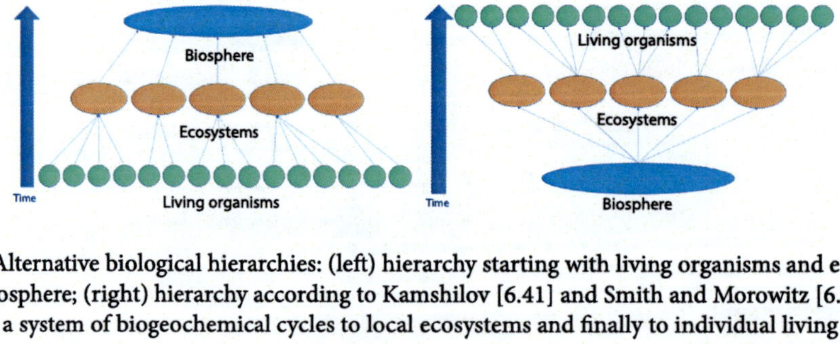

Figure 6.3 Alternative biological hierarchies: (left) hierarchy starting with living organisms and ending with the whole biosphere; (right) hierarchy according to Kamshilov [6.41] and Smith and Morowitz [6.89] from the biosphere as a system of biogeochemical cycles to local ecosystems and finally to individual living organisms.

we are now. Nevertheless, we can already see some significant signs of the emergence of the noosphere: globalization in the forms of the internet, social media, crowd thinking, etc.

If we arrange the times corresponding to each hierarchical step (the 3rd column in Table 6.2) on the time scale from the Big Bang to our days (and a little bit into the future), we get a rather regular and smooth curve (Figure 6.4). It includes both branches of the main hierarchy from Table 6.2. The Cosmic one began with the Big Bang and continued until the formation of galaxies. The Substantive branch began with the appearance of interstellar dust in the form of heteroatomic molecules and continues until now.

Within the Substantive branch, molecules gradually became more complex (most likely on the surface of the planets), combined into heteropolymers (macromolecules) and then formed the next hierarchical levels of life, prokaryotes, eukaryotes, etc. The timing of each of these steps is based on the best approximation from the aforementioned sources. The only

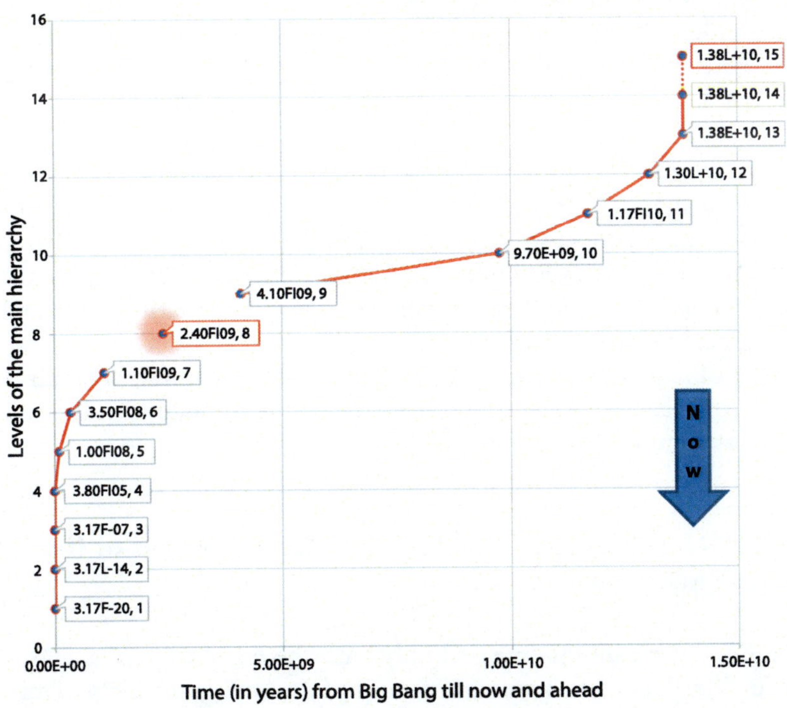

Figure 6.4 Dynamic of the successive steps of the main material hierarchogenesis of the universe (see Table 6.2) since the Big Bang. The X-axis is the timeline of evolution of the universe in gigayears, and the Y-axis is the sequential series of hierarchical events. The scale of Y-axis is arbitrary, and the distances between any two neighboring events are the same. Points: 1–quarks, 2–hadrons, 3–nuclei, 4–atoms, 5–stars and planet systems, 6–galaxies, 7–heteroatomic molecules, 8–heteropolymers/macromolecules, 9–living droplets, 10–prokaryotic cells, 11–unicellular eukaryotic organisms, 12–multicellular organisms, 13–agroecosystems, 14–nations/states, 15–noosphere (?). Step 8 (macromolecules) cannot be dated with certainty, and the number referring to this point is an interpolation. Point 15 (noosphere?) refers to the future, and the dotted line between points 14 and 15 describes a probable prediction. As will be shown below, the lower left part of the curve (points 1-4) corresponds to absence of clear meanings, its central part (points 5-9) to potential meanings, and its rightmost upper part to actual meanings.

exception was the time of appearance of macromolecules (point 8), which was interpolated because, at least for the moment, it cannot be estimated even approximately. But the most interesting and important are points 9 and 10, which differ from all others by their approximately central position relative to the symmetry of the curve. At the same time, this period corresponds to one of the major transitions in the evolution of the universe — the transition from abiotic to biotic systems endowed with the function of self-reproduction, since the simplest inheritance mechanism is the key feature of living system [6.68]. These functions are characteristic of the most ancient, most primitive agents [6.88]. Thus, the emergence of the first agents and agency as such coincided with the most important turning point in the evolution of the universe.

The curve in Figure 6.4 is approximately odd, i.e., it is symmetrical relative to its center. The symmetry results from the similarity of the patterns of slowing down during cosmic/chemical evolution (the 1st phase of general evolution) and acceleration during biological/human evolution (its 2nd phase). Earlier, I proposed a simple mathematical model [6.59] based on the semantic approach, which approximates this curve.

It should be noted that this sequence of 15 hierarchogenetic events is not definitive and may be slightly changed [6.39] [6.40] [6.105]. Jagers op Akkerhuis [6.39] [6.40] did not include levels of nuclei, macromolecules, live droplets, agroecosystems, and nations/states, while Tyler Volk [6.105] also did not include macromolecules and life, but added animal social groups and tribal metagroups, which I have not considered because they do not meet condition 3 of the hierarchogenetic event (Table 6.1). In addition, both authors ignore the last two stages of the main, cosmic branch of the evolution of the universe, i.e., stars and galaxies. On the whole, however, Table 6.2 with its 15 lines gives us an approximate number of hierarchogenetic events and demonstrates the general regularity of hierarchogenesis as the main stages in the evolution of matter.

This scheme of sequential hierarchogenesis reflects the main rates and direction of matter evolution. And each of its steps represents the merger of many fundamentally different systems into a system of a higher hierarchical level.

6.3 Potential Meanings During the Abiotic Period of the Evolution of the Universe

Actual (manifested) meanings emerged only with the emergence of agents, i.e., with the emergence of life. However, potential meanings (meanings-in-themselves) could have existed earlier, during the first phase of hierarchogenesis, the deceleration phase (Figure 6.4). They could have, but did they really exist?

Any potential meanings in the very early universe could only be probabilistic, because at that time the universe consisted exclusively of quantum objects (quanta of fields, elementary particles and, finally, simplest atoms), obeying laws of quantum mechanics. And all these objects were both waves and particles at the same time. Since meanings of "wave" and "particle" are actually opposite, meanings of quantum objects as fundamental "bricks" of the original universe were internally inconsistent and very blurred. Accordingly, all other potential meanings (such as mass, energy, momentum, etc.) in that period could not be other than very fuzzy and indistinct.

This follows directly from the "uncertainty principle" originally formulated by Heisenberg in 1927 [6.82], according to which, values of position and linear momentum, energy and time, speed and mass-momentum and so on cannot be exactly measured for a quantum particle simultaneously. In other words, the values of all these conjugate variables at a given time can only be measured approximately, because a precise measurement of one of them leads to complete uncertainty in the value of the other. And without measurements, not only both conjugate variables of a particle remain uncertain, but even its reality is in question. Martin Macháček [6.54] wrote:

> The particle, if it is not measured, could be mathematically described by a function assigning probabilities to possible different values of physical quantities belonging to the particle. That means the measurement—or the interpretation of probability—itself is responsible for its chosen state and could be then responsible for its reality.

In other words, without measurement we cannot definitely speak about reality of anything but quanta of fields in the very early universe. At the same time, such measurement is possible only as a result of wave packet reduction by an "observer" (or physical instrument, measurer). And most importantly, only macroscopic (i.e., non-quantum-mechanical) objects can be such instruments or "measurers." Since there were no such objects in the initial quantum-mechanical universe, there were no clear potential meanings, either (see the last column in Table 6.2).

On the end of this period (the Dark Ages), the universe consisted almost exclusively of evenly distributed atoms of rarefied hydrogen (with a small admixture of helium), and its entropy was close to maximum. Accordingly, the universe was very close to thermodynamic equilibrium and looked quite stable. However, such homogeneous states close to equilibrium are in fact stable only in the absence of gravity.

Sir Isaac Newton [6.65] already in the 17th century noticed that such states are gravitationally unstable, and this instability may result from the slightest deviations from homogeneity in the density distribution. Such inhomogeneities will steadily increase under the action of gravitational forces. This is exactly what happened at the very end of the Dark Ages: density fluctuations grew larger, and hydrogen atoms within these increasingly dense inhomogeneities collided more frequently. As a result, this led to enormous gravitational increase of pressure and temperature inside aggregations of atoms and, as a consequence, to the reaction of fusion of hydrogen atoms, forming helium atoms. The Dark Ages ended, the "Cosmic Dawn" came, and the first stars were lit [6.14].

Emerging stars were not only the first things (macroscopic objects) but also the first *evolving* things that "were born," evolved, and finally "died" in some way or another. Such evolving macroscopic objects can be called "*eventities*," which is a shorthand for "evolving entity," and simultaneously means an entity that exists because of the events that constitute its evolution. And stars are the first and most important of these *eventities* in the universe.

Their appearance led to the emergence of classical Newtonian mechanics with its strict (not probabilistic) regularities and laws. At the same time, thermonuclear fusion is possible only between particles, not waves. Thus, the reduction of the proton field wave function to individual particles played the role of a quantum "observer" or "measurer" [6.77]. The ignition of fusion in protostars actually "measured" whether their density and temperature were sufficient to initiate this process. And this "measurer," which in quantum-physical terms is synonymous with an "observer," has allowed our world to go beyond its quantum existence.

About 30 years ago, Asher Peres [6.73] stated that "observers" in quantum physics, like "observers" in special relativity, do not imply the actual presence of humans. These virtual physicists can also be inanimate automata, if programmed accordingly. For myself, I would add that they can also be physical potential meanings and fundamental laws of nature that correlate these meanings to each other. Below we will discuss this hypothesis in more detail. John Bell once [6.9] asked: "Was the wave function waiting to jump for thousands of millions of years until a single-celled living creature appeared? Or did it have to wait a little longer for some highly qualified measurer … with a PhD?" My answer might be: "No, it only had to wait about 100 million years for the first thermonuclear fusion."

Stars were the first local and macroscopic sources of free energy to appear in the universe, and they dramatically changed the direction of its evolution away from quantum mechanical. This free energy became a kind of fuel for the subsequent macroscopic evolution, accompanied by a steady growth of complexity determined by the laws of nature [6.62]. In the beginning, these laws were only the laws of classical Newtonian mechanics. Later, after supernova explosions, the formation of planets, and the appearance of chemical elements heavier than hydrogen and helium, additional laws of nature came into play, such as chemical, geological, hydrodynamic, and many others.

All these laws, including the original quantum-mechanical ones, conceal a certain mystery. First of all, it is not quite clear how they came to be in the first place. Then, what seems amazing is the ability of these fundamentals of the universe to be later formulated in the language of mathematics, originating from counting cattle (arithmetic) and dividing arable land into individual fields (geometry). Finally, their supposed stability for at least 13 billion years [6.91] [6.92] in our world, where everything is constantly changing, either degenerating or evolving, seems truly incredible.[1]

Each of these mysteries cannot be fully explained within the framework of natural sciences. The first one is largely epistemological, and its discussion is beyond the scope of this book. The second has been discussed by Eugene Wigner [6.112], but without apparent success. As for the latter, it probably has to do with the semantic aspects of these laws as mathematical statements and will be considered later.

If we return to the laws of classical Newtonian mechanics, they clearly make sense, since they describe, for example, periodic events in any stellar system, such as the motion of paired stars, the motion of planets, their satellites, comets, asteroids, etc. Since there are no meaningful statements consisting of meaningless elements, their elements (atomic meanings), such as mass, force, velocity, acceleration, distance, also have meanings. And all these meanings in the absence of agents at the time were, of course, only potential.

Such a situation occurred about three hundred million years after the Big Bang and, accordingly, about 13.5 billion years ago (see the last column in Table 6.2). If we consider the semantic triangle during this period, we will find that it was reduced to one of its sides

[1] Although this constancy of the basic laws of nature is confirmed by observation of the deep cosmos and follows from the equivalence principle, one of the cornerstones of the theory of relativity, we cannot rule out that the fine structure constant has changed throughout the history of the universe [6.108]. However, this has not yet been proven, and if it were proven, all our ideas about space and time would undergo radical changes. Obviously, all this is beyond the scope of this book.

with the vertices "referent" and "potential sense." Or, better said, this side of the semantic triangle was its original form, which was later developed to its full three-vertex version. Of course, such side or, more precisely, a facet of the future semantic prism is actually a parallelogram rather than a line segment because each potential meaning has, obviously, a multitude of referents in the universe. On the other hand, referents have more than one potential meaning depending on "context" — a word that should probably be put in quotes when the meanings are potential. For example, the proton has very different potential meanings in stars, in a hydrogen atom, in molecules H_2 or H_2O, or as an element of molecular energy transport in living cells.

Symbols in the context of the semantic triangle, as well as actual meanings, simply did not exist in the absence of agents. Ludwig Wittgenstein [6.116] wrote "each sign by itself seems dead." But a "dead" sign is actually not a sign, just as a corpse is not a person, carrion is not an animal, and hay is not grass. Thus, although it can sometimes be said that potential meaning itself is related to something like signs (e.g., light as a sign of a confluence of stars), such a "sign" in the absence of agents is dead and therefore not a real sign. As for "referents," they were represented by concrete things or their characteristics. For example, for mass, they could be stars, planets, and other space objects. For acceleration, specific angular accelerations of planets or satellites. For distance, the perigee and apogee values of those planets and satellites. And so on.

"Meaning" as the vertex of the semantic segment (the reduced triangle) is a different story. At that distant time, it was represented by the generalization of all existing (and maybe all possible) referents. This is because the law of universal gravitation or Newton's second law has been applied to any mass, regardless of its size, location or what it is made of. This is just an example, but it shows the relationship between meanings and referents for all other laws of nature and all their elements.

Here it is impossible not to mention that according to the idea of perceptual evolution proposed by C. Christopher Martin and Richard Gordon [6.55], meaning can also be placed in the context of perception and action. Moreover, this context makes it possible to clearly define meaning as *a perception that is comprehended by the agent and prompts it to action.* In addition, such a context forms another perceptual triangle, with perception, action and meaning as its vertices, and its sides corresponding to "comprehension of perception," "meaning of action" and "cause of action." Martin and Gordon [6.55] understand "perception" very broadly, attributing this property to the simplest living organisms, such as bacteria and archaea, and even in a sense to physical systems. This, in turn, implies the presence of simplest meanings in their actions. As for abiotic physical objects, we can talk about the perception of one electron by another through the forces of electrical repulsion. The action in this case will be curvature of their trajectories. However, the meaning of such perception and the action generated by it will be, in the absence of agents that could detect electron trajectories, only potential [6.85]. In a sense, such potential meanings in the early universe can be seen as a kind of Platonic forms or eidoi of their referents [6.113]. Such an interpretation can be seen as a weighty argument in favor of realism in its centuries-long polemic with nominalism [6.20] [6.21]. For it means that not only things, but also abstract objects (such as mass, velocity, distance, time, etc.) arose long before man with consciousness and even before the emergence of life. Possibly, Plato, Aristotle and Thomas Aquinas were closer to the truth than William of Ockham or John Buridan [6.13].

It must be said that the laws of nature in this context should be understood quite broadly and, in particular, should include the laws of mathematics, which relate to nature only indirectly. But mathematics allows us to describe other laws of nature and to do so in a language that is immanent to most of them: the language of mathematics. This language is immanent to all quantitative laws, because only it can ensure the exceptional accuracy of their predictions, like the discovery of the planets "at the tip of a pen."

This is the principal difference between the laws of mathematics and the laws of logic, which describe the laws of human thought, and which could not exist not only in the early universe, but much later, up to the appearance of *Homo sapiens*. For example, the commutative law of multiplication is actually an integral part of the law of universal gravitation. This is the only reason why the Earth attracts the Moon with the same force with which the Moon attracts the Earth, and, strictly speaking, it is not the Moon that revolves around the Earth, but both of them actually revolve around a center located about six times closer to the Earth between them. Since mathematics, as we have seen above, is a way of representing all the laws of nature, including its own, it could not have appeared later than these laws themselves.[2]

However, the laws of nature and its elements are not the only objects that exist outside of time in the sense that they retain their identity throughout the existence of the universe and time itself. Another type of such objects is the world constants, e.g., the speed of light in a vacuum, **c**; the Planck constant, **h**; the elementary charge, **e**; the Avogadro constant, $\mathbf{N_A}$; the Boltzmann constant, $\mathbf{k_B}$; etc. These world constants can be viewed as peculiar laws of nature, perhaps as a kind of law of conservation. For example, for the speed of light, the equivalent law can be formulated as follows: "The speed of light in vacuum is a constant that equals to 299,792,458 $m{*}s^{-1}$" or mathematically "$C = 299{,}792{,}458\ m{*}s^{-1}$".

We cannot say when these constants appeared or whether their values have changed. Strictly speaking, we can assert their constancy for the period of the last 13 Gy, during which our universe can be directly observed, because 13 Gy ago (800 My after the Big Bang) there was a gamma-ray burst 090423, which so far is the oldest event we have observed in the universe [6.67] [6.46]. Of course, it is possible to assume that the world constants could have appeared sometime in the early universe, i.e., during the first 800 My, or changed their values during this period. However, the models themselves, which describe the early evolution of the universe before it became observable, assume that these constants, like the laws of nature, already existed and had the same values as they do now. At the same time, the conclusions of these models do not contradict the parameters and trends of the subsequent evolution of the universe. Thus, they can be considered as the best possible representation of the initial period of our universe's existence. Consequently, we have no reason to assume that the fundamental physical laws and world constants have not existed or have not remained stable in the universe we cannot observe.

Somewhere between these world constants and Newton's laws of nature are the laws of conservation. While the former ones relate quantities to their values, and the latter relate two or more quantities together, the laws of conservation state that some quantity is a constant in any given closed system. And for the whole universe, which is one such system, the laws of conservation, as I have already noted, actually state the existence of additional

---

[2] I am not referring here to mathematics in its symbolic representation as used by humans, but as the natural relations that later became represented by symbols in mathematics.

world constants. So far in physics there are six such laws: conservation of energy, momentum, angular momentum, charge, baryonic and lepton number, but there may be other yet unknown conservation laws [6.38].

This means that our universe arose and existed in two incarnations: the material world located in the space-time continuum and the semantic (or perhaps quasi-semantic, given that originally it was inhabited by potential rather than actual meanings) world, which exists outside of space and time in the sense that it does not change from point to point in space and from moment to moment in time, at least for the last 10 Gy (Figure 6.5), which has been repeatedly confirmed (see, for example [6.63]). These two worlds not only coexist but also interact closely, albeit asymmetrically. The semantic world of laws dictates permissible behavior to the material world of nature, while the natural world fills the laws of the semantic world with concrete content and allows them to realize themselves through natural forces and processes.

Perhaps "dictates" is not the best word here and should be replaced with "controls." Howard Pattee was probably right when he wrote:

> "The consistency with physical laws means that everything is dependent on the laws — none of biological or mental processes is inconsistent with any physical law. However, … this does not mean that everything is determined by the laws." [6.71], p. 151.

Or, better said, following Seth Lloyd [6.51], that "natural systems speak a language whose grammar consists of the laws of physics." But in any case, the natural world is controlled by the world of laws, not *vice versa*.

Another fundamental difference between these two worlds is the difference in the nature of their objects. If one understands evolution very broadly, including chaotization due to the 2nd law of thermodynamics, then almost any material macroscopic object (except, perhaps, very stable crystals such as diamonds) can be regarded as an *eventity*. However, the laws of nature and the world constants, by virtue of their eternity, can in no way be called

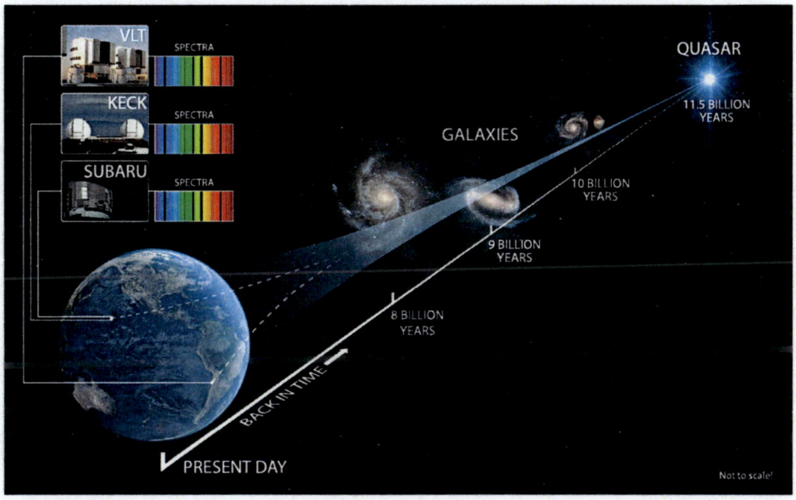

Figure 6.5 Three telescopes testing nature's laws in 3 distant galaxies. (Image credit: Swinburne Astronomy Productions; from [6.63]).

eventities. Perhaps they can be called *eternities* (from *eternal entities*), but they clearly do not evolve in any sense.

In addition, unlike the colossally huge (almost 10^{11} light years in diameter) size of the material world, the semantic world is so small (or rather, concise, since "small" has a spatial connotation) that, according to Seth Lloyd [6.52], it can be placed on the back of an ordinary T-shirt. That said, the material world looks very complex, diverse, and is evolving on all time scales. A rough estimate of the amount of information created and processed after the Big Bang is 10^{92} bits [6.50]. The amount of information existing in the semantic world is immeasurably smaller. Nevertheless, it controls not only the evolution of the material world, but also conditions of its very existence.

The point is that if the world constants were different, the universe could not have existed for a long time. But even in those highly improbable scenarios in which it would have existed for billions of years, living things could not have appeared in it [6.24]. Over a huge range of possible values, the world constants must have almost exactly defined values for life to be possible. Lee Smolin [6.90] calculated that the probability of randomly falling numbers compatible with life is 1 in 10^{229}. Assuming that the world constants got their values by chance, this negligible probability means that a universe like ours simply cannot exist. It should also be considered that Smolin made his calculations based on the laws of nature, including mathematical laws, which may be quite different in other possible universes. Therefore, in fact, the probability that random values of the world constants turned out to be compatible with life, let alone reason, is actually still much lower, although 10^{-229} is already equal to zero from all possible points of view.

Thus, we can summarize that the accidental emergence and existence of the laws of nature and the world constants (semantic world) and, as a consequence, our material world as it exists, was impossible. This means that there is some reason for the existence of the semantic and material worlds. Such a reason must exist in the semantic world, for where else could it exist? After all, it is the semantic world that dictates the laws to the material world, not vice versa. Besides, neither forces, nor events, nor phenomena in the material world of space-time can be the cause of eternal entities. At least, no one has observed anything like that.

It seems obvious that this reason has a meaning. Moreover, it is probably the biggest and most important meaning in our universe. Unfortunately, we are not able to understand it, at least not yet. Perhaps because we do not yet know (i.e., have not discovered) all the laws of nature. Most likely, we know only a tiny part of them, as Richard Hutchin [6.38] wrote about the laws of conservation. After all, at the end of the 19th century, it seemed to most scientists that the basic laws of nature had already been discovered, and only minor refinements remained. Subsequent discoveries in physics, including, above all, both theories of relativity and quantum theory, have completely dashed those hopes. So, I am afraid there is not enough room on Lloyd's T-shirt to fully describe the semantic world.

At the same time, the semantic world as we know it now has its own structure, and this structure is hierarchical. Such hierarchy of *eterentities* has 3 levels: *atomic potential meanings* (e.g., mass, velocity, or force); *laws and constants*, perhaps groups of laws and constants (e.g., mechanical, quantum mechanical, chemical, electrodynamic, thermodynamic, etc.); and the *Code of Laws of Nature* in general.

Some physicists are thinking of a Theory of Everything (TOE) that could fully explain and tie together all the physical concepts of the universe (e.g., [6.34] [6.110]). It is not yet clear whether such a TOE exists, but we already have an outline of a COE (Code of Everything) as a set of laws of nature discovered so far, which, while clearly incomplete, demonstrates an internal interconnection and coherence. A weighty argument for the existence of COE as a coherent text where words and phrases correlate with each other, in contrast to a vocabulary where there is no such correlation, is the emergence of life, which, as discussed above, would not be possible without the extremely fine tuning of both the laws of nature and world constants.

To conclude this section, it is worth repeating that all the meanings we have considered so far were not actual but only potential meanings (the last column in Table 6.2). Semantic triangles did not exist at that time because there were no symbols. Nor were there agents who could create these symbols and actualize potential meanings. All this came into existence only after the emergence of life, which we will examine in the next section of this chapter.

6.4 Evolution of Meanings in the Biological Systems

We know little about the earliest biological or, better said, pre-biological organisms that existed before LUCA (Last Universal Common Ancestor) [6.96]. There are many different hypotheses about them, such as "gene-first" (e.g., [6.22] or [6.64]; or "metabolism-first" [6.124]), which we will not consider here because of their speculative nature and lack of tangible evidence. Let us start right away with LUCA, which was a prokaryote and already had a fully developed protein synthesis programmed by genes [6.87].

The emergence of LUCA and other prokaryotic organisms, which used the simplest versions of genetic codes in the form of DNA and RNA with the processes of transcription and translation, dramatically changed the semantic structure of the emerging systems. These systems were obviously *eventities* in all senses since cell divisions and evolution were a prerequisite for their existence. However, the emergence of the genetic code added a third vertex ("symbol" or "sign") to the single side of the reduced semantic triangle that existed throughout nonbiological evolution with its two vertices as "referent" and "potential meaning". If an amino acid is a "referent," and its eidos, realized in its function, is a "meaning," then the role of a "symbol" closing the semantic triangle is played by a DNA or RNA codon consisting of three nucleotides. In this case, the meaning of the amino acid ceases to be potential and becomes actual (Table 6.2), and the enzyme consisting of them becomes a simple functional measuring device [6.69]. And a prokaryotic cell synthesizing a protein based on the information encoded in the sequences of nucleotides can be considered a primitive agent. This opened a new chapter in the material evolution of *eventities*, a chapter of agents that have begun to accelerate the evolution that was previously slowing down (Figure 6.4 and Table 6.2). The emergence of agents has created an "epistemic cut" [6.70] that separates emergent genetic symbols from the dynamics of construction that they control. This made this chapter of evolution and especially the subsequent ones much more interesting than the previous ones [6.101].

Figuratively speaking, the very first pages of this chapter were written by prokaryotes, which were the first organisms to live on Earth. Whether they were bacteria or archaea is not entirely clear, so we will consider them as a whole, especially since archaea, like bacteria, have had two major perceptual systems—chemotaxis and photoreception [6.55]—and that they probably split much later [6.123].

Anyway, the emergence of the simplest prokaryotes in the form of LUCA led to the appearance of symbols separated from referents and meanings, which is a necessary condition of the "epistemic cut" and the triadic relation represented by the semantic triangle [6.72]. The processes of mRNA transcription, translation, and degradation in these organisms occur simultaneously, although, of course, on different mRNA molecules [6.76]. This does not allow us to speak of the presence of any temporal structure as an ordered sequence of events embedded in changes in their spatial structures, except for cell divisions, which, however, are not accompanied by the formation or transformation of any spatial structures.

At the same time, the spatial structure of the prokaryotic coding system has a quite obvious hierarchy consisting of three levels: codon – gene – genome or phenotype. This spatial hierarchy finds its expression in the corresponding semantic hierarchy related to the embodiment of genetic code meanings in the structure and, respectively, in the function of proteins. Moreover, the lowest level of such semantic hierarchy – the atomic meaning ("word") – corresponds to the codon, the next level – "phrase" – to the gene, and the highest level – "text" – is represented by the genome and the phenotype as its actualization.

In eukaryotes, which, as we have seen above, are the next, 11th step of the general hierarchy (Table 6.2), the semantic hierarchy is much more complex. One of the most fundamental differences between eukaryotes and prokaryotes is the appearance of a temporal structure in the form of a mitotic cycle with a strict sequence of its phases. And the duration of each such phase is more or less constant in the sense that it fluctuates around the same values. However, the duration of interphase and prophase can be quite long. This temporal structure closely interacts with intracellular spatial compartmentalization, generating the orchestration of all cellular processes [6.120]. This orchestration includes semantic processes of "reading" the genetic code, such as transcription and translation.

Despite this peculiarity, eukaryotes have the same three levels of semantic hierarchy as prokaryotes in their organelles, primarily in mitochondria. However, they have three additional levels, which begin where the three levels common to them end. The lowest level of the specifically eukaryotic hierarchy, representing its temporal structure, is the eukaryotic phenotype as a system of phenotypes of the nucleus, mitochondria, chloroplasts (in green plants) and other intracellular organelles.

Such a phenotype, or in other words, the state of a eukaryotic cell, at a given moment of time can be considered as an atomic meaning ("a word") at the fourth (first additional) level of the structural-semantic hierarchy of eukaryotes. Accordingly, this or that stage of the mitotic cycle — prophase, prometaphase, metaphase, anaphase, and telophase [6.18] — can be considered as a kind of "phrase" at the next hierarchical level. Such a "phrase" is nothing but a sequence of "words," that is, events in the form of distinguishable states of the eukaryotic cell as *eventity*. Finally, the mitotic cycle as a whole can be viewed as a "text" describing the complete temporal structure of eukaryotes at the sixth level, the highest level of the structural-semantic hierarchy for eukaryotes.

The next, 12th step of the basic hierarchogenesis (Table 6.2) was the emergence of multicellular organisms with their complex temporal structure manifested in cell differentiation, morphogenesis, and ontogenesis in general. This temporal structure added three more levels of structural-semantic hierarchy, preserving the six existing, i.e., prokaryotic and eukaryotic ones. The role of the first additional level (the seventh level in the general hierarchy) is played by the spatial structure of the embryo at any given moment of time as a conglomerate of spatial structures of all its cells, which can be considered as a "word." The next level ("phrase") as a sequence of distinguishable states of embryo (morphogenetic events) can be related to the appearance of new elements of form and/or new cell types in this *eventity*.

According to Natalie and Richard Gordon [6.30], this is the result of differentiation waves (contraction and expansion) that are generated by the cell state splitter organelle. And the whole ontogenesis, again, can be interpreted as a "text" consisting of "phrases" referring to successive stages of embryogenesis and then to the developmental stages of larval and/or juvenile stages to the adult stage. Table 6.3 presents this whole nine-level semantic hierarchy of multicellular organisms.

It should be noted that relations in this whole nine-level semantic hierarchy of living *eventities* are not limited to relations between neighboring hierarchical levels. Interactions between distant levels also take place. The most obvious example is that ontogenetic development is determined by the information encoded in the genotype (nuclear DNA) of the fertilized zygote. Here we see how the genotype, i.e., the third hierarchical level, not only influences, but also determines, albeit indirectly, the ninth level—the level of ontogenesis.

Another example is the effect of differentiation waves on gene expression in the genotype. If an embryonic cell has experienced contraction because of the propagation of a differentiation wave, its cytoskeleton sends one biochemical signal to the nucleus, and if the cell has experienced expansion, a completely different signal is sent [6.31]. These signals then alter gene expression and thereby regulate the fate of the embryonic cell. This means that each

Table 6.3 Three triplets of the spatiotemporal semantic hierarchy of multicellular living organisms and its individual hierarchical levels in each of the triplets (it should be noted that each subsequent triplet includes all the hierarchical levels of the previous ones).

Position in the general spatiotemporal hierarchy	Position of hierarchical level in the triplet	Biological spatial or spatiotemporal structures that represent the corresponding hierarchical level of *eventity*
Prokaryotic triplet	"Word"	Codon
	"Phrase"	Gene
	"Text"	Genotype/Phenotype
Eukaryotic triplet	"Word"	State (phenotype) of a cell in each moment
	"Phrase"	Phase of mitotic cycle
	"Text"	Entire mitotic cycle
Multicellular triplet	"Word"	State ("phenotype") of an embryo in each moment
	"Phrase"	Sequence of distinguishable states of the embryo
	"Text"	All the successive stages of embryogenesis

differentiation wave (morphogenetic event), being a part of the ontogenetic "phrase," i.e., the eighth level of the general semantic hierarchy, can control changes at the sixth level, the level of eukaryotic genotype, changing the active genotype by expression of certain genes.

One more example can be observed in the study of mitochondrial DNA and its role in the human aging process. It has been shown [6.118] that metabolic rewiring caused by mutations in mitochondrial DNA drives clonal expansion and, consequently, leads to cellular aging and, eventually, to aging of the whole organism. Here, changes in mitochondrial genotype and, consequently, phenotype (the 3rd level of the general hierarchy) affect the 9th level in the form of terminal stages of ontogenesis.

This whole hierarchical structure of 9 (3 × 3) levels (Table 6.3) is based on codes, which are systems of signs or symbols. The simplest, most fundamental, and well-known of these is the genetic code proposed by George Gamow [6.107] and then confirmed by numerous biochemical studies. Later, Gordon Tomkins [6.97] proposed the idea of a metabolic code used for hormone-mediated intercellular communication. He viewed such a code as a relationship between extra- and intracellular events. At the same time, metabolic symbols are structurally unrelated to the molecules that contribute to their accumulation during a nutritional or metabolic crisis.

Then new biological codes, including splicing codes, histone codes, tubulin codes, compartment codes, and many others, sprang out of the horn of plenty. Marcello Barbieri [6.5] [6.6] [6.7] organized all this diversity of biological codes and developed his original semantic theory of evolution based on them [6.8]. He has convincingly shown that there is ample evidence that biological codes are an integral part of the living world and participate in a wide variety of biological processes.

If we consider each of the triplets of our 9-level hierarchical structure (Table 6.3) in terms of such codes, we can see that each of them corresponds to one or more specific biological codes. The first, prokaryotic triplet, which belongs to the period between the 10th and 11th steps of the basic hierarchy (Table 6.2), corresponds to the genetic code. The second, eukaryotic triplet (between the 11th and 12th steps) corresponds to the metabolic, splicing, compartmental and, possibly, some other biological codes. The third, multicellular triplet (after the 12th step), corresponds to the differentiation code introduced by Natalie and Richard Gordon [6.30], p. 717:

> We can regard the differentiation code, the mapping between the sequence of differentiation waves cells experience and their resultant cell types, as one of these biological codes. Codes mean that objects (like tRNA) or processes (like differentiation waves) act as symbols for either other objects or processes. Thus, codons and differentiation waves take on meanings.

This semantic hierarchy, in which the division into triplets is based on codes and the levels within each triplet are based on structural hierarchy, is not the only possible one. At least one more, related to the behavior of organisms, can be distinguished.

Usually, the concept of behavior is associated with animals, but here we will understand it more broadly, including various taxes and tropisms in prokaryotes and plants. Among them are such phenomena as chemotaxis and thermotaxis of bacteria [6.106] [6.27] or archaea [6.75], phototropism in the form of turning stems and leaves of plants toward the light [6.33], more intensive growth of roots on the side where the concentration of water and nutrients is higher [6.23], and so on. All these phenomena can be seen as "decision-making" that is pretty close to the concept of "basal cognition" [6.49]. This concept applies cognition that is usually considered a uniquely human capacity to all the living organisms, from single cells and up [6.26]. Such basal cognition, as well as decision-making at the cellular level, has received both experimental [6.122] and theoretical confirmation through mathematical modeling [6.56].

However, any decision-making at any level is impossible without the information that external stimuli contain [6.45]. Accordingly, any behavior, if it is not random, is controlled either by stimuli or by intentions and desires. The former are primarily characteristics of the simplest organisms, and the role of the latter increases as biological organization grows. However, there is one most important intension common to all living organisms without exception. It is the Darwinian "struggle for life," i.e., the desire to survive and pass on its DNA to the greatest number of descendants. This desire makes obvious sense: to survive as an individual and as a species. And this meaning, the main meaning of life as such, is actual, not potential, as were the meanings that preceded the emergence of life. In other words, although we do not know exactly when the first agents appeared, the most primitive modern bacteria or archaea can certainly be considered agents. As Søren Brier wrote:

> "Biosemiotics encompasses all living systems from the cell, over bacteria, fungi, plants and animals to humans as sign producers and interpreters" [6.15].

As for stimuli, physical and chemical gradients were the first that affected living organisms and they remain the most common in bacteria and archaea, i.e., all the prokaryotes. These include temperature, salinity, humidity, solar radiation, concentration of nutrients and inhibitors, etc. And as soon as a certain gradient changes the behavior of an organism (in the case of a prokaryotic cell it can only be the direction or perhaps the speed of its movement), it becomes clear that this gradient is essential or has a meaning for it.

In the absence of any gradient in a completely uniform environment, the movement of prokaryotic cells loses its direction and becomes random. At first glance, such chaotic movement does not have any meaning. However, for prokaryotic cells, as for many other organisms, this is not entirely true. The point is that this movement is nothing more than a random search for nutrient gradients, allowing the bacterium to find locations with higher concentrations of nutrients. This random search is an optimization method that does not require a gradient and is used in artificial neural network for hyper-parameter optimization [6.10].

In addition, bacterial cells can communicate with each other by sending and receiving signals. Most often these signals are chemical, by means of signaling molecules called autoinducers [6.58]. However, there is another option: physical signals in the form of bioluminescence. In this case, bacteria produce bioluminescent proteins that emit light [6.93]. Typically, bacteria use these signals to regulate their population density [6.12]. This process has been called "quorum sensing" [6.4] and is quite widespread. The slowdown in the rate of cell divisions resulting from this quorum sensing is a meaningful change in the behavior of the bacteria.

The main feature that distinguishes any signal from noise is its meaning. This transfers such signal-based communication, like the aforementioned phenomena of taxis, into a semantic plane. At the same time, this semantics is behavioral in nature. Unlike structural semantics, which in prokaryotes contains three levels, behavioral semantics of prokaryotes has only two hierarchical levels. And the meanings constituting the lower level can be considered elementary (i.e., not further decomposable). This is very close to the idea of logical atomism introduced by Russell [6.78] and Wittgenstein [6.115], but in relation to meanings rather than facts.

Considering the structural-semantic hierarchy, we have distinguished the initial hierarchical levels ("words") in each of the hierarchical triples. But generally speaking, they are not elementary meanings since each of them refers to some spatial or spatiotemporal material structure. And if a codon meaning can still be considered elementary (in biological context), then the phenotype of a eukaryotic cell or the spatial structure of an embryo at a certain time point are, of course, composite meanings consisting of many elementary ones. The behavioral hierarchy is another matter. It deals with sensations or feelings, not matter. And if the meaning of behavior is primary, i.e., is not the result of other behavioral meanings, then it is really elementary and indecomposable.

Examples of such elementary meanings at the lowest level of the behavioral hierarchy are *desire to survive*, *feeling hungry*, *uncomfortable* or *there are too many of us*. However, even prokaryotes have another higher level of such hierarchy — the level of judgments, which combines two or more elementary meanings and generates new meaning(s). In the case corresponding to the "quorum sensing," the simplest judgment might be *since there are too many of us, I, as a cell, should stop or at least slow down my divisions*. Such a judgment is realized almost automatically: an increase in the concentration of autoinducers or an intensification of bioluminescence inhibits the division of bacterial cells.

We cannot, of course, "step into the shoes" of a bacterium and assume what it is judging. Probably it does not even "know" that there is any outside world. To us external observers, however, it looks like a kind of judgment. Moreover, natural selection eliminated those bacteria that were incapable of making these judgments because they conferred a huge selection advantage. Therefore, I will continue to use the term "judgment," recognizing its obvious conventionality.

There are other variants of such conditional judgments. They might be as follows: *if I'm hungry, I should move toward increasing nutritional concentration*. Or *if I am experiencing discomfort, I should move in the direction of reducing the discomfort factor*. Or more complex judgments, such as *if I am hungry and there is no nutrient gradient, I should move indiscriminately, changing direction until I find food*. The main difference between these judgments and "quorum sensing" judgments is the desire to survive as the main driving factor. The prokaryotic cell strives to find food because it "wants" to survive. Of course, this is not as deliberate an intention as our desire to drink a cold beer on a hot summer day. On the other hand, neither is it a "quorum sensing" judgment, which does not imply desire at all. Such prokaryotic desires lie somewhere in the middle and should probably be the subject of a separate study. In any case, even primitive organisms such as bacteria or archaea have two levels of the semantic hierarchy of behavior and are capable of making simplest judgments not only automatically, due to chemical or physical signals, but also deliberately, due to the desire *to survive* inherent in all living things.

We stayed on prokaryotes for so long and in such detail because they are the simplest living organisms available for direct observation, and I wanted to demonstrate their capabilities as the simplest agents that exist in the world. Naturally, more complex, more organized animal and even plant species recognize more of these elementary meanings and can combine them into more complex judgments.

Such additional meanings include the notions of *hot, dark, suffocating, this can be eaten, this can eat me, female/male,* etc. These elementary meanings can be combined into judgments ranging from the simplest, such as *getting colder – need to breathe more often; enemy approaching – danger!; hungry – need to find food;* to more complex, such as *little food – need to migrate to a new place; female/male close – have to move in that direction; enemy close – what to do: run, hide or defend yourself?*

In humans, these kinds of judgments are the most complex, including such logical structures as enthymemes, epicheiremas, and complete syllogisms. However, the simplest logical structures can also be found in the behavior of higher animals. A hare, for example, is able to construct the simplest syllogism: at the level of unconditioned reflexes, it knows that it should run away from large predators (major premise). When it sees a wolf or a fox, it knows that it is a large predator (small premise) and decides to run (conclusion).

Beavers show behaviors that indicate the use of more complex judgments: enthymemes and epicheiremas. Here is an example of three such judgments:

1. Trees that fall across a small river or stream (natural dams) allow me (the beaver) to build my lodge with an underwater entrance, because the dam raises the water level and makes that level relatively stable.
2. I (the beaver) can create artificial dams myself because I can fell trees, including those falling across streams.
3. I (the beaver) build my lodge anywhere along the stream I want because I can create artificial dams.

The first two judgments (1 and 2) represent enthymemes as incomplete syllogisms, and the 3rd is an example of epicheirema, where the conclusion is made on the basis of enthymemes.

However, this interaction between different types of judgments cannot be regarded as an act of hierachogenesis because it does not create a new level of hierarchy. If we recall the four characteristics of hierarchogenesis (Table 6.1), we will see that the simple interactions of judgments exhibited by higher animals do not correspond to the 3rd characteristic of hierarchogenesis. After all, the components of the compound sentences whose examples we have looked at are not "radically different from each other" as required by this condition.

The situation began to change when our ancestors realized that instead of running through steppes and forests in search of food, they could grow edible plants and breed edible animals right on the spot. *Homo sapiens* was the first and only species to create systems of a fundamentally new level of the main hierarchy of matter (Figure 6.4 and Table 6.2) in the form of artificial ecosystems or agroecosystems. Such systems consisted of many different species of multicellular organisms, united into a single whole by one of them, us. There are other species that are in a commensalism relationship with other species. For example, leaf-cutting ants, or termites, which bring home leaves to feed their mushroom gardens, ants that milk aphids, or fungi that cultivate algae in lichens. However, none of these organisms form complete multi-species ecosystems with multilevel food chains and sufficiently closed biogeochemical cycles. Agroecosystems of individual households and their conglomerates with their neighbors created a kind of artificial environment in which humans spent most of their lives, occasionally venturing out into the natural environment for hunting or fishing.

As a result, we, as a central species in agroecosystems, not only obtained a reliable source of food, but also got rid, at least partially, of competitors using scarecrows, bird deterrent wind-ratchets, domesticated dogs, etc. This happened about 10,000 years ago [6.3]. Our ancestors moved to a sedentary way of life and got both the time and the need to radically complicate their behavior. But this is already the topic of the next section.

6.5 The Evolution of Meanings in Human Societies and the Relationship between Hierarchies of Substance and Semantics

The transition to a settled way of life with cultivation of plants and breeding of animals instead of hunting and gathering was accompanied by an extraordinary complication of human behavior, the creation of new meanings and their intensive exchange through communication. Although human biological evolution came to an almost complete halt after the Agricultural Revolution, the rate of behavioral evolution became extremely high. Humans are like all other biological species, and even to a greater extent, are *eventities*. And this is due not only to their behavioral evolution itself, but also to the role that events have come to play in their lives. Each person's personality is first and foremost a set of events that have happened to him or her and are preserved in memory. Heraclitus wrote that our conscious individual "self" is nothing more than the memory in your brain. Howard Pattee even spread this statement to all the individualities, where we can include such higher animals as horses, dolphins, elephants, dogs, etc. He wrote:

> "I would say that any 'complex individuality' is defined only by its memory, and therefore such memory-controlled individuals would be capable of evolving or learning many forms of behavior" [6.72], p. 319.

However, our memory is obviously the most developed and complicated in the living world. Consequently, we humans, more than anyone else, in every sense justify the name of *eventities*.

This unusually rapid evolution of behavior was the result of the positive feedback between improvements in agriculture and the development of articulate speech, divided into words and phrases. We do not know when such a spoken language arose, but the oldest writing in Sumer and Elam, which originated from the token system, dates from about 5,500 [6.119] years ago. Considering the huge gap between oral and written languages, we can assume that the former appeared not later than the formation of agroecosystems.

After all, if collective hunting required only a few exclamations like "Come in from the left!", "Come in from the right!", "Don't move!", "Run away!", etc., not to mention gathering, which is mainly an individual activity, then successful farming requires both complex coordination of efforts and transmission of the accumulated experience to the next generations. Neither is possible without a developed oral language to formulate complex instructions and explain their importance.

If primitive people, like other animals, taught their offspring only by their own example, such memetic copying is not sufficient for organizing agricultural activities, because here it is necessary to perform different actions depending on the situation, time of the year, the harvest in the previous season, and so on. In other words, the new generation must learn not only a set of actions, but also the rationale for each action, which is impossible without a rather complex spoken language. Thus, the increasing complexity of agricultural practice stimulated the increasing complexity of oral communication and, consequently, of language, and the increasing complexity of language opened new possibilities for better agriculture. Such interrelationships, defining positive feedback between the two, led to an explosive development of both. Moreover, developed oral language made it possible to formulate judgments and ideas for interlocutors and, as a consequence, to discuss and compare them. This, in turn, led to the artificial selection of such judgments and ideas in addition to the artificial selection of domesticated animal and plant species.

As a matter of fact, with the emergence of such a complex language, the problem of meaning moved into the realm of semantics proper, not biosemantics and especially not potential semantics in the agentless world. Therefore, we will not consider this stage of the evolution of meanings in detail and will focus only on the hierarchogenetic events of this evolution as the very last steps of the entire hierarchogenesis of meanings.

The next hierarchical level is a coherent system of judgments, including syllogisms, enthymemes, or epicheiremas. Thomas Kuhn [6.47] called such systems *paradigms*, referring primarily to scientific theories such as evolutionary theory, molecular-kinetic theory, quantum mechanics, Big Bang theory, etc. Since then, however, the term has become widespread and filled with much more content. This allows us to use it, perhaps with the prefix *proto-*, to denote both the system of judgments that unites ancient agronomy and the judgments that constitute ancient zootechnics.

The third such proto-paradigm was fishing, but it stands apart from the first two. Having emerged as a form of hunting long before the agricultural revolution, it did not share the fate of gathering and hunting, which had long ago become a hobby for most peoples. Fishing is still an important source of food (fish and other seafood) for people in many countries. Fish farming and aquaculture in general, which came much later than agriculture, have not, however, replaced fishing itself, which still supplies an important part of our diet.

We should also mention the domestication of insects, especially bees, which developed out of the collection of wild honey as a type of gathering. Beekeeping is a completely separate proto-paradigm with its own judgments, rules, and traditions, unknown to most people. We can also mention the breeding of silkworms, which are larvae of moths, but it developed mainly in China and is not widespread throughout the world.

One way or another, about 10,000 years ago the next hierarchy level of cognitive meanings emerged in the form of paradigms, represented at first by primitive agronomy, animal husbandry, fish farming, beekeeping, and some others. Over time, these paradigms became more and more sophisticated. New ones also emerged by logical development and integration of previously existing ones. They encompassed all new spheres of human activity: family life, cults, traditions, laws, polity, military art, technology, and, finally, science.

However, the level of paradigms is not the only new level of the hierarchy of meanings that humans have created. Almost simultaneously with the appearance of the first proto-paradigms, our ancestors tried to combine them into a more or less coherent system, which can be called a *worldview*, and which allowed them to comprehend the world in which they live as a single whole. In the beginning, such worldviews took the form of very primitive religions. But as human society developed, religions became more sophisticated, philosophy emerged, and finally the scientific worldview became dominating and began to be taught in most schools around the world. This whole hierarchy is represented in Figure 6.6 (top).

Although worldview in the full sense of the word is primarily characteristic of humans, we can talk about its emergence in the form of "Umwelt" [6.48] in other organisms. The term was introduced by Jakob von Uexküll [6.100], who defined it as the phenomenal world (or world of the self) of living beings. Accordingly, the "epistemic cut" we discussed above [6.35] [6.70] runs between this Umwelt and the environment for living beings and between the worldview and the real world for humans.

At the same time, this hierarchy of meanings, as in the case of the biological hierarchy, is not the only possible one. The fact is that elementary meanings can be combined not only into judgments, but also into abstract concepts (Figure 6.6, middle).

For example, a set of specific horses you have met can be grouped into the concept of a "horse." It is commonly believed that only humans are capable of such abstractions. However, such a concept as "food," which unites all objects that can be eaten, is rather abstract and yet makes sense for a highly organized animal, since it generates similar actions regardless of what it is going to eat. But people are also capable of combining these abstract concepts into larger concepts. "Horse," "cow" and "chicken" can be combined into the concept of "domestic animal." "Wheat," "corn" and "apple tree" can be combined into the concept of a "domestic plant." Domestic and wild animals and plants can be combined in the concept of "living things," etc. As the highest level of such hierarchy, we can consider some general picture of the world as an interconnected set of abstract concepts and paradigms, i.e., the "worldview" which we have already considered above.

It is difficult to say anything definite about the evolution that formed this second hierarchy. It is even possible that, according to the idea of Kamshilov [6.41] and Smith and Morowitz [6.89] mentioned above, this evolution could go in the opposite direction: from a vaguely perceptible "worldview," available to any living being to separate abstract concepts — "food," "reproduction," "struggle for life" — and to their referents in the form of concrete objects (Figure 6.6, bottom).

In conclusion, let us try to imagine the next coming hierarchogenetic step in the evolution of meanings in human society. Perhaps it will be a kind of conglomerate of different worldviews, religions, and philosophical systems, which so far as we can see, are mostly fighting each other. If humanity can get on the path of cooperation in the field of religions and philosophical systems, and the different worldviews complement each other, looking at reality from different angles, then something will possibly emerge, for which there is yet no suitable word. It could probably be called *super-understanding*, *super-knowledge* or *omnilogy* (not to be confused with omniology).

There is a suspicion that such *super-understanding* may arise not in a separate consciousness, but only in the entire noosphere, when and if it arises due to the next hierarchogenetic event in the material world. If so, then this next hierarchical step will be the same for both semantic and both biological evolutions, as it was predicted by Teilhard de Chardin [6.95] for the Ω point.

I already mentioned in Section 6.1 that knowledge is a kind of information killer. Accordingly, the simpler the system, the less information is needed to describe it, and the better knowledge works. As for complex systems, scientific knowledge is far from being fully understood. Almost two centuries ago the Russian poet Fyodor Tyutchev [6.99] wrote in his poem "Silentium":

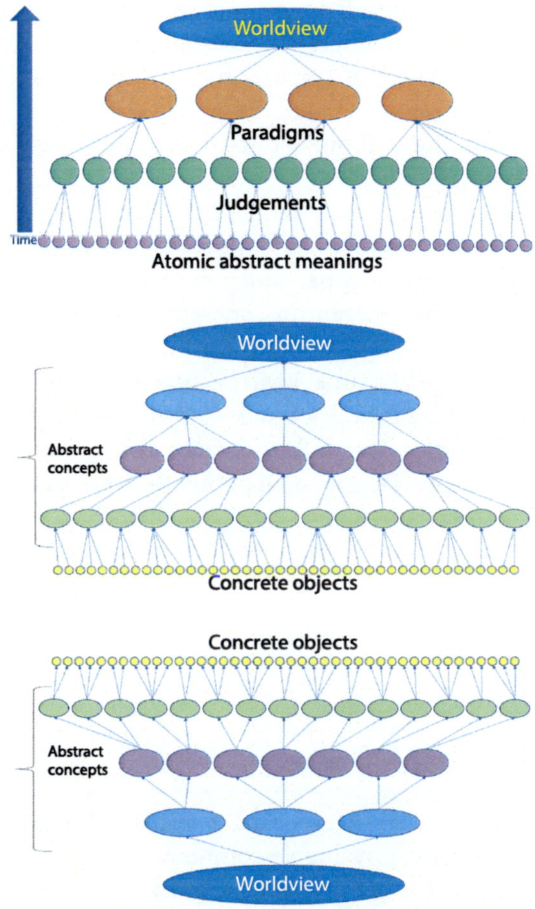

Figure 6.6 Hierarchies of human semantics: (top) such hierarchy beginning with atomic abstract meanings organized into paradigms and then into an integrated worldview; (middle) alternative hierarchy of abstract concepts (Plato eidoi) from concrete observable objects to the worldview as a whole (explanations in the text); (bottom) Inverse alternative hierarchy of abstract concepts, beginning with an obscure worldview, through increasingly more and more detailed concepts and up to the formulation of atomic abstract meanings as denotations of concrete things.

> How can a heart expression find?
> How should another know your mind?
> Will he discern what quickens you?
> A thought once uttered is untrue...

Indeed, any fundamentally new inner thought is always an insight, a flash, free from any context or, rather, containing all possible contexts, including contradictions to itself. Verbalizing this thought, we have to squeeze it into the Procrustean bed of this or that context. This makes it "untrue," because it is an oversimplification, an understatement of the truth. Poetry, however, is an exception. Using poetic imagery, rhythms, the music of verse, and so on, the poet tries to express his thoughts close to their inner true source. When the poet succeeds, the poems are brilliant. And Tyutchev's poem is one example. Incidentally, scientists try to do the same, presenting their inner thought in the form of a long article or even book, where it is discussed in different contexts, presented visually as a set of figures, compared with the ideas of other scientists or with experiments, etc. But few scientists manage to do this well.

Tyutchev, of course, in his line "A thought once uttered is untrue," did not mean scientific knowledge. He was referring to knowledge in the broadest sense. This line describes in an extremely concise and succinct manner the general relationship between meanings and symbols, or, in a broader sense, between semantics and semiotics.

However, only the attainment of super-understanding or super-knowledge in Ω may help to avoid the simplifications inherent in concrete knowledges (e.g., in scientific theories). Hopefully, at this point a complete superposition and unification of all this concrete knowledge (even contradicting each other) will become possible.

The hypothesis presented here is only a crude approximation, discarding the lion's share of information about the world. The author is fully aware that reality is immeasurably brighter, richer, and more unexpected than all of the ideas presented above.

Declarations

I declare no conflict of interest.

Acknowledgments

The author would like to thank Dr. Alexei Sharov and Dr. Richard Gordon for their discussion of this chapter and the very valuable comments they made. In addition, I am very grateful to Katerina Lomis for her help in preparing my text.

References

[6.1] Alicea, B. (2009) *Hierarchies of Biocomplexity: modeling lifes energetic complexity*. Cornel University, arXiv. DOI: https://doi.org/10.48550/arXiv.0810.4547

[6.2] Allday, J (2016) *Quarks. Leptons and the Big Bang*, 3rd ed. Boca Raton: CRC Press, Taylor & Fransis Group.

[6.3] Anissimov, M. (2022) *How did Agriculture Begin?* PublicPeople.org, July 02. www.publicpeople.org/how-did-agriculture-begin.htm

[6.4] Antunes, L.C.M, Ferreira R.B.R, Buckner M.M.C and Finlay B. B. (2010) Quorum sensing in bacterial virulence. *Microbiology*, 156, 2271–2282 DOI: https://doi.org/10.1099/mic.0.038794-0

[6.5] Barbieri, M. (2003) *The Organic Codes*. Cambridge University Press, Cambridge, UK.

[6.6] Barbieri, M. (2014) From Biosemiotics to Code Biology. *Biological Theory*, 9(2). DOI: https://doi.org/10.1007/s13752-013-0155-6

[6.7] Barbieri, M. (2015) *Code Biology. A New Science of Life*. Springer, Dordrecht .

[6.8] Barbieri, M. (2020) *The Semantic Theory of Evolution*. Routledge. London & NY.

[6.9] Bell, J.S. (1981) Quantum Mechanics for Cosmologists. In: *Quantum Gravity 2: A Second Oxford Symposium*. C. J. Isham, R. Penrose and D.W. Sciama (eds.) Oxford, Clarendon Press: 611-637.

[6.10] Bergstra, J. and Bengio, Y. (2012) Random search for hyper-parameter optimization. *Journal of Machine Learning Research.* 13, 281–305.

[6.11] Blackstone, N.W. (2013) Why did eukaryotes evolve only once? Genetic and energetic aspects of conflict and conflict mediation. *Philos. Trans. R. Lond. B Biol. Sci.* 368(1622), #20120266.

[6.12] Blackwell, H.E. and Fuqua, C. (2011) Introduction to Bacterial Signals and Chemical Communication. *Chem. Rev.*, 111(1), 1–3. https://doi.org/10.1021/cr100407j

[6.13] Borghini, A. (2021) Understand the Philosophical Theories of Nominalism and Realism. *ThoughtCo.* https://www.thoughtco.com/nominalism-vs-realism-2670598

[6.14] Bowman, J.D., Rogers, A.E.E., Monsalve, R.A, Mozdzen, T.J. and Mahesh, N. (2018) An absorption profile centered at 78 megahertz in the sky-averaged spectrum. *Nature* 555, 67–70.

[6.15] Brier, S. (2006) *Biosemiotics. In Encyclopedia of Language & Linguistics* (2nd Ed.), Keith Brown (ed.), Elsevier, 31-40.

[6.16] Butterfield, N.J. (2009) Modes of pre-Ediacaran multicellularity. *Precambrian Res.* 173, 201–211.

[6.17] Carlson, E.K. (2019) The oldest discovered cluster of galaxies is revealing the early universe. *Astronomy.* https://astronomy.com/news/2019/09/the-oldest-discovered-cluster-of-galaxies-is-revealing-the-early-universe

[6.18] Cheeseman, I.M., and Desai, A. (2008) Molecular architecture of the kinetochore-microtubule interface. *Nature Reviews Molecular Cell Biology* 9, 33–46. DOI: https://doi.org/10.1038/nrm2310

[6.19] Chopra, A. and Lineweaver, C.H. (2018) The cosmic evolution of biochemistry. In: *Habitability of the Universe before Earth*, Gordon, R., Sharov, A.A. (eds.) Elsevier B.V.: Amsterdam, The Netherlands: 75-78.

[6.20] Christiansen, P.V. (1993) Peirce as Participant in the Bohr-Einstein Discussion. In: *Charles S. Peirce and the Philosophy of Science: Papers from the Harvard Sesquicentennial Congress.* E. C. Moore (ed.) The University of Alabama Press, Tuscaloosa, London: 223–232.

[6.21] Conti, A. (2009) Realism. In: *The Cambridge History of Medieval Philosophy.* R. Pasnau (ed.), Cambridge University Press, Cambridge: 647-660. DOI: https://doi.org/10.1017/CHOL9780521762182.010

[6.22] Copley, S.D., Smith, E. and Morowitz, H.J. (2007) The origin of the RNA world: co-evolution of genes and metabolism. *Bioorg. Chem.* 35 (6), 430–443.

[6.23] Coutts, M.P., Nielsen, C.C.N. and Nicoll, B.C. (2000) The development of symmetry, rigidity and anchorage in the structural root system of conifers. In: *The Supporting Roots of Trees and Woody Plants: Form, Function and Physiology.* Stokes, A. (eds), Developments in Plant and Soil Sciences, vol 87. Springer, Dordrecht: 3-17. https://doi.org/10.1007/978-94-017-3469-1_1

[6.24] Davis, P. (2007) *Cosmic Jackpot: Why Our Universe Is Just Right for Life.* Houghton Mifflin, Boston.

[6.25] DK. (2015) *Timelines of History* Penguin Random House, London.

[6.26] Dodig-Crnkovic, G. (2022) Cognition as Morphological/Morphogenetic Embodied Computation In Vivo. *Entropy* 24(11), #1576. https://doi.org/10.3390/e24111576

[6.27] Eisenbach, M. (2011) Bacterial Chemotaxis. Wiley Online Library, *eLS.* https://doi.org/10.1002/9780470015902.a0001251.pub3

[6.28] Erwin, D.H. (2015) Early metazoan life: divergence, environment and ecology. *Philos. Trans. R. Soc. B Biol. Sci.* 370(1684), #20150036. https://doi.org/10.1098/rstb.2015.0036

[6.29] Fisher, D. (2019) *Which English Word Has the Most Definitions?* The Spruce Crafts. https://www.thesprucecrafts.com/which-word-has-the-most-definitions-4077796

[6.30] Gordon, N.K. and Gordon, R. (2017) *Embryogenesis Explained*. World Sci. Publ., Singapore.

[6.31] Gordon, R. (1999) *The Hierarchical Genome and Differentiation Waves: Novel Unification of Development, Genetics and Evolution*. World Scientific & Imperial College Press, Singapore, London. http://www.worldscientific.com/worldscibooks/10.1142/2755

[6.32] Gordon, R. and Mikhailovsky, G.E. (2021) There were plenty of day/night cycles that could have accelerated an origin of life on Earth, without requiring panspermia. In: *Planet Formation and Panspermia: New Prospects for the Movement of Life through Space*. B. Vukotić, J. Seckbach and R. Gordon, (eds.) Astrobiology: Perspectives on Life of the Universe.Wiley-Scrivener, Beverly, Massachusetts, USA: 195-206.

[6.33] Goyal, A., Szarzynska, B. and Fankhauser Ch. (2013) Phototropism: At the crossroads of light-signaling pathways. *Trends in Plant Science* 18(7), 1120-1127. DOI: https://doi.org/10.1016/j.tplants.2013.03.002

[6.34] Hawking, S.W. (2006) *The Theory of Everything: The Origin and Fate of the Universe*. Jaico Books Publ., Mumbai.

[6.35] Hoffmeyer, J. (2000) Code-duality and the epistemic cut. *Ann N Y Acad Sci*. 901, 175-86. DOI: https://doi.org/10.1111/j.1749-6632.2000.tb06277.x.

[6.36] Hoffmeyer, J. (2008) From thing to relation. On Bateson's bioanthropology. In: *A legacy for living systems. Gregory Bateson as precursor to biosemiotics*. J. Hoffmeyer (ed.) Springer, Dordrecht, New York: 27-44.

[6.37] Hole, F.A. (1984) Reassessment of the neolithic revolution. *Paléorient*. 10, 49–60.

[6.38] Hutchin, R.A. (2015) The Unknown Conservation Laws, *Journal of Modern Physics*, 6, 729-732. DOI: http://dx.doi.org/10.4236/jmp.2015.66078

[6.39] Jagers op Akkerhuis, G.A.J.M. (2010) *The Operator Hierarchy: A Chain of Closures Linking Matter, Life and Artificial Intelligence*. Ph.D. Thesis, Radboud University Nijmegen, Nijmegen, The Netherlands.

[6.40] Jagers op Akkerhuis, G.A.J.M. (2018) Why on theoretical grounds it may be likely that 'life' exists throughout the universe. In: *Habitability of the Universe Before Earth*. Gordon, R. and Sharov, A.A., (eds.) Elsevier B.V.: Amsterdam, The Netherlands: 491-505.

[6.41] Kamshilov, M.M. (1974) *The evolution of the biosphere*. Mir Publ., Moscow.

[6.42] Kauffman, S. A. (1993) *Origins of Order: Self-Organization and Selection in Evolution*. Oxford University Press, New York, Oxford.

[6.43] Kauffman, S. A. (2000) *Investigations*. Oxford University Press, New York, Oxford.

[6.44] Knoll, A.H. (2011) The multiple origins of complex multicellularity. *Ann. Rev. Earth Planet. Sci*. 39, 217–239.

[6.45] Kobayashi, T.J. and Kamimura, A. (2012) Theoretical Aspects of Cellular Decision-Making and Information-Processing. In: *Advances in Systems Biology*. I. Goryanin and A. Goryachev(eds). Advances in Experimental Medicine and Biology, 736. Springer, New York: 275-291 https://doi.org/10.1007/978-1-4419-7210-1_16

[6.46] Krimm, H. *et al*. (2009) "GRB 090423: Swift detection of a burst". *GCN Circulars*. 9198, 1.

[6.47] Kuhn, T.S. (1970) *The Structure of Scientific Revolutions* (2nd ed.) University of Chicago Press, Chicago, London.

[6.48] Kull, K. (1998) On semiosis, Umwelt, and semiosphere, *Semiotica* 120(3-4), 299-310.

[6.49] Levin, M., Keijzer, F., Lyon, P. and Arendt, D. (Eds) (2021) Basal cognition: Multicellularity, neurons and the cognitive lens. *Philosophical Transactions of the Royal Society B: Biological Sciences*, 376 (1821).

[6.50] Lloyd, S. (2002) Computational Capacity of the Universe. *Phys. Rev. Lett*. 88, #237901.

[6.51] Lloyd, S. (2007) Excerpt from *Programming the Universe*. https://www.penguinrandomhouse.ca/books/102977/programming-the-universe-by-seth-lloyd/9781400033867/excerpt

[6.52] Lloyd, S. (2013) On the Spontaneous Generation of Complexity in the Universe. In: *Complexity and the Arrow of Time*. C.H. Lineweaver, P.C.W. Davis, and M. Ruse (eds.), Cambridge University Press, Cambridge, UK: 80–112.
[6.53] Lyons, N.A. and Kolter, R. (2015) On the evolution of bacterial multicellularity. *Curr. Opin. Microbiol.* 24, 21–28.
[6.54] Macháček, M. (2019) A Peircean Interpretation of Probability in Quantum Mechanics. In: *Semiotics 2018: Resilience in an Age of Relation*. G.R. Owens and E. Katić (eds.): 233-241. DOI: https://doi.org/10.5840/cpsem201817
[6.55] Martin, C.C. and Gordon, R. (2001) The evolution of Perception. *Cybernetics and Systems: An International Journal*, 32(3), 393-409. DOI: https://www.tandfonline.com/doi/abs/10.1080/01969720151033553
[6.56] Merlevede, A., Legault, E.M., Drugge, V., Barker R.A. and Drouin-0uellet J. (2021) A quantitative model of cellular decision making in direct neuronal reprogramming. *Sci Rep*. 11, 1514. https://doi.org/10.1038/s41598-021-81089-8
[6.57] Michalowski, M.J. (2015) Dust production 0.7–1.5 billion years after the Big Bang. *PTA Proc.* 123, 51–56.
[6.58] Michiko, E.T. and Bonnie, L.B. (2003) Chemical communication among bacteria. *PNAS*, 100(suppl_2), 14549-14554 https://doi.org/10.1073/pnas.1934514100
[6.59] Mikhailovsky, G.E. (2018) From Identity to Uniqueness: The Emergence of Increasingly Higher Levels of Hierarchy in the Process of the Matter Evolution. *Entropy*, 20, 533- 551. http://www.mdpi.com/1099-4300/17/7/4863
[6.60] Mikhailovsky, G.E. and Gordon, R. (2018) Symbiosis: Why was the transition from microbial prokaryotes to eukaryotic organisms as a cosmic gigayear event? In: *Habitability of the Universe before Earth*. R. Gordon, and A.A. Sharov (eds.) Elsevier B.V., Amsterdam: 355-405
[6.61] Mikhailovsky, G.E. and Gordon, R. (2020) Shuffling type of biological evolution based on horizontal gene transfer and the biosphere gene pool hypothesis. *BioSystems*, 193-194, #104131, 1-12.
[6.62] Mikhailovsky, G.E. and Levich, A.P. (2015) Entropy, information, and complexity or which aims the arrow of time? *Entropy*, 17(7), 4863-4890. https://www.mdpi.com/1099-4300/20/7/533
[6.63] Murphy M. (2014) *Three eyes tracking nature's laws*. https://astronomy.swin.edu.au/~mmurphy/2014/09/three-eyes-tracking-natures-laws/
[6.64] Neveu, M., Kim, H.J. and Benner, S.A. (2013) The "strong" RNA world hypothesis: Fifty years old. *Astrobiology*, 13(4), 391–403. DOI: https://doi.org/10.1089/ast.2012.0868
[6.65] Newton, I. (1684) *Philosophiae Naturalis Principia Mathematica*. Royal Society, London.
[6.66] Ogden, C.K. and Richards, I. A. (2013) *The Meaning of Meaning: A Study of the Influence of Language Upon Thought and of the Science of Symbolism*. Martino Fine Books.
[6.67] Olivares, H. *et al.* (2009) GRB 090423: GROND detection and preliminary photo-z. *GCN Circulars*. 9215(1).
[6.68] Pattee, H.H. (1969) How does a molecule become a message? *Developmental Biology Supplement* 3, 1–16.
[6.69] Pattee, H.H. (1971) Can life explain quantum mechanics? In: *Quantum Theory and Beyond — Essays and Discussions Arising from a Colloquium*. Bastin, Ted (ed.), Cambridge University Press, London, 307–319.
[6.70] Pattee, H.H. (2001) The Physics of Symbols: Bridging the Epistemic Cut. *Biosystems*. 60, 5-21.
[6.71] Pattee, H.H. (2008) Physical and functional conditions for symbols, codes, and languages. *Biosemiotics* 1(2), 147–168.

[6.72] Pattee, H.H., and Kull, K. (2009) A biosemiotic conversation: Between physics and semiotics. *Sign Systems Studies*, 37(1-2), 311-331. https://casci.binghamton.edu/publications/pattee/pattee.html

[6.73] Peres, A. (1995) *Quantum Theory: Concepts and Methods.* Springer

[6.74] Puiu, T. (2018) NASA zooms in on one of the oldest galaxies with the help of a cosmic magnifying glass. *ZME Sci.* https://www.zmescience.com/space/old-far-away-galaxy-0423523/

[6.75] Quax, T., Albers, S. V., and Pfeiffer, F. (2018) Taxis in archaea. *Emerging topics in life sciences*, 2(4), 535–546. https://doi.org/10.1042/ETLS20180089

[6.76] Ralston, A. (2008) Simultaneous gene transcription and translation in bacteria. *Nature Education* 1(1), 4.

[6.77] Rieffel, E.G., and Polak, W. H. (2011) *Quantum Computing: A Gentle Introduction (Scientific and Engineering Computation).* The MIT Press, Boston.

[6.78] Russell, B. (1918) *The Philosophy of Logical Atomism: A Centenary Reappraisal.* Springer Int. Publ.

[6.79] Salthe, S.N. (1985) *Evolving Hierarchical Systems.* Columbia Univ. Press: New York, NY, USA.

[6.80] Salthe, S.N. (2012) Hierarchical Structures. *Axiomathes* 22, 355–383. https://doi.org/10.1007/s10516-012-9185-0

[6.81] Seigel, E. (2011) Found: The First Atoms in The Universe!!!, *ScienceBlogs*, November 2011. https://scienceblogs.com/startswithabang/2011/11/14/found-the-first-atoms-in-the-u

[6.82] Sen, D. (2014) The Uncertainty relations in quantum mechanics. *Current Science.* 107(2), 203–218. https://www.currentscience.ac.involumes/107/02/0203.pdf

[6.83] Sharov, A. A. (2010) Functional information: Towards synthesis of biosemiotics and cybernetics. *Entropy*, 12(5), 1050-1070.

[6.84] Sharov, A. A. (2018) Mind, agency, and biosemiotics. *Journal of Cognitive Science*, 19(2), 195-228.

[6.85] Sharov, A.A. (2021) personal communication.

[6.86] Sharov, A.A. and Gordon R. (2013) Life before Earth. *arXiv*, #1304.3381. https://doi.org/10.48550/arXiv.1304.3381

[6.87] Sharov, A.A. and Gordon, R. (2018) Life before Earth. In: *Habitability of the Universe before Earth.* R.Gordon and A.A.Sharov (Eds.) Elsevier, Amsterdam: 265-296.

[6.88] Sharov, A. A. and Tønnessen, M. (2021) *Semiotic agency. Science beyond mechanism* (Biosemiotics vol. 25). Springer , Dordrecht.

[6.89] Smith E., and Morowitz, H.J. (2016) *The Origin and Nature of Life on Earth: The Emergence of the Fourth Geosphere.* Cambridge Univ. Press, NY.

[6.90] Smolin, L. (1999) *The Life of the Cosmos.* Oxford University Press, NY.

[6.91] Swartz, N. (1999) Laws of Nature In: *Internet Encyclopedia of Philosophy.* https://iep.utm.edu/lawofnat

[6.92] Swartz, N. (2003) *The concept of physical law* (2nd Ed) Cambridge Univ. Press, NY.

[6.93] Tanet, L., Tamburini, C., Baumas, C., Garel, M., Simon, G. and Casalot L. (2019) Bacterial bioluminescence: Light emission in *Photobacterium phosphoreum* is not under quorum-sensing control. *Frontiers in Microbiology*, 10(4), 365. https://www.frontiersin.org/article/10.3389/fmicb.2019.00365

[6.94] Teilhard de Chardin, P. (1967) *The vsion of the past.* Harper and Row Publ., NY.

[6.95] Teilhard de Chardin, P. (1976) *The phenomenon of man.* Harper and Row Publ., NY.

[6.96] Theobald, D. L. (2010) A formal test of the theory of universal common ancestry. *Nature.* 465 (7295), 219–222. DOI: https://doi.org/10.1038/nature09014

[6.97] Tomkins, M.G. (1975) The metabolic code. Biological symbolism and the origin of intercellular communication. *Science* 189, 760–763. DOI: https://doi.org/10.1126/science.169570

[6.98] Tønnessen, M. (2015) The Biosemiotic Glossary Project: Agent, agency. *Biosemiotics*, 8(1), 125-143.

[6.99] Tyutchev, F.I. (1830) *Silentium*! (translated to English by V. Nabokov) https://allpoetry.com/poem/8537755-Silentium-by-Fyodor-Ivanovich-Tyutchev

[6.100] Uexküll, J. von (1957) A Stroll Through the Worlds of Animals and Men: A Picture Book of Invisible Worlds. In: *Instinctive Behavior: The Development of a Modern Concept*. Claire H. Schiller, C.H. (ed. and translator) International Universities Press, NY, 5–80.

[6.101] Umerez, J. (2001) Howard Pattee's theoretical biology--a radical epistemological stance to approach life, evolution and complexity. *Biosystems*, 60(1-3), 159-77. DOI: https://doi.org/10.1016/s0303-2647(01)00114-9

[6.102] Verdier, N. (2005) Hierarchy: A short history of a word in Western thought. In: *Hierarchy in Natural and Social Sciences*; Pumain, D. (ed.) Springer: Dordrecht, The Netherlands, 1-17.

[6.103] Vernadsky, V. I. (1991) *Scientific thought as a planetary phenomenon*. Moscow: Nauka (in Russian).

[6.104] Vernadsky, V.I. and Ross, J.A. (2014) *150 Years of Vernadsky*. Create Space Independent Publ., North Charleston

[6.105] Volk, T. (2017) *From Quarks to Culture*; Columbia Univ. Press, NY, USA.

[6.106] Wadhams, G and Armitage, J. (2004) Making sense of it all: bacterial chemotaxis. *Nat Rev Mol Cell Biol* 5, 1024–1037. https://doi.org/10.1038/nrm1524

[6.107] Watson, J. D. (2002) *Genes, Girls, and Gamow: After the Double Helix*. Random House, NY.

[6.108] Webb J. (2003) Are the laws of nature changing with time?, *Physics World*, 33-38. http://www.nat.vu.nl/~wimu/Varying-Constants-Papers/WEBB-Fund-constants.pdf

[6.109] *Webster's College Dictionary* (1997) Random House, Inc.

[6.110] Weinberg, S. (2011) *Dreams of a Final Theory: The Scientist's Search for the Ultimate Laws of Nature*. Knopf Doubleday Publishing Group.

[6.111] Weiss, A. (2006) Big bang nucleosynthesis: Cooking up the first light elements. *Einstein Online*. https://www.einstein-online.info/spotlights/BBN.html

[6.112] Wigner, E. P. (1960) The unreasonable effectiveness of mathematics in the natural sciences. Richard Courant lecture in mathematical sciences delivered at New York University, May 11, 1959. *Communications on Pure and Applied Mathematics*. 13, 1–14.

[6.113] Wikipedia (2022) Platonism *Wikipedia*. https://en.wikipedia.org/wiki/Platonism

[6.114] Wikipedia (2023) Human history. *Wikipedia*. https://en.wikipedia.org/wiki/Human_history#Ancient_history_(3000_BCE_to_500_CE)

[6.115] Wittgenstein, L. (1922) *Tractatus Logico-philosophicus*. Harcourt, Brace, Inc., NY; Kegan, Trench, Trubner, Ltd., London.

[6.116] Wittgenstein, L. (1953) *Philosophical investigations*. Blackwell, Oxford.

[6.117] Witze, A. (1922) JWST spots some of the most distant galaxies ever seen. Premier observatory offers a glimpse of the early Universe. *Nature News* DOI: https://doi.org/10.1038/d41586-022-03751-z

[6.118] Wolf, A.M. (2021) MtDNA mutations and aging—not a closed case after all? *Sig Transduct Target Ther* 6, 56. https://doi.org/10.1038/s41392-021-00479-6

[6.119] Wright, J., (Ed.) (2014) *International Encyclopedia of Social and Behavioral Sciences*. Elsevier. https://sites.utexas.edu/dsb/tokens/the-evolution-of-writing/

[6.120] Yao, X. (Ed.) (2021) Orchestration of cell plasticity by phase separation *Journal of Molecular Cell Biology*, 13(4), 237–238. https://doi.org/10.1093/jmcb/mjab029

[6.121] Zackrisson, E. and Sahlén, M. (2021) The First Stars and Galaxies. *Uppsala Univ., Dept. Physics and Astronomy*. https://www.physics.uu.se/research/astronomy-and-space-physics/research/galaxies/first-stars-galaxies/

[6.122] Zainab R., Perdon S., Wang X., Ha T., Harley B., and Leckband D. (2016) Nanoscale mechanics guides cellular decision making, *Integrative Biology*, 8(9), 929–935. https://doi.org/10.1039/c6ib00113k

[6.123] Zhu, Q., Mai, U., Pfeiffer, W. *et al.* (2019) Phylogenomics of 10,575 genomes reveals evolutionary proximity between domains Bacteria and Archaea. *Nat. Commun.* 10, 5477. DOI: https://doi.org/10.1038/s41467-019-13443-4

[6.124] Zhou, R., Basu, K., Hartman, H., Matocha, Ch.J., Sears, S.K., Vali, H. and Guzman, M.I. (2017) Catalyzed Synthesis of Zinc Clays by Prebiotic Central Metabolites. *Scientific Reports*, 7, 522. DOI: https://doi.org/10.1038/s41598-017-00558-1

[6.125] Zimmer, C. (2009) On the origin of eukaryotes. *Science*, 325, 666–668.

7

Semiotics of Potential Meanings

Alexei A. Sharov

Elixirgen Scientific, Baltimore, Maryland, USA

Abstract

In biosemiotics, meaningful things, processes, or relations are those that have significance for organisms as indicated by either behavioral choice or regulation of innate habits and processes. In organisms, meanings are communicated by signs, such as words in human speech, animal calls, neuron firing, or hormone signaling. But in this chapter I discuss potential meanings — those that are not communicated currently or in the past. An example of potential meaning is a natural resource that initially has not been utilized by organisms because of the lack of either organs necessary to perceive and manipulate it or "knowledge" of how to do it. These problems can be resolved by adaptive evolution of organs and/or by learning new habits, hence potential meanings can turn into actual meanings. Because intrinsic properties of resources remain the same before and after actualization, it makes sense to consider such potential meanings exist even before they become actual. In physics, potentiality is described as a field which is not observed unless we introduce a measuring device. Similarly, I interpret potential meaning as a semiotic field that can be detected by using agents as measuring devices. Potential meanings can propagate and evolve via physicochemical activities that can inadvertently produce more complex local environments enriched with non-redundant potential meanings. Following Deely, I assume that such complex environments can ignite the origin of life or provide resources for already existing life. Emergence of living organisms and their evolution led to actualization of some already existing potential meanings, and these actualized natural meanings together with new artificial meanings became encoded by organisms in hereditary signs and memory for the purpose of reusing them when necessary within the lifespan and beyond (in the progeny). Memory in organisms is a repository of potential meanings (i.e., passive memory units that can be activated to communicate their meaning) that enables organisms to adapt on demand by converting potential meanings into actual meanings. Once actualized, meanings tend to spread among semiotic agents via interaction or communication and further evolve at longer time scales.

Keywords: Biosemiotics, agency, potential meaning, semiotic field, memory, adaptation, criteria of meaning objectiveness

Email: sharov@comcast.net

7.1 Introduction

Semiotics is the study of various kinds of signs and sign processes (semiosis) such as interpretation and production of signs by humans and non-human organisms [7.70] [7.73]. The specific feature of signs is their capacity to communicate something, generally called meaning, to the organism that interprets the sign. This process is known as *signification* [7.47]. As a sign is perceived, its meaning is retrieved or generated by the interpreter based on its previous experience in dealing with this kind of signs and in the context of external and internal conditions.

To conceptualize semiosis, it is important to define its boundaries. In particular, there were multiple attempts to delineate "nonsigns" that constitute the non-semiotic world. According to Saussure, everything outside human language is non-semiotic: "nothing is distinct before the appearance of language" [7.14]. In contrast, Peirce considered that (nearly) the whole universe as semiotic: "The entire universe [. . .] is perfused with signs, if it is not composed exclusively of signs" [7.63], CP 5.448. Biosemiotics assumes an intermediate stance, where the semiotic world is coextensive with life, whereas non-living world is non-semiotic [7.2]. Sharov and Tønnessen [7.83] suggested adding to the sphere of semiosis autonomous agents capable of using signs even if these agents are not organisms. Examples of such agents are cell subagents (ribosomes, proteins), human artifacts (robots, computers), viruses, and multi-organism agents (colonies, populations, human organizations).

Despite my long interest in biosemiotics, in this chapter I decided to focus on non-signifying entities that are potentially useful for agents. The first kind of such entities are natural resources, which included just a few kinds of molecules and radiation particles at the origin of life, and gradually expanded to other kinds of non-organic and organic resources of various sizes. Historically, each kind of resource was initially not utilized by any living systems, and then at some point became used by some species. From a semiotic point of view, this means that a resource was initially not meaningful for living systems, and then became meaningful in the sense that organisms started selectively interacting with the resource, and possibly gained the capacity to recognize it and selectively manipulate it. Across the evolutionary time-span, a resource remained the same, which means that its useful properties existed long before they were utilized by life. These useful properties can then be considered as *potential meanings*, which were later "discovered" by organisms. After the resource became assimilated by organisms, its meaning was actualized (i.e., turned into a proper meaning) (Table 7.1). The notion of potential meaning is similar to the idea of non-occupied ecological niche in ecology. It can be applied to such problems as assessing the probability of invasion of alien organisms or the possibility of the origin of life.

The second kind of entities that carry potential meanings but do not signify are what I call *potential signs*, following Nöth [7.57], p. 66. These are not proper signs because they do not signify; but they have a chance to signify if they become engaged with certain agents or subagents (Table 7.1). For example, virus capsids carry virus DNA or RNA but remain inactive unless they get a chance to infect a host cell and signify by making a new generation of viruses. Other examples of potential signs are units of heredity (genes), long-term memory, or communication (pheromones, books, internet blogs). These signs can be activated by agents on demand and start signifying.

Table 7.1 Features of proper meanings and signs and potential meanings and signs.

Proper meaning	**Potential meaning**	**Non-meaning**
Entity* that is currently significant for some semiotic agency in a specific way	Entity that can be significant for some semiotic agency in a specific way, as follows from past observations or models	Unknown entity or entity without known significance
Proper sign	**Potential sign**	**Non-sign**
Entity that currently signifies meaning of something else to semiotic agency	Entity that carries meaning of something else, and can signify it (based on past observations or models) but does not signify now	Unknown entity or entity without meaning of something else

* Entity means a thing, feature, event, relation, or process, which can be external to agency or internal.

Relatively little attention in semiotics is given to things that do not signify. In this respect, Queiroz *et al.* [7.67], p. 81, differentiate between *effective and potential semiosis*:

> We understand potential semiosis as a triadically-structured process that could take place, but that is not effectively taking place at a given time t. Effective semiosis, in turn, concerns a sign in effective action, i.e., a sign that, by its being actualized, has an actual effect on the interpreter.

Thus, potential semiosis is only a possibility because signification is not present. The notion of "form," which is a cornerstone of the philosophy of objective idealism, is a predecessor of potential meaning because things are perceived and utilized by agents according to their form. Thus, the form is the set of constraints on how each thing can be perceived and used even if it is currently not perceived or contemplated. Thus, I start the chapter with an overview of the notion of form as it appears in ancient philosophy and semiotics. However, the notion that forms exist on their own outside of time and space (e.g., Plato's world of ideas) has already lost its charm; thus, it makes sense to ground meanings/ideas in the function of organisms. Such grounding became possible thanks to recent developments in biology, including physiology, embryology, cognitive science, and molecular biology.

Thus, I turn to biosemiotics, which is a synthesis of biology and semiotics [7.82], on the basis that meanings are grounded in the functional activity of agents. After defining semiotic agency and signs, I discuss the notion of potential meaning, where semiotic potentiality is compared to the field theory in physics. The existence of a field (e.g., electric, magnetic, or gravitational) is justified by the assumption that we can always place a specific particle (e.g., electron in the electric field) and measure the field by observing particle's motion. In the same way, we can justify the existence of potential meaning associated with a thing (i.e., semiotic field) by a possibility of adding a competent agent/interpreter and examining its response to this thing at various functional levels.

Potential meanings include predicted physical conditions in the past that facilitated the origin of life as well as conditions on the surface of exoplanets, which are expected to be meaningful for life or potential life. Simple environments such as homogeneous gas clouds certainly cannot support life. Following Deely [7.16], I suggest that complexity of the environment is a necessary condition for the origin of life, and that complexity can be quantified by the number of non-redundant intrinsic potential meanings relevant for projected primordial agents. Some potential meanings can propagate via spontaneous activity of things (e.g., by catalysis or crystallization), which may contribute to a local increase of physical complexity. Deely (ibid.) suggested that such increase of complexity may ignite the origin of life.

Potential meanings can be actualized by organisms if they integrate them into their own goal-directed activities and encode this integration in hereditary or memory signs. Such information storage allows organisms to retrieve the content of genome or memory on demand in the future, and re-act the process of meaning actualization by, for example, consumption of a resource or any other kind of useful activity. Information storages in agents are built as a network of actual and potential signs. Actual signs are represented by continuously sensed stimulation from the environment or other agents, but most of these stimuli are filtered out as noise. Only specific patterns of stimuli serve as pointers[1] to meaningful potential signs that represent memory content. Thus, memory content can be retrieved on demand by subagents with the help of pointer signs.

7.2 Form in the Philosophy of Idealism

The nature of human thinking and communication has been among the most disputed problems in philosophy since ancient times. According to Plato, knowledge is attained by grasping unchangeable ideas or forms, and the world of things is only an imperfect copy of the "real" world of ideas. In contrast, Aristotle assumed that forms exist only in substances, where they are combined with some matter. Then, sensing is a process of change within an organism mediated by sense organs, and this process leads to acquiring knowledge about the form of a thing. Aristotle thought that the form is actually present in the object no matter whether the object is perceived or not, and sense organs are capable of receiving the form of the object (without matter) in a similar way as wax takes the form of a signet ring [7.3], 424a17.

The importance of human thinking in obtaining and keeping knowledge was firmly recognized in the philosophy of Enlightenment, although there was no consensus on the source of true beliefs. According to rationalists, such as Descartes and Spinoza, truth can be attained by mind even without sensory experience. Descartes believed that humans were born with innate knowledge endowed by God, and Spinoza thought that substances and attributes were based on self-evident truths illuminated by God. In contrast, empiricists, such as Bacon, Locke, and Hume, rejected innate ideas and assumed that knowledge comes only or mostly from sensory experience. In addition, Kant noticed that experience alone is not sufficient and may require reason to process it into a coherent thought.

---

[1] The term "pointer" comes from programming languages where it means a digital address in computer memory. Here, however, I use it in a more abstract sense as any relation or correlation (e.g., index, icon, or both) that can help semiotic agents to reach the target sign carrying the content or memory which can be retrieved on demand.

The merit of empiricism is in the notion that human thinking is strongly constrained by interaction with nature itself, and thus, knowledge is mostly *discovered in reality* rather than delivered from God or constructed from self-evident axioms. The development of empiricism, nominalism, and materialism resulted in the decline of idealism in secular philosophy, and in the shift of interest towards empirical studies that stimulated the development of experimental science. However, the nature of knowledge is not fully explained by the dominating contemporary schools of philosophy. According to the still popular correspondence theory of truth, the proposition "snow is white" is true if and only if the real snow is indeed white [7.88], as if we evaluate object properties directly without sensing and perception, which are mind-dependent processes. Propositions in natural language, such as "snow is white," are interpreted by the mind and thus carry some elements of subjectivity,[2] which cannot be ignored.

Semiotics attempted to rescue many notions from the idealistic philosophy, such as "form" and "relation." Hjelmslev considered form as the central notion of semiotics: "semiotics is a science of forms (or structures) and not of substances" [7.57], p. 66. In contrast, Charles Peirce considered "relation" as the core of his semiotics doctrine. He wrote: "A sign therefore is an object which is in relation to its object on the one hand and to an interpretant on the other in such a way as to bring the interpretant into a relation to the object corresponding to its own relation to the object" [7.64], p. 390. Queiroz, Emmeche, and El-Hani used the notion of form to reformulate Peirce's triadic sign relation (Sign – Object – Interpretant) as "a triadic process of communication of a form from the object to the interpretant through sign mediation" [7.67], p. 79. John Deely [7.16], p. 46, explains sign relations:

> So the sign is not a thing as such, not a physical element merely existing in the environment, like a rock or a toad, but something as doubly related: A stands for B in context C. Yet to say 'doubly related' is not to say 'two relations', but to say rather that one 'thing' is related to two other 'things' at one and the same time by one single relation.

He clarifies that context C is what Peirce called "interpretant," and that triadic relations exist even in seemingly dyadic cause-effect relations. The interpretant as a third element of a sign relation is a "thematic factor" that ensures the consistency of sign effects even in the absence of interpreting mind. In particular, Peirce considered sign relations actually existing even when signs are not communicated [7.63], CP 8.332. Perhaps he meant sign types rather than tokens. Some interpretations of Peirce included potentiality in the form of incomplete (i.e., degenerate) sign relations where the interpretant could be latent [7.13] [7.43].

Deely [7.16] considers a sign as a certain kind of cause, where the effect is not intrinsic because a sign needs to stand for something *other than itself*. In particular, material cause and intrinsic formal cause (e.g., development of embryo from an egg) are not signs, according to Deely. Associative drift of ideas when one thing leads to another "cannot by itself constitute the process of semiosis" because it is potentially endless and does not have an "external check" (ibid., p. 53). Eventually, Deely comes to the notion that signs represent "specificative

---

[2] Subjectivity here means subject-specific semiotic processes, which can be studied by objective scientific methods.

causality" which is "extrinsic formal causality of the specificative or 'objective' type" (ibid., p. 56). An example of specification is a dinosaur's bone: "The dinosaur, long dead, is present in the fossil bone as its extrinsic specifier" (ibid., p. 51). Deely further explains that a fossil bone is not a sign of a dinosaur if discovered by a gardener who knew nothing of paleontology. Because the interpretant is missing, the sign relation is incomplete. However, if the bone is spotted by a paleontologist, his professional training (interpretant) would be activated by the sight of a bone, and he would recognize the bone as a sign of a dinosaur.

If a sign is communicated to mind, Deely [7.16], p. 58-60 wrote, then it is based on the interpretant (or thirdness) of experience, and if a sign is not communicated, then the interpretant is based on the laws of nature. In other words, Deely thinks that physical causation is a kind of interpretation mediated by the laws of nature. The fossilized bone is of course not the same bone that was part of a dinosaur because its organic components have been replaced over a long time by external mineral compounds. However, the form of the bone is still preserved and indicates to mind that it was a part of a dinosaur skeleton. Deely describes it as *virtual semiosis* that includes physical forces (i.e., secondness):

> In this way the interaction is a virtual semiosis — that is, a series of interactions at the level of Secondness that, at the same time, provides an actual pathway through time whereby it is possible that what happened long ago might be partially understood. Indeed, from such a standpoint, the present might be regarded as a mosaic of traces from the past, each providing, for a sufficiently knowledgeable present observer, the starting point of a journey into what used to be. Notice that the possibility in question need not be actualized in order for it to be virtually present; nor is it merely "possible" in some abstract, conceptual sense [7.16], p. 58–59.

Deely (ibid., p. 60) also assumes that fossilization of a bone is similar, in a semiotic sense, to the projection of the image of an object to the retina in the eye because both processes transfer the form to a new medium, and in both cases, the form can signify and stimulate the mind. He followed Peirce's idea that physical laws are semiotically equivalent to habits of organisms, and applied the notion of virtual semiosis to the whole physical world: "Yet it is clear that, within our experience, the world of stones and stars precisely exists as a constitutive part of the semiotic web which our experience is, however that world may further or previously exist as well" (ibid., p. 55). In particular, he used the notion of virtual semiosis to explain the evolution of the physical world: "Virtual semiosis is also at work in the ways present interactions anticipate future conditions radically different from what presently obtains. In other words, present effects are virtual signs prospectively as well as retrospectively" (ibid., p. 60). Deely also thought that virtual semiosis at some point created such conditions in the universe that facilitated the origin of life: "For, over and above the individual interaction of bodies, there is a macroformation of the universe that takes place, over-all, directionally, toward the establishment of conditions under which virtual semioses can give place to actual ones. Physiosemiosis, let us say, moves in the direction of biosemiosis" (ibid., p. 61).

The idea of virtual semiosis (or physiosemiosis) is generally consistent with the traditional practice in semiotics, where signs are considered meaningful even in the absence of interpreters (e.g., natural signs) or when meanings are unknown (e.g., symptoms of unknown health condition or writings in unknown language). But it is inconsistent with traditional biosemiotics which assumes that semiosis occurs only in living organisms [7.2].

Below I propose a weaker version of biosemiotics, where semiosis is coextensive with semiotic agency (a generalization of organism) rather than with life, and can be studied in actual and potential form. In particular, potential meaning can replace Deely's notion of virtual semiosis.

7.3 Semiotic Agency in Biosemiotics

Biosemiotics brought the idea that sign processes (semiosis) are coextensive with life, which makes a living organism the essential part of semiosis. Thomas Sebeok [7.71], p. 69, wrote: "a full understanding of the dynamics of semiosis" may, "in the last analysis, turn out to be no less than the definition of life." Biosemiotics further integrates semiosis with organisms, which interpret signs, produce signs for communication with other organisms (exosemiosis) and with organism parts (endosemiosis), and preserve memories for future use (mental memory and heredity). However, organisms are not the only systems that are capable of semiosis because interpretation and production of signs also takes place in organism parts and in multi-organism systems such as colonies, populations, and ecological consortia. Thus, it makes sense to use the notion of *semiotic agency as a generalization of organism in semiotic context*. Sharov and Tønnessen [7.83], p. 153, defined semiotic agents (or simply "agents") as systems with "a capacity for acting purposefully and using signs to make informed choices." Examples of agents are organisms, cells, autonomous cell components, colonies, species, ecological consortia, and human organizations. Autonomous human artifacts are also semiotic agents because they are produced for helping humans, and functionally are similar to ribosomes that are produced by cells for making proteins. Although some semiotic agents are not alive, they all depend by origin and/or by function on living organisms, and therefore, the notion of semiotic agency does not contradict the main assumption of biosemiotics that semiosis is coextensive with life.

Sharov [7.79] considered organisms, individual cells, and colonies as *primary agents* because of their capacity of self-reproduction, whereas ribosomes, other functional cell components, and organs in multicellular organisms do not self-reproduce and were called *secondary agents*. Secondary agents usually self-organize from parts produced by other kinds of agents or subagents, but primary agents always appear at the root of the agential production tree. Artificial agents, such as computers and robots, are secondary agents produced by humans and human-made machines, and humans appear as primary agents at the root of the production tree.

Being an agent is a necessary condition to perform semiotic functions. It is associated with such physical features as non-equilibrium dynamics, free energy to perform work, multiple accessible modes of structure and function (e.g., dynamic switches) [7.58]. However, these physical features are just means used by agents to perform their functions, whereas the design and origin of functions is emergent and cannot be deduced from physical laws. The key feature of semiotic agency, which is goal-directedness, is not physical just because the methodology of physics is to study means, not the goals. The notion of goal stands for historically emerged specific integration of all processes in the agency that is self-supporting, propagating, and competitive.

It can be argued that many biological structures and processes have optimal physical properties under certain constraints [7.68], which is often presented as proof that physics explains biological function. However, the optimality of biological structures is not a physical law but rather a result of goal-directed activity of organisms and lineages of organisms. Classical physics is focused on universal laws of change which take place in real situations with specific constraints (e.g., initial and boundary conditions). These constraints are measured in biology and biophysics and used as parameters in equations, but they are not explained by physics.

Agents are not just systems with goals, but systems whose mere existence is a consequence of their goal-directed behavior. This means that goals reshape the way agents exist. As a result, agents have an *extended ontological status* because their existence is coupled with goal-directed meaning-making in a positive feedback loop [7.83]. The idea that observation and interpretation are intrinsically integrated with activity of organisms (e.g., in functional circles) lies at the core of both biosemiotics and enactivism [7.5] [7.79].

Agents differ from merely physical things because (i) with the exception of primordial agents,[3] all agents are produced only by other agents with at least comparable level of complexity [7.77], (ii) agents have the power to control some physical processes in their bodies and in the environment, also known as autonomy [7.31] [7.83], and (iii) advanced agents develop internal models of the world [7.92] [7.93].

Semiotic agents often have multiple levels (or layers) of identity. Long-term identity of organisms is species-specific and it is preserved by hereditary signs mostly in the genome and chromatin. This identity is not affected by the life cycle and is the same in a fertilized egg, embryo, larval stage (if present), and adult stage. Other levels of identity correspond to individual stages of development, sex, and special physiological states related to such factors as stress, pregnancy, disease, or specific environment. These additional levels of identity emerge via developmental or physiological change regulated by various epigenetic processes such as gene expression, chromatin change, secretion of growth factors and hormones, and signaling in the immune system and nervous system.

The majority of semiotic agents have composite nature and consist of many subagents, some of which are morphologically identifiable (such as organs, tissues, cells, organelles), whereas others are integrated functionally via distal connections (e.g., functional subnets of neurons, and immune system). Subagents have their own identity, and collectively, the identity of subagents contributes to the identity of the whole organism. Some subagents are phylogenetically different from the higher-level agent such as symbionts or parasites. All subagents of either similar or different origins persist together in a partially cooperative and partially antagonistic way. But in general, cooperation is more common between closely related subagents whereas antagonism is more common between unrelated subagents

[3] Primordial agents are first semiotic agents with long-term heredity capable of autonomous goal-directed activity and open-ended evolution. Primordial agents were able to appear spontaneously because they were very simple, and their emergence was a very rare event. All subsequent agents were more complex and they reproduced with a steady regularity. It is impossible to point out which system was the "first" primordial agent because its detection depends on selected thresholds of minimum complexity and minimum probability of producing offspring with comparable complexity and similarity. These thresholds should not be too low, so that they are not crossed by non-living systems that cannot evolve. And they should not be too high, because otherwise they will be crossed by systems that have been evolving by natural selection long before we detect them.

(e.g., between a parasite and its host organism).[4] In summary, an agent is both a unity and a community of smaller subagents. But the hierarchy of subagents has a finite number of levels because smallest subagents (e.g., functional protein domains in living cells) do not have autonomous semiotic parts.

7.4 Signs and Meanings as Semiotic Tools

In traditional semiotics, meanings are viewed as mental representations that are separated from objects and events in the real world. Howard Pattee called it the "epistemic cut" — a boundary that separates the observer and the observed, the knower and the known [7.60]. Biosemiotics bridges the epistemic cut by exploring how organisms encode their meanings in their genomes, epigenetic systems, and neural systems. In other words, semiosis appears to be a part of the real world of biology and becomes represented by the functional organization of agents. For example, Kobus Marais [7.45], p. 6, described meanings without anthropocentric connotations as "extended and embodied cognitive work, done to infer the relevance or significance or consequences of any observed or conceived phenomenon for the interpreter." He clarified that meanings are always normative: "I explained elsewhere how significance is related to the interest of the organism and, therefore, always normative, and how this significance can grow in complexity, from a choice to eat or being eaten, to a large and complicated cognitive system like 'the law'" (ibid., p. 7).

Following agency-centered approach in biosemiotics [7.83], meaning can be defined as a capacity of things to be significant (positively or negatively) for agents in certain situations or context.[5] Some meanings are *intrinsic* to things, such as gravity, chemical valence, crystallization, catalysis, or aggregation of hydrophobic molecules in water. Other meanings are *associative* if they are established arbitrarily[6] by agents for certain purposes and functions. Examples of associative meanings are (i) pairing of complementary nucleotides in the DNA, (ii) genetic code that associates triplets of nucleotides with amino acids in protein synthesis on ribosomes, (iii) simple reflexes, and (iv) conditioning in behavior. Life emerged from natural intrinsic meanings, which became supplemented with associative meanings. In complex organisms, meanings are organized into functional networks at multiple structural and functional levels of subagents.

Semantics and pragmatics are two interdependent aspects of meaning [7.75]. Semantics answers the "what" question; it is what kind of entity is significant, or what entity is associated

[4] Closely-related subagents, such as cells within a multicellular organism, tend to cooperate due to centralized self-reproduction at the level of whole organisms. In contrast, closely-related conspecific organisms reproduce independently of each other, and thus, can benefit from intensive competition. But even at the level of organisms, relatedness is a factor that tends to be correlated with cooperation and altruism due to kin selection and other forms of group section.

[5] Significance can be conscious (e.g., simple preference or motivated preference) or unconscious (e.g., innate habits at all functional levels: metabolic, physiological, developmental, and behavioral). Significance often correlates with usefulness because of natural selection; however, evolutionary adaptations can be delayed, resulting in lower levels of correlation. Usefulness can be modulated by interaction of organisms. For example, sexual selection in deer populations resulted in the increase of horns in males, but long horns reduce mobility and therefore increase the risk of males being killed by predators.

[6] Arbitrariness is considered within limits of availability and functionality.

with a sign (reference). In addition, pragmatics answers the "why" question: why an entity is significant. It can be a relation between the shape, structure, and dynamics of an object and properties of other object(s) in the body or environment that enables a beneficial process such as catalysis, form change, or mobility. In biology, such relation or matching of parts to the whole is called "adaptation" or "fittedness" [7.38]. Note that function or adaptation is understood here in a wide sense and is not limited to the interaction of organism with its environment. It is equally important to consider internal functions/adaptations such as selecting a more efficient metabolic pathway or establishing coordination between two gene regulatory networks that were previously independent. In other words, adaptations include both internal coupling of subagents and coupling with the environment. Adaptations arise at multiple time scales in evolution, and development, physiology, and behavior can be improved by internal regulatory capacities, natural selection, and/or learning at multiple structural and functional levels. However, neither natural selection nor reflex would be possible without system-level preferences and interpretations that have formed historically in previous long-term evolution and/or recursive learning [7.83].

Following this understanding, meanings are relational components of the world inhabited by semiotic agents. However, they are not observer-independent entities as in physics but relations linked to individual agents or agent types. In particular, different agents (e.g., animals from different species) endow the same physical thing with different meanings. According to Jakob von Uexküll [7.93], the same meadow with flowers has different meanings for a girl, ant, cow, and cicada-larva that bores a tunnel in the grass stem. Humans can inhibit their own goal-directedness and attempt to reconstruct meanings of other organisms. But even in this case, inferred meanings of organisms do not necessarily coincide with the real meanings of studied organisms.

Meanings are context-dependent because in some conditions they are beneficial for an organism and in other conditions they are either useless or even harmful. Thus, organisms need to perceive signs to know when and where to initiate physiological processes or behaviors. Thus, signs appear meaningful because they guide organisms in making decisions on which actions to select. Here, sign is seen as a material thing (sign vehicle) that is interpreted by an organism or another kind of semiotic agent. Such material understanding of sign does not preclude deeper-level elaborations, such as viewing sign as also a type, relation, or role in certain semiotic processes. Such multilayer terminology allows a deeper understanding of meanings.

Signs and meanings are considered in biosemiotics in a wider sense than in linguistics, and include non-mental signification such as by genes encoding proteins or molecular messengers that transfer signals from membrane receptors to the nucleus of a cell. A biosemiotic "sign is a tool that (thanks to its form rather than matter or energy) prompts agents to perform functional activity" [7.83], p. 344, or change, which tends to be beneficial to the agent itself or associated agents (progeny, super-agents).

Although biosemiotics is compatible with the theory of Charles Peirce, the emphases are different. Peirce emphasized the universality of human logic and the objective nature of interpretation. Because most humans (within the same cultural tradition) interpret signs similarly, the role of interpreter can be reduced to logic. Thus, Peirce attributed interpretation activity to signs rather than to the interpreting mind. John Deely explained Peirce's semiotics as "the study of [...] the behavior or action peculiar to signs" [7.16], p. 48. Peirce's semiotics is largely anthropocentric because he mostly explored human signs that represent distinct theoretical types of sign relations. In contrast, biosemiotics stems from the Umwelt theory of Jakob von Uexküll [7.93], who emphasized the subjective nature of meaning: species of organisms differ in their sense organs and habits, and thus, interpret the same stimuli and objects in the environment differently according to their specific needs and capacities. Because organisms control the way they interpret signs, the notion of agency appears more important than the notion of sign. Thus, biosemiotics attributes activity to whole organisms and their subagents rather than to signs [7.83].

The importance of agents or interpreters in semiosis was emphasized by Charles Morris [7.53] [7.54]. With the rise of biosemiotics, the scope of semiotic theory was substantially broadened, and the notion of agency became firmly established as a part of semiotic theory [7.83] [7.89]. Sharov and Tønnessen wrote that the difference between signs and agents is that the role of signs is to signify, and the role of agents is to interpret signs and act. In some cases these roles are combined as, for example, kinase proteins transfer signals by activating their substrates, and therefore signification is not separated from action. Although very few signs are real sign-agent hybrids (as kinase proteins), all signs prompt agents for certain activity, physical or mental. This, however, does not imply that the sign itself is active; instead, it is the complex of agent + sign (or subagent + sign) that becomes active. For example, the light image on the retina of the eye interacts with photoreceptors, and the result of this interaction is a generated pattern of neural impulses that travels to the visual cortex in the brain. The pattern of impulses is processed by neurons in the brain, and if it matches the visual image of a familiar object (e.g., food) then some neural circuits get activated to initiate specific habits associated with this object (e.g., feeding, salivation).

Another example of signification is a glucose molecule that binds the outer part of a chemoreceptor in the membrane of a bacterium. After glucose binding, the chemoreceptor changes its conformation (i.e., identity) and activates a specific kinase protein inside the cell. This action initiates a cascade of signaling that eventually affects the rotation of flagella to move the bacterium towards increasing abundance of glucose, a phenomenon known as chemotaxis [7.66]. Although glucose on its own is a passive sign, it changes the conformation of chemoreceptor, which results in a change of receptor identity. A complex of receptor + glucose becomes a subagent within bacterial cell that mediates chemotaxis. These examples show that Aristotelian formal causation is not sufficient to describe semiosis. In addition, a specific interaction with agent is needed for signification, and most agents interact with signs cautiously to avoid possible deception. Agents keep some level of control over their interaction with signs, and usually do not let signs act on their own.

7.5 Conceptualizing Potential Meanings and Signs

The assumption of traditional biosemiotics that signs and meanings exist only in living systems [7.2] seems contradictory to the existence of natural signs, many of which are not produced by organisms (e.g., position of the sun, gravity, temperature). The existence of natural signs is accepted in most semiotic theories [7.29] [7.49] [7.72], and thus it needs to be explained in biosemiotics. The simplest way to reconcile biosemiotics with the existence of natural signs is to emphasize that sensing and interpreting signs requires organism activity, and thus the result of interpretation is organism-dependent. However, Peirce insisted that natural signs are mind-independent; he wrote:

> Of this sort are all natural signs and physical symptoms. The index has no generality in itself. It does not depend on a mental association, but upon a real reaction between the mind and the external world at the moment when the index acts [7.63], CP 3.361.

Now we know that there is no direct reaction between the mind and the external world; instead, this reaction is mediated by receptors, neurons, and synapses which are not the same in different organisms. Thus, although the external world has its objective properties, sensing and interpreting of natural signs is species-specific and can also be organism-specific. In particular, indexes are not universal across species: bees or ants would not understand the meaning of a pointed finger or directed human gaze.

Peirce's notion of mind-independent interpretation of natural signs cannot explain the existence of additional natural meanings. For example, for most humans, a bright spot in the night sky means a star. But people familiar with astronomy can also differentiate between stars and planets. Moreover, an astrophysicist will add that the bright light of a star indicates very high temperature, and the energy that heats the star comes from the nuclear fusion reaction. Thus, being a true star (not a planet), high temperature, and nuclear fusion are additional meanings of the same natural sign (bright spot in the sky) that are not available to uneducated people. These are *potential meanings* because there is a possibility of their actualization. In this example, observers have a possibility to gain knowledge about stars, and then potential meanings will become actualized for them. Similarly, in the Deely's [7.16] example (see above), a fossil dinosaur's bone in the ground means only a weird-shaped rock for the gardener. The potential meaning of being a dinosaur's bone can be revealed by the gardener if he learns about paleontology and fossil bones. The advantage of education is in converting potential meanings into actual ones, which increases the semiotic competence of the agent.

There is a long tradition in science to account for potential events or relations, such as, for example, field theory in physics or valence in chemistry. Thus, I think that the notion of potential meaning fits this tradition in a legitimate way. In particular, potential meaning can be considered as a semiotic field that changes the activity of competent agents. As electric field is measured by placing a charged object into it, potential meaning of a thing can be evaluated by placing an agent nearby and observing its response. In such experiment, meaning is actualized because it becomes interpreted by a semiotic agent. An important difference between physical and semiotic fields is that a physical field is expressed in the magnitude of force, whereas in semiotic field, the pattern of forces often appears more important than the magnitude. Semiotic impacts include some forces; they may be weak but should be sufficient to activate receptors. Then, these impacts are selectively transformed and amplified by agents, and selection criteria can include matching patterns or successful interaction. For example, protein chemoreceptors recognize the shape of incoming molecules by tight binding to a matching 3D protein binding site. Photoreceptors detect photons within a specific range of wavelength. Animals are able to recognize large-scale spatial and temporal patterns (visual, acoustic), as well as associations between different stimuli.

An additional complication is that a charged particle would immediately move in the electric field, whereas an organism may continue staying in place even after perceiving a stimulus. The behavior of an animal is more complex and does not always reflect perceived meanings. Additional factors to consider are subagents of the organism (e.g., sense organs and receptor cells), as well as the context of prior sensations and historically accumulated mental models of the world. Such an agency-based approach to potential meanings is different from Deely's notions of "virtual semiosis" or "physiosemiosis," which assume that interpretation is grounded in physical causation alone (e.g., fossilization of dinosaur's bone).

To claim the existence of reality with its intrinsic differences it is necessary to find reproducible responses of many different semiotic agents (and subagents) to the same semiotic field. Examples of such experiments are mazes for detecting smell, sound, light, or color preferences in animals where immediate and lagging behavioral and physiological responses are recorded [7.65] [7.74] [7.95]. Such experiments generally prove that the differences between meaningful physical conditions are real and reproducible. Simultaneously, they allow researchers to classify animals into those that do (or do not) respond to these differences, and find causes of these differences. Experiments with animals are, however, difficult because animal responses depend on such factors as species, sex, genotype, memory, and physiological state. Thus, physicists prefer using external sensor instruments to minimize subjectivity in detecting reality. Because such sensors are produced by humans, and observation results are recorded and interpreted by humans, sensors can be treated as external organs or subagents of humans. Because of human involvement, there is a possibility that human knowledge is species-specific. However, most scientists believe that scientific data is not species-specific and generally represents the reality that could have been confirmed by non-human rational agents if they were present. Such a belief is based on well-established scientific practice that uses reproducibility, statistics, and consistency with other phenomena as criteria of objectivity. Scientific practice also implies a critical stance towards already accumulated human knowledge and detachment from innate intuition, cultural norms, and folk beliefs.

The notion of potential meaning can be further applied to hypothetical things that have not been observed before; for example, composition of rocks or atmosphere on an exoplanet. In this case, it is we (humans) who do not have sufficient knowledge to uncover potential meanings, such as habitability of the planet or usability of sources of energy. Some potential meanings can be uncovered from better models of planet formation, however logic alone is generally not sufficient for obtaining new knowledge. Stuart Kauffman [7.34], p. 7, formulated this idea: "[t]he world is not a theorem," and explained further: "we claim that the morphological and behavioral evolution of the biosphere, besides the impossibility of being entailed [...] is inherently not mathematizable in terms of sets." Thus, empiricists were deeply right in the primacy of experimental research. All model-based predicted truths need to be validated before we trust them.[7] Experiments on projected exoplanet conditions can be conducted by human researchers with their instruments, but the results may appear incomplete and some potential meanings can be missing. Thus, it is better to employ other organisms for evaluating novel environments for habitability. For example, various species of bacteria can be grown in simulated exoplanet conditions. It may happen that bacteria will find new metabolic solutions to survive in such conditions, and educate us on how to exploit natural resources on the exoplanet. In such experiments, new potential meanings are revealed by competent semiotic agents (bacteria) used as measurement devices.

---

[7] Most animals use predominantly innate models in their behavior, and keep trying the same strategy even if it failed repeatedly. Cautious use of acquired models of behavior is found mostly in conscious animals (vertebrates, some arthropods and cephalopods) [7.28].

If we recognize the existence of natural meanings in the physical world, then nature appears potentially cognizable. For example, the past can be inferred from the present, as advocated by geologist Charles Lyell [7.42]. Knowledge of causation relations allows us to reconstruct possible causes of present conditions. Similarly, future states of the world can be predicted from trends at the present state (e.g., climate change). We can hypothesize that natural objects even have such features that we are not aware of. Richard Feynman wrote: "But see that the imagination of nature is far, far greater than the imagination of man. No one who did not have some inkling of this through observations could ever have imagined such a marvel as nature is" [7.24], p. 10. But these unknown features are not yet meanings because we don't know what they are.

The idea of potential meanings supports realism in philosophy because it assumes that object types, their properties, and relations are not just arbitrary constructs of mind but they are found or discovered in the real world [7.51]. Such realist position was defended by Peirce [7.11] and Deely [7.17] [7.18]. However, semiotic realism can be based on supernatural explanations, for example by assuming the existence of global universal mind (in panpsychism or pansemiosis), or by inferring religious doctrines. I avoid such interpretations and ground the objectiveness of human knowledge in scientific practices, such as reproducibility and statistics. Thus, meanings are not objective unless they are explored thoroughly.[8] Then, after validation, objective meanings can be extrapolated to the world, where observer agents are not present, including the past and future. In particular, it is reasonable to consider things that carry objective meanings as signs if they signify to agents, or as potential signs if they do not signify now but have a capacity to signify later.

Such ontology of meanings is agency-dependent, similar to radical constructivism [7.91], but in addition, meanings are pragmatic and become corrected through repetition of the corresponding functional cycle. It is initially subjective but can spread via interaction and communication and become objective within the community of agents. Such collective knowledge is updated with the emergence of each new kind of agency: first, actual meanings are updated in space and time proximity of these new agents, and second, potential meanings are updated in the whole universe, as well as in the reconstructed past, and in the projected future.

Deely [7.18], p. 29-30, proposed that human semiosis is qualitatively different from animal semiosis because humans can reach the objective status of things, whereas animals rely on phenomenology where objectivity is incomplete. I agree that humans have substantially greater semiotic capacity than animals, but this difference is not as sharp as portrayed by Deely because human knowledge is also incomplete and is partially based on phenomenology. High level of objectivity is achievable in such spheres of human knowledge as classical physics. But studies of complex systems, both natural and artificial, are less objective because these systems are highly uncertain and allow multiple partially incompatible explanations. For studying such systems, phenomenological approaches may appear more useful than causation-based physical knowledge and should not be discarded. Humans routinely use phenomenological knowledge in their everyday activities without any noticeable reduction of success as compared to robots controlled by causation-based mathematical models of physical motion. Knowing animal habits is sufficient for most human interactions with

[8] This criterion, however, does not restrict using the term "meaning" in relation to individual cognition.

them; knowing exact molecular-level mechanisms of these habits and evolutionary mechanisms that supported them is not that important, except for specific research projects.

One of the factors that contribute to the partial limitations of meaning is the ambiguity of human language. This is how Sebeok and Danesi describe the uncertainty of human language and compare it with the Heisenberg's uncertainty in quantum mechanics:

> Suppose that a scientist reared and trained in New York observes a physical event that s/he has never seen before. Curious about what it is, she takes out a notebook and writes down her observations in American English. At the instant that the American scientist observes the event, another scientist, reared and trained in the Philippines who speaks only the indigenous Tagalog language, also sees the same event. The Philippine scientist similarly takes out a notebook and writes down her own observations in Tagalog. Now, the question is: To what extent will the contents of the observations, as written in the two notebooks, coincide? The answer, of course, is that the two sets of observations will not coincide completely. The reason for this is not due, clearly, to the nature of the event, but rather to the fact that the observers were different people, and that the representational systems used (English vs. Tagalog) provided each scientist with different verbal forms for characterizing the event. So, as Heisenberg aptly suggested, the true nature of the event is indeterminable, although it can be investigated further, paradoxically, on the basis of the notes taken by the two scientists [7.73], p. 8.

Even a simple statement that the Earth circles around the Sun can be misinterpreted in many different ways depending on the beliefs and level of education. For people who do not know that the Earth and Sun are cosmic bodies, this statement is meaningless. Others may consider that the trajectory of Earth around the Sun is a perfect circle, which is wrong because the real trajectory is an ellipse. Nevertheless, prolonged communication between people with different cultural backgrounds may lead to better mutual understanding that bridges semantic gaps. Does this mean that human knowledge can be understood correctly by equally intelligent aliens from other stellar systems? We don't know, but it is conceivable that such communication is possible if the interaction is peaceful and mutually encouraged. These considerations support the notion that human knowledge tends to be objective rather than being narrowly species-specific.

Because the objectiveness of meanings is only partial, there is always some degree of *uncertainty*. The most simple type of uncertainly comes from misidentification or inaccurate measurement of things. For example, from a large distance, a mocking bird can be misidentified as a dove, or measles can be misidentified by its symptoms as flu. In these examples, it is possible to repeat identification with alternative or modified methods, such as coming closer to the observed bird or using the PCR test to identify the virus. In other cases, identification cannot be updated because the object is either unreachable (e.g., Schrödinger's cat) or it has changed after the initial identification procedure (e.g., Heisenberg's uncertainty). Required measurement may also appear not feasible (e.g., measuring membrane depolarization in all neurons of the brain), resulting in semantic uncertainty. Expressing results of measurements in human language, transmitting them to others, or translating to another language may add another layer of uncertainty to meanings.

Despite partial uncertainty, potential meanings gain objectiveness via the spread of actual meanings (i.e., knowledge, in a wide sense) between semiotic agents. As I wrote above, potential meanings of the physical world are grounded in the human knowledge, but this scientific knowledge emerges first in a single head or single lab. Then it starts spreading: in the community of scientists of one country, of many countries, and then among the general public. Knowledge obtains the status of objectiveness only after sufficient spread. Potential meanings were first discovered by non-human organisms (e.g., photosynthesis, or a capacity to swim), and then this knowledge of combined natural and associated potential meanings has spread to other organisms of the same population, and then to other species. The pathways of spread are various: heredity, horizontal gene transfer, physiological regulation, symbiosis, imitation, communication, and teaching. Humans inherited a large set of these meanings from hominid ancestors, and then generated an entirely new set of cultural meanings supported by language, communal work, and later by means of experimental science: physics, chemistry, and biology.

When potential meanings are actualized and then spread among agents, they change due to the contribution of new context and new goals. Thus, the spread of potential meanings is not a simple copy-paste process but a translation from one system of meanings (an analogue of language) to another [7.39] [7.44] [7.45] [7.80]. According to Yuri Lotman, translation is one of the central processes that drives the dynamics of semiosphere,[9] working not only between languages, but also within the same language [7.41] [7.52]. Lotman argues that translation helps to unify common knowledge (homogenizing effect), and also to define smaller domains in the semiosphere where new contexts and new knowledge starts crystallizing (heterogenizing effect). Jesper Hoffmeyer [7.31], p. 5, proposed a wider notion of semiosphere that encompasses sign processes of all living organisms: "[...] life on Earth manifests itself in a global and evolutionary semiosphere, a sphere of sign processes and elements of meaning that constitute the frame of understanding within which biology must work."

Because of the spread of knowledge, meanings once actualized keep existing and evolving. Although extinctions happen, the semiosphere as a whole, preserves the majority of meanings that once appeared in a single population. But not all meanings are preserved in their original context because most of them went through multiple rounds of translation. Moreover, meanings can spread beyond planetary limits because some simple organisms, such as bacteria, can be ejected into space with rocks after collision of planets with comets or asteroids [7.22]. Bacteria can survive long cosmic travels within asteroids or rogue planets that do not belong to any star system. There is indirect evidence that Earth was contaminated with microbial life originated from other star systems — a hypothetical phenomenon of panspermia [7.33] [7.81].

The main conclusion from this analysis of potential meanings is that knowledge is carried by all semiotic agents, not just humans. Through spread and translation to other agents, knowledge has a tendency to become objective, but it is always in the making and is never completed. The core areas of knowledge within the semiosphere appear relatively stable and homogenic, whereas more heterogeneity and semiogenesis appears in newly created

---

[9] Lotman [7.41], p. 205, wrote "Semiosphere is the [human cultural] semiotic space, outside of which semiosis cannot exist." "The concept of semiosphere is linked to a definite semiotic homogeneity and individuality" (ibid., p. 208).

domains. But the problem of meaning change or semiogenesis [7.83] is beyond the scope of this chapter.[10]

7.6 Potential Meanings and Complexity of the Non-Living World

It is well known that life on Earth evolved through billions of years towards higher complexity. Prokaryotes evolved into much more complex eukaryotic cells, which eventually turned into multicellular animals and plants [7.7]. The genomes of organisms became longer with time and harbored more genes and regulatory elements [7.46] [7.76]. Ecosystems also increased their complexity because organisms keep creating and occupying new ecological niches. Naturally, there is a question whether such trend of increasing complexity is unique to life or may also exist to a minor extent in the non-living world?

Physics does not support the increase of complexity in non-living systems. In particular, the second law of thermodynamics tells us that closed systems change only towards degradation as their entropy increases irreversibly. New hopes for the possibility of the progressive increase of complexity in the non-living world came into sight after the discovery of self-organization in open thermodynamic systems. The phenomenon is also known as "order from noise" [7.90] and "order from fluctuations" [7.55]. Examples of self-organization in non-living systems include crystal growth and dissipative structures. At cosmic scales, self-organization includes formation of galaxies, stars, and planets. In geological formations and atmosphere, self-organization processes are exemplified by continental drift, crust recycling, hydrological cycles, hurricanes, and tornados. These phenomena depend on the continuous supply of energy and resources and disappear after exhausting available resources.

The scenario of progressive increase of complexity in non-living systems has been repeatedly advocated by John Deely as a possible explanation of the origin of life:

> Out of cosmic dust, stellar systems form through subatomic, atomic, and molecular interactions. At various stages of the process, even as now on earth we can in laboratories bring into being a few elements not existent in nature itself, new elements not previously given in nature precipitate from the natural interactions. These new elements, in turn, prove essential to the formation in planetary systems of the conditions under which living beings become possible, and these beings, in turn, further modify the planetary conditions so that successive generations of living beings are incompatible with the original conditions of life [7.16], p. 61.

Later he compared the origin of life with recurrent attempts to light a cigarette:

> Those intermittent gradual steps whereby a virtual semiosis latent in the purely physical interactions of the components of the universe moved the universe here and there closer and closer to the possibility of life I might compare to the repeated unsuccessful attempts of a smoker to light his cigarette [7.18], p. 38.

[10] According to Brenner and Igamberdiev [7.9] [7.10], forms of existence (or "ontolons") exhibit potential dynamic multiplicity in living and even in physical systems, or, in other words, they work as "seeds" to produce derivative forms. From the perspective of Kauffman [7.35], these derivative forms represent the "adjacent possible" — a manifold of future directions of change.

Similar ideas were developed later by George Mikhailovsky [7.50], who assumed that chemical complexity of the universe increased before the origin of life and led to the emergence of the first living systems as a result of megaevolution or hierarchogenesis.

To explore the complexity of physical systems, it is first necessary to define what complexity is. The complexity of organisms has been evaluated by the number of its non-redundant functions or functional components, such as nucleotide sequences, genes, or proteins [7.1] [7.62]. In physical systems, however, complexity is more difficult to quantify because physical systems have no functions (in the biological sense), and thus it is not clear what elements to count. Here I propose to quantify physical complexity by the number of *non-redundant and stable potential meanings* related to possible primordial systems that started evolution towards full-scale life in a specific location (e.g., planet, geological formation, lake, or hydrothermal vent). Stability, whether it is static or dynamic (as in dissipative systems), is important because non-stable potential meanings may vanish without a trace, and thus do not contribute to the subsequent evolution of physical complexity. With this approach, the complexity of a physical system is composed of the number of types of atoms, molecules, and larger associations between molecules, such as separation of phases, magnetic domains, or lipid membranes. To exclude redundancy in the set of possible meanings, I suggest counting only those differences between them that make a difference for possible primordial agents, following Bateson's approach to information [7.6], p. 459. If two things (or stimuli) cannot be distinguished by projected primordial agents, then their difference does not contribute to physical complexity.

This semiotic approach to physical complexity requires that primordial agents can distinguish between potential meanings and use them selectively as resources or signs if brought into contact with them. The possibility of distinguishing potential meanings by these agents and exploiting them as resources may also depend on the cost-benefit balance. For example, neutrino is not likely to be used as a resource or signal because neutrinos are rare and difficult to detect. Similarly, it is impractical to use unstable atoms or molecules as resources. Thus, physical complexity is a pragmatic category that needs to be weighted by some reasonable level of effort or cost.

I think that Deely's hypothesis of physical evolution towards higher complexity has a chance of gaining theoretical support if complexity is measured as described. Another suggestion is to use this approach at substantially larger spatial and temporal scales than those used in typical laboratory or simulation studies. Many non-living things have a capacity to produce other things via such processes as gravitational attraction, collision, sedimentation, self-organization, oxidation, catalysis, or nuclear reaction. One of these processes — the accumulation of heavier elements, including those that are needed for life (C, O, N, P, S), via nuclear fusion in stars — has been well studied at cosmic scales [7.48]. After these elements are produced, they persist for billions of years and become dispersed by supernovae explosions. Eventually they are recycled in new cosmic bodies, including planets. An important source of complex organic molecules that may contribute to the origin of life is the interstellar medium (ISM) [7.15] [7.21]. In particular, polycyclic aromatic hydrocarbons (PAHs), which may serve as catalysts or sources of carbon, are abundant in the ISM, especially around carbon-rich stars [7.20].

All production processes in the universe change the balance of potential meanings in each cosmic object (e.g., planet, star, or comet) or in distinct parts of these objects (e.g., surface of a planet, sea bottom, or lake). Thus, it seems possible that new elements, molecules, and aggregates of molecules accumulate in certain locations. If the concentration of unusual atoms and molecules reach thresholds for catalysis or crystallization, the next generation of new components may arise adding more complexity to the entire chemical "community." It is also conceivable that local pockets of complexity can split, move around, and keep their identity for some time. Most of these pockets disappear, but some persist, and may generate second-order pockets of somewhat higher complexity. This kind of physicochemical evolution is not semiosic yet because there are no hereditary signs in the physical world and no signification.[11] The rates o f complexity increasing in such locations are most likely low, and possibly, there is an upper limit of complexity that can be reached, because most locations with increased complexity eventually deteriorate or get destroyed by larger-scale disturbances.

Increased physical complexity of local environment is likely to facilitate the origin of life, as suggested by Deely [7.16] [7.18]. However, this does not mean that the most difficult task in the origin of life is performed by physical factors. In particular, there is no evidence of the existence of primordial soup, where all necessary organic molecules are present at physiological concentrations [7.78]. The only abundant organic molecules in the universe that can be used as resource for emerging life are liquid hydrocarbons and PAHs. Other organic molecules specific to life, such as carbohydrates, amino acids, nucleic acids, and phospholipids, are present in space at extremely low concentrations and thus cannot be effectively captured by primordial agents as resources. Thus, these molecules became locally abundant only after the origin of life, when the biochemical pathways towards their synthesis emerge via natural selection [7.78].

7.7 Actualization of Potential Meanings by Organisms

7.7.1 The Origin of Life

When life emerged, natural potential meanings became assimilated by semiotic agents as components of their metabolism and self-production activity. As a result, potential meanings turned into actual meanings, and carriers of potential meanings became sign vehicles (or shortly – signs). Applying Deely's interpretation of Peirce's triadic sign relation, a material sign vehicle (e.g., a resource molecule) got associated with a specific object (e.g., a product of reaction or interaction) that appears significant for the agent in relation to its goal-directed activity, which constitutes thirdness (or interpretant) [7.16].

Let us consider how such actualization of potential meanings happened at the origin of life. Because primordial agents have a chemical nature, we are looking for molecules that are able to produce other molecules, such as catalysts that support fusion or modification of substrate molecules. Before the origin of life such catalysis had not been goal-directed. To achieve goal-directedness it is necessary to transform catalysis into self-production. But this task is too complex for a single catalyst molecule. The only way to succeed is to establish a sort of "cooperation" between several molecule types and molecular aggregates that modulate the actions of each other. Assuming that one kind of molecule managed to achieve

[11] Although this process is not semiosic, studying potential signs and meanings should be a part of semiotics as a discipline.

self-production in such cooperative environment, it becomes a replicator that exploits some already existing potential meanings, such as natural production capacities of matter (e.g., catalysis, specific binding, and scaffolding) for its own functional "needs." The evolution of such system starts when small changes in replicators and/or their local conditions become amplified if they facilitate the recruitment, self-production, and/or spread of these replicators. When formed, such circular causation units can function as hereditary signs capable of supporting a primordial version of Darwin's natural selection.

As described in the chapter "Chemical Origins of Life, Agency, and Meaning," first primordial replicators were not genes and even not heteropolymers[12]; instead, most likely they were simple catalytically-active molecules resembling coenzymes, vitamins, or amino acid [7.78]. Such coenzyme-like replicators (CLR) require carbon and energy for their self-production. Carbon-containing gas molecules (methane or carbon dioxide) cannot be effectively captured and utilized by CLRs as a source of carbon. Thus, the most likely source of carbon for first replicators was oil (hydrocarbon) droplets in water. It can be hypothesized that CLRs managed to attach to the surface of oil droplets and use oil as both a habitat and a source of carbon. Also, oil droplets adsorb hydrophobic aromatic hydrocarbons,[13] which can capture energy by photoactivation. Hence, CLRs can use this energy to facilitate various chemical reactions. In particular, this energy could have been used for oxidation of the outer layer of oil, converting hydrocarbons into fatty acids and other amphiphilic molecules. Amphiphilic molecules reduce surface tension in oil droplets, which in turn can facilitate self-production of replicators, and increase the chances of oil droplet of splitting into smaller parts, which is analogous to cell reproduction. Descendent oil droplets (after division) carrying CLRs can merge with clean oil droplets, which is analogous to capturing new resource [7.78].

Such system with non-polymeric replicators that depend on cooperation with several other molecule types is similar to Eigen and Schuster's hypercycle [7.23], although it is a simpler system with only one or few replicators (CLRs), whereas a hypercycle has many polymeric replicators. A single replicator can establish cooperative relations via *niche construction*, where the local niche (e.g., oil droplet) provides cooperating components. Components of such agent-niche coupling are discussed by Howard Pattee, who assumes that the meaning of first molecular messages depends not only on their own properties, but also on the properties of their ecosystem serving as context or analogue of "language" [7.59]. However, Pattee does not mention niche construction. In another paper, he proposes that primordial molecular symbols are grounded on folding of proteins and non-coding RNA [7.61]. Although folding of macromolecules undoubtedly participates in primitive semiosis (protosemiosis) in living cells [7.84], it is irrelevant to the problem of the origin of life because nucleic acids and proteins did not exist before the origin of life and appeared much later in evolution powered by natural selection [7.78].

[12] The main reason why traditional scenarios of the origin of life (e.g., RNA world, protein interaction world, and lipid world) are not acceptable is that these scenarios assume abundant and versatile organic molecules (amino acids, sugars, nucleodites). The hypothesis of "primordial soup" containing these molecules at physiological concentrations is unrealistic and misleading.

[13] For example, polycyclic aromatic hydrocarbons (PAHs) are highly abundant in the interstellar medium [7.20] [7.21] and in carbonaceous meteorites [7.15].

In this hypothetical scenario, catalysts-replicators play the role of semiotic agents because they enhance their activity (self-production and catalytic oxidation of hydrocarbons) by molecular recognition (e.g., specific binding to the surface of oil droplets modified by oxidation, and capacity to become activated by energy-rich molecules), that is functionally comparable to sensing. Self-production of CLRs is another kind of semiotic process that preserves variation either of molecule structure or its interaction with helper components of the evolving system. This enables natural selection among competing lineages of oil droplets populated by CLRs, which leads to improved functions and faster propagation. Thus, CLRs simultaneously play the role of hereditary signs and self-interpreting agents.

An important component of the system is local environment (oil droplet), which is modified by replicators to their own benefit. Thus, the local environment (and its components) becomes another hereditary sign, which signifies when the modified environment facilitates survival, self-production, and spatial propagation of descendants.

Individual components of this system may have existed before the formation of agents and hereditary signs, but then they were only potential signs. After integration into a self-amplifying functional circle, potential signs started signifying on a regular basis and turned into actual signs. Only specific meanings were actualized in this process—those that became components of the functional circle. Thus the process of actualization of potential signs is simultaneously the process of semiotic selection. Non-actualized potential meanings can be carried by evolving agents for a long time without being utilized. And then some of them can become actualized at a later evolutionary event when an agent manages to make them useful for a new function.

7.7.2 Potential Signs in Evolution

As primitive organisms expand their metabolic repertoire, they actualize an increasing set of potential meanings. For example, the simultaneous presence of hydrogen sulfide and carbon dioxide in water allowed living systems to develop chemosynthesis to produce sugars and amino acids [7.86], p. 179. The presence of carbon dioxide and radiation from the Sun were important factors for the emergence of photosynthesis. At later stages of evolution, the presence of land above the surface of the ocean prompted the migration of organisms out of the water to the land, which occurred independently in many lineages of organisms, both animals and plants. Similarly, the presence of a sufficiently dense atmosphere was necessary for the development of flight in insects, reptiles, birds, and even some mammals. Whenever an organism develops a new adaptation to the environment or to internal factors, it converts some potential meanings into actual meanings (i.e., new functions).

Besides assimilating natural capacities of things (potential meanings), organisms started making new artificial components that never existed before. Some of these components are produced unintentionally[14] as a byproduct of other activities; these have no internal functions when they first emerge[15] but they can become meaningful as signs for other organisms. Other artificial things are produced by organism on purpose (intentionally)[6] if they

[14] Here "intentional" means goal-directed communication, where goals can be organism-related or lineage-related. In this sense, intention is not necessarily mental, but includes evolutionary and developmental adaptations.

[15] However, these natural signs can later acquire functions for some other species of organisms.

are products of learning, development, or evolution. The reason why purposeful things are produced is because they are meaningful for an organism, which "expects" (not necessarily in the psychological sense) them to be useful for improving certain physiological or behavioral functions. Some of these artifacts are resources, tools, or scaffolds that are necessary for various physiological functions (e.g., digestion, locomotion, perception). Others are signs that regulate the function of subagents (e.g., growth factors or neural signals) or used for communication—to modify physiology or behavior of other organisms (e.g., pheromones, communication sounds, or mimicry).

Production of potential signs by organisms is always associated with a risk of functional uncertainty because several factors may affect the success of interpretation. First, a potential sign may fail to encounter a competent interpreting agent or subagent. Second, the sign can be intercepted by an enemy and used against the producer of the sign. For example, pheromones of bark beetles attract predators [7.4]. And third, it may happen that a sign is properly interpreted but this interpretation failed to bring anticipated benefits because of changed conditions, both external and internal.

Some kinds of signs produced by cells signify immediately after their production (e.g., hormones or cytokines), and thus, they act as signals and do not require memory. Fast signification can be supported by abundant receptor proteins in target cells that interpret these signals. Another mechanism of immediate signification is based on the activity of signaling molecules themselves. For example, kinase proteins in cells specifically bind to their target molecules and phosphorylate them. Thus, kinase molecules combine two roles: sign and agent, which appear inseparable because signification depends on the catalytic activity of the molecule. This mechanism is especially important for secondary messengers that transfer signals from membrane receptors to the nucleus of cells.

7.7.3 The Phenomenon of Memory

Most regulatory signs in organisms, however, remain inactive and therefore they are potential signs that are reserved for future use; these kinds of signs are known as *memory*. Because memory signs are normally inactive, they require an interface composed of actual signs capable of activating them. Memory includes at least two levels of signs: normally inactive potential content-signs and a network of pointer-signs that provide links to content-signs, so that *following pointer signs, agents or subagents can activate content-signs and retrieve their meaning*. By pointing I mean an efficient and specific method by which a subagent can reach the target content-sign, which then can be followed by revealing the content. Such method can be represented by a physical connection comparable to Peirce's index (e.g., a synapse that connects neurons). Another method of pointing is by using a 3-dimensional pattern of the protein as bait that binds specifically to the target molecule (e.g., protein or nucleic acid). This method of pointing is dominant among the intracellular signaling pathways, and resembles Peirce's icon because of its pattern-matching mechanism. Some pointing methods combine multiple mechanisms, both indexical and iconic.

The crucial feature of memory is that potential content-signs can be activated for interpretation (i.e., linked with interpreting agents) on demand. This job is done by subagents when they follow pointer-signs. Examples of mediating subagents in cells are transcription factors, transporter proteins, and translation factors. In the nervous system, mediating subagents are neurons. Some pointer signs (e.g., protein receptors) are signifying continuously, but other more specialized layers of pointer signs are activated on demand and communicate processed signals further down the network. In simple organisms, content signs execute and coordinate specific functions, such as duplication of DNA, secretion of hormones, or muscle contraction. In conscious animals, content signs can support categorization of outside objects (e.g., as food item or predator). Mental signs often combine content and pointer roles; this makes the memory structure much more complex than in simple organisms.

Let us consider three examples of memory at low, medium, and high levels of complexity. The first example is regulation of sensitivity of bacterial chemoreceptors, where chemoreceptors are methylated after their binding to attractant (e.g., glucose). After repeated bindings to glucose, the receptor gets heavily methylated, which affects the physical conformation of the protein and reduces the sensitivity of receptor [7.37]. In other words, chemoreceptors keep memory of their previous sensing activity to adjust their sensitivity. This mechanism has many useful functions. It prevents overstimulation of receptor, and makes it more sensitive to an increase of glucose concentration, allowing bacterial cells to move across the gradient of concentrations. To avoid over-methylation of the receptor, there is also a stimulus-independent process of demethylation (removing methyl groups), which restores the sensitivity of receptor. In this case, methylations of receptor are syntax-signs, and the change of protein receptor conformation is their interpretation. The sensitivity of receptor is a content-sign that affects the flow of signals from the receptor to cell nucleus. The chemoreceptor memory is simple because: (i) it is only quantitative and no qualitatively distinct patterns can be retained; (ii) it is short-term; and (iii) the effect of memory is local because memory signs are interpreted by the same molecule instead of being transmitted.

The second example is the genome in eukaryotes which represents multi-generation long-term memory of a medium-complexity level. Genes are fragments of very long DNA molecules that form chromosomes together with histone proteins. The core meaning of genes is the sequence of nucleotides in the "coding region," which is eventually translated into the synthesis of a specific protein on ribosomes. In addition, the genome carries regulatory regions which are needed for DNA replication (copying), maintaining chromosome integrity, and regulating the activity of each gene or a complex of genes. Most genome components are passive potential signs that can activate cell subagents (e.g., transcription factors, RNA polymerase and DNA polymerase complexes) which happen to interact with them. The genome requires regular copying and repair. Every cell division is accompanied with replication of the entire genome. Repair of DNA includes correction of mismatching nucleotides, repair of single-strand and double-strand breaks, resolving knots and other abnormal structures. Regulation of gene activity (expression) is modulated by proteins called transcription factors, which bind specifically to relatively short motifs in the DNA sequence and activate or repress the activity of RNA polymerase. Copying and repair of the DNA as well as regulation of gene expression are syntactic operations that ignore the content of gene sequence. In comparison to chemoreceptor memory, the genome has the following advanced features: it carries long unique sequences of nucleotides, rather than uniform modifications of chemoreceptor proteins; it supports long-term preservation of genes; and the sequence of nucleotides in genes is interpreted far away, in the cytoplasm of the cell. Communication with cytoplasm is achieved by making an RNA-copy of the gene (mRNA) and transmitting it to the cytoplasm, where the mRNA controls the synthesis of polypeptides—precursors of functional proteins. However, the genome has many functional limitations: it is mostly a read-only memory; the dominant type of regulation is based on local events such as binding of regulatory proteins to promoters or enhancers of genes; and the meaning of genes is expressed only via protein synthesis, which is performed by ribosomes. These organelles convert mRNA nucleotide sequences into sequences of amino acids in proteins. In other words, genomic memory regulates living cells mostly via the informational bottleneck of ribosomes.

Finally, the most sophisticated type of biological memory is neural, which is performed by the nervous system of animals. Its molecular mechanisms are not fully uncovered, but biologists know that neural memory is maintained by cells called neurons. These cells have many long appendages, including dendrites that collect information from other neurons or sensory cells and one axon that sends signals to muscles, other organs, or to other neurons. Neurons communicate mostly via electric signals that travel along the membrane surface of dendrites and axons and pass these signals to other cells via synapses—small chambers where membranes of two cells come in close contact and can exchange chemicals called neurotransmitters. Advantages of the neural memory in comparison to genomic memory are the following. First, neural memory is a multicellular system that is serviced by neurons—the most complex cells—in contrast to genomic memory, which is a part of a single cell and maintained by small molecular-scale subagents that are specialized in simple biochemical functions. Second, neural memory is designed by evolution to control complex animal organisms; it is fast, contextual, and reaches every organ, whereas genomic memory controls functions of a single cell. Third, neural memory is writable and editable. Fourth, it is a distributed memory that is not localized in a single neuron. Such distributed architecture supports robustness of function even in conditions of stress or damage. Fifth, neural memory has enormous associative capacity because each neuron in the human brain has on average ca. 7,000 synaptic connections with other neurons [7.30]. Moreover, new synapses can be formed on demand. Finally, sixth, there are strong indications of deep compartmentalization of neuronal activity that combines dendritic computation within individual neurons with interneuron communication via synapse connections [7.27] [7.40]. Dendritic computation can be mediated by G protein-gated ion channels and it was hypothesized to provide modules to resolve the scaling problem of the brain [7.56]. The scaling problem is that doubling the number or recognized objects requires ca. 500-fold increase in computational resources to be implemented. Dendritic computation allows parallel computation at sub-neuron scales, and hence has a potential to overcome these resource limitations.

7.8 Potential Signs and Meanings in Human Life

The process of assimilation of natural potential signs and meanings and production of new signs and meanings has been accelerated with the origin and evolution of human species and human civilization. Non-human animals often use outside objects as resources or scaffolds (e.g., for nest-building), but only a few species can use external objects as *tools* to make nests [7.32], acquire food or water, make a shelter, or use tools for defense [7.25] [7.85]. Hominids started making stone tools more than 3 million years ago [7.26], a capacity that has not been observed in other animals. Tools made of stone, animal bones, wood, and clay have been diversified and improved in quality during the Neolithic period [7.8] [7.19]. Several species of plants became cultivated in the Neolithic period. Some animal species were also domesticated at that time. Intensive use of tools and farming gave rise to many new meanings that were stored not only in the memory of individuals but also in the joint *social memory* supported by such activities as imitation, training, and reciprocated communication.

An important historical change that amplified the use of potential signs by humans was the progress in language and communication. The characteristic feature of human language is symbolic interpretation [7.28], where words are associated with abstract meanings such as concepts. Some animals (e.g., dogs and chimps) can understand the meaning of human words, and this indicates that symbolic interpretation is not limited to humans. Nevertheless, in comparison to animals, human language is a qualitatively new form of communication in terms of its complexity, rich context, and range of generalization. The use of language requires large memory and more connectivity between neurons, which supports more potential signs as part of what we call "human knowledge."

The emergence of written languages and number systems created a new channel of communication mediated by written signs, which made direct human-to-human talk optional. Written communications units are potential signs because they are used on demand and are not interpreted immediately. The major advantage of writing is in the increased reach of disseminated messages, when communication does not require direct human involvement. Another advantage is the increased functional life of messages and increased accuracy of information transfer. Thus, most important human knowledge was transferred to long-lasting media, initially on rock surfaces, clay tablets, animal skin, dried plant leaves (papyrus), and later on paper. These artifacts comprise hierarchically organized potential signs that are interpreted by humans via complex rules that include the alphabet, phonetics, vocabulary (thesaurus), grammar, and syntax. In addition, each written document can be reasonably well understood only on condition of general mental knowledge that is accumulated by individuals through their lifetime. Those written documents that are widely accessible to most people in a society become an important component of social memory within each culture. Initially, such documents were limited to religious books, but eventually most cultures evolved towards open access to the cultural heritage offered at libraries and other institutions.

Each book is a potential sign which turns into an actual sign when it is read and interpreted by a competent human being. In the process of interpretation, the initial agent, which can be called naïve (in respect to the contents of the book), becomes a different kind of agent, such as, for example, a specialist in a new profession.

Human potential signs usually have a specific purpose for which they are produced. For example, a stop sign is placed at the intersection of roads to regulate the passage of cars in various directions and avoid collision. Similarly, a textbook on algebra is produced to disseminate knowledge on operations with numbers and solving equations. However, the meaning and purpose of potential signs can be modified after they are produced, and these signs can acquire additional meanings that were not intended by producers or authors. For example, sacred texts like the Bible or Koran are continuously reinterpreted in various ways depending on the evolving religious and cultural practices.

An important class of human potential signs is quantity, which is an abstract category that generalizes stable relations between things subject to permitted manipulations. The simplest example of a quantity is the number of apples in a basket, which is determined by counting. According to Campbell [7.12], quantities can be expressed by numbers if algebraic operations among numbers represent empirical relations among quantities. For example, the number of apples combined from two baskets equals the sum of apple counts in each original basket. Counting apples is an example of measurement procedure, where numbers are assigned to quantities.

The traditional "realist" philosophy assumes that the result of measurement should be stable if repeated or if the object is moved or disassembled and then assembled back. It is also assumed that the same quantity can be measured by different methods (or devices) and then the results would match if calibrated from one method to another. For example, temperature can be measured by different thermometers (mercury, alcohol, bimetallic) but the result can then be converted to a standard scale (Kelvin or Fahrenheit degrees). Among different scales for measuring the same quantity (or magnitude) the preferred scale is the one that is additive in respect to physical operations. For example, the length of a two concatenated rods equals the sum of lengths of individual rods. This approach has been generalized in the Representational Theory of Measurement (RTM) proposed initially by Patrick Suppes [7.69] [7.87]. According to Krantz *et al.* [7.36], p. 12, measurement is a mapping from the actual relational structure into a certain numerical relational structure that is invariant to acceptable transformations.

Development of human knowledge has been accompanied by the evolution of technology towards higher levels of automation and control. As a result, human society became increasingly dependent on the technosphere, which is the functional extension of social agency. In the XXth century, human technology was already capable of manipulating potential signs, such as messages and quantities, without human intervention. That was the birth of cybernetics—the science of communication in animal and machine [7.94]. The main achievement of that time was the theory of information that specified conditions upon which messages can be reliably transmitted over communication channels independent of their meanings. This result indicated that all human knowledge (in the form of potential signs) can be digitized and become converted into sequences of zeroes and ones.

Progress in cybernetics and computation technology revolutionized all human operations and communication, but it failed to replace the human mind as the main interpreter of potential signs. Despite all automation and developments in artificial intelligence (AI), the human brain remains the bottleneck of semiosis. There is certainly some progress in AI, exemplified by advanced robotics, self-driving cars, and game projects such as computers playing chess and AlphaGO. But AI still has a long road to go towards development of expert-type software programs that can make meaningful guesses outside of terminological clichés. I believe that biosemiotics can be useful for understanding why educated humans are still more competent in interpreting potential signs than computers.

Declaration

I declare no conflict of interest.

Acknowledgments

I thank Kalevi Kull, Donald Favareau, Richard Gordon, George Mikhailovsky, Anton Sukhoverkov, and an anonymous reviewer for their helpful discussion of this manuscript.

References

[7.1] Adami, C., Ofria, C. and Collier, T.C. (2000) Evolution of biological complexity. *Proc Natl Acad Sci USA* 97(9), 4463-4468.

[7.2] Anderson, M., Deely, J., Krampen, M., Ransdell, J., Sebeok, T.A. and von Uexküll, T. (1984) A semiotic perspective on the sciences: Steps toward a new paradigm. *Semiotica* 52(1/2), 7-47.

[7.3] Aristotle (2016) *De Anima*. Oxford University Press, Oxford.

[7.4] Aukema, B.H. and Raffa, K.F. (2005) Selective manipulation of predators using pheromones: Responses to frontalin and ipsdienol pheromone components of bark beetlesin the Great Lakes region. *Agricultural and Forest Entomology* 7, 193–200.

[7.5] Barandiaran, X.E. (2017) Autonomy and enactivism: Towards a theory of sensorimotor autonomous agency. *Topoi* 36, 409–430.

[7.6] Bateson, G. (1987 [1972]) *Steps to an ecology of mind. Collected essays in anthropology, psychiatry, evolution, and epistemology*. Jason Aronson, Northvale, NJ.

[7.7] Benton, M.J. (2008) *The history of life. A very short introduction*. Oxford Uniuversity Press, Oxford.

[7.8] Bower, J. (1991) The Pastoral Neolithic of East Africa. *Journal of World Prehistory* 5(1), 49-82.

[7.9] Brenner, J.E. and Igamberdiev, A.U. (2019) Philosophy in reality: Scientific discovery and logical recovery. *Philosophies* 4(22), 1-45.

[7.10] Brenner, J.E. and Igamberdiev, A.U. (2021) *Philosophy in reality: A new book of changes*. Springer Nature.

[7.11] Burch, R. (2022) Charles Sanders Peirce. <https://plato.stanford.edu/archives/sum2022/entries/peirce/>

[7.12] Campbell, N.R. (1920) *Physics: The elements*. Cambridge University Press, London.

[7.13] Christiansen, P.V. (1993) Peirce as participant in the Bohr-Einstein discussion. In: *Charles S. Peirce and the philosophy of science: Papers from the Harvard Sesquicentennial Congress*. E.C. Moore, (ed.) The University of Alabama Press, Tuscaloosa and London: 223–232.

[7.14] de Saussure, F. (1959) *Course in general linguistics*. Philosophical Library, New York.

[7.15] Deamer, D., Dworkin, J.P., Sandford, S.A., Bernstein, M.P. and Allamandola, L.J. (2002) The first cell membranes. *Astrobiology* 2(4), 371-381.

[7.16] Deely, J. (1992) Semiotics and biosemiotics: Are sign-science and life-science coextensive? In: *Biosemiotics. The semiotic web 1991*. T.A. Sebeok and J. Umiker-Sebeok, (eds.) Mouton de Gruyter, New York: 45-75.

[7.17] Deely, J. (2001) Umwelt. *Semiotica* 134(1/4), 125-135.

[7.18] Deely, J. (2014) Semiotic entanglement: The concepts of environment, Umwelt, and Lebenswelt in semiotic perspective. *Semiotica* 199(1), 7-42.

[7.19] Diamond, J. and Bellwood, P. (2003) Farmers and their languages: the first expansions. *Science* 300(5619), 597-603.

[7.20] Ehrenfreund, P. and Cami, J. (2010) Cosmic carbon chemistry: from the interstellar medium to the early Earth. *Cold Spring Harb Perspect Biol* 2(12), a002097.

[7.21] Ehrenfreund, P., Rasmussen, S., Cleaves, J. and Chen, L. (2006) Experimentally tracing the key steps in the origin of life: The aromatic world. *Astrobiology* 6(3), 490-520.

[7.22] Ehrlich, H.L. and Newman, D.K. (2009) *Geomicrobiology*. CRC Press, Taylor & Francis, Boca Raton, FL, 5th.

[7.23] Eigen, M. and Schuster, P. (1979) *The hypercycle, a principle of natural self-organization*. Springer-Verlag, Berlin, New York.

[7.24] Feynman, R.P. (1998) *The meaning of it all: Thoughts of a citizen-scientist*. Perseus Books Group, New York.

[7.25] Finn, J.K., Tregenza, T. and Norman, M.D. (2009) Defensive tool use in a coconut-carrying octopus. *Curr Biol* 19(23), R1069-1070.

[7.26] Foley, R.A. and Lahr, M.M. (2011) The evolution of the diversity of cultures. *Philos Trans R Soc Lond B Biol Sci* 366(1567), 1080-1089.

[7.27] Francioni, V. and Harnett, M.T. (2022) Rethinking Single Neuron Electrical Compartmentalization: Dendritic Contributions to Network Computation In Vivo. *Neuroscience* 489, 185-199.

[7.28] Ginsburg, S. and Jablonka, E. (2019) *The evolution of the sensitive soul. Learning and the origin of consciousness*. MIT Press, Cambridge, London.

[7.29] Guillemette, L. and Cossette, J. (2006) The semiotic process and the classification of signs. *Signo, Louis Hébert (dir.)*.

[7.30] Herculano-Houzel, S. (2009) The human brain in numbers: A linearly scaled-up primate brain. *Front Hum Neurosci* 3, 31.

[7.31] Hoffmeyer, J. (2008) *Biosemiotics: An examination into the signs of life and the life of signs* University of Scranton Press, Scranton, PA.

[7.32] Hölldobler, B. and Wilson, E.O. The evolution of communal nest-weaving in ants: Steps that may have led to a complicated form of cooperation in weaver ants can be inferred from less advanced behavior in other species. *American Scientist* 71(5), 490-499.

[7.33] Joseph, R. (2009) Life on Earth came from other planets *The Journal of Cosmology* 1, 1-56

[7.34] Kauffman, S. and Roli, A. (2021) The world Is not a theorem. *Entropy* 23(1467), 1-9.

[7.35] Kauffman, S.A. (2000) *Investigations*. Oxford Univ. Press, Oxford, New York.

[7.36] Krantz, D.H., Luce, R.D., Suppes, P. and Tversky, A. (1971) *Foundations of measurement v.1: Additive and polynomial representations*. Academic Press, San Diego and London.

[7.37] Krembel, A., Colin, R. and Sourjik, V. (2015) Importance of multiple methylation sites in *Escherichia coli* chemotaxis. *PLoS One* 10(12), e0145582.

[7.38] Kull, K. (2022) The biosemiotic fundamentals of aesthetics: Beauty is the perfect semiotic fitting. *Biosemiotics* 15(1), 1-23.

[7.39] Kull, K. and Torop, P. (2003) Biotranslation: Translation between Umwelten. In: *Translation, translation*. S. Petrilli, (ed.) Rodopi, Amsterdam, New York: 315-328.

[7.40] London, M. and Hausser, M. (2005) Dendritic computation. *Annu Rev Neurosci* 28, 503-532.

[7.41] Lotman, J. (2005) On the semiosphere. *Sign Systems Studies* 33(1), 205–229.

[7.42] Lyell, C. (2010 [1832]) *Principles of geology. An attempt to explain the former changes of the Earth's surface, by reference to causes now in operation*. Cambridge University Press, Cambridge, MA.

[7.43] Macháček, M., 2018: (2019) A Peircean interpretation of probability in quantum mechanics. *Semiotics (SSA)* 2018, 233–241.

[7.44] Marais, K. (2019) *A (bio)semiotic theory of translation: The emergence of social-cultural reality*. Routledge, New York.

[7.45] Marais, K. (2023) What does it mean to translate In: *Translation beyond translation studies*. K. Marais, (ed.) Bloomsbury Academic, London, New York: 1-14.

[7.46] Markov, A.V., Anisimov, V.A. and Korotayev, A.V. (2010) Relationship between genome size and organismal complexity in the lineage leading from prokaryotes to mammals. *Paleontological Journal* 44(4), 363-373.

[7.47] Martinelli, D. (2010) *A critical companion to zoosemiotics. People, paths, ideas*. Springer, Dordrecht, New York.

[7.48] Mason, P.A. and Biermann, P.L. (2018) Astrophysical and cosmological constraints on life. In: *Habitability of the Universe before Earth*. R. Gordon and A.A. Sharov, (eds.) Elsevier, Amsterdam: 89-126.

[7.49] Meier-Oeser, S. (2011) Medieval semiotics. <https://plato.stanford.edu/archives/sum2011/entries/semiotics-medieval/>
[7.50] Mikhailovsky, G. (2018) From identity to uniqueness: The emergence of increasingly higher levels of hierarchy in the process of the matter evolution. *Entropy* 20(533), 1-18.
[7.51] Miller, A. (2021) Realism. <https://plato.stanford.edu/archives/win2021/entries/realism/>
[7.52] Monticelli, D. (2019) Borders and translation: Revisiting Juri Lotman's semiosphere. *Semiotica* 2019(230), 389-406.
[7.53] Morris, C.W. (1971) *Writings on the general theory of signs*. Mouton, The Haague, Paris.
[7.54] Morris, C.W. (1977) *Foundations of the theory of signs*. University of Chicago Press, Chicago.
[7.55] Nicolis, G. and Prigogine, I. (1977) *Self-organization in nonequilibrium systems: From dissipative structures to order through fluctuations*. Wiley, New York.
[7.56] Nikolić, D. (2023) Where is the mind within the brain? Transient selection of subnetworks by metabotropic receptors and G protein-gated ion channels. *Comput Biol Chem* 103, 107820.
[7.57] Nöth, W. (1990 [1985]) *Handbook of Semiotics*. Indiana University Press, Bloomington, IN.
[7.58] Pattee, H.H. (1967) Quautum mechanics, heredity and the origin of life. *Journal of Theoretical Biology* 17, 410-420.
[7.59] Pattee, H.H. (1969) How does a molecule become a message? *Developmental Biology Supplement* 3, 1-16.
[7.60] Pattee, H.H. (2001) The physics of symbols: Bridging the epistemic cut. *Biosystems* 60(1-3), 5-21.
[7.61] Pattee, H.H. (2023) Symbol grounding by folding: The primary biosemiosis. *Open Semiotics* 4, 97-198.
[7.62] Patthy, L. (1999) Genome evolution and the evolution of exon-shuffling-a review. *Gene* 238, 103-114.
[7.63] Peirce, C.S. (1931–1958) *Collected Papers (eds. Hartshorne and Weiss; Burks)*. Harvard University Press., Cambridge, MA. 1-8.
[7.64] Peirce, C.S. (1958) *Charles S. Peirce: Selected Writings (Values in a Universe of Chance)* (ed. Philip P. Wiener). Dover Publications, New York.
[7.65] Pleskacheva, M.G. (2009) Behavior and spatial learning in radial mazes in birds. *Neurosci Behav Physiol* 39(8), 725-739.
[7.66] Porter, S.L., Wadhams, G.H. and Armitage, J.P. (2011) Signal processing in complex chemotaxis pathways. *Nat Rev Microbiol* 9(3), 153-165.
[7.67] Queiroz, J., Emmeche, C. and El-Hani, C.N. (2008) A Peircean approach to 'information' and its relationship with Bateson's and Jablonka's ideas. *American Journal of Semiotics* 24(1-3), 75-94.
[7.68] Rosen, R. (1967) *Optimality principles in biology*. Springer Science+Business Media, LLC.
[7.69] Scott, D. and Suppes, P. (1958) Foundational aspects of theories of measurement. *Journal of Symbolic Logic* 23(2), 113–128.
[7.70] Sebeok, T.A. (1976) *Contributions to the doctrine of signs*. Indiana University Press, Bloomington, IN.
[7.71] Sebeok, T.A. (1985 [1968]) *Contributions to the doctrine of signs*. Indiana University Press, Bloomington, IN.
[7.72] Sebeok, T.A. (2001) *Signs: An introduction to semiotics*. University of Toronto Press, Toronto, London, 2nd.
[7.73] Sebeok, T.A. and Danesi, M. (2000) *The forms of meaning. Modeling systems theory and semiotic analysis*. Mouton de Gruyter, New York.
[7.74] Sharma, S., Rakoczy, S. and Brown-Borg, H. (2010) Assessment of spatial memory in mice. *Life Sci* 87(17-18), 521-536.

[7.75] Sharov, A.A. (1992) Biosemiotics: Functional-evolutionary approach to the problem of the sense of information. In: *Biosemiotics. The semiotic web 1991.* T.A. Sebeok and J. Umiker-Sebeok, (eds.) Mouton de Gruyter, New York: 345-373.

[7.76] Sharov, A.A. (2006) Genome increase as a clock for the origin and evolution of life. *Biology Direct* 1, 17.

[7.77] Sharov, A.A. (2010) Functional information: Towards synthesis of biosemiotics and cybernetics. *Entropy* 12(5), 1050-1070.

[7.78] Sharov, A.A. (2016) Coenzyme world model of the origin of life. *Biosystems* 144, 8-17.

[7.79] Sharov, A.A. (2018) Mind, agency, and biosemiotics. *Journal of Cognitive Science* 19(2), 195-228.

[7.80] Sharov, A.A. (2023) Biology of translation: The role of agents. In: *Translation beyond translation studies.* K. Marais, (ed.) Bloomsbury Academic, London, New York: 63-79.

[7.81] Sharov, A.A. and Gordon, R. (2018) Life before Earth. In: *Habitability of the Universe before Earth.* R. Gordon and A.A. Sharov, (eds.) Elsevier, Amsterdam: 267–298.

[7.82] Sharov, A.A., Maran, T. and Tønnessen, M. (2015) Towards synthesis of biology and semiotics. *Biosemiotics* 8(1), 1-7.

[7.83] Sharov, A.A. and Tønnessen, M. (2021) *Semiotic agency. Science beyond mechanism.* Springer, Dordrecht. 25.

[7.84] Sharov, A.A. and Vehkavaara, T. (2015) Protosemiosis: Agency with reduced representation capacity. *Biosemiotics* 8(1), 103-123.

[7.85] Shumaker, R.W., Walkup, K.R. and Beck, B.B. (2011) *Animal tool behavior: The use and manufacture of tools by animals.* Johns Hopkins University Press, Baltimore.

[7.86] Srivastava, S., Ghosh, N. and Pal, G. (2013) Metagenomics: Mining environmental genomes. In: *Biotechnology for Environmental Management and Resource Recovery. Springer. 2013.* R.C.K.a.A. Singh, (ed.) Springer, Heidelberg: 161-190.

[7.87] Suppes, P. (1951) A set of independent axioms for extensive quantities. *Portugaliae Mathematica* 10(4), 163–172.

[7.88] Tarski, A. (1983 [1935]) *Logic, Semantics, Metamathematics.* Hackett, Indianapolis, 2nd.

[7.89] Tønnessen, M. (2016) Agency in biosemiotics and enactivism. In: *Meaning, mind and communication. Explorations in cognitive semiotics.* J. Zlatev, G. Sonesson and P. Konderak, (eds.) Peter Lang GmbH, Internationaler Verlag der Wissenschaften, Frankfurt am Main, Bern, New York: 69-82.

[7.90] von Foerster, H. (1960) On self-organizing systems and their environments. In: *Self-organizing systems.* M.C.Y.a.S. Cameron, (ed.) Pergamon Press, London: 31–50.

[7.91] von Glasersfeld, E. (1984) An Introduction to radical constructivism. In: *The invented reality.* P. Watzlawick, (ed.) Norton, New York: 17-40.

[7.92] von Uexküll, J. (1957) A stroll through the worlds of animals and men: A picture book of invisible worlds. In: *Instinctive behaviour: The development of a modern concept.* C.H. Schiller, (ed.) International Universities Press, Inc., New York: 5–80.

[7.93] von Uexküll, J. (1982 [1940]) The theory of meaning. *Semiotica* 42(1), 25-82.

[7.94] Wiener, N. (1948) *Cybernetics or control and communication in the animal and the machine.* Wiley & Sons Inc., New York, 2d.

[7.95] Zhang, S., Si, A. and Pahl, M. (2012) Visually guided decision making in foraging honeybees. *Front Neurosci* 6, 88.

8

A Constructivist Approach to Meanings in the Universe

Alexander Kravchenko

Theoretical and Applied Linguistics, Baikal State University, Irkutsk, Russia

Online resources:
https://isea.academia.edu/AlexanderKravchenko
https://www.researchgate.net/profile/Alexander-Kravchenko-4
https://loop.frontiersin.org/people/189470/bio

Abstract
The problem of meaning and its "existence" in the universe is addressed from an epistemological perspective. It is argued that the problem cannot be resolved within the dominant paradigm of external realism as "objectivity without parentheses" because of the epistemological trap of language that sets limits to our knowledge. Instead, a radical constructivist approach is used, based on Humberto Maturana's systems approach to language and cognition as biological phenomena. Such an approach highlights the crucial role of the observer in constructing a universe as "objectivity with parentheses," the domain of "things" brought forth through the operations of distinction made in language. Viewed from this angle, signs, meanings, and other objects of investigation in biosemiotics appear as "things" in the observer's idiosyncratic universe with their properties defined by the operations of distinction. However, the distinctions we make in language affect theory adequacy, and because the observer's universe is an image of the observer's language, we tend to reify things to which words as names "refer" and make endeavors to find them, identify, and analyze. In this chapter, an alternative view is suggested, positing an intrinsically anthropocentric nature of the key biosemiotic concepts and a necessity to distinguish between qualitatively different levels of interpretation of semiotic phenomena in the human and non-human cognitive domains.

Keywords: Constructivism, sign, signification, meaning, interpretation, observer, language

Email: sashakr@hotmail.com

8.1 Do We Ask the Right Kinds of Questions?

What is meaning? Why is this question so important that the quest for a definitive answer never ends as scholars from various fields of research try to come to grips with the concept? It seems intuitively transparent and understandable but, nonetheless, eludes a rigorous uncontroversial definition. We all use the concept freely in our daily lives, speaking of something as meaningful or meaningless, asking someone to clarify their meaning or explaining to others what we meant when we said or did something. And of course, many of us have felt, at one time or another, an urgent need to understand the meaning of life. But what is the meaning of "meaning"? Addressed a century ago in what became a classic book on semantics [8.55], this seemingly tautological question still taunts philosophers, linguists, psychologists and other researchers in the human sciences. No wonder it has become the focal point in biosemiotics as the study of signs and meaning in living organisms and systems.

Philosophers speak, basically, of two types of "theory of meaning," a semantic theory dealing with the semantics of linguistic expressions aka symbols, and a foundational theory of meaning dealing with the origins of meaning as the semantic content of linguistic expressions [8.70]. As pointed out by Quine [8.59], p. 3, "conceptualization on any considerable scale is inseparable from language, and our ordinary language of physical things is about as basic as language gets." Thus, "meaning" appears to be an anthropocentric notion, something inherent in the human enlanguaged world, in which we exist and with which we interact, adapting to the environment to sustain our unity as living systems, *Homo loquens* [8.23]. By contrast, in biosemiotics all living creatures are viewed as semiotic systems; therefore, the notion of meaning is not restricted to the enlanguaged world of humans, and this, as argued in [8.31], becomes a major epistemological problem, because language matters in anything and everything we do as human beings, science being no exception [8.35].

In a very general sense, human adaptive behavior may be viewed as a purposeful, end-directed activity constitutive of agency as a capacity to generate such behavior [8.40]. However, we are not alone; we share this world with other non-human living organisms, and what can be said about humans in terms of their purposeful behavior (agency) can also be said about non-human organisms. Seeing agency as a capacity shared by human and non-human living organisms presupposes a common mechanism for purposeful behavior, such as making sense of the environment and using it to achieve a particular goal [8.65] by adaptively regulating its operation and its relation to the environment [8.13], p. 33. By and large, sense-making depends on semiosis as a process in which perceived objects and phenomena function as signs — something perceptually present that an agent interprets as indicating something perceptually absent [8.53] [8.32]. Grounded in the lived experience of an organism as a history of its encounters with various aspects of the environment preserved in memory, such an interpretation is at the basis of the organism's purposeful (orientational, adaptive) behavior, which a human observer describes as its meaningful interactions with the environment.

In the literature, meaning is often described as something "created" or "made" [8.56] [8.52] [8.15] [8.80], inter alia, that is, a kind of thing intentionally brought into existence by an agent. If, however, meaning is viewed as a value-based relationship between the organism and its world of experience [8.79], it is not "created" but emerges in the complex dynamics of the organism-environment system as a unity of interactions. This begs the question, "Once meaning, as a value-based relationship in an organism's world of experience, has emerged, does it continue to exist in the universe?" An illuminating answer to this question is possible only when we have a coherent answer to another question, "What is a universe?"

Our view of the universe depends on the kind of questions we ask in our quest for knowledge. In philosophy of external realism as the still-dominant paradigm in contemporary epistemology, the world is seen as largely made up of objects that are mind-, language- and theory-independent [8.3]. Therefore, the object of knowledge — what we know about the world and its properties — also appears to be independent of the knower. Radical constructivism [8.17] [8.21] [8.72], as an alternative epistemology, takes a different stance on the issue of knowledge [8.49]. As argued by Maturana [8.44] in his seminal "Biology of Cognition," the question "What is the object of knowledge?" becomes meaningless once we understand that, although we are used to talking about reality orienting each other through linguistic interactions to what we deem are sensory experiences of concrete entities, these entities, just like thoughts and descriptions, are states of relative activity between neurons that generate new descriptions.

Language is our existential domain in which we happen and become what we are. As Foerster [8.18], p. 71, put it, "[t]he world is an image of language. Language comes First, and the world is a consequence of it." Because we as humans "happen" in language as our immediate environment (the relational domain of cooperative interactions), and because what we know of the world comes through the agency of the body in action [8.12], to us, as observers, language literally creates the world through the fundamental operation of distinction as the specification of an entity operationally cleaved from a background:

> that which results from an operation of distinction and can thus be distinguished, is a thing with the properties that the operation of distinction specifies, and which exists in the space that these properties establish. Reality, therefore, is the domain of things, and, in this sense, that which can be distinguished is real. Thus stated, there is no question about what reality is: It is a domain specified by the operations of the observer. [8.45], p. 55.

As pointed out by Foerster [8.18], p. 80, "it is language that, because of its denotative aspect, seduces us and makes us look for the properties of reality "out there" instead of within ourselves." We begin to believe that there is one universe in which signs, meanings, goals, and so forth, along with the universe itself, exist as entities independent of the observer, as something "out there," whose origin, nature, and constitutive properties can be identified, analyzed, and explained; hence the "What is (an) X?" question researchers in various fields endeavor to answer, overlooking the subject-dependent nature of science that is not a manner of revealing an independent reality as "objectivity without parentheses" [8.47] but a manner of bringing forth a particular reality "bound to the conditions that constitute the observer as a human being," or "objectivity with parentheses":

> objectivity with parenthesis entails accepting that existence is brought forth by the distinctions of the observer, that there are as many domains of existence as kinds of distinctions the observer performs: objectivity in parenthesis entails the *multiversa*, entails that existence is constitutively dependent on the observer, and that there are as many domains of truths as domains of existence she or he brings forth in her or his distinctions [8.47], p. 11.

The importance of this epistemological tenet in construing a theory of something as part of the "objective world" has been well illustrated by Vaz [8.76] in his constructivist analysis of how immune activity is seen and conceptualized in mainstream immunology and why almost all attempts to invent new vaccines fail. This, Vaz shows, is the result of overlooking the fact that immunoglobulins and antibodies are different entities that belong to two different non-intersecting domains in which a system is simultaneously described, one operational (system as a collection of components) and the other phenomenological (system as a singular entity in interaction with a medium which contains the system and makes it possible). Specific antibodies and immunoglobulins arise in separate domains of description: "antibodies are functional entities created (named) by the immunologist [...], while immunoglobulins are considered components of the structural dynamics of the organism, without the inherent intention of reacting with (any) foreign materials" [8.76], p. 72. Failure to see this aspect (operations of distinction) of our constructing the world in language results in a major conceptual confusion that cannot but tell on theory adequacy — immunology being just one example. Another ready example is a similar conceptual confusion in linguistics that fails to see the difference between language as a structured system of vocalizations (signs) produced by human organisms (the hallmark of structural linguistics) and languaging as human-specific behavior in the phenomenal domain [8.35]. Similarly, biosemiotics, in its attempts to understand and describe meanings as components of the universe, may well be in the process of drawing a picture not unlike in immunology or orthodox linguistics, when a theory called upon to empower us with knowledge comes short of doing the expected job.

Therefore, before we attempt to come up with a meaningful answer to the question, "What is meaning?" (excuse the pun), we should ask the kind of questions that would illuminate our initial epistemological assumptions [8.16]. Do signs "signify," or do they "mean"? If they do both, how is "meaning" different from "significance" and "signification"? Do signs "have" meanings as their properties observed and described by the observer? Or is meaning an entity brought forth by an operation of distinction done in language? Do the properties of this entity specified by the operation of distinction make it appear as a "thing," "process," or "relation"? If the latter, what aspect of the organism-environment system dynamics is distinguished by the observer as "meaning" and why? Can "meaning" be an inferential notion based on the interpretation of causal relationships in the living system's end-directed dynamics, and if so, can we speak of meaning in non-living systems such as artificial intelligence agents?

Much depends on how these and other similar questions are answered. However, it is not my ambition to attempt to give such answers even as a first approximation. Rather, I invite the reader to take a closer look at some concepts and notions we often take for granted because we don't realize how the language we speak sets a dangerous trap on our way to knowing.

8.2 The Epistemological Trap of Language

By and large, biosemiotic research into the nature and origins of meaning is driven by the belief that signs and meanings, either actual or potential, exist in the universe, in the activity of living organisms. This belief is a direct descendant of Descartes' famous "*Cogito, ergo sum*" and reflects our propensity for hypostatizing, supported by the common belief in the instrumental function of language, the hallmark of linguistic orthodoxy. This and many other epistemological beliefs, such as the existence of time, information, knowledge, and, ultimately, mind and (self-)consciousness, are existentially pregiven in the enlanguaged world of humans as an image of language. Their seemingly pregiven existence is a necessary consequence of "objectivity with parentheses," something brought forth in language by the observer who arises in language as "a manner of living in recurrent interactions in a flow of coordinations of coordinations of consensual behaviors" [8.50], p. 30; cf. [8.60].

While Heidegger [8.22], p. 239, called language our "house of being," another metaphor is more appropriate: language is an epistemological trap [8.30] of which we don't seem to be aware. We are used to seeing the world through the prism of language in which the world is brought forth by the observer, forgetting that it is very often the archaic world view made by a prehistoric observer who was the ultimate point of reference and "the measure of all things" in the operations of distinction he made in language. Heidegger's metaphor itself is a good illustration of how we are entrapped in language as something "interpretatively terminal" [8.42], p. 705, that cannot be described "externally" [8.19]: as a building for human habitation, "house" is an external description of an observed artifact purposefully designed and created by the observer as part of his adaptive interactions with the environment from which the observed ("a house") is operationally cleaved. Language, however, may not be a "house of being" in which we dwell because language, *pace* some cognitive scientists (e.g., [8.15]), is not an artifact invented to meet some pragmatic need, such as an instrument for communication, we cannot leave this "house" and observe it from some vantage point external to it even if we wanted to. Language cannot be operationally cleaved from the background of the environment because, to us as living systems, it *is* the environment constitutive of the human organism-environment systems, and "our being occurs in languaging" [8.48], p. 196.

As an eigenform, language sets the limits to what we can know about the world [8.32], not unlike in the famous movie *The Truman Show*, and we take for granted what our lived experience tells us. We are forever in the epistemological trap of language, but unlike in *The Truman Show*, we cannot escape from its settings. Much, if not all, of what we think we, as observers, know about the world, is the result of our operations of distinction we make in language, or naming—specification of an entity by operationally cleaving it from a background. Once such a distinction is made, a thing with the properties that the operation of distinction specifies is added to the enlanguaged world. Reality, or the universe, is a domain specified by the operations of the observer [8.45], and there are as many realities as there are observers. This radical constructivist tenet has an important epistemological implication.

Different observers construct their observer-dependent universes based on their ontogenetic history of structural coupling with the medium as a flow of coordinations of coordinations of consensual behaviors (languaging). Every observer is unique in their experiences; therefore, their interpretations of the observed are also unique (idiosyncratic). To different observers speaking the same language the value of their relationships with similar aspects of the environmental medium will be different inasmuch as their histories of structural coupling with the medium are different, thus affecting their interactions in the flow of languaging as coordinations of coordinations of behaviors, that is, understanding. This is especially true about semiotics, biology, and cognitive science, when it comes to the different concepts of meaning developed in these fields, semiotic and non-semiotic: meaning as a sign-function and meaning as relevance, organization, or configuration [8.69]. However, such a divide is questionable if semiosis is understood as meaning-making based on interpretation. It takes us back to the question, "What is the meaning of 'meaning'?", not in the general sense of an entity that exists in the universe prior to a distinction (naming) made by the observer, but in the sense of the distinction itself and the properties of the "object-with-parentheses" this distinction specifies, because "to 'speak' about 'speaking', to signify signification or to communicate about communication cannot but influence the universe of speaking, signification and communication" [8.14], p. 29.

Every time we engage in linguistic interactions, or languaging, we display behavior uniquely characteristic of humans. This behavior includes a whole range of things that we do as we communicate. Counter to the orthodox view, language (used in the trivial sense, as, for example, when we routinely refer to our natural communicational activity) is not just verbal patterns, utterances, vocalizations, wordings, etc. It is always a complex of closely integrated processes which span perception, sensory-motor activities (including the production and perception of speech), emotional states, interoception, proprioception, memory, etc. In other words, language is embodied as the manifestation of a specific kind of bodily activity and enacted in the "here-and-now" of the communicative interaction.

When we engage in languaging as natural human activity, we don't reflect on what exactly it is that we do when we speak to or hear one another — just as we don't ask ourselves what exactly happens in the relevant parts of our body when we step over an obstacle in our path or sit on a bench in the park to rest a little and enjoy the view and the weather. What matters is that in the process of linguistic interactions with others we exploit the affordances provided by our consensual domain to monitor, modify and, if necessary and, indeed, possible, change the behavior of others and self in such a way as to sustain, remedy, or enhance our ecological equilibrium with the niche. It is in this sense that language may be described as "natural"; it is something that happens naturally, like breathing, to which no second thought is given. However, if we begin to experience problems with normal breathing (whatever the cause may be), it becomes the focus of our attention, and we try to deal with the problem looking for ways to bring our respiratory function back to normal. This is an adaptational response of our body to the changed environment as the organism mobilizes all the available resources to restore the balance with its niche. Such a response may involve a variety of physical and/or physiological activities that go beyond the respiratory process as such, rising to a higher level of coordination of various bodily functions that pertain to breathing as a merely physiological process. Something like this happens when we start thinking and talking about language and the meaning of linguistic expressions, thus engaging in metalinguistic discourse.

Pointing to these two different aspects of language as communicational activity, Love [8.41] introduced the concept of orders of language as cognitively distinct kinds of such activity. First-order language is a communicative process in which we engage as interacting agents, something that we do without thinking about how we do it. For example, when I ask a male colleague at work (who I know might have some ideas on how to deal with a particular problem I've been working on), "Would you like to join me for lunch?" and get, as a response, something like "Yeah, sure," I engage in an interaction with that person with the goal to coordinate our separate behaviors in the shared time-space of the physical context of our environment so that we find ourselves at the same place (e.g., at a table in the cafeteria) at the same time (e.g., 12 a.m.), which could give me a good opportunity to unobtrusively pick my colleague's brain over a shared meal. This is first-order language (or, rather, languaging).

If, however, my invitation to join me for lunch were addressed, say, to a young attractive female from another department who has recently joined the workforce of our company and with whom I've been only cursorily acquainted, in response to my invitation I might hear something like, "And what could that possibly mean?" It is not the case that my female colleague, while being a speaker of English, does not understand the words that constitute the utterance "Would you like to join me for lunch?" What makes her ask the question is her uncertainty about the orientational purport of my utterance — Eco's [8.14] p. 29, "indeterminacy principle" — caused by the lack of sufficient interactional experience in our mutually shared domain of interactions that would allow her to interpret my invitation at face value, so to speak, or as an instance of first-order language. Considerations of this sort take us to another issue, that of the "epistemological threshold" in semiotics [8.14] and the problem that arises with the view of semiotics as being about knowledge [8.10] [8.38]. What, in that case, is the role of "non-semiotic meaning" in what we call "knowledge"?

The very way the question is formulated is a good illustration of Foerster's observation above about the "seductive" role of language that makes us look for the properties of reality "out there" instead of within ourselves. Once "non-semiotic meaning" becomes a distinction made in language by an observer (a semiotician, in our case), a thing with properties specified by this distinction is brought forth in language as our cognitive domain, and we attempt to identify its position and function in the universe relative to other things and their functions, such as "signs," "semiotic meaning," "signification," "knowledge," "concept," etc. This is, basically, what a theory of semiotics is about, and because different observers operate in their own idiosyncratic universes as their unique experiential domains, it is often hard, if not impossible, to reach a consensus on what a unified theory of biosemiotics should be.

Moreover, even within an idiosyncratic "universe" of a particular (bio)semiotician, things that "exist" in it are often not specified well enough, they are fuzzy and, therefore, hard to distinguish from one another. Thus, Sonesson [8.69], p. 150, observes:

> Even though semiotics is not exclusively concerned with signs, but is also required to attend to meanings of other kinds, the concept of sign remains crucial, and semiotic inquiry still has to start out from a distinction between signs and other meanings.

Here, "signs" are distinguished from "meanings of other kinds," which poses the question of whether signs are, indeed, meanings, or whether meanings are signs, while both appear to be things "out there." In a similar fashion, Kull and Lotman [8.39], p. 11, speak of protoconcepts that "include all distinguishable and repeatedly usable signs as results of categorization," while on the next page we learn that most signs include "protoconcepts and concepts in everyday language" [8.39], p. 12. Sharov and Tønnessen [8.65], p. 10, speak of "meaningful use of signs" as semiotic tools, thus implying that signs exist "out there" (in the universe) and their use by a living system can be meaningless despite the role of signs, which is to signify. Deely [8.10], p. 72, defines semiotics as the study of "the action of signs" that provides "the basis for the whole of human knowledge," while Morehouse [8.52], p. 32, defines meaning as "a relationship between two sorts of things: signs and the kinds of things they intend, express or signify," although it is hard to imagine how a sign, as a thing, can act — that is, pose in the guise of an agent — in one case or intend anything in the other. Examples of such obviously dubious explanations of semiotic phenomena and concepts found in the biosemiotic literature are numerous, which may be one of the reasons why biosemiotics still cannot be described as a mature science with a noticeable effect on the human praxis of living. This somewhat awkward situation can change once biosemioticians realize the perils of the epistemological trap humans unwittingly set for themselves as the kind of living systems that they are—human organisms as observers that arise in language as their mode of living.

8.3 The Observer, the Observed, and the Problem of Interpretation

Living organisms adapt to their environments by behaviorally reacting (as a result of interpretation) to the perceptually present aspects of their niche we call "signs" — something that in the lived experience of the organism is associated with something perceptually absent that may be beneficial, detrimental, or neutral to the unity of the organism-environment system. A sign is always a sign to someone who interprets it as such, and as distinctions made in language signs belong exclusively to the cognitive domain of the observer constructing an idiosyncratic universe as an image of the observer's language. The constitution of such a universe depends on the ontogenetic history of fine structural coupling of the observer with the medium, and because for different observers such histories are different, their idiosyncratic universes also differ: what may appear as a sign to one observer is not necessarily interpreted as a sign by another. But what is sign interpretation as a process?

A sign is something that allows the observer/interpreter to take into account what is not perceptually present, and this is its semiotic function. In other words, "a recognized *absence*" of something drives semiosis as the living process [8.37], p. 20. We know that something is a sign based on our experience of previous interactions with the environment retained in memory as a dynamic network of states of relative neuronal activity as representations of physical interactions [8.29], p. 99. This experience allows us to associate something perceptually present with what is perceptually absent but retained in memory, thereby categorizing it as a sign, a "mark" of something else that is of significance in the observer's experience of orientational (adaptive) interactions with the environment because of the value of what the sign signifies. This significance is not an intrinsic property of that which is perceptually present; it emerges in the flux of the lived experience of an organism in the process of Hebbian learning and is mediated by the organism's behavioral response to the aspect of the environment perceived and interpreted as a sign.

According to Deacon [8.9], p. 294, to interpret something as a sign is to interpret immediate, perceptually present physical conditions "as representing other displaced or as yet unrealized possible conditions." The two types of conditions in this definition, while both referring to something not perceptually present, refer to two different, although intersecting, ontological domains. "Displaced conditions" are conditions that the observer experienced in his past interactions with the environment; they are retained in memory as representations in the sense described above — states of relative neuronal activity caused by the organism's interactions with the environment. In this sense, "displaced conditions" are "real" in that they are part of the observer's lived experience. "Unrealized possible conditions" are conditions that may be obtained in the observer's interactions with the environment in the (near) future but are not yet part of his lived experience, and in this sense, they are not "real," they are potentialities—something that may happen but does not necessarily have to. But how can perceptually present physical conditions represent (in the trivial sense, that is, "present again") unrealized conditions as potentialities that have not yet become part of the lived experience? This is possible in the universe as an image of our language in which we exist and operate as observers interpreting and describing the observed.

Because "potentiality" is a noun, and nouns usually denote things in the observer's universe, and because we cannot free ourselves from the epistemological trap set for us by the very language we speak, we are tempted to view, describe, and treat potentialities as something that exists and may, therefore, become perceptually present in our future interactions with the environment. We categorize our experience as a flux moving along the time axis from the past through the present and into the future as temporal domains that exist in the universe as "objectivity without parentheses." However, as has been shown by Simsky, Kravchenko and Druzhinin [8.68], the construction of a temporal sequence that unites what is and what already or still is not, is possible only in linguistic semiosis, that is, language: "past" and "future" are mental constructs that belong to the present just as any other act of thinking. In other words, time appeared with the appearance of language [8.37], p. 22. When we speak about something as a potentiality, we unwittingly confuse the world we construct in language, or "objectivity with parentheses," with the "real" world as the flux of experience in our specious present, or "objectivity without parentheses."

In the world of the living that we, humans, strive to cognize, only an observer—a languaging human organism — is capable of making operations of distinction, filling the universe with things that result from such operations. In this sense, signs exist, but as objects brought forth in language they exist in the observer's universe and have properties specified by the observer as a system capable of descriptions (linguistic semiosis). Because of this, signs cannot be viewed as part of a non-human organism's universe in the sense that non-human, non-languaging organisms are not observers capable of descriptions, and to recognize something as a sign a human is needed [8.37], p. 13. As I argued in [8.33], p. 128, we must be aware of the epistemic cut between the observed adaptive response of an organism and the observer describing the observed behavior as if it were the result of the acquisition by the organism of some features of the environment — that is, the organism's perception and categorization of something as a "sign." The concept of sign, just as any other concept, belongs to the cognitive domain of the observer as part of his description of his cognitive domain in which he exists and operates as an organism-environment system, but in no possible sense can it be part of the cognitive domain of the observed organism such as a cell or an animal. We often assume that at least some animals have concepts because we interpret their observed interactional behavior as adequate, overlooking how the measuring device (the conceptual structure "expressed" in the language we "use") becomes the measured (an organism's adaptive behavior).

When we say that signs signify, in the universe as an image of our language they become nothing less than agents capable of actions directed at something else that becomes (if our lived experience is sufficient enough for that) associated with the sign on a more or less permanent basis. Thus, a relationship of signification is established, but because signs in our universe become agents (they possess a capacity to "signify"), we begin to speak of the meaning of a sign, even though "meaning" is not the same as "signification." Importantly, something becomes a sign only as a result of establishing a relationship of signification between what is perceptually present and what is not. A relationship is not a physical entity or phenomenon that can be perceived by a higher animal, either human or non-human. It is the result of projecting what is perceptually present (or, rather, its neuronal representation) onto that part of the animal's lived experience (a stable state of relative neuronal activity as a representation of the animal's previous interactions with the environment) that appears analogous to the current experience of the animal as an organism-environment system. This is the animal's interpretation of the flux of current experience that may trigger a behavioral response depending on the value (positive, negative, or zero) such an interpretation "assigns" to the relationship between the animal and that perceptually absent aspect of the environment as part of the animal's lived experience of which the currently experienced physical conditions are a sign. Note that, because of the neural uncertainty, there may be an "error" in interpretation, and the animal's behavioral response may turn out to be maladaptive. We can say that a dog once beaten with a stick "knows," upon seeing a man with a stick, what the stick signifies by relating it to its lived experience retained in memory as a dynamic network of states of relative neuronal activity. The dog's behavioral response upon seeing an approaching man armed with a stick is triggered by its interpretation of the observed (granted that there is no error in interpretation); however, to the dog, the stick does not and cannot become a sign, because the dog is not an observer capable of making a corresponding distinction in language.

Maritain's [8.43], p. 53, contention that all animals "make use of signs without perceiving the relation of signification," also shared by Deely [8.10], p. 73, appears counterintuitive and is an example of how the epistemological trap of language affects our interpretation and understanding of the observed. As succinctly put by Foerster [8.18], p. 69,

> [w]hen a verb becomes a noun, it suddenly creeps into us as if it were a thing. When you have a process that can become a thing, you are on a bad track. Many of the difficulties we encounter in understanding stem from the fact that we are constantly dealing with things that actually are processes.

Signification is a process that leads to the emergence of a sign, it is a distinction made by the observer in the flux of his experience of interacting with, and adapting to, the environment. But because "signification" is a noun — and nouns are, typically, names of things as objects in the universe that can be perceived and handled or acted upon — it becomes a thing, either perceived or not. However, in the universe of non-human animals, experientially established relationships are not things that an animal perceives and describes, neither do signs as relational concepts that emerge and are distinguished in the flux of an animal's experience exist as entities whose function is to signify. *Sensu stricto*, it does not make much sense to speak of animals as "using signs" unless a reservation is made, such as "animals use what to humans appear as signs."

It should be noted that such a reservation does not imply that signs exist as objective natural phenomena without any reference to the specifically human interpreter, as suggested, for example, by Vehkavaara [8.77]. In his arguments for the naturalization of semiotic concepts that should be based "solely on the objects of external experience," Vehkavaara claims that the semiotic concepts are definable independently of our internal conceptions and intuitions about them, and speaks of two stances in semiotic analysis, from the point of view of "object-agent" and from the point of view of "meta-agent (observer/researcher)" [8.77], p. 551 ff. This idea, shared by some researchers (cf. [8.6] [8.64]) and refuted in [8.32], is a *bona fide* example of "objectivity without parentheses" posited by philosophy of external realism. As such, it is incompatible with Humberto Maturana's constructivist approach to language and cognition: "things arise in language when the observer arises. Therefore, [...] all that exists exists in language as consensual coordinations of consensual actions of observers, the observer included" [8.46], p. 3. Because it emerges in the experientially grounded process of interpretation, and the lived experiences of different organisms may considerably differ, the significance of what the observer observing the behavior of a particular non-human animal categorizes as a sign is determined by the animal's history of fine structural coupling with the medium, it is idiosyncratic rather than universal. What appears, from the point of view of the observer, as a significant aspect of the perceived environment to an adult dog, may not be a "mark" of anything in the limited lived experience of a pup. Moreover, "glitches" in interpretation do occur, when something perceptually present is interpreted by the animal as bearing on some relevant part of its overall experience even though that is not the case. Whatever its actual outcome, the crucial point is the inherent indeterminacy of interpretation as a cognitive process in which signs emerge. If, for example, speaking about the feeding behavior of ticks, we assume that they interpret the smell of butyric acid as a sign, we must also assume the possibility of an error in interpretation — but what are the chances for a tick to "misinterpret" the smell of butyric acid?

Awareness of the problem revealed by questions of this kind resulted in the concept of protosemiosis [8.58] as contrasted with eusemiosis [8.66]: while in the former signs/signals are used for automated control of actions, in the latter signs point to objects interpretable by mind [8.63], p. 216. However, the suggested distinction between protosemiosis and eusemiosis, while drawing an explanatorily useful line, raises at least two relevant questions, one about the conceptual difference between signs and signals, and the other about the nature and function of mind as the interpreter of signs and its relationship with agency.

Clearly, the notion of interpretation at the different levels of life poses a conceptual problem in biosemiotics [8.33], and one way to deal with it is to posit different levels of meaning that may be described either in terms of semiotic hierarchy [8.78] [8.79] or semiotic threshold zones that separate vegetative, animal, and cultural semiosis [8.37]. However, as notions, both semiotic hierarchy and semiotic thresholds emerge as a result of particular operations of distinction made in language, and as such, they belong to and become part of the universe as an image of language. In other words, they are intrinsically anthropocentric, just like any other notion or concept used, not just in biosemiotics, but in any scientific discourse. If we agree, for example, that any semiosis assumes some kind of memory, biosemioticians must and do speak of memory in cells, plants, and animals, even though, as a distinction made in language, *memory* (from Lat. *memor* "mindful") implies the existence of mind (it is not, therefore, accidental that Peirce spoke of semiosis as rules of mind). This is not to say that biosemiotics posits the existence of mind in cells or plants, but it is, again, a good reminder of the epistemological trap of language in which we dwell, and of how theory adequacy depends on the kinds of distinctions made by scientists.

In our attempts to understand and describe biosemiosis, qualitatively different levels of interpretation of the environmental aspects in the animal world and the enlanguaged world of humans (cf. [8.1]) should be assumed, based on the double view of systems that may be described simultaneously — and, therefore, confusingly — in two different domains mentioned in Section 8.1, operational (system as a collection of components) and phenomenological (organism-environment system as a singular entity capable of agency). Focusing on the interpretation of interpretation as a cognitive process would be a good step towards bringing greater explanatory clarity into biosemiotics as the study of signs and meaning in the living systems based on the "use" of second-order language.

8.4 Meaning and the Mindful Observer

Etymologically, to "mean" is "to have in the mind, intend," while the noun "mind" is related to the Anglo-Saxon "munan," "to think." Therefore, the problem of meaning is inseparable from the problem of mind — another thing that "exists," along with thought and (self-) consciousness, in the human universe as described by the observer, while the problem itself, compressed in the question "What is mind?", is seen by many as intractable [8.75].

As I see it, this intractability stems from the failure of cognitivists, philosophers, psychologists, etc., to use a non-reificatory approach to such "phenomena" as mind, (self)consciousness, knowledge, information, and the like. I use the word "phenomena" in quotation marks to show that, strictly speaking, these are not phenomena as something observable. Even if we view them in terms of first-order experience, we should not forget that the nature of this experiential domain is radically different from the experiential domain of "objects." It is the domain of languaging — the relational domain of coordinations of coordinations of behavior in a (second-order) consensual domain, and as such it is constitutive of our humanness.

Because our universe is an image of the language we speak, and, as our domain of existence, it is filled with objects that the observer brings forth through the operations of distinctions in language, reification of the unobservable becomes mother's milk to us as languaging cognizers, and we are very seldom aware of this. For example, according to *Webster's Encyclopedic Unabridged Dictionary of the English Language*, mind is "(in a human or other conscious being) the element, part, substance or process that reasons, thinks, feels, wills, perceives, judges, etc." — that is, mind is something within a conscious living being that generates purposeful behavior, thereby becoming an agent, and in this, human and non-human animals do not differ as they both appear mindful and conscious. However, this is a delusion. Mind is neither an element/part nor substance or process in a human being; it is not something capable of purposeful behavior (agency). There is no such thing as "the mind" in the operation of the nervous system, and "the mind" is nothing but an explanatory notion [8.51]. Similarly, "consciousness" ("awareness of being aware") is an explanatory notion, too, and does not refer to any material substrate or physical phenomenon (just like "love," "hatred," "kindness" and various other unobservables). It is a distinction made by the observer in the relational domain of language in which the observer himself emerges; it helps us orient others and self in our cognitive domain of linguistic interactions, and

> what happens in language (in all the dimensions that arise through the recursion of the linguistic behavior) also becomes, as part of the relational space, part of the domain of transformation of the human nervous system, giving rise to what appears as mind/body mutual modulations [8.51], p. 24.

This brings up the issue of the "kinds of minds" [8.11]. Do non-human animals, whose behavior in an observed situation appears to us as making sense, think because they have minds? And if we answer the question positively, would that mean that the mental processes, say, in a dog or a dolphin, are similar to the mental processes in humans? Or are these processes essentially different?

Neuroscience as the study of the nervous system and its function has a special focus on these and many other questions about the nature of mind and consciousness. However, we should not forget that neuronal processes that take place in the brain are not properties of the brain as a material substrate of the autonomous mind [8.28]. The brain is a complex organ that regulates the many diverse processes within the body of an organism, but organisms do not exist in a vacuum. Every organism is immersed in a plethora of multiple connections and interactions with the world of which the organism itself is a part, and its neuronal processes are representations of its interactions with the world and self. Learning what kinds of interactions are represented by what specific neuronal processes helps us understand how the brain functions as a complex organ, but it does not answer the question of the nature of mind and thought. Any attempts at getting a coherent answer to this question, as well as the question whether the mental processes in human and non-human animals are basically similar or radically different, must start with another, much simpler question, "What makes our biological species, *Homo sapiens,* so radically different from any other species, and could this difference be at the biological root of sapience?" And the answer, of course, is "Language."

The fundamental mechanism at the basis of many complex "phenomena" pertaining to our humanness — mind, thought, (self-)consciousness, etc. — is linguistic semiosis as a biological adaptation, the ability, in our interactions with the environment, to take into account what is perceptually not present. This taking into account what is perceptually not present is a process of abstraction that lays the neuro-physiological ground for (abstract) thought as non-physical interactions with states of relative neuronal activity as representations of what happens in the observer's cognitive domain, the relational domain of language. This process — what we call "thinking" — is an essential part of the human organism-environment system's dynamics, and the observable changes in these dynamics appear as mind/body mutual modulations driven by meaning — the process of establishing a correlation between the transformations of the human nervous system, caused by what happens in the second-order consensual domain of language in all its dimensions, and the organism's inductive behavior. As such, meaning is an exceptional feature of the enlanguaged world of humans; it is the default of human existence [8.27].

There is a view that humans "create" meanings out of their experiences as they move toward a goal in the present moment; as meaning-makers, "they are making sense of their experiences by placing them within a context and anticipating how present actions move them toward a set of goals" [8.52], p. 32; cf. [8.71], p. 1026. However, meaning is not a thing we "create"; in the enlanguaged world of humans it is an aspect of the organism-environment system's dynamics, including the internal process of correlating our inductive behavior with the goals we pursue as observers in the universe we create in language as our mode of living. In short, meaning emerges in, and comes with, thinking, which is neither a passive nor detached activity. Thinking involves an interaction between the person and the environment [8.26]. As a distinction made in language, "thinking" is

> a manner that the observer has of indirectly referring to the internal operation of the nervous system as it participates in the generation of behavior. [...] every animal has a brain that thinks according to the manner of living that it lives. The human brain thinks in language [8.51], p. 24.

This is the crucial point in approaching the problem of meaning in biosemiotics, because, as was succinctly put by Müller [8.54], p. 340, "the one great barrier separating brute and man is Language. Man speaks, and no brute has ever uttered a word. Language is our Rubicon, and no brute will dare to cross it."

Unfortunately, the role of language as "our Rubicon" in the evolution of humans — a major evolutionary transition [8.73] — is often overlooked because of the conceptual confusion of language with communication that we owe to the orthodox structuralist view of language as a communication tool. Another firmly established belief is that, evolutionarily, sapience precedes language. Together, these two drive many researchers' attempts, biosemioticians being no exception, to find and explain continuity between animal and human behavioral capacities [8.67] [8.62] [8.4] inter alia, particularly the gradual transition from apes to humans as far as their cognitive powers go [8.61]. However, as argued by Penn, Holyoak and Povinelli [8.57], p. 110, "[t]he profound biological continuity between human and nonhuman animals masks an equally profound functional discontinuity between the human and nonhuman mind." As I argued in [8.34], this profound discontinuity is rooted in the emergence of language as the pivoting point in the evolution of the human brain, laying the basis for abstract thought. Language and the human brain co-evolved [8.8], and we are cognitively superior to all other animals because we can talk [8.20].

Remarkably, Jaroš and Maran [8.24] claim that the different takes on the issue of (dis) continuity in animal and human cognitive development are not supported by empirical evidence but are based on the ontological stances of the researchers from the two camps and can be reconciled. Such a reconciliation, they suggest, can be achieved with the help of what they call a "3C" model for the analysis of the similarities and differences between human and animal capacities for cognition, culture, and communication as different "faculties," and validated "by quotes from fundamental works of individual researchers" (ibid., p. 383) — including, of course, Darwin. For space considerations, I am not going to analyze Jaroš and Maran's position in detail. However, there are two things I would like to comment on.

First, as the saying attributed to Seneca the Elder reminds us, "*Errare humanum est,*" and when Darwin [8.7], p. 35, insisted that there was "no fundamental difference between man and the higher mammals in their mental faculties," he was certainly wrong, because, as emphasized above, "every animal has a brain that thinks according to the manner of living that it lives," and, unlike all other mammals, "the human brain thinks in language." The point is that quotes from fundamental works cannot serve as a means for validating a theoretical claim.

Second, when human-animal relations are analyzed in three domains, cognition, culture, and communication ("3C"), it only increases the conceptual confusion as a consequence of our epistemological entrapment in language. "Culture" is a distinction made in language and, as opposed to "nature," in our universe it marks the difference between the way of life of humans as languaging and, therefore, sapient animals and other social non-languaging — and, therefore, non-sapient — animals, such as primates. It is an epistemological mistake to extend the inherently anthropocentric notion of culture to non-human animals

— something not uncommon in the literature (cf. [8.5]). Similarly, it is a mistake to view language as a specific communication system of humans, in which case the biological (orientational) function of language, identified by Morris [8.53], is obscured: while communication is the process of establishing common ground for perceptually guided interactions between organisms regardless of the degree of their biological complexity, language is a uniquely human orientational behavior *based on taking into account what is not perceptually present*. All living beings can communicate, but only humans can talk, and this feature is central to their biology [8.25]; it marks the qualitative difference between human and animal cognition because, in humans, cognition and language are structurally coupled [8.2]. Failure to understand language as human life in semiosis has far-reaching epistemological consequences that can be overcome only with a systems approach to the study of our species [8.36] and our universe as an image of our language, our mode of living as observers in search of the Holy Grail of meaning.

In biosemiotics, there is a strong tradition to treat meaning as a result of an organism's action that "imprints its meaning on the meaningless object and thereby makes it into a subject-related meaning-carrier in the respective Umwelt (subjective universe)" [8.74], p. 31. However, when we speak about signs, interpretation, meanings, etc., in the non-human living world, we must bear in mind that, doing so, we extend the analysis of human experience to non-human subjects through empathy [8.79], but that does not mean that non-human animals have the kind of mind and think the kind of thoughts humans have and do. Their manner of living is not in language as a relational domain of coordinations of coordinations of behavior.

Since both "meaning" and "mind" are things brought forth by the observer in language, and because the human brain thinks in language, meanings are not something hidden in the universe as potentialities, they are the product of the operations of distinction made by a mindful observer in the relational domain of language. Grounded in radical constructivist epistemology, such an understanding of cognition and language lays a new, heretofore almost unexplored, pathway to the origin and evolution of meanings in the observer's universe.

References

[8.1] Brier, S. (2015) Cybersemiotics and the reasoning powers of the universe: philosophy of information in a semiotic-systemic transdisciplinary approach. *Green Letters: Studies in Ecocriticism* 19(3), 280–292.

[8.2] Bunnell, P. (2015) Dancing with ambiguity. *Cybernetics & Human Knowing* 22(4), 101–112.

[8.3] Button, T. (2013) *The Limits of Realism*. Oxford University Press, Oxford.

[8.4] Clowry, G. J. (2014) Seeking clues in brain development to explain the extraordinary evolution of language in humans. *Language Sciences* 46, 220–231.

[8.5] Cowley, S. and Kuhle, A. (2020) The rise of languaging. *Biosystems* 198, #104264, doi.org/10.1016/j.biosystems.2020.104264

[8.6] Cuffari, E. C. (2020). On life-language continuity. *Constructivist Foundations* 15(2), 149–151.

[8.7] Darwin, C. (1871) *The Descent of Man, and Selection in Relation to Sex*. John Murray, London.

[8.8] Deacon, T. W. (1997) *The Symbolic Species: The Co-evolution of Language and the Human Brain*. Penguin Books, London.

[8.9] Deacon, T. W. (2015) Steps to a science of biosemiotics. *Green Letters: Studies in Ecocriticism* 19(3), 293–311.

[8.10] Deely, J. (2015). What semiotics is. *Language and Semiotic Studies* 1(1), 63–94.

[8.11] Dennett, D. C. (1997) *Kinds of Minds: Toward an Understanding of Consciousness*. Basic Books, New York.

[8.12] Di Paolo, E. (2005) Autopoiesis, adaptivity, teleology, agency. *Phenomenology and the Cognitive Sciences* 4(4), 429–452.

[8.13] Di Paolo, E. A., Cuffari, E. C. and De Jaegher, H. (2018). *Linguistic Bodies: The Continuity Between Life and Language*. The MIT Press, Cambridge, MA.

[8.14] Eco, U. (1976) *A Theory of Semiotics*. Indiana University Press, Bloomington, IN.

[8.15] Evans, V. (2015) *The Crucible of Language: How Language and Mind Create Meaning*. Cambridge University Press, Cambridge.

[8.16] Foerster, H. von (1972) *Notes on an Epistemology for Living Things*. BCL Report No. 9.3 (BCL Fiche No. 104/1). University of Illinois, Urbana.

[8.17] Foerster, H. von (1973) On constructing a reality. In: *Environmental Design Research, Vol. 2*. W. F. E. Preiser (ed.) Dowden, Hutchinson & Ross, Stroudsburg, PA: 35–46.

[8.18] Foerster, H. von (2002) Vision, language and knowledge: The double blind. In: *New Paradigms, Culture, and Subjectivity*. D. F. Schnitman & J. Schnitman (eds.) Hampton Press, Cresskill NJ: 65–81.

[8.19] Gasparyan, D. (2020) Semiosis as eigenform and observation as recursive interpretation. *Constructivist Foundations* 15(3), 271–279.

[8.20] Gentner, D. (2003) Why we're so smart. In: *Language in Mind: Advances in the Study of Language and Thought*. D. Gentner & S. Goldin-Meadow (eds.) The MIT Press, Cambridge, MA: 195–235.

[8.21] Glasersfeld, E. von (1995) *Radical Constructivism: A Way of Knowing and Learning*. Falmer Press, Taylor & Francis Inc., Bristol, PA.

[8.22] Heidegger, M. (1998) Pathmarks, ed. by W. McNeill. Cambridge University Press, Cambridge (Letter on "Humanism", pp. 239–276, trans. by F. A. Capuzzi).

[8.23] Herder, J. (2002) Treatise on the origin of language. In: *Herder: Philosophical Writings*. M. Forster (ed.) Cambridge University Press, Cambridge, 65–164.

[8.24] Jaroš, F. and Maran, T. (2019) Humans on top, humans among the other animals: Narratives of anthropological difference. *Biosemiotics* 12, 381–403.

[8.25] Jennings, R. and Thompson, J. (2012) The biological centrality of talk. In: *Cognitive Dynamics in Linguistic Interactions*. A. Kravchenko (ed.) Cambridge Scholars Publishing, Newcastle upon Tyne: 33–63.

[8.26] Kellogg, R. T. (1994) *The Psychology of Writing*. Oxford University Press, Oxford.

[8.27] King, L. A. (2014) The commonplace experience of meaning in life. *International Journal of Existential Psychology & Psychotherapy* 5(1), 9–17.

[8.28] Kravchenko, A. V. (2007) Whence the autonomy? A reply to Harnad and Dror. *Pragmatics & Cognition* 15(3), 587–597.

[8.29] Kravchenko, A. V. (2008) *Biology of Cognition and Linguistic Analysis: From Non-realist Linguistics to a Realistic Language Science*. Peter Lang, Frankfurt/Main etc.

[8.30] Kravchenko, A. V. (2016) **Эпистемологическая ловушка языка** [The epistemological trap of language]. *Tomsk State University Journal of Philology* 3(41), 14–26.

[8.31] Kravchenko, A. V. (2020a) Codes, interpretation, meaning: why language matters. *Constructivist Foundations* 15(2), 159–163.

[8.32] Kravchenko, A. V. (2020b) Linguistic semiosis and human cognition. *Constructivist Foundations* 15(3), 285–287.

[8.33] Kravchenko, A. V. (2020c) A critique of Barbieri's code biology. *Constructivist Foundations* 15(2), 122–134.

[8.34] Kravchenko, A. V. (2021) Approaching linguistic semiosis biologically: implications for human evolution. *Rivista Italiana di Filosofia del Linguaggio* 15(2), 139–158 (Special Issue *Languages, languaging and biosemiotics*).

[8.35] Kravchenko, A. V. (2022a) The distinctions we make as observers affect theory adequacy. *Constructivist Foundations* 18(1), 83–85.

[8.36] Kravchenko, A. V. (2022b) The Maturanian turn: Good prospects for the language sciences. *Constructivist Foundations*18(1), 30–41.

[8.37] Kull, K. (2009) Vegetative, animal, and cultural semiosis: The semiotic threshold zones. *Cognitive Semiotics* 4, 8–27.

[8.38] Kull, K. (2017) On the limits of semiotics, or the thresholds of/in knowing. In: *Umberto Eco in his Own Words*. T. Thellefsen and B. Sørensen (eds.) De Gruyter Mouton, Berlin, Boston: 41–47.

[8.39] Kull, K. and Lotman, M. (2016) Theses on semiotic concepts. In: *Concepts for Semiotics*. C. J. Rodríguez Higuera and T. J. Bennett (eds.) University of Tartu Press, Tartu: 11–16.

[8.40] Kull, K., Deacon, T., Emmeche, C., Hoffmeyer, J. and Stjernfelt, F. (2009). Theses on biosemiotics: Prolegomena to a theoretical biology. *Biological Theory* 4(2), 167–173.

[8.41] Love, N. (1990) The locus of languages in a redefined linguistics. In: *Redefining Linguistics*. H. G. Davis and T. J. Taylor (eds.) Routledge, London: 53–118.

[8.42] Love, N. (2007) Are languages digital codes? *Language Sciences* 29(5), 690–709.

[8.43] Maritain, J. (1957) Language and the theory of sign. In: *Language: An Enquiry into Its Meaning and Function*. R. N. Anshen (ed.) Harper & Bros., New York: 86–101.

[8.44] Maturana, H. R. (1970) *Biology of Cognition*. Biological Computer Laboratory (BCL) Research Report BCL 9.0. University of Illinois, Urbana.

[8.45] Maturana, H. R. (1978) Biology of language: The epistemology of reality. In: *Psychology and Biology of Language and Thought: Essays in Honor of Eric Lenneberg*. G. Miller and E. Lenneberg (eds.) Academic Press, New York: 27–63.

[8.46] Maturana, H. R. (1986) Some reflections. Continuing the conversation. *ASC Newsletter* 2(5), 2–3.

[8.47] Maturana, H. R. (1988) Ontology of observing: The biological foundations of self-consciousness and of the physical domain of existence. In: *Texts in Cybernetic Theory: An In-depth Exploration of the Thought of Humberto R. Maturana, William T. Powers, and Ernst von Glasersfeld*. R. E. Donaldson (ed.). American Society for Cybernetics, Felton, CA: 1–54.

[8.48] Maturana, H. R. (2008) Virtual realities and human existence. Appendix 5 in: H. R. Maturana and G. Verden-Zöller. *The Origin of Humanness in the Biology of Love*. Ed. by P. Bunnell. Imprint Academic, Exeter: 196–203.

[8.49] Maturana, H. R. and Varela, F. J. (1987) The Tree of Knowledge: The Biological Roots of Human Understanding. Shambhala, Boston, MA.

[8.50] Maturana, H. R. and Verden-Zöller, G. (2008) The Origin of Humanness in the Biology of Love. Ed. by P. Bunnell. Imprint Academic, Exeter.

[8.51] Maturana, H. R., Mpodozis, J. and Letelier, J. C. (1995) Brain, language, and the origin of human mental functions. *Biological Research* 28, 15–26.

[8.52] Morehouse, R. E. (2012) Beginning Interpretive Inquiry: A Step-by-Step Approach to Research and Evaluation. Routledge, London & New York.

[8.53] Morris, C. W. (1938). *Foundations of the theory of signs. In: International Encyclopedia of Unified Science*. Vol. 1, Part 2. O. Neurath, R. Carnap, C. W. Morris (eds.) The University of Chicago Press, Chicago: 1–59.

[8.54] Müller, F. M. (1861). Lectures on the Science of Language. First Series. Longman and Roberts, London.
[8.55] Ogden, C. K. and Richards, I. A. (1923) The Meaning of Meaning: A Study of the Influence of Language upon Thought and of the Science of Symbolism. Harcourt, Brace & World, Inc., New York.
[8.56] Park, C. L. (2010) Making sense of the meaning literature: An integrative review of meaning making and its effects on adjustment to stressful life events. *Psychological Bulletin* 136(2), 257–301.
[8.57] Penn, D. C., Holyoak, K. J. and Povinelli, D. J. (2008) Darwin's mistake: explaining the discontinuity between human and nonhuman minds. *The Behavioral and Brain Sciences* 31, 109–130; discussion 130–178.
[8.58] Prodi, G. (1988) Material bases of signification. *Semiotica* 69(3/4), 191–241.
[8.59] Quine, W. (1960) Words and Objects: An Inquiry into the Linguistic Mechanisms of Objective Reference. John Wiley & Sons; The MIT Press, New York; London.
[8.60] Raimondi, V. (2014) Social interaction, languaging and the operational conditions for the emergence of observing. *Frontiers in Psychology* 5, #899. https://doi.org/10.3389/fpsyg.2014.00899.
[8.61] Savage-Rumbaugh, S. and Lewin, R. (1994) *Kanzi: The Ape at the Brink of the Human Mind.* Wiley, New York.
[8.62] Schoenemann, P. T. (2009) Evolution of brain and language. *Language Learning* 59(Suppl. 1), 162–186.
[8.63] Sharov, A. A. (2018) Mind, agency, and biosemiotics. *Journal of Cognitive Science* 19(2), 195–228.
[8.64] Sharov, A. A. (2020). Levels of interpretation of signs. *Constructivist Foundations* 15(2), 147–148.1.65. Sharov, A. A. and Tønnessen, M. (2021) *Semiotic Agency: Science Beyond Mechanism*. Springer, Cham, Switzerland.
[8.66] Sharov, A. A. and Vehkavaara, T. (2015) Protosemiosis: agency with reduced representation capacity. *Biosemiotics* 8(1), 103–123.
[8.67] Sherwood, C. C., Subiaul, F. and Zawidzki, T. W. (2008) A natural history of the human mind: tracing evolutionary changes in brain and cognition. *Journal of Anatomy* 212, 426–454.
[8.68] Simsky, A. D., Kravchenko, A. V. and Druzhinin, A. S. (2021) Action-thoughts and the genesis of time in linguistic semiosis. *Slovo.ru: Baltic Accent* 12(2), 7–28.
[8.69] Sonesson, G. (2006). The meaning of meaning in biology and cognitive science: A semiotic reconstruction. *Sign Systems Studies* 34(1), 135–213.
[8.70] Speaks, J. (2021) Theories of Meaning. The Stanford Encyclopedia of Philosophy (Spring 2021 Edition), E. N. Zalta (ed.). https://plato.stanford.edu/archives/spr2021/entries/meaning/.
[8.71] Staver, J. R. (2012) Constructivism and realism: Dueling paradigms. In: *Second International Handbook of Science Education.* B. J. Fraser, K. Tobin and C. J. McRobbie (eds.) Springer, Dordrecht: 1017–1028.
[8.72] Steffe, L. P. (2007) Radical constructivism: A scientific research program. *Constructivist Foundations* 2(2/3), 41–49.
[8.73] Szathmáry, E. and Maynard Smith, J. (1995) The major evolutionary transitions. *Nature* 374, 227–232.
[8.74] Uexküll, J. von (1940/1982) The theory of meaning. *Semiotica* 42(1), 25–82.
[8.75] Uttal, W. R. (2005) *Neural Theories of Mind: Why the Mind-Brain Problem May Never Be Solved.* Lawrence Erlbaum Associates, Mahwah, NJ.

[8.76] Vaz, N. M. (2022) Maturanian observer-dependent immunology. *Constructivist Foundations* 18(1), 69–77.
[8.77] Vehkavaara, T. (2003) Natural self-interest, interactive representation, and the emergence of objects and Umwelt. *Sign Systems Studies* 31(2), 547–587.
[8.78] Zlatev, J. (2002) Meaning = life (+ culture) An outline of a unified biocultural theory of meaning. *Evolution of Communication* 4(2), 253–296.
[8.79] Zlatev, J. (2018) Meaning making from life to language: The Semiotic Hierarchy and phenomenology. *Cognitive Semiotics* 11(1), #20180001. https://doi.org/10.1515/cogsem-2018-0001
[8.80] Zlatev, J., Steffensen, S. V., Harvey, M. I. and Kimmel, M. (2018) Introduction to the special issue "Meaning making: Enactive, participatory, interactive, symbolic." *Cognitive Semiotics* 11(1), #20180006.

Part II
MEANINGS IN THE EVOLUTION OF LIFE

9

Chemical Origins of Life, Agency, and Meaning

Alexei A. Sharov

Elixirgen Scientific, Baltimore, Maryland, USA

Abstract

Traditional models of the origin of life from primordial soup to nucleic acids and formation of primitive cell are misleading because primordial soup never existed. The problem is to find the simplest functional agency that does not require rare organic resources (amino acids or nucleotides) and evolves in a semi-Darwinian way towards greater robustness and complexity. Such agents could emerge only in rare locations with relatively high physical and chemical complexity with liquid phase separation and catalysis. Lipid world scenarios with surface-bound catalysis appear most promising. The first replicating agents could have been catalysts resembling coenzymes which colonized the surface of oil droplets in water. The key event in the origin of life was the emergence of agency-niche coupling, where agents modified and sustained specific properties of the niche (e.g., surface tension of oil droplets) that was beneficial for agent survival and self-production. Circular causation establishes primordial functional cycles in oil droplets that play the role of hereditary signs, which are passed to descendent oil droplets via two channels: information channel that passively preserves properties of the niche, and agential channel represented by active replicators that signify by their functional activity. The proposed two-channel heredity model assumes that novelty appears in the form of agency-niche coupling, which is self-interpretation of replicators in the context of previously constructed niche. Agency-niche coupling is the simplest form of abduction — finding a way to replicate in a historically constructed environment. Reproduction is equivalent to sensing the environment because successful reproduction confirms that the environment is favorable, and the functional circle becomes closed. In subsequent evolution, replicators lost their agential properties and turned into passive genes, whereas agency was largely transferred to non-replicating functional components (e.g., proteins) and to reproducers (e.g., cells and multi-cellular organisms).

Keywords: Catalytic agents, replicators, lipid world, surface metabolism, circular causation, two channels of heredity, agency-niche coupling

9.1 Conflicting Paradigms in Studying the Origin of Life

Studies of the origin of life are valuable mostly because of methodological significance: they represent our understanding of the phenomenon of life in general. Among the most disputed is the question whether organisms are fully controlled by molecules as proposed by Richard

Email: sharov@comcast.net

Dawkins in his "selfish gene" doctrine, or organisms and single cells are autonomous agents capable of goal-directed and context-dependent activity with the help of genes that store intergenerational memory. The latter view is held by the Extended Evolutionary Synthesis and biosemiotics. Another controversy concerns the likelihood of the origin of life. One scenario assumes high probabilities of the origin of life on Earth via self-organization and catalysis of molecules within the primordial soup, whereas another view is that there was no soup, and instead, the primitive form of natural selection resulted in the local production of functional organic molecules. The latter scenario assumes that the origin of life is not only a rare event, but also a *long process of gradual evolution* that possibly took billions of years to reach the level of the last universal common ancestor (LUCA)[1] [9.32] and started long before Earth [9.101].

Historically, past theories of the origin of life closely followed the discoveries in biology and chemistry. Haldane and Oparin [9.79] proposed that primordial life was based on chemical reactions that yield organic materials from non-living resources. Discovery of nucleic acids and their replication led to the RNA-world hypothesis which assumed that RNA molecules appeared by chance before life and then started duplicating with the help of other RNA molecules that functioned as catalysts [9.41]. Studies of catalytic functions of proteins resulted in the notion that metabolism appeared before heredity in so-called "protein interaction world" [9.5]. Self-organization of lipid molecules via non-covalent interactions was the basis for developing the lipid-world scenario of the origin of life [9.94] [9.103]. Although researchers often disagree on which chemical compounds were most important for primordial systems, there is a consensus that life is a chemical system capable of Darwinian evolution [9.54].

Theoretical foundations of the origin of life have been shaped by the dominating evolutionary synthesis in the second half of the 20th century known as Neo-Darwinism with its gene-centric approach. In particular, Richard Dawkins [9.22] considered genes as units of natural selection, called replicators. In contrast, organisms and groups of organisms do not qualify as units of selection because they have much shorter lifespan. Dawkins called them *vehicles* or *survival machines* because their function is just to carry genes/replicators. He wrote about genes: "They are in you and in me; they created us, body and mind; and their preservation is the ultimate rationale for our existence." Dawkins also proposed that accidental appearance of replicators (either nucleic acids or other units capable of self-copying) was the key event in the origin of life [9.22], p. 17–20. According to his theory, evolution results from the combination of spontaneous mutations and passive copying of genes.

This mechanistic doctrine in biology has been challenged by the extended evolutionary synthesis [9.86], and more recently by biosemiotics [9.49] [9.102]. Biosemiotics offers a radically different explanation of life by assuming that living organisms are not zombies/vehicles controlled by genes but semiotic agents capable of "acting purposefully and using signs to make informed choices" [9.102], p. 153. This capacity is also present in subagents such as organs, cells, and intracellular autonomous units, such as ribosomes and functional proteins. DNA sequences are passive hereditary signs that

[1] List of abbreviations: CLR – coenzyme-like replicator, CoA – coenzyme A, GARD – graded autocatalysis replication domain model, ISM – interstellar medium, LUCA – last universal common ancestor, mRNA – messenger RNA, NAD – nicotinamide adenine dinucleotide, PAH – polycyclic aromatic hydrocarbons.

are copied and interpreted by intracellular subagents, such as polymerase enzymes and ribosomes. In other words, genes are information tools used by organisms to support their existence and identity. Naturally, it is difficult to overturn biological terminology after more than a century of mechanistic interpretations. Thus, I will occasionally use terms "replicator" and "natural selection" as metaphors stripped of their underlying reductionist assumptions.

Initially it was assumed that the origin of life was a single-step accident where typical organic compounds self-assembled into a primitive cell-like living system [9.39] [9.45] [9.79]. However, there was no success in demonstrating that such self-organizing event is possible. Then, Dawkins [9.22] proposed that life may have started from a single self-replicating macro-molecule, called a replicator, assuming that there was a mechanism for its copying and a large supply of monomers. This hypothesis matches well to the RNA-world scenario of the origin of life [9.41]. Later, Szathmáry [9.111] discussed alternative types of replicators that could have a higher chance to emerge spontaneously and support the origin of life: autocatalytic sets, lipid vesicles, and membranes. In addition, it is necessary to find potential pathways of evolutionary change of such replicators towards more familiar living systems, which may include a long sequence of additional steps. Such multistep scenarios of a sequential emergence of various biological functions are already discussed [9.82] [9.91]. Because small steps of change are more likely to happen than giant one-step emergence, the multistep scenario overall appears more plausible.

This idea of a long multistep evolution of primordial systems prompts us to reconsider many assumptions that were treated as obvious in earlier scenarios of the origin of life. First, these systems require an internal mechanism of adaptive change, which most likely was natural selection. Second, if natural selection started operating at the initial step of the origin of life, then heredity has to be supported not by nucleic acids (as assumed previously) but by much simpler molecules that were likely not polymers. Such kind of heredity is often called "compositional inheritance" [9.96] or "compositional genome" [9.119]. But this type of heredity does not resemble digital strings in a computer or human text; and thus, there may be doubts if it is stable enough to support accumulation of adaptive traits. However, this hypothetical type of heredity is in fact digital. Digital information requires strong domains of attraction that make spontaneous cross-domain switches very unlikely. In this respect, covalent bonds in organic molecules are strongly digital, as emphasized by Schrödinger [9.93]; thus, self-producing organic molecules connected by covalent bonds with sufficient energy barriers can support digital heredity. Self-production is needed to compensate for natural loss of molecules due to dissipation or damage. Such molecules can carry hereditary information as long as they keep self-producing (e.g., if their numbers do not fall below the threshold that is needed for self-production). The functionality of compositional heredity was demonstrated by computer simulations [9.103] [9.105].

And third, there is no need to hypothesize the stereotypical "primordial soup" because evolving chemical catalytic systems can start making new organic molecules on demand. These changes of basic assumption seem radical enough to talk about the shift towards a new paradigm in studies of the origin of life.

The emergence of first forms of evolving replicators with limited heredity, natural selection, and sufficient evolvability can be treated as the groundbreaking event in the origin of life, because these features provide a real chance for replicators to develop adaptations that eventually yield unlimited heredity and evolvability,[2] which is the hallmark of the extant forms of life. It would be naïve to expect that primordial systems immediately acquired unlimited heredity via template copying. Thus, it makes sense to consider three periods in the origin of life: (1) physicochemical evolution before the origin of evolving replicators, (2) emergence of first replicators and their evolution towards unlimited heredity, and (3) evolution towards basic forms of life on Earth as we know it, such as LUCA [9.32] [9.115]. In this chapter, I briefly discuss physicochemical evolution and then focus on the second transitional period, in which life properties became gradually established. I describe scenarios concerning how first replicators emerged by establishing agency-niche coupling where agents/replicators used niche construction to initiate and sustain their self-reproduction. Then replicators that happen to share their niche were forced to cooperate due to group selection and stochastic corrector model [9.109] [9.112]. Pascal *et al.* [9.82], p. 2, proposed calling such transitional systems as partially alive and evolving towards the increase of "aliveness," which culminated with the establishment of membrane-bound cells with unlimited heredity.[3] In subsequent evolution, most replicators lost their agential properties, which were largely transferred to reproducers, and passive DNA became the major repository of hereditary signs.

9.2 Physicochemical Evolution

Physicochemical evolution is a hypothetical long-term trend of complexity increase in some local physical environments in the Universe, which inadvertently increased the chances of the emergence of life in these conditions. Classical thermodynamics indicates that the increase of complexity is a highly improbable event in an isolated system of particles that do not multiply and retain their identity. However, dynamic chemical systems with autocatalysis and cross-catalysis can generate and retain their metastable states with increased level of complexity if supplied by energy [9.80] [9.81] [9.88]. Obviously, life cannot emerge in environments that lack sufficient physicochemical diversity, such as in the interstellar hydrogen gas clouds. Thus, self-organization processes are needed to prepare environments suitable for the emergence of replicators. Such processes exist at various spatial and temporal scales in the Universe, from the formation of galaxies, stars, and planets at the large scale to continental drift, hydrological cycle, hurricanes, and tornados at the scale of individual planets or continents. Self-organization also occurs at smaller scales in such processes as crystallization, sedimentation, filtration, and thermal convection.

The increase of complexity can also be supported by production of extremely stable particles, such as heavy elements produced by thermonuclear synthesis in stars. Thus, "[t]he cycle of birth and death of stars constantly increases the abundance of heavy elements in

---

[2] According to Szathmáry and Maynard Smith [9.114], unlimited heredity requires template copying of replicators. It can support a vast number of possible types of hereditary molecules that far exceeds the number of physically existing hereditary molecules [9.110].

[3] The notion of life does not have an exact definition and is used according to language habits. But in relation to the origin of life, I prefer associating life with the start of sustained adaptive evolvability driven by heritable variation and selection. Thus, degrees of "aliveness" discussed by Pascal belong to already emerged life in my understanding.

the interstellar medium, a crucial prerequisite for terrestrial (rocky) planet formation and subsequently for the origin of life" [9.37], p. 493.

The increase of physicochemical complexity has a subjective component because organisms differ in their capacity to establish specific interactions with different components of their environment. Thus, the same physical conditions may appear complex and heterogeneous for one kind of proto-life system and simple and homogeneous for another kind of system. A naïve approach of measuring complexity by counting all known types of particles (elementary and composite) as well as all possible types of interaction would not work because most of these physical entities are not suitable for life. Thus, selective counting is needed, where elements are weighted based on their stability, reactivity with other entities, and context-dependency. Because we don't know any alternative types of life, I focus on the origin of only those lineages of life that are represented on Earth. The bias associated with this approach can be later corrected if we encounter alien forms of life. It is also possible that the bias is relatively small if all possible forms of life in the Universe are chemically similar, showing just minor variations of the same principles of structure and function. Such similarity can result from two mutually compatible scenarios. The first scenario is that life originated just once in each accessible portion of the galaxy and then spread by microbial panspermia, and the second scenario is that many forms of life evolved in parallel and converged into similar chemical types as they followed comparable constraints and bottlenecks in their evolution.

Following this approach, physicochemical complexity before the origin of life can be evaluated in relation to hypothetical primordial replicators; and thus, the notion of habitat complexity (see above) appears to merge with habitability. Gowanlock and Morrison [9.44] list the following main components of habitability: (1) source of energy, (2) range of major chemical elements (C, H, N, O, P, S) necessary for building biomolecules, (3) liquid water as a medium to facilitate chemical interactions. To this we can add the self-organization capacities discussed above. Conditions that are favorable for self-organizations have been explored by modeling of kinetic barriers: "At moderate temperatures, allowing for the presence of liquid water, and for timescales expressed in seconds to years, systems based on covalent bonds were inferred as the ones more likely to support self-organization" [9.82], p. 4.

Although conditions of habitability seems reasonably simple, it appears that very few planets have habitable environments because of various physical constraints such as atmosphere conditions, orbit of planet around the star, and position of the star inside the galaxy [9.44]. However, despite generally low levels of complexity of non-living environments, there may be isolated pockets of increased complexity that persist and possibly even expand due to natural self-organization. Component substances of these pockets can be transferred by natural processes (e.g., currents, winds) to other locations and serve as seeds of increased complexity if they can support self-organization processes similar to those at parental sites.

Physicochemical complexity of non-living environments generally is not stable and can fluctuate in space and time. However, because the origin of life is a rather long evolutionary process, the important constraint is the *lower limit* of these fluctuations. Thus, I focus here on local stable microhabitats, such as oil droplets in water that can keep increased physicochemical complexity over relatively long times.

9.3 Hypothetical Primordial Soup

Most scenarios of the origin of life assume the existence of "primordial soup" — a diverse mixture of organic molecules in water that can self-organize into proto-cells. The idea of soup comes from the famous Miller's experiment [9.70], where electric sparks generated various organic molecules from inorganic gases and water. The soup hypothesis is assumed to generate resources of organic molecules that are later used by emerging self-reproducing functional polymers such as nucleic acids in the RNA-world scenario [9.41], polypeptides in the collectively autocatalytic sets of polymers [9.55], or various kinds of hydrophobic or amphiphilic polymers in the lipid world scenario [9.94]. According to Bernstein [9.11], p. 1689, "all non-magical notions of how life emerges are predicated on the basis of the presence of organic compounds, as well as some other substances that we know make up living things today." What Bernstein misevaluated is that the primordial soup is magic by itself, as follows from this discussion.

Supporters of the soup hypothesis argued that some amino acids, carbohydrates, and nucleotides can be synthesized without protein-based catalysis [9.8] [9.61] [9.87] [9.92]. Hydrothermal vents were long considered as a source of various organic molecules that can contribute to primordial soup [9.8] [9.50] [9.57]. One of the hypothetical mechanisms includes synthesis of formaldehyde from carbon dioxide in ZnS-enriched compartments of hydrothermal vents, as the first step [9.72] [9.73] [9.74]. Subsequent steps are supposed to produce carbohydrates via the formose reaction. However, this scenario is not realistic because the concentrations of formaldehyde in hydrothermal vents are in the nanomolar range [9.47]. Formaldehyde is a simple molecule, and indeed, its concentration can be 10-fold higher in some special conditions. But this is not sufficient to reach the threshold for the formose reaction, which is a billion-fold higher [9.69] [9.78]. I am not aware of experimental systems mimicking the origin of life that include the formose reaction. Thus, carbohydrates and more complex derivative molecules cannot emerge in this way.

Recent research has drawn attention to extraterrestrial sources of organic molecules: interstellar clouds and carbonaceous meteorites [9.15] [9.18] [9.28] [9.36] [9.37]. Although meteorites indeed contain some amino acids, nucleobases, and sugars, the concentration of these molecules is extremely low. In the Murchison meteorite, nucleobases and sugars are present at concentrations of approximately ~1 ppm [9.15] [9.18]. It was suggested that concentrations can increase by such mechanisms as diffusion across microporous minerals [9.19] [9.30], evaporation cycles [9.21], or absorption on clay [9.9]. But it has not been shown that any of these mechanisms are efficient enough to increase the concentration of biomolecules to physiologically-relevant levels [9.27], p. 373-4.

Specific polymers, such as RNA or peptides, can persist only on condition of unlimited local supply of monomers, a condition that has not been demonstrated by the primordial soup hypothesis. Thus, the scenarios of the RNA-world or peptide world contradict with the fact that the concentrations of nucleotides or amino acids are far below the limits that allow polymerization [9.27] [9.59] [9.60]. The most problematic are various versions of RNA-world scenarios that assume the existence of nucleic acids in the primordial soup. Nucleic acids clearly show features of previous adaptive evolution: exactly spaced hydrogen bonds, fixed orientation along the sugar-phosphate backbone, and chirality of sugars. Except for a divine origin, such perfect structure and function can be explained only by heritable variation and natural selection. This means that even if the RNA-world systems were indeed precursors of living cells, they appeared much later than the origin of the first molecular agents capable of adaptive evolution. Freeman Dyson [9.34] proposed that the first evolving primordial systems had no nucleic acids and supported their heredity in a non-genetic way. He assumed that nucleic acids appeared later via polymerization of ATP-like molecules, and later caused a "genetic takeover" when earlier non-genetic forms of life were wiped out and replaced by cells with nucleic acids. This line of thinking is now supported by some researchers [9.10] [9.27] [9.91], who agree that life originated long before nucleic acids, and hereditary functions were initially performed by other means. Bernhardt argued that RNA is too complex to have arisen prebiotically, and in addition, it is highly unstable, and generally, a weak catalyst. These features are not compatible with the RNA-world hypothesis.

Instead of considering primordial soup with amino acids, nucleobases, carbohydrates, and phospholipids, I focus on self-aggregation of organic compounds that are indeed plentiful in the non living conditions. The most important candidate is oil (i.e., a mixture of liquid hydrocarbons), which is the most abundant non-gaseous organic resource in the universe [9.26] and is abundant on terrestrial planets [9.65]. Oil molecules can be converted to fatty acids, which opens a possibility to make other lipids. Such "oily" version of the primordial soup is known as the lipid world [9.94]. The hypothesis of the lipid world emerged from the discussion on the origin of chemical complexity in pre-biotic environments. Shenhav *et al.* [9.106] argued that two levels of microscopic complexity can be found in living cells: the first one is based on covalent molecular structures, whereas the second one involves non-covalent assemblies such as lipid membranes. Most theories of the origin of life discuss the first kind of complexity, and the second kind of complexity is largely ignored in this respect. The specific feature of non-covalent assemblies is that their complexity is represented by combinations of various types of molecules that are bound together. The dynamics of such systems was explored with a GARD (Graded Autocatalysis Replication Domain) computer simulation model [9.94] [9.95] [9.104]. The model assumes that amphiphilic and hydrophobic molecules (e.g., lipids, amino acids, and peptides)[4] aggregate into droplets in water and control the rates of association and dissociation of other molecules. Droplets grow by accumulation of molecules from the environment and may split after reaching a critical size, which resembles self-reproduction. The term "autocatalysis" in this model means selective non-covalent association with molecules of the same type rather than changing covalent bonds[5].

Simulations demonstrated that such self-reproducing vesicles in the GARD model tend to produce progeny vesicles with a similar composition of components to parental vesicles, which was interpreted as a primitive kind of compositional heredity that does not require nucleic acids [9.96]. However, computer simulations of the evolutionary dynamics of such systems showed that they do not have sufficient evolvability [9.118]. The main factor that limits their evolvability is the lack of covalent bond change to generate novel kinds of molecules.

[4] Concentration of amino acids in oil droplets before the origin of life is expected to be extremely low. Thus, it is not realistic to assume the presence of peptides.

[5] This definition does not match the traditional understanding of autocatalysis. But it is consistent with my model of niche-dependent self-production of molecules (see below).

A modified version of the GARD model assumes polymerization of simple molecules into short oligomers via stable covalent bonds [9.103]. According to computer simulations, primordial systems described by this modified model (called P-GARD) have increased evolvability because newly synthesized molecules can be unique and stable due to covalent bonds, and hence, can better perform hereditary functions. The appearance of a new molecule type or a loss of a molecule type is equivalent to a mutation. These "mutations" can affect the rates of survival and self-reproduction of oil vesicles, and then the numbers of most successful systems would increase by natural selection [9.51]. P-GARD is a hybrid model that combines two kinds of chemical complexity that involves both non-covalent and covalent interactions.

Although P-GARD models show evolvability in computational experiments, it is still not clear how comparable real systems could have emerged in nature. Similar to the RNA-world and protein interaction world, this model is based on the wrong assumption of rich primordial soup. Although experiments showed that some amphiphilic lipids can be synthesized from organic molecules detected in the interstellar medium (ISM) by UV radiation [9.27], there is no evidence that lipids can accumulate in the oil droplets in water and persist long enough to support the origin of life. Polymerization activity assumed in P-GARD scenario implies something like protein synthesis, which is not realistic in conditions where amino acids are extremely rare. Even amphiphilic lipids were likely not sufficiently abundant to form micelles or membranes in water prior to the origin of life. Lipids in meteorites have very low concentrations and could be artifacts of heating during the passage of meteorite through the atmosphere. For example, fatty acids are rare components of oil droplets in water because the typical content of oxygen in petroleum is only ~1% [9.116]. Phospholipids are unlikely to become abundant before cellular life because they are complex molecules that easily degrade. Deamer *et al.* [9.27], p. 375, wrote: "Although phospholipids are universal components of cell membranes today, it seems improbable that such complex molecules were available on the prebiotic Earth." Thus, it is more likely that phospholipids emerged via adaptive evolution for the purpose of making stable and elastic membranes.

9.4 Life Emerged in the Form of Molecular Agency

Self-organization is necessary for life as it allows a system to persist despite partial degradation and dissipation, but it is not sufficient: "Physical self-organized systems eventually stop functioning without affecting any future self-organizing events, whereas living systems produce the next generation of living systems of the same kind" [9.102], p. 127. Thus, life is more than just self-organization, such as crystal growth or dissipative structures [9.35] [9.55]. Instead, primordial life emerged in the form of agency capable of capturing resources, self-assembly, and heritable self-reproduction, which are necessary conditions for the Darwinian evolution. Therefore, the problem of the origin of life is in finding such context/niche where organic molecules may become functional agents.

Elements of the agential approach to the origin of life can be found in the writings of Alexander Oparin [9.79] and John Haldane [9.45]. Both of them stressed the importance of two functional levels: molecular and proto-cellular in the organization of primordial systems. Oparin thought that cells formed via a liquid phase separation process, whereas Haldane already knew that cells are enclosed in a membrane.

Recent versions of the agential scenario includes self-replicating virus-like nucleic acids within mineral compartments of hydrothermal vents [9.56] [9.57], autogen model of Deacon *et al.* [9.23] [9.25], lipid world [9.94], and coenzyme world [9.100]. Koonin [9.56] assumed that chemical agents, such as self-replicating nucleic acids, emerged within mineral compartments of hydrothermal vents. It is a compartmentalized version of the RNA-world model with more functionality and cross-catalytic reactions. Deacon *et al.* [9.25] developed a model called *autogen* that includes several peptide components with cross-catalytic activity. The autogen includes peptides that can get assembled into a sphere resembling a viral capsid, which encloses several catalysts. According to Deacon, this structure can support long-term survival under unfavorable conditions. If the enclosure is broken, then catalysts are released and help to repair the enclosure. The advantage of these agential models is that they are focused on functions and evolution rather than on individual types of molecules. But they also have serious weaknesses. First, both models (hydrothermal enclosures and autogen) do not explain the supply of monomers (i.e., nucleotides for the former model and amino acids for both models). And second, they assume the existence of a rather complex system where parts are already well fitted to the whole without explaining how such high complexity and perfect fitness of parts emerged in the first place. For example, the latest version of Deacon's autogen includes cross-catalysis between two types of replicators [9.24], but the origin of each individual replicator type was not explained. An additional problem is that RNA would degrade faster than if it is replicated without protein enzymes and with a sparse supply of monomers.

Because changes of covalent bonds appear essential for the evolvability of primordial systems [9.119], we need to understand the origin of catalysts which could have facilitated the emergence of such systems. Primordial catalysts have to be different from catalysts in extant living cells such as proteins or non-coding RNA because there was no supply of activated amino acids or nucleotides at the origin of life (see the above subsection on primordial soup). An additional constraint is that proteins can be strong catalysts only if the order of amino acids is controlled by nucleic acids; but nucleic acids could not have been present at the origin of life. In particular, the random protein synthesis proposed in [9.76] generates only weak catalysts and the proportion of molecules with catalytic activity is small. Thus, random protein synthesis is prohibitively wasteful in conditions where amino acids are scarce resources. Mineral catalysts cannot self-reproduce, and thus, their role in the origin of life was limited. The only option left is that first catalysts were relatively small organic molecules, similar to coenzymes, which managed to support their own synthesis and/or assembly in specific local conditions; I called this scenario *coenzyme-world* [9.98] [9.100]. Because many coenzymes (e.g., ATP, NAD, and CoA) contain or are similar to nucleotides, coenzyme-like replicators (CLRs) can be considered predecessors of nucleotides, as it was suggested previously [9.58] [9.124]. Other kinds of CLRs could have been similar to amino acids or vitamins.

9.5 Lipid World as a Habitat for Self-Reproducing Catalytic Agents

Now let us consider potential habitats in which first CLRs were most likely to colonize. Simple homogeneous habitats, such as 3-dimensional volume of gas or water, are not favorable for catalysis, because molecules encounter each other faster in two dimensions than in three dimensions [9.2]. This is why surfaces are so important in living cells [9.48]. For example, phospholipid membranes have multiple functions such as separating cells from the environment, preventing the dissipation of cell components, holding receptor proteins to sense the outside world, and carrying tools for cell propulsion.

Among surfaces available in physical environment, two kinds are the most likely places for the origin of life: water–mineral and water–oil surfaces. These two kinds of surfaces are selected for discussion because they both include water, which is indispensable for most chemical reactions in living cells. These surfaces are also relatively stable in comparison to highly unstable air-water surfaces. Here I focus mostly on water-oil surfaces because (1) oil self-aggregates in water, forming fine dynamic droplets that can divide or merge, (2) oil-bound molecules avoid the dilution that occurs with water-soluble organic molecules [9.27], p. 374, (3) oil-bound molecules float on the surface and can interact with other oil-bound molecules, (4) oil can be used as a source of carbon, and (5) oil droplets can turn into a phospholipid membrane (see below). Mineral-water surfaces do not have these features (except feature #2), and are therefore less likely to support the emergence of primordial self-reproducing molecular networks. Nevertheless, minerals can be helpful in supporting catalysis in primordial systems [9.122], especially if covered by oil film or submerged in oil droplets in water.

The next question is what could have been the source of carbon for making these CLRs? There are two main options for producing carbon-based molecules: fixation of carbon-based gaseous components (CO_2, CO, CH_4), or decomposition of oil (hydrocarbons). The first option was studied by Hartman [9.46] and Morowitz *et al.* [9.71], who assumed that carbon was assimilated from carbon dioxide by the Calvin cycle, which works in green plants. But the Calvin cycle requires protein catalysts and a large amount of energy in the form of ATP, which were not available before the origin of life. The computational analysis of molecular network structure suggests that the Calvin cycle has a negligibly low probability of realization by pure chance [9.125]. Another scenario of the origin of life by carbon dioxide fixation assumed that this pathway of carbon fixation was realized in micropores and microcompartments with inorganic walls containing catalytic Fe(Ni)S minerals in alkaline hydrothermal vents [9.108]. The authors argued that "the difference in pH across these thin barriers produced natural proton gradients with equivalent magnitude and polarity to the proton-motive force required for carbon fixation in extant bacteria and archaea" (p. 1). They further hypothesized that the most ancient pathway of carbon fixation involved acetyl CoA. This is just another kind of "metabolism-first" model of the origin of life that ignores the crucial question of primordial heredity, without which life could not have emerged. The idea is that physical conditions in alkaline hydrothermal vents are favorable for the emergence of CO_2 fixation pathways in already existing and evolving primordial living systems. But it did not happen at the very beginning of life, when there was no glycerol, no lipid membrane, and no other organic molecules (e.g., amino acids, nucleobases, or ATP) at physiological concentrations to support the hypothetical carbon fixation pathway.

Thus, it is highly unlikely that life started from the assimilation of carbon dioxide by Calvin cycle, which leads us to an alternative scenario that the first primordial agents were heterotrophs decomposing oil. Deamer [9.29] wrote: "It seems likely that primitive cells incorporated lipid-like molecules from the environment as a nutrient, rather

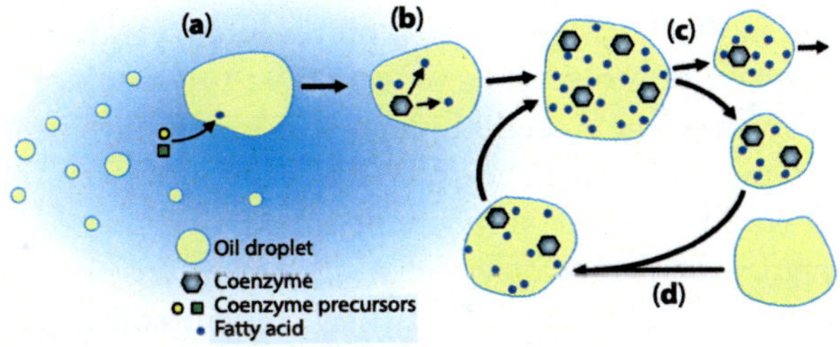

Figure 9.1 Colonization of the surface of oil droplets by coenzyme-like replicators (CLRs). (a) Precursors anchor to the oil droplets via rare fatty acid molecules and become assembled into a CLR. (b) CLR oxidizes oil to fatty acids, which provides additional sites for making new CLRs. (c) Fatty acids make oil surface hydrophilic, which increases the chance of a droplet to split into parts. (d) Small droplets can capture additional oil as a resource.

than undertaking the much more complex process of synthesizing complex lipids by an enzyme-catalyzed process."[6]

Here is a hypothetical scenario of how CLRs could have colonized the surface of oil droplets in water. Consider simple precursors of CLRs that are dissolved in water but cannot anchor to the hydrophobic oil surface. But they can bind to rare fatty acid molecules with hydrophilic ends and assemble into functional CLRs (Figure 9.1a). This CLR can now catalyze the oxidation of outer ends of oil molecules in the droplet and make more fatty acids, where additional CLRs can be assembled from precursors (Figure 9.1b). Some components of CLRs could be taken from the oil droplet rather than from the water. These may include microelements and hydrophobic organic molecules, such as aromatic hydrocarbons. A CLR is therefore a simple agent that performs two functions: self-reproduction (autocatalysis) and niche construction, which makes the oil surface more hydrophilic and thus more accessible to the binding of CLR precursors.

High-energy habitats are generally not favorable for the origin of life because their destructive effects (e.g., hydrolysis) overshadow potential benefits. Thus, primordial chemical agents had to rely mostly on sparsely-dispersed energy sources and spend most of the time in the inactive state, waiting for accidental energy sources. The most likely source of energy for reactions catalyzed by CLRs is photoactivation catalyzed by organic pigments that could have been present in oil droplets. It appears that polycyclic aromatic hydrocarbons (PAHs) and their derivatives absorb light (UV and blue) and can function as primitive pigments [9.28]. PAHs are among the most abundant organic gas-phase molecules in space [9.37], especially in circumstellar envelopes of carbon-rich stars [9.36], they are also abundant in the carbonaceous meteorites [9.27]. PAHs can arrive on planets with meteorites and cosmic dust, and because of their hydrophobic nature, they can be incorporated into oil droplets. Besides capturing energy, organic pigments may protect other functional components from damaging UV radiation [9.27], p. 374.

Upon activation, CLRs can function as catalysts that help in oxidizing oil molecules and producing new CLRs. This pattern of activity comprises niche-dependent self-reproduction of CLRs and niche construction, closing the self-reinforcing functional circle by increasing

---

[6] Cited from https://ucmp.berkeley.edu/education/events/deamer1.html.

the probability of a droplet to split (Figure 9.1c). Later, small droplets can coalesce with other oil droplets, and in this way capture new oil resources (Figure 9.1d).

The first CLRs were simple molecules and were relatively weak catalysts if isolated from their environment. However, their catalytic capacities could have been facilitated by interaction with other compounds around them. We already discussed their potential interaction with PAHs that could function as pigment supporting photoactivation. In addition, CLRs could have augmented their catalytic capacity via interaction with metals or minerals. This hypothesis can be supported indirectly by comparison of CLRs with some amino acids that readily interact with heavy metals such as cobalt, nickel, or zinc, and these complexes show catalytic activity [9.33]. Many inorganic compounds are natural contaminants in petroleum, and thus they were possibly captured by the oil droplets at the origin of life. Besides oil droplets, CLRs can also colonize mineral surfaces covered with a film of oil, where more mineral particles can be mixed with oil than in free-floating oil droplets. An increased abundance of mineral particles can boost catalytic activity of CLRs, increasing their rate of self-reproduction and spread along mineral surfaces. Because CLRs can move from oil droplets to oil-covered surfaces of oleophilic minerals, mineral surfaces and oil droplets could host one and the same population of CLRs.

In summary, this model of catalytically active replicators (CLRs) on the surface of oil droplets or oil-covered minerals differs from the lipid-world scenario in the following details. First, the lipid-world model assumes the existence of a primordial soup with abundant lipids, amino acids, and other diverse organic molecules. In contrast, the CLR model does not require abundant organic molecules except liquid oil and PAHs, which are known to be abundant. Second, the original lipid-world model did not include change in covalent bonds, whereas CLRs are active catalysts capable of changing covalent bonds. And third, the CLR model assumes that primordial metabolism occurred on the surface of oil droplets, which provides a physical scaffold for catalysis and molecular interaction, whereas the lipid-world hypothesis ignores the spatial location of molecules.

Oil droplets (and possibly, oil-covered minerals) with CLRs create a two-level hierarchical system that includes self-reproduction at both levels: (1) catalytically active molecules on the surface and (2) oil droplets (and minerals) hosting populations of surface-bound CLRs. Self-reproduction at both levels is facilitated by CLRs because they modify their local environment (i.e., surface properties of oil droplets) and this change enables them to capture resources, self-reproduce, and propagate to other oil droplets (Figure 9.2). I see this kind of replicator-induced niche construction as the key event in primordial evolution that creates systems with a chance to expand their functional capacities via natural selection and eventually reach unlimited heredity. The *replicator-niche coupling model* presented here is a simpler, and therefore, more realistic model than earlier proposed models which included two or more mutually coupled replicators: hypercycle [9.38], autocatalytic set [9.55], or autogen [9.24].

The roots of the idea of agency-niche coupling can be found in the writings of Howard Pattee, who noted that the origin of life cannot be explained by studying candidates of initial coding molecules; it is equally important to learn what the first ecosystem was in which these coding molecules appeared to be functional and meaningful. He wrote "it is reasonable to consider the hypothesis that the first 'messages' were expressed not in the highly integrated and precise genetic code that we find today, but in a more global set of geophysical and geochemical constraints, which we could call the primeval 'ecosystem

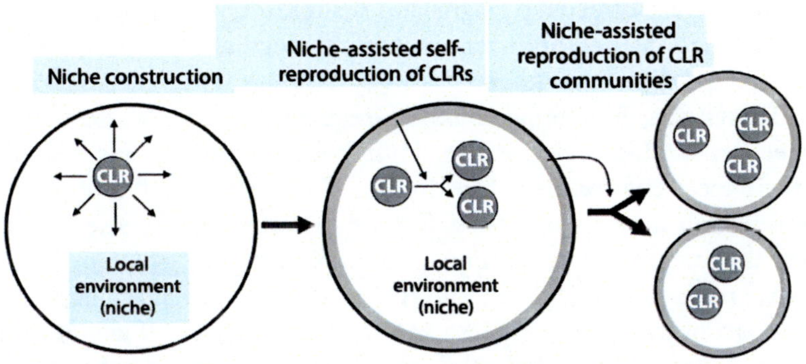

Figure 9.2 Replicator-niche coupling model: Coenzyme-like replicators (CLR) initiate niche construction (e.g., oxidation of the outer layer of oil droplets shown by shading), and the modified niche facilitates self-reproduction of CLRs and reproduction of niche-bound communities of CLRs.

language'" [9.84], p. 3. However, Pattee did not write about the importance of niche construction at the origin of life, because the idea of niche construction was developed much later [9.77].

9.6 Semiotics of Heredity

Heredity means passing organisms' properties from parents to their descendants in the form of signs and scaffolds that are interpreted by the organisms and their parts. Signs are interpreted by organisms and their subagents as something else, whereas scaffolds are preserved as they are with necessary repair, replacement, or expansion. The same material thing can function as a sign in one kind of interaction and as a scaffold in another kind of interaction. For example, DNA functions as a sign when it mediates binding of transcription factors or synthesis of a messenger RNA (mRNA), which controls the synthesis of proteins in the cytoplasm. But in addition, DNA functions as a scaffold when it is replicated or repaired. Other examples of scaffolds in the cell are lipid membranes, cytoskeleton fibers, and chromatin structures. Scaffolds can be treated as degenerate signs with a reflexive meaning, as in the case where a map is identical to the territory.

Signs and scaffolds comprise the informational channel of heredity, but in addition, heredity requires the agential channel, which is the activity of agents guided by interpretation of signs. Semiotic agency was defined as "a capacity for acting purposefully and using signs to make informed choices" [9.102], p. 153. Agency is a generalization of the notion of organism that is applicable, besides organisms, to autonomous organism parts (e.g., organs, cells, organelles, and functional proteins) and to integrated multi-organism systems (e.g., colonies, populations, and symbiotic consortia). The properties of organisms are passed from parents to their descendants by the passive informational channel and the active agential channel [9.99]. An important feature of life is the *functional coupling* between the informational and agential channels of heredity, which means that signs and scaffolds preserve and enhance the quantity and quality of catalytic agents in various ways, and catalytic agents, in turn, copy, produce, and repair signs and scaffolds. The quantity of agents is increased via more efficient synthesis of their components and improved self-assembly when these processes are directed by signs and scaffolds. Qualitative changes may include increased functional robustness and/or more accurate regulation (e.g., by feedback and better timing).

In primordial systems, functional coupling emerges via natural selection at two levels: individual selection of CLRs and group selection between populations of CLRs, each associated with their local environment (oil droplet). Potential benefits of a combination of two levels of selection may come from a faster adaptation due to selection of individual CLRs, whereas the integrity of interactions can be sustained by group selection, such as in the stochastic corrector model [9.112]. The core of the primordial self-reproducing system is a minimal subset of irreplaceable CLRs and their interactions. If any of these core components or interactions is lost entirely, then it cannot be recovered based on other components, which may result in a decline of the whole community of CLRs.

In particular, CLRs play the role of hereditary signs because they encode the surface properties of oil droplets (phenotype) and transfer the code to the next generation of colonized oil droplets in the form of additional copies of the same molecule kind [9.98]. Because of their self-reproduction capacity, oil droplets with a specific community of CLRs on their surface are higher-level agents that carry passive hereditary signs (amphiphilic molecules) and support the function of catalytic agents (CLRs). This can be seen as the beginning of semiosis in these systems. To support evolvability, CLR molecules with hereditary functions should not exist as resources in the environment (otherwise primordial life would never change beyond the initial attraction domain).

The origin of heredity is an important event in global evolution because it marks the origin of agency and meaning in the universe. Before this event, the universe was knowable but it has not been known. Objects and relation between them had only potential meanings, such as types of objects, part-whole relations, or causation. With the origin of agents and heredity, some potential meanings were converted into actual meanings if they were used by agents in a specific and heritable way. Potential meanings (e.g., object types, correlations, causations, catalysis) are not associated with goals or values, but actual meanings that include a pragmatic value in relation to the specific kind of agents. According to the umwelt-theory of Jakob von Uexküll [9.121], animals associate external objects with specific functions, such as eating, hiding, or courting. Similarly, simple agents interpreted molecular-level interactions as signals for initiating activity (e.g., catalysis).

First agents were new kinds of objects in the universe that did not exist before. Moreover, they have constructed their own way of existence by establishing a unique agency-habitat coupling, which is a kind of "natural abduction" [9.4]. Charles Peirce used the term "abduction" for mental creative processes that generate a new explanatory hypothesis [9.64]. However, it can also be applied to mindless organisms such as earthworms [9.63]. Magnani [9.63], p. 5, wrote: "Like humans, worms build external artifacts endowed with precise roles and functions, which strongly affect their lives in various ways, and of course their opportunity to 'know' the environment." He attributed this adaptation to natural selection. The abductive nature of the agency-habitat coupling is comparable in worms and first primordial agents, except that habits of worms are substantially more complex than the habits of first agents. Thus, first agents already carried some meaningful representation of their environment, which is the start of semiosis.

9.7 Evolvability of Molecular Agents

The major criterion of progress in biological macroevolution is the increase of functional complexity [9.97]. Thus, the increase in the number of non-redundant components and relations in molecular communities is the main path towards full-fledged life. The increase of complexity cannot be limited to the information channel only because each kind of agent has only a limited number of capacities, which cannot increase no matter how complex the information input is. The primacy of the information channel in evolution is a part of the gene-centric doctrine of neo-Darwinism, which assumes that mutations are the primary causes of adaptive evolutionary change. This idea is supported by von Neumann's model of self-reproducing automata, which relies on the existence of a universal constructor, a computational agency that can create any possible digital pattern in a cellular automaton [9.120]. In this model, all evolution is restricted to the change of digital sequences that describe the pattern to be constructed.

Although limited evolution of cellular automata is possible via universal constructor model, the scope and speed of such evolution are both very poor, as follows from computer implementations of von Neumann's universal constructor [9.13] [9.85]. Notoriously, von Neumann's model cannot depict the origin of life because that would require an accidental single-step magical appearance of the universal constructor, which is a quite complex mechanism. Thus, the progress in biological evolution should consider both components of complexity: informational and agential. In particular, the emergence of new catalytic agents and new regulatory relations within the agential network are important evolutionary events towards advanced primordial systems.

Fixation of evolutionary changes is possible in compositional hereditary systems despite the absence of template-based replication; however, it takes the form of addition or subtraction rather than substitution that is typical for genetic alleles. New self-producing CLRs can appear and support additional catalytic changes in the system. Also, some old types of self-producing CLRs can be lost irreversibly due to competition. But initial steps of evolution of systems with compositional hereditary could have been largely random and non-accumulative. Only natural selection (including stochastic corrector mechanism [9.109]) can drive the accumulation of adaptive changes in such systems.

New kinds of CLRs can appear by modification of already existing CLRs if this modification happens to be self-amplifying. For example, consider CLRs of type A that are already produced in the systems; then accidental phosphorylation of A to A_p can result in the establishment of a new kind of CLR (A_p) if the reaction of $A \rightarrow A_p$ is autocatalytic. A chain conversion of A into A_p may initially be harmful due to a decrease in the number of molecules A, and weakening of functions supported by these types of molecules. However, the molecular network may eventually adapt to shield some molecules A from phosphorylation or inhibit the catalytic activity of A_p if it gets too abundant. As a result, the primordial system acquires a new kind of CLR, which can self-reproduce and support a new chemical function (i.e., phosphorylation). This function can be inherited by descendants if they get A_p molecules after the division of an oil droplet.

Computational analysis of evolvability in primordial systems is usually done by checking if new quasi-species of replicators can emerge in a homogeneous environment [9.119]. This approach, however, does not account for the role of environmental hereditary factors in early evolution. Because first replicators were simple monomers, their information capacity as heredity carriers was limited, and they alone could not support a large number of potential evolutionary changes. However, evolution could initially rely on the variability of environmental hereditary factors. As replicators start spreading to neighboring areas, they can encounter different local environments, such as oil droplets with a different composition of hydrocarbons and contaminants (e.g., metals, minerals, and PAHs). Depending on local conditions, natural selection will drive the evolution in somewhat different directions, and eventually, these changes can be irreversible as supported by modified covalent bonds within replicators (CLRs) or in downstream molecules.

After diversification of replicators in various semi-isolated environments, replicators from different lineages may happen to meet again after the next waves of spatial dispersion. Consider that oil droplet carrying replicators of type *A* can merge with oil droplet with replicators of type *B* originating from a different lineage. An oil droplet carrying two types of CLRs, *A* and *B*, can produce new downstream molecules if, for example, catalytic activity of *B* is applied to the products of catalyst *A* or vice versa. The increase in the diversity of downstream molecules may generate some beneficial effects for the community of replicators (i.e., oil droplets with both kinds of replicators may reproduce faster than oil droplets with just one kind of replicator), and the combination of replicators *A* and *B* will become stabilized by natural selection.

Initially, replicators were monomers; and being not connected they were transferred to offspring oil droplets in different combinations at random. The probability of inheriting the full set of replicators by pure chance can be problematic, especially if the copy number of some replicators is small. If oil droplets with the full set of replicators have selective advantage (i.e., a higher rate of survival and reproduction) then their relative proportion in the population will increase over time. This is known as "stochastic corrector" mechanism [9.112], which is a primordial version of the purifying selection.

New replicators can also become established by polymerization of non-polymeric CLRs [9.98]. The initial role of polymers could have been structural rather than hereditary. For example, a CLR of type *A* can support the polymerization of CLRs of type *B*, making long polymers *B-B-B*..., which could be useful in stabilizing the surface of oil droplets. However, some polymers may develop additional self-supporting function that stabilize their specific features, such as geometry (e.g., circular nucleic acids that support synthesis of similar circular molecules) or composition (e.g., *A-B-A-B-A-B*... simple repeat polymers can support the synthesis of other polymers with the same order of monomers).

Long-term coexistence of multiple kinds of replicators inevitably leads to their coadaptation out of necessity, because they share their local niche (oil droplet), on which they depend functionally, and participate in niche construction. The dependency on the niche creates conditions for group selection that favors cooperative activities of replicators, because a failure to cooperate may result in the deterioration of niche properties and depletion of resources. As a result, such weakened networks of replicators will have reduced chances to survive and propagate. This mechanism of group selection differs from the stochastic corrector model [9.112] because it enhances coadaptation between replicators, rather than just acting as purifying selection that eliminates groups with non-matching components.

There are two kinds of co-adaptations between catalysts: direct and indirect. Direct coadaptation occurs when one kind of catalyst facilitates reactions that either produce another kind of catalyst or protect it from degradation. In contrast, indirect coadaptation is mediated by the local environment, when one kind of catalyst modifies the properties of the niche which appears beneficial for the production or function of another kind of catalyst. Such benefits can be achieved by generating resources, scaffolds, or signs that are needed for the second kind of catalysts. Indirect co-adaptations mediated by the local environment (e.g., by the surface properties of oil droplet or abundance of resources) seem especially important for evolvability. Each kind of CLR performs some chemical function, such as capturing energy or resource molecules; and these effects could be useful for many types of replicators that share the same oil droplet. Another potential function of CLRs may be catalytic degradation of oil and its derivatives. In particular, some organic molecules could be derived from fatty acids by pathways similar to the beta-oxidation pathway.

9.8 Emergence of Unlimited Heredity and Cell Membrane

The primordial evolution was slow and fragile due to the absence of universal methods for producing new replicators. Transformations of CLRs by adding functional groups or polymerization were important pathways for deriving new replicators, but they were not universal. This problem has been solved in a radical way only with the emergence of template-based replication, where each monomer is added to the newly constructed polymer sequentially based on a uniform rule applied to a neighbor monomer in another linearly aligned polymer that serves as a template [9.109]. This method of polymer copying is universal because the rule of replication works for any sequence of monomers from the given set of monomer kinds.

The template-based replication did not appear before the origin of life as assumed by the RNA-world hypothesis. Instead, it emerged by the process of adaptive evolution driven by a kind of natural selection in primordial agents with compositional-type heredity. The main prerequisite for the emergence of template-based replication was the existence of polymeric replicators whose functions depended on the order of monomers. To be functional, the order of monomers was preserved even before the emergence of template-based replication. An example of a hypothetical mechanism that helped keep the order of monomers is the capacity of the last monomer in the chain to control the choice of the next monomer to be added. Naturally, this mechanism allowed errors and supported only a few types of repetitive patterns. But it could keep polymers functional, which is necessary for setting evolution on the path towards template-based replication.

It is possible that some hetero-polymers tended to bind side by side and get aligned on the surface of oil droplets. Such binding could have been beneficial due to increased mechanical stiffness of the surface. Then the shorter aligned polymer in a pair of sequences can be elongated by adding monomers that weakly matched the longer polymer [9.100]. This mechanism of neighbor-controlled elongation could have evolved from the more primitive mechanism of self-controlled elongation (see above). In subsequent evolution, natural selection increased the fidelity of template-based copying. The first replicating polymers were probably similar to nucleic acids; but it is unlikely that they had the sugar-phosphate backbone (as in RNA or DNA) that does not anchor to the oil surface. The more likely candidates are peptide nucleic acids (PNAs) [9.75], whose pseudopeptide backbone can bind to the lipid-water surface [9.123].

The emergence of template-based replication was the turning point in the evolutionary history, when primordial systems acquired high evolutionary potential [9.52] [9.98]. Szathmáry [9.110] called this type of replication "unlimited heredity" and considered it one of the major transitions in evolution [9.113].

Another important event in the origin of life was the emergence of cell membrane. Although initial metabolism of replicators on the surface of oil droplets provided advantages during early stages of the evolution of life (access to water-dissolved resources, anchoring of replicators), it led to serious problems later in evolution. The first problem was the limited supply of oil in the environment. If all available oil was already captured by surface catalytic networks, the rates of evolution would have declined due to resource shortage. The oil droplet is a very inefficient way of using resources by replicators, because oil is utilized only at the surface. In contrast, a cell membrane is a much more efficient way of using oil because the membrane is thin. In other words, oil from one droplet would be enough to make membranes for a hundred cells. The second problem was the small surface of oil droplets that did not provide enough space to store resources for the primordial agency. With the emergence of cell membrane, resources could be stored more efficiently in the big volume of cytoplasm. The third problem was the limited mobility of polymers in a two-dimensional surface of oil droplets because long molecules cannot easily pass each other on a surface. When polymers moved to the cytoplasm, this problem was also resolved.

A hypothetical model of the emergence of cell membrane assumes that water gets inside an oil droplet after mechanical agitation [9.100]. Such membranes easily arise in oil emulsion, but they are unstable. The cell membrane is unlikely to offer any functional advantage to a cell if it breaks so easily. Membranes can be stabilized by replacing fatty acids with lipids, and especially, phospholipids. Lipids, however, require glycerol, which is not available in the environment. Thus, the formation of stable cell membrane should have been closely associated with the origin of metabolic pathways leading to internal synthesis of glycerol [9.100]. Although glycerol has many functions in the cell, it is conceivable that the synthesis of glycerol evolved by natural selection for the purpose of making phospholipids as building blocks of stable cell membrane.

It is not clear whether template-based replication appeared before or after the formation of a stable cell membrane. But in any case, these two evolutionary events were needed for the existence of membrane-bound cells with mixed coenzyme-based and ribozyme-base catalysis. The increase of complexity within communities of singular and polymeric replicators is limited if the copy number of replicators is not controlled during cell division. If some replicators have low copy number, then after uncontrolled cell division, one of the daughter cells may be left without a single copy of some specific kind of replicators that is important for the survival of the entire functional network. Thus, mechanisms of cell division control (e.g., aggregation of replicators into chromosomes and separation of cells together with chromosomes) appeared early in the evolution of cellular forms of life.

Subsequent evolution towards a typical prokaryote cell was characterized by a gradual transformation of active RNA replicators into passive hereditary signs encoded by DNA, whereas active catalytic and regulatory functions were mostly transferred to proteins or to non-reproducing non-coding RNA. This process was accompanied by increased activity of the whole cell. In bacteria, this activity is mediated by various appendages: flagella, which support locomotion and sensing, fimbriae and pili, which are capable of selective adhesion, immune response, invasion of other cells, secretion, sensing, pulling, and DNA transport

[9.17] [9.53]. Bacterial cells developed elaborate sensing organelles: membrane receptors responsible for chemotaxis, photoreceptors, and even magnetosensing. Signals from sensors are integrated via complex signal transduction network and control the activity of other cell components, such as activation and repression of genes, regulation of metabolism, rotation of flagella, and modification of a primitive cytoskeleton. Many bacteria even have a primitive immune system which protects them against viruses [9.6]. Thus, biological agency evolved from simple monomeric catalysts (replicators), to template copied ribozymes, and finally to whole cells and organisms, whose activity is regulated by hereditary signs and sensorial information.

9.9 How Old is Life?

Most researchers assume that life emerged on Earth soon after its formation. Traces of early life date from 3.77 to 4.28 billion years ago [9.31], and some scholars argue that the origin of LUCA happened ca. 4.5 billion years ago [9.12]. An alternative hypothesis is that microbial life is abundant in the Universe and is more likely to be transferred physically[7] between planets and possibly between star systems within a galaxy, a process called panspermia, than to originate anew [9.101]. Now it is well established that microbial life is abundant deep in the Earth's crust [9.62] [9.89], which makes it reasonable to predict that living bacteria are present on exoplanets, including rogue planets [9.1] [9.7] that do not belong to any star system [9.40] [9.42] [9.117].

The age of life can be evaluated by examining the macro-evolutionary trend of genome complexity increase [9.97]. Initial comparative studies of genome complexity change stalled after discovering that the total amount of DNA in some single-cell organisms substantially exceeds the amount of DNA in a human cell [9.83]. Sequencing of full genomes of many organism species showed that the combined length of DNA in all chromosomes is also a poor measure of genome complexity because DNA includes many repetitive regions that are either not functional as information carriers or redundant. Many repetitive segments of DNA participate in non-informational functions (e.g., maintenance of the nuclear matrix, chromosome condensation, and regulation of mitosis and meiosis) [9.16] [9.90]. Thus, it makes sense to measure genome complexity not by the total DNA length but by the length of functional and non-redundant genome segments [9.3].

To project the macro-evolutionary trends in the change of genome complexity, I plotted the combined length of functional and non-redundant genome segments of major phylogenetic lineages against the time of their origin (Figure 9.3) [9.97] [9.102]. Functional genome segments are those that include exons or regulatory regions of genes such as promoters and enhancers. Most functional segments are highly conserved in evolution. However, in mammals and humans, the evolution of regulatory regions was continuing at faster rates, which resulted in lower levels of evolutionary conservation. Thus, I use an additional criterion of functionality suggested by Simons [9.107], which is the lack (or strong depletion) of transposons, which are virus-like DNA sequences capable of hopping to new genome

[7] Crick and Orgel's hypothesis [9.20] that life was transferred to Earth by intelligent beings is highly unlikely, as discussed in [9.101].

locations. It appeared that about 15% of the mammalian genome is nearly transposon-free and is likely functional [9.97].

Log-transformed genome complexity in major lineages of eukaryotes increased with time (Figure 9.3), which indicates an exponential growth with a doubling time of ca. 340 million years [9.101]. The rate of genome increase was later checked by Markov *et al.* [9.66], and their estimate was close enough (360 million years).

The regression line reaches zero log-complexity at about 9.7 billion years ago, which can be interpreted as expected origin of life (Fig. 9.3) [9.97] [9.101]. This point corresponds to a primordial system with just one type of CLR. This replicator was not a nucleotide, but its descendants after long-term evolution lost catalytic activity and became passive sequences of nucleotides (i.e., genes). The first two points of the data used for regression (prokaryotes and single-cell eukaryotes) are the most uncertain and can affect the slope of regression. Thus, sensitivity analysis was used to evaluate the uncertainty if these two points change in a reasonably selected range (±0.3 By and ±0.3 log10 bp). Within these conditions, the predicted time of the origin of life changed by ca. ±2 By [9.97]. In particular, the predicted age of life is still greater than the age of Earth (4.5 By).

Marzban *et al.* [9.68] have reanalyzed the regression data shown in Figure 9.3 using the measurement error model [9.14] that assumes a uniform frequency distribution of errors at all data points. The confidence interval of the age of life estimated by Marzban *et al.* included 4.3 billion years. However, the model assumes a uniform frequency distribution of errors, and this simplified assumption does not apply to our data, where more recent data points are much more accurate than data points representing events separated from the present by several billion years. For this reason, the measurement error model used by Marzban *et al.* is not justified [9.101].

One of the limitations of this analysis is that it reflects the increase of genome size mostly in eukaryotes. However, a recent extensive study of genome length distributions in bacteria and archaea indicates that genome length depends on the phylogeny [9.67]. Besides obvious dependencies that parasitic bacteria tend to have smaller genomes compared to free-living species, it appeared that more ancient taxonomic groups (e.g., *Firmicutes*) have smaller average genomes than more advanced groups (e.g., *Halobacteria, Proteobacteria, Bacteroidota,* and *Cyanobacteria*). This fact indicates that the evolution of genome size

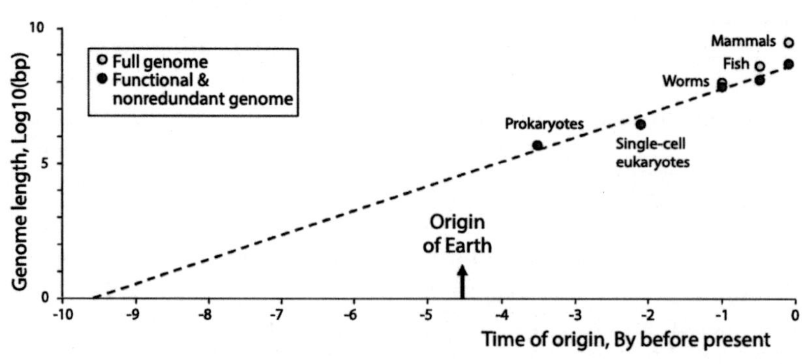

Figure 9.3 The increase of genome complexity of organisms with time (billions of years before the present). The projected age of life is 9.7 By (modified from [9.97]).

in prokaryotes may also have been progressive but with a slower rate of increase than in eukaryotes. Another limitation of the model is that I did not account for potential acceleration of evolution due to shorter day/night cycle on early Earth [9.43].

This analysis does not provide direct evidence that life is older than Earth, and thus, only the direct analysis of extraterrestrial life would be conclusive. Nevertheless, the hypothesis of microbial panspermia should not be rejected based solely on beliefs and scientific traditions. If the panspermia hypothesis is true, then most extraterrestrial living organisms in our galactic region likely originated from the common evolutionary root, and should have very similar biochemical organization (nucleic acids, proteins, carbohydrates, lipids, metabolic cycles, and gene regulation pathways). This prediction can be tested when microbial organisms are discovered on planets, satellites, and possibly other cosmic bodies within the Solar system. It is also likely that all extraterrestrial organisms within the Solar system have genetic similarity with terrestrial bacterial species. Considering these predictions, panspermia can be tested experimentally, and I hope that the waiting period will not be too long.

9.10 Conclusions

Agency and life have physical and chemical roots, but this does not mean that physics or chemistry holds the secret of life. Chemistry and physics can provide *means* for the emergence of agency and life but they do not explain it. Agency is a higher-level phenomenon that is above physics and chemistry and includes interpretation of signs. The origin of life is a turning point in the evolution of matter when and where chemical and physical processes start playing new semiotic roles by supporting the existence and propagation of a larger evolving whole, a system that eventually becomes an organism. This turning point can be recognized by *agency-niche coupling*, where agency interprets its local environment (niche) as a medium for life and reproduction, and simultaneously it modifies its niche in a way that facilitates survival and propagation of future generations (niche construction). Agency-niche coupling is the simplest form of abduction — finding a way to replicate in a historically constructed environment. Reproduction combines sensing of the environment (successful reproduction confirms that the environment is favorable), and action (allocation of progeny organisms into this environment), which closes the functional circle.

Acknowledgments

I want to thank Dr. Richard Gordon, Dr. George Mikhailovsky, and two anonymous reviewers for their helpful comments and suggestions on this manuscript.

Declaration

The author declares no conflict of interest.

References

[9.1] Abbott, D.S. and Switzer, E.R. (2011) The Steppenwolf: A proposal for a habitable planet in interstellar space. *Astrophysical Journal Letters* 735(2), L27.

[9.2] Adam, G. and Delbrück, M. (1968) Reduction of dimensionality in biological diffusion processes. In: *Structural Chemistry and Molecular Biology*. A. Rich and N. Davidson, (eds.) W.H. Freeman, New York: 198-215.

[9.3] Adami, C., Ofria, C. and Collier, T.C. (2000) Evolution of biological complexity. *Proc Natl Acad Sci USA* 97(9), 4463-4468.

[9.4] Andrade, E. (2011) Natural abduction: The bridge between individuals' choices and the production of evolutionary innovations. *Signs* 5, 112-146.

[9.5] Andras, P. and Andras, C. (2005) The origins of life -- the 'protein interaction world' hypothesis: protein interactions were the first form of self-reproducing life and nucleic acids evolved later as memory molecules. *Med Hypotheses* 64(4), 678-688.

[9.6] Arroyo-Olarte, R.D., Bravo Rodriguez, R. and Morales-Rios, E. (2021) Genome Editing in Bacteria: CRISPR-Cas and Beyond. *Microorganisms* 9(4).

[9.7] Badescu, V. (2011) Constraints on the free-floating planets supporting aqueous life. *Acta Astronautica* 69(9-10), 788-808.

[9.8] Benner, S.A., Kim, H.J. and Carrigan, M.A. (2012) Asphalt, water, and the prebiotic synthesis of ribose, ribonucleosides, and RNA. *Acc Chem Res* 45(12), 2025-2034.

[9.9] Bernal, J.D. (1951) *The physical basis of life*. Routledge and Kegan Paul, London.

[9.10] Bernhardt, H.S. (2012) The RNA world hypothesis: The worst theory of the early evolution of life (except for all the others). *Biol Direct* 7, 23.

[9.11] Bernstein, M. (2006) Prebiotic materials from on and off the early Earth. *Philos Trans R Soc Lond B Biol Sci* 361(1474), 1689-1700; discussion 1700-1682.

[9.12] Betts, H.C., Puttick, M.N., Clark, J.W., Williams, T.A., Donoghue, P.C.J. and Pisani, D. (2018) Integrated genomic and fossil evidence illuminates life's early evolution and eukaryote origin. *Nat Ecol Evol* 2(10), 1556-1562.

[9.13] Buckley, W.R. (2008) Signal crossing solutions in von Neumann self-replicating cellular automata. In: *Proceedings Automata 2008*. R.A.-S. A. Adamatzky, A. Lawniczak, G. J. Martinez, K. Morita, T. Worsch, (ed.) Luniver Press: 453–503.

[9.14] Buonaccorsi, J.P. (2010) *Measurement error: Models, methods, and applications*. CRC Prress, Taylor & Francis Group, Boca Raton, FL.

[9.15] Callahan, M.P., Smith, K.E., Cleaves, H.J., 2nd, Ruzicka, J., Stern, J.C., Glavin, D.P., House, C.H. and Dworkin, J.P. (2011) Carbonaceous meteorites contain a wide range of extraterrestrial nucleobases. *Proc Natl Acad Sci U S A* 108(34), 13995-13998.

[9.16] Cavalier-Smith, T. (2005) Economy, speed and size matter: evolutionary forces driving nuclear genome miniaturization and expansion. *Annals of Botany* 95(1), 147-175.

[9.17] Chang, C.Y. (2018) Surface Sensing for Biofilm Formation in *Pseudomonas arugenosa*. *Front Microbiol* 8, 2671.

[9.18] Cooper, G., Kimmich, N., Belisle, W., Sarinana, J., Brabham, K. and Garrel, L. (2001) Carbonaceous meteorites as a source of sugar-related organic compounds for the early Earth. *Nature* 414(6866), 879-883.

[9.19] Copley, S.D., Smith, E. and Morowitz, H.J. (2007) The origin of the RNA world: Co-evolution of genes and metabolism. *Bioorg. Chem.* 35(6), 430-443.

[9.20] Crick, F.H.C. and Orgel, L.E. (1973) Directed panspermia. *Icarus* 19, 341-346.

[9.21] Damer, B. (2016) A field trip to the Archaean in search of Darwin's warm little pond. *Life (Basel)* 6(2), 21.

[9.22] Dawkins, R. (1976) *The selfish gene*. Oxford University Press, New York.

[9.23] Deacon, T.W. (2006) Reciprocal linkage between self-organizing processes is sufficient for self-reproduction and evolvability. *Biological Theory* 1(2), 136–149.
[9.24] Deacon, T.W. (2021) How molecules became signs. *Biosemiotics* 14(3), 537–559.
[9.25] Deacon, T.W., Srivastava, A. and Bacigalupi, J.A. (2014) The transition from constraint to regulation at the origin of life. *Front Biosci (Landmark Ed)* 19, 945-957.
[9.26] Deamer, D. (2011) *First life: Discovering the connections between stars, cells, and how life began.* University of California Press, Berkley, CA.
[9.27] Deamer, D., Dworkin, J.P., Sandford, S.A., Bernstein, M.P. and Allamandola, L.J. (2002) The first cell membranes. *Astrobiology* 2(4), 371-381.
[9.28] Deamer, D.W. (1992) Polycyclic aromatic hydrocarbons: primitive pigment systems in the prebiotic environment. *Adv Space Res* 12(4), 183-189.
[9.29] Deamer, D.W. (1999) How did it all begin? The self-assembly of organic molecules and the origin of cellular life. In: *Evolution: Investigating the evidence.* D.A. Springer and J.G. Scotchmoor, (eds.) Paleontological Society, Pittsburgh, PA. 9.
[9.30] Deamer, D.W. and Georgiou, C.D. (2015) Hydrothermal conditions and the origin of cellular life. *Astrobiology* 15(12), 1091-1095.
[9.31] Dodd, M.S., Papineau, D., Grenne, T., Slack, J.F., Rittner, M., Pirajno, F., O'Neil, J. and Little, C.T. (2017) Evidence for early life in Earth's oldest hydrothermal vent precipitates. *Nature* 543(7643), 60-64.
[9.32] Doolittle, W.F. (2000) Uprooting the tree of life. *Sci Am* 282(2), 90-95.
[9.33] Dutta, A., DuBois, D.L., Roberts, J.A. and Shaw, W.J. (2014) Amino acid modified Ni catalyst exhibits reversible H2 oxidation/production over a broad pH range at elevated temperatures. *Proc Natl Acad Sci U S A* 111(46), 16286-16291.
[9.34] Dyson, F.J. (1999) *Origins of life.* Cambridge University Press, Cambridge.
[9.35] Egel, R. (2012) Primal eukaryogenesis: on the communal nature of precellular States, ancestral to modern life. *Life (Basel)* 2(1), 170-212.
[9.36] Ehrenfreund, P. and Cami, J. (2010) Cosmic carbon chemistry: from the interstellar medium to the early Earth. *Cold Spring Harb Perspect Biol* 2(12), a002097.
[9.37] Ehrenfreund, P., Rasmussen, S., Cleaves, J. and Chen, L. (2006) Experimentally tracing the key steps in the origin of life: The aromatic world. *Astrobiology* 6(3), 490-520.
[9.38] Eigen, M. and Schuster, P. (1979) *The hypercycle, a principle of natural self-organization.* Springer-Verlag, Berlin, New York.
[9.39] Gánti, T. (2003 [1971]) *The principles of life.* Oxford University Press, Oxford, New York.
[9 .40] Gibson, C.H. (2000) Turbulent mixing, viscosity, diffusion, and gravity in the formation of cosmological structures: The fluid mechanics of dark matter. *Journal of Fluids Engineering-Transactions of the Asme* 122(4), 830-835.
[9.41] Gilbert, W. (1986) Origin of life: The RNA world. *Nature* 319(6055), 618.
[9.42] Gladman, B. and Chan, C. (2006) Production of the extended scattered disk by rogue planets. *Astrophysical Journal* 643(2), L135-L138.
[9.43] Gordon, R. and Mikhailovsky, G. (2021) There were plenty of day/night cycles that could have accelerated an origin of life on Earth, without requiring panspermia. In: *Planet formation and panspermia: New prospects for the movement of life through space.* J.S.a.R.G. B. Vukotić, (ed.) Wiley-Scrivener, Beverly, Massachusetts, USA: 195-206.
[9.44] Gowanlock, G. and Morrison, I.S. (2018) The habitability of our evolving galaxy. In: *Habitability of the Universe before Earth.* R. Gordon and A.A. Sharov, (eds.) Elsevier, Amsterdam: 149-171.
[9.45] Haldane, J.B.S. (1929) Origin of life. *The Rationalist Annual* 148, 3-10.
[9.46] Hartman, H. (1998) Photosynthesis and the origin of life. *Orig Life Evol Biosph* 28(4-6), 515-521.

[9.47] Herschy, B., Whicher, A., Camprubi, E., Watson, C., Dartnell, L., Ward, J., Evans, J.R. and Lane, N. (2014) An origin-of-life reactor to simulate alkaline hydrothermal vents. *J Mol Evol* 79(5-6), 213-227.

[9.48] Hoffmeyer, J. (1998) Surfaces inside surfaces. On the origin of agency and life. *Cybernetics and Human Knowing* 5(1), 33-42.

[9.49] Hoffmeyer, J. (2008) *Biosemiotics: An examination into the signs of life and the life of signs* University of Scranton Press, Scranton, PA.

[9.50] Ikehara, K. (2016) Evolutionary steps in the emergence of life deduced from the bottom-up approach and GADV hypothesis (top-down approach). *Life (Basel)* 6(1).

[9.51] Inger, A., Solomon, A., Shenhav, B., Olender, T. and Lancet, D. (2009) Mutations and lethality in simulated prebiotic networks. *J Mol Evol* 69(5), 568-578.

[9.52] Jablonka, E. and Szathmáry, E. (1995) The evolution of information storage and heredity. *Trends Ecol Evol* 10(5), 206-211.

[9.53] Jonson, A.B., Normark, S. and Rhen, M. (2005) Fimbriae, pili, flagella and bacterial virulence. *Contrib Microbiol* 12, 67-89.

[9.54] Joyce, G.F. and Orgel, L.E. (1998) The origins of life — a status report. *American Biology Teacher* 60, 10–12.

[9.55] Kauffman, S.A. (1986) Autocatalytic sets of proteins. *J. Theor. Biol.* 119(1), 1-24.

[9.56] Koonin, E.V. (2009) On the origin of cells and viruses: primordial virus world scenario. *Ann N Y Acad Sci* 1178, 47-64.

[9.57] Koonin, E.V. and Martin, W. (2005) On the origin of genomes and cells within inorganic compartments. *Trends Genet* 21(12), 647-654.

[9.58] Kritsky, M.S. and Telegina, T.A. (2004) Role of nucleotide-like coenzymes in primitive evolution. In: *Origins: Genesis, evolution and diversity of Life*. J. Seckbach, (ed.) Kluwer, Dordrechtt, The Netherlands: 215-231.

[9.59] Lane, N., Allen, J.F. and Martin, W. (2010) How did LUCA make a living? Chemiosmosis in the origin of life. *Bioessays* 32(4), 271-280.

[9.60] Lane, N., Martin, W.F., Raven, J.A. and Allen, J.F. (2013) Energy, genes and evolution: Introduction to an evolutionary synthesis. *Philosophical Transactions of the Royal Society B-Biological Sciences* 368(1622), 20120253.

[9.61] Levy, M., Miller, S.L., Brinton, K. and Bada, J.L. (2000) Prebiotic synthesis of adenine and amino acids under Europa-like conditions. *Icarus* 145(2), 609-613.

[9.62] Li, J., Mara, P., Schubotz, F., Sylvan, J.B., Burgaud, G., Klein, F., Beaudoin, D., Wee, S.Y., Dick, H.J.B., Lott, S., Cox, R., Meyer, L.A.E., Quemener, M., Blackman, D.K. and Edgcomb, V.P. (2020) Recycling and metabolic flexibility dictate life in the lower oceanic crust. *Nature* 579(7798), 250-255.

[9.63] Magnani, L. (2007) Animal abduction: From mindless organisms to artifactual mediators. In: *Model-based reasoning in science, technology, and medicine*. L. Magnani, Li, P., (ed.) Springer, Heidelberg: 3-38.

[9.64] Magnani, L. (2009) *Adaptive cognition. The epistemological and eco-cognitive dimensions of hypothetical reasoning*. Springer, Berlin, Heidelberg.

[9.65] Marcano, V., Benitez, P. and Palacios-Pru, E. (2003) Acyclic hydrocarbon environments n-C18 on the early terrestrial planets. *Planet. Space Sci.* 51(3), 159-166.

[9.66] Markov, A.V., Anisimov, V.A. and Korotayev, A.V. (2010) Relationship between genome size and organismal complexity in the lineage leading from prokaryotes to mammals. *Paleontological Journal* 44(4), 363-373.

[9.67] Martinez-Gutierrez, C.A. and Aylward, F.O. (2022) Genome size distributions in bacteria and archaea are strongly linked to evolutionary history at broad phylogenetic scales. *PLoS Genet* 18(5), e1010220.

[9.68] Marzban, C., Viswanathan, R. and Yurtsever, U. (2014) Earth before life. *Biol Direct* 9, 1.

[9.69] Michitaka, T., Imai, T. and Hashidzume, A. (2017) Formose reaction controlled by a copolymer of N,N-dimethylacrylamide and 4-vinylphenylboronic acid. *Polymers* 9(11), 549.

[9.70] Miller, S.L. (1953) A production of amino acids under possible primitive earth conditions. *Science* 117(3046), 528-529.

[9.71] Morowitz, H.J., Kostelnik, J.D., Yang, J. and Cody, G.D. (2000) The origin of intermediary metabolism. *Proc. Natl. Acad. Sci. U. S. A.* 97(14), 7704-7708.

[9.72] Mulkidjanian, A.Y. (2009) Origin of life in the Zinc World: 1. Photosynthetic, porous edifices built of hydrothermally precipitated zinc sulfide (ZnS) as cradles of life on Earth. *Biology Direct* 4, 26.

[9.73] Mulkidjanian, A.Y. and Galperin, M.Y. (2007) Physico-chemical and evolutionary constraints for the formation and selection of first biopolymers: Towards the consensus paradigm of the abiogenic origin of life. *Chem. Biodivers.* 4(9), 2003-2015.

[9.74] Mulkidjanian, A.Y. and Galperin, M.Y. (2009) On the origin of life in the Zinc world. 2. Validation of the hypothesis on the photosynthesizing zinc sulfide edifices as cradles of life on Earth. *Biology Direct* 4, 27.

[9.75] Nelson, K.E., Levy, M. and Miller, S.L. (2000) Peptide nucleic acids rather than RNA may have been the first genetic molecule. *Proc. Natl. Acad. Sci. U. S. A.* 97(8), 3868-3871.

[9.76] Oba, T., Fukushima, J., Maruyama, M., Iwamoto, R. and Ikehara, K. (2005) Catalytic activities of [GADV]-peptides. Formation and establishment of [GADV]-protein world for the emergence of life. *Orig Life Evol Biosph* 35(5), 447-460.

[9.77] Odling-Smee, F.J., Laland, K.N. and Feldman, M.W. (2003) *Niche construction. The neglected process in evolution.* Princeton University Press, Princeton, NJ.

[9.78] Omran, A., Menor-Salvan, C., Springsteen, G. and Pasek, M. (2020) The Messy Alkaline Formose Reaction and Its Link to Metabolism. *Life (Basel)* 10(8).

[9.79] Oparin, A.I. (1953 [1924]) *The origin of life*. Dover Publications, Mineola, NY.

[9.80] Pascal, R. (2012) Suitable energetic conditions for dynamic chemical complexity and the living state. *Journal of Systems Chemistry* 3, 3.

[9.81] Pascal, R. (2013) Life, metabolism and energy. In: *Astrochemistry and astrobiology: Physical chemistry in action.* C.S.C. I. W. L. Smith, S. Leach, (ed.) Springer, Heidelberg: 243–269.

[9.82] Pascal, R., Pross, A. and Sutherland, J.D. (2013) Towards an evolutionary theory of the origin of life based on kinetics and thermodynamics. *Open Biol* 3(11), 130156.

[9.83] Patrushev, L.I. and Minkevich, I.G. (2008) The problem of the eukaryotic genome size. *Biochemistry (Moscow)* 73(13), 1519-1552.

[9.84] Pattee, H.H. (1969) How does a molecule become a message? *Developmental Biology Supplement* 3, 1-16.

[9.85] Pesavento, U. (1995) An implementation of von Neumann's self-reproducing machine" (PDF), , MIT Press, 2 (4): 337–354. *Artificial Life* 2(4), 337–354.

[9.86] Pigliucci, M. and Müller, G.B. (eds.) (2010) *Evolution - the extended synthesis.* MIT Press, Cambridge MA, London.

[9.87] Powner, M.W., Gerland, B. and Sutherland, J.D. (2009) Synthesis of activated pyrimidine ribonucleotides in prebiotically plausible conditions. *Nature* 459(7244), 239-242.

[9.88] Prigogine, I. (1980) *From being To becoming*. Freeman, San Francisco.

[9.89] Robinson, J.M. and Mikucki, J.A. (2018) Occupied and empty regions of the space of extremophile parameters. In: *Habitability of the Universe before Earth.* R. Gordon and A. Sharov, (eds.) Academic Press, Cambridge, MA, London: 199-230.

[9.90] Rollins, R.A., Haghighi, F., Edwards, J.R., Das, R., Zhang, M.Q., Ju, J. and Bestor, T.H. (2006) Large-scale structure of genomic methylation patterns. *Genome Res* 16(2), 157-163.

[9.91] Ruiz-Mirazo, K., Briones, C. and de la Escosura, A. (2017) Chemical roots of biological evolution: the origins of life as a process of development of autonomous functional systems. *Open Biol* 7(4).
[9.92] Sanchez, R.A., Ferris, J.P. and Orgel, L.E. (1968) Studies in prebiotic synthesis. IV. Conversion of 4-aminoimidazole-5-carbonitrile derivatives to purines. *J Mol Biol* 38(1), 121-128.
[9.93] Schrödinger, E. (1940) *What is life? The physical aspect of the living cell.* Cambridge University Press, Cambridge.
[9.94] Segré, D., Ben-Eli, D., Deamer, D.W. and Lancet, D. (2001) The lipid world. *Orig. Life Evol. Biosph.* 31(1-2), 119-145.
[9.95] Segré, D., Lancet, D., Kedem, O. and Pilpel, Y. (1998) Graded autocatalysis replication domain (GARD): Kinetic analysis of self-replication in mutually catalytic sets. *Orig Life Evol Biosph* 28(4-6), 501-514.
[9.96] Segré, D., Shenhav, B., Kafri, R. and Lancet, D. (2001) The molecular roots of compositional inheritance. *J Theor Biol* 213(3), 481-491.
[9.97] Sharov, A.A. (2006) Genome increase as a clock for the origin and evolution of life. *Biology Direct* 1, 17.
[9.98] Sharov, A.A. (2009) Coenzyme autocatalytic network on the surface of oil microspheres as a model for the origin of life. *International Journal of Molecular Sciences* 10(4), 1838-1852.
[9.99] Sharov, A.A. (2014) Evolutionary constraints or opportunities? *Biosystems* 123, 9-18.
[9.100] Sharov, A.A. (2016) Coenzyme world model of the origin of life. *Biosystems* 144, 8-17.
[9.101] Sharov, A.A. and Gordon, R. (2018) Life before Earth. In: *Habitability of the Universe before Earth*. R. Gordon and A.A. Sharov, (eds.) Elsevier, Amsterdam: 267–298.
[9.102] Sharov, A.A. and Tønnessen, M. (2021) *Semiotic agency. Science beyond mechanism.* Springer, Dordrecht. 25.
[9.103] Shenhav, B., Bar-Even, A., Kafri, R. and Lancet, D. (2005) Polymer GARD: computer simulation of covalent bond formation in reproducing molecular assemblies. *Orig Life Evol Biosph* 35(2), 111-133.
[9.104] Shenhav, B. and Lancet, D. (2004) Prospects of a computational origin of life endeavor. *Origins of Life and Evolution of the Biosphere 34: , 2004.* 34, 181–194.
[9.105] Shenhav, B., Oz, A. and Lancet, D. (2007) Coevolution of compositional protocells and their environment. *Philos Trans R Soc Lond B Biol Sci* 362(1486), 1813-1819.
[9.106] Shenhav, B., Segrè, D. and Lancet, D. (2003) Mesobiotic emergence: Molecular and ensemble complexity in early evolution. *Advances in Complex Systems* 6(1), 15-35.
[9.107] Simons, C., Pheasant, M., Makunin, I.V. and Mattick, J.S. (2006) Transposon-free regions in mammalian genomes. *Genome Res* 16(2), 164-172.
[9.108] Sojo, V., Herschy, B., Whicher, A., Camprubi, E. and Lane, N. (2016) The Origin of Life in Alkaline Hydrothermal Vents. *Astrobiology* 16(2), 181-197.
[9.109] Szathmáry, E. (1999) The first replicators. In: *Levels of selection in evolution*. L. Keller, (ed.) Princeton University Press, Princeton: 31-52.
[9.110] Szathmáry, E. (2000) The evolution of replicators. *Philos Trans R Soc Lond B Biol Sci* 355(1403), 1669-1676.
[9.111] Szathmáry, E. (2006) The origin of replicators and reproducers. *Philos. Trans. R. Soc. Lond. B Biol. Sci.* 361(1474), 1761-1776.
[9.112] Szathmáry, E. and Demeter, L. (1987) Group selection of early replicators and the origin of life. *J Theor Biol* 128(4), 463-486.
[9.113] Szathmáry, E. and Maynard Smith, J. (1995) The major evolutionary transitions. *Nature* 374, 227-232.
[9.114] Szathmáry, E. and Maynard Smith, J. (1997) From replicators to reproducers: The first major transitions leading to life. *Journal of Theoretical Biology* 187, 555-571.

[9.115] Theobald, D.L. (2010) A formal test of the theory of universal common ancestry. *Nature* 465(7295), 219-222.

[9.116] Turgeon, A. and Morse, E. (2022) Petroleum. *National Geographic.*

[9.117] Vanhamaki, H. (2011) Emission of cyclotron radiation by interstellar planets. *Planetary and Space Science* 59(9), 862-869.

[9.118] Vasas, V., Fernando, C., Szilagyi, A., Zachar, I., Santos, M. and Szathmary, E. (2015) Primordial evolvability: Impasses and challenges. *J Theor Biol* 381, 29-38.

[9.119] Vasas, V., Szathmary, E. and Santos, M. (2010) Lack of evolvability in self-sustaining autocatalytic networks constraints metabolism-first scenarios for the origin of life. *Proc Natl Acad Sci U S A* 107(4), 1470-1475.

[9.120] von Neumann, J. (1966) *Theory of self-reproducing automata.* University of Illinois Press, Urbana.

[9.121] von Uexküll, J. (1957) A stroll through the worlds of animals and men: A picture book of invisible worlds. In: *Instinctive behaviour: The development of a modern concept.* C.H. Schiller, (ed.) International Universities Press, Inc., New York: 5–80.

[9.122] Wächtershäuser, G. (1988) Before enzymes and templates: Theory of surface metabolism. *Microbiol Rev* 52(4), 452-484.

[9.123] Weronski, P., Jiang, Y. and Rasmussen, S. (2007) Molecular dynamics study of small PNA molecules in lipid-water system. *Biophys. J.* 92(9), 3081-3091.

[9.124] Yarus, M. (2011) Getting past the RNA world: The initial Darwinian ancestor. *Cold Spring Harb Perspect Biol* 3(4).

[9.125] Zubarev, D.Y., Rappoport, D. and Aspuru-Guzik, A. (2015) Uncertainty of prebiotic scenarios: the case of the non-enzymatic reverse tricarboxylic acid cycle. *Sci Rep* 5, 8009.

10

Evolution of Biomolecular Communication

Gustavo Caetano-Anollés

Evolutionary Bioinformatics Laboratory, Department of Crop Sciences and Carl R. Woese Institute for Genomic Biology, University of Illinois, Urbana, Illinois, USA

Online resources:
https://cropsciences.illinois.edu/directory/gca
https://scholar.google.com/citations?user=xBqljbEAAAAJ&hl=en
https://www.linkedin.com/in/gustavo-caetano-anolles-29a25620

Abstract

Unfolding complexity in the living world requires an evolving biomolecular communication that responds to environments that are internal and external to biological systems. Here I explore the diversity and evolution of macromolecular structures and functions, which are encoded by a molecular-level "vocabulary" of modules. A genomic census of protein structure, ontological terms of molecular functions, and molecular components of ribosomes supports the ideas of gradual growth of complexity, grammar dependency, and statistical laws arising from principles of information compression. Comparative genomic and phylogenomic data are consistent with a biphasic model, in which the first phase that unifies parts of a system is followed by a second phase that diversifies the resulting modules. The predicted emergence of both modularity and hierarchy is supported by tracing the evolution of biological networks. The chronology and organization of evolving networks reveal an evolutionary trend towards optimized economy, flexibility, and robustness in organisms from superkingdoms Archaea, Bacteria and Eukarya, and viruses. Results support the proposa l that learning-like processes in evolution occur when biomolecular modules interact or combine with other modules to generate hierarchical layers of genetic "vocabularies" in an entangled network of biocommunication clustered around the ribosome. The evolution of biomolecules is driven by persistence of autonomous agents, which use their previous historical and contextual experiences to adapt to new environments and extend the life of living systems. The proposed theoretical framework integrates the biphasic model, agency in biology, Peirce's triadic semiotic system, Shannon's theory of communication, Chomsky's hierarchy of formal languages, von Neumann's universal constructor, arguments of computational intelligence, and the four-dimensional theory of identity and persistence of perdurantism.

Keywords: Complexity, hierarchy, language, modularity, persistence, lexicon, syntax, semantics and pragmatics

---

Email: gca@illinois.edu

10.1 Introduction

There is a general agreement among evolutionary biologists that the complexity of the living world unfolds both in time and space constrained by the environment. This idea was already a part of evolutionary thinking quite early in the history of science. In fact, it was Lamarck — not Darwin — who suggested over 200 years ago that two complementary forces were at work in evolution, *Le pouvoir de vie* (the "force of life" that increased complexity) and *L'influence des circonstances* (a contextual "force of adaptation" imposed by environmental change) [10.40]. These two forces were not only used to explain biological complexity, but also diversification by branching descent — a problem that vexed Lamarck's classification studies of invertebrates. Lamarck, however, viewed *L'influence des circonstances* as a direct effect that the environment had on the use and disuse of biological makeup. This would require that physiological novelties somehow transformed into genetic novelties, something he could not address. While epigenetics now supports "soft inheritance" phenomena of the type envisaged by Lamarck, a general direct mechanism that would link varying environments (internal and external to biological systems) to information stored in genomes still awaits dissection by systems biology research. Meanwhile, Darwinian thinking and the extended modern evolutionary synthesis was able to explain evolutionary change as a population-level phenomenon of natural selection, furthering the *L'influence des circonstances* theme. However, this theoretical framework failed to explain phenomena related to Lamarck's *Le pouvoir de vie* theme, including the rise of novelties responsible for modern biochemistry, how molecular novelties transformed into phenotypic novelties, and why processes of change occurred at different rates. It is here, where systems biology may offer a new evolutionary synthesis [10.9], and evolutionary genomics provide the tools with which to answer these central evolutionary questions.

Remarkably, the *Le pouvoir de vie* theme was already advanced by the Milesian natural philosophy of Anaximander and the pluralists Anaxagoras and Empedocles, all of whom were concerned about explaining the dynamics of the living world. Note that pre-Socratic philosophy has been mostly supported by doxographic evidence, especially because the great majority of Greek and Roman philosophical writings have been irrevocably lost. Fortuitously, the discovery of a set of 52 papyrus fragments in the ancient city of Panopolis in Upper Egypt in 1904 has turned out to be of extraordinary significance. The papyrus scroll was translated in 1992. The papyrologist Alain Martin attributed the text to Empedocles, and together with Oliver Primavesi, published the *editio princeps* seven years later [10.43]. A biological reinterpretation of the direct scribal transmission of Empedocles' philosophy now suggests that evolutionary and network thinking were already quite developed ~2,400 years before Lamarck, Darwin and systems biology [10.16]. A thematic indexing of the poem revealed that it carefully described natural selection, evolving lineages and networks, natural history and systematization of organisms, evolutionary growth, accretion, functional modules, origin of life, life cycles, and fossil remnants. The central thesis of the Strasbourg papyrus, however, is a "double tale" that describes two opposing forces, one of *unification* capable of growing things together from "many" (Love, Φιλότης), and the other of *change* capable of growing things "apart" (Strife, Νεῖκος). The double tale coherently explains the living world with a network paradigm of accretion and diversification in which modules and fractal-like patterns of complexification arose at all levels of biological organization.

Empedocles' philosophy can be an entry point to understand biomolecular communication. Here, I start by describing our exploration of the birth of modules and the establishment of hierarchical patterns of biological organization with the tools of evolutionary genomics and bioinformatics. Modules are structured entities, sets of integrated parts that functionally coordinate with each other and cooperate in performing tasks [10.30]. They have a greater degree of interaction among themselves than with other parts or modules of the system. Module components are integrated by the performance of functions, and thus, modules become evolutionary units. They have features that are "persistent" and spread in evolution, for example, via cellular lineages or viruses. Examples of modules are motifs in protein and nucleic acid sequences, macromolecular domains, protein subunits, components of cellular organization, cell types, and organs. In turn, the hierarchy of modules implies a nesting relation of parts or subsystems, which can be either compositional (scalar hierarchy) or taxonomical (subsumption hierarchy) [10.55] [10.56]. Compositional hierarchies are ordered by the "is-a-part-of" relationship types, while taxonomical hierarchies are ordered by the "is-a-kind-of" relationships. In a complementary manner, Simon [10.63] considers that hierarchies can be described by either *"states"* defining observables (the world as sensed) or *"processes"* (the world as acted upon), which prompts the system to act *"purposefully upon its environment."* The two logical forms of hierarchy have been integrated into a *model hierarchy*, which is also described by two partial hyperscalar hierarchies focusing on nesting and its context [10.26].

My premise is that both modules and hierarchy in biological organization enable "evolutionary learning."[1] Hierarchy imposes constraints on modules, making them structural, functional and evolutionary units of communication. I assume that these modular units combine with other units in evolution to generate novel "molecular vocabularies"[2] which are interlinked to previously generated "vocabularies" through a hierarchy of semiotic experiences. Because languages establish organized systems of signs that create and transmit meaning, molecular modules may also follow Peirce's triadic system [10.50], CP 1.540. Molecular units of communication must act as "signs" (representamen, encodings) of semiotic "objects," interpretations (embodiments as icons, indices and symbols), and "interpretants" (i.e., decoders that assign meaning to signs). Within this biosemiotics framework [10.8] [10.38], which applies logic to the real world, biomolecular communication occurs through numerous interactions between molecular modules in the crowded and contextually diverse environment of the cell. This new perspective is informed by the arguments of von Neumann, Pattee, and Sebeok (reviewed in [10.8]), who emphasized the causal power of signs and meaning in all types of biological communication with, for example, genotype being a symbolic representation of phenotype. Module interactions in different contexts enable a multiplicity of functions, and it is through these functions that communication occurs. In other words, modules may mean different things for cellular communication, all depending on details of the molecular makeup, function and physiological context.[3] I

---

[1] Evolutionary learning is not equivalent to mental learning, although they share some common features, such as a capacity for adaptation.

[2] The term "vocabulary" does not imply equivalence to human language. Instead, I consider a molecular vocabulary as a primitive semiotic system.

[3] Similarly (but not equivalently), words mean different things for human communication depending on the details of syntax, semantics and pragmatics.

show that there is significant evidence supporting this module-centered molecular view of biocommunication, and that levels of organization constrain each other in evolution in such a way that lower levels become increasingly granular (cohesive and interlinked), as higher levels of organization add complexity to the evolving biological system. My focus is to explore how communication "learns to learn"[4] in biological evolution through previous historical and circumstantial experiences by establishing cohesive, compressed and hierarchical language-like structures associated with the emergent modules.

10.2 A Biphasic Model of Evolutionary Growth

Accretion is a process of evolutionary growth, in which biological parts are added in a piecemeal way to an evolving system [10.21]. Accretion fulfills the principle of space-time continuity, which assumes that novelties arise and integrate with previous novelties in a sequential way. This process represents the mereological aspect of a biological system, which unfolds at different time scales, persists despite of the loss of parts, and sustains the molecular makeup and integrity responsible for cellular complexity. Epistemologically, accretion aligns with "constructivism" in many of its inceptions (e.g., as multilevel semiotic networks [10.61]), but also with temporal parts and "perdurantism," the four-dimensional theory of identity and persistence [10.19].

A decade ago, we proposed a phylogenomic-based *biphasic model* of module creation that explained the evolutionary growth of biological systems [10.44]. This model predicted both the evolutionary establishment of nested arrangements (hierarchies) of modules, and the formation of tightly linked modular groups by convergence under optimization or selection. These modular groups can diversify later into new levels of biological organization. The biphasic model creates novelty via two evolutionary phases that are often asynchronous. In Phase 1, parts associate in different combinations but are weakly linked to each other. These initial loose associations compete with each other and are selected for performance. The emerging interactions constrain how associations are structured, causing parts to self-organize into modules with tighter links. In Phase 2, evolutionary variants of the newly constructed modules diversify and become new parts of modules populating higher-level organization.

In Figure 10.1, the model is illustrated using two paradigms: the first is a tree paradigm showing evolving lineages, and the second is a time series of evolving networks. In the tree paradigm, the roots of the tree embody the initial Phase 1 construction phase, with root tips representing evolving parts (Figure 10.1a). The coalescence of roots towards the trunk of the tree illustrates the piecemeal assembly of parts that yields the modular structure. This structure later forms the trunk of the tree. In Phase 2, the branches emerging from the trunk represent the initial diversification of the constructed module and/or its interaction with other modules to form more complex structures. Here, branches represent variants of the diversifying module and the resulting leaves become diversified modules or novel module assemblies embodying higher organizational levels. A number of events can illustrate how parts become assembled together to form a whole (roots converging to the trunk) (Figure 10.1b). Once parts have formed the module, then the module diversifies in sequential or

[4] See footnote 1.

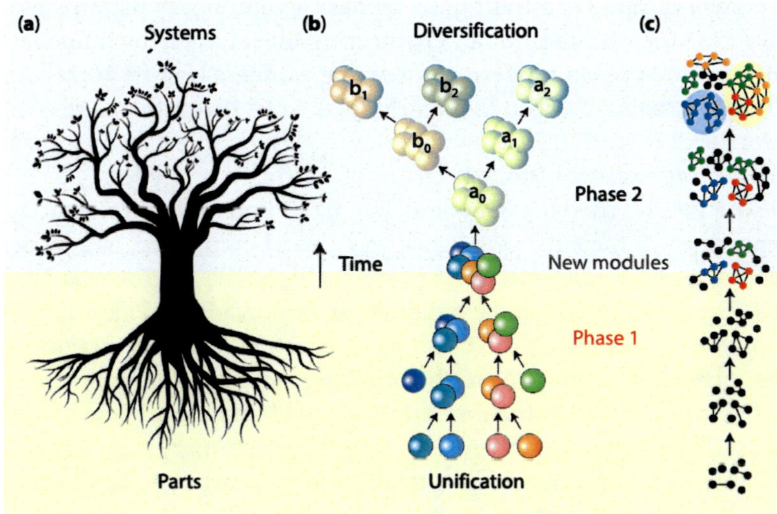

Figure 10.1 The biphasic model of evolution explained in relation to various paradigms: (a) In the tree paradigm, roots, trunk and branches are pathways of evolutionary change as time flows in the direction from the roots to the shoots. The roots represent the rise of modules from primordial components, the trunk represents the world of new modules, and the branches of the crown are pathways of diversification of the newly formed modules, akin to phylogenies. (b) The evolutionary change can be explained by a time series of events of accretion in which individual parts (spheres) come together to form a modular whole (Phase 1). The module then diversifies in sequential (modules a_0, a_1 and a_2) or branching (modules b_0, b_1 and b_2) manner in the absence or presence of other evolving modules (Phase 2). (c) The network paradigm describes two evolutionary phases with a time series of an evolving graph, in which nodes and links describe parts and interactions in an evolving biological system, respectively. The rise of hierarchy and modularity during Phase 1 produces highly connected sets of nodes and links (communities). Once interactions stabilize in the emerging subnetworks, communities become modules. In Phase 2, modules coalesce into new communities, generating higher level network substructures.

branching manner, or both, either isolated from other modules (which is unlikely to happen) or in full interaction with them. In the time series of networks paradigm, competitive optimization results in the emergence of network communities (modules) in a dynamic process of system innovation (Figure 10.1c). Nodes and links of the network describe parts and interactions, respectively. The larger the number of links, the more cohesive and stable is the structure of the evolving system. During Phase 1, the emergence of highly connected communities generates subnetworks in the form of modules that further stabilize the overall network structure. In Phase 2, modules become new parts, which often coalesce into higher level network substructures or, in other cases, engage in a combinatorial interplay that brings together different network parts in a landscape of increasing organization. This second phase lays down the foundation for hierarchy, as newly formed higher-level parts subsume lower-level modular parts. Note that the network paradigm of the biphasic model makes explicit the central prediction that both modularity and hierarchy emerge in system's evolution. This prediction has been verified through experimentation across timescales and levels of complexity, for example, by measuring the degree of modularity and hierarchy of evolving networks in proteins with statistical descriptors, including the Newman-Girvan index, the Fast Greedy Community index, and the clustering coefficient [10.18].

There are many examples illustrating the biphasic evolutionary patterns in biology. But here I first describe the evolution of an engineering object made by humans, the Empire State Building — one of seven modern engineering wonders (Figure 10.2a). The high-rise building with its typical Art Deco architecture was erected in less than a year in Manhattan during the year 1930. Figure 10.2a shows a series of photographs taken as the building was constructed. They represent a historical record of retrodiction. By June 20, 1930, the supporting steel structure reached the 26th floor, by July 27 half of the steel structure was completed, by September 10 the entire steelwork was almost completed, and by September 19 the steel structure reached 1,048 feet and 102 stories high. The antenna was an afterthought added in 1953. The construction of the building can be considered Phase 1, in which different building parts were added to form a complete structure, the structural module. Phase 2 embodies a number of changes done to the building, including the reconstruction of floor 76 following the crash of a B-25 bomber into it in 1945 and a complete modernization of elevators in 2011. They can be described as a series of evolutionary steps (similar to steps a_0, a_1 and a_3 in Figure 10.1b) that gradually changed the building core. In a somewhat similar way, a protein complex is gradually built by stepwise addition of its protein component parts. Figure 10.1b, for example, depicts the evolution of the F_1/F_0 ATP synthase complex, which acts as proton or sodium ion pump and generates most ATP to power cell activities. The complex has two rotors connected by a 2-protein axle, and a stator that modulates rotation speed. One of the rotors (F_0) is associated with the membrane and modulates ion pumping. Phylogenomic data-driven analysis allows us to reconstruct the time of origin to each of its protein components [10.17]. Their origins span most of the geological time scale, i.e., ~ 3 billions of years (Gy), showing a protracted history of improvements that increased

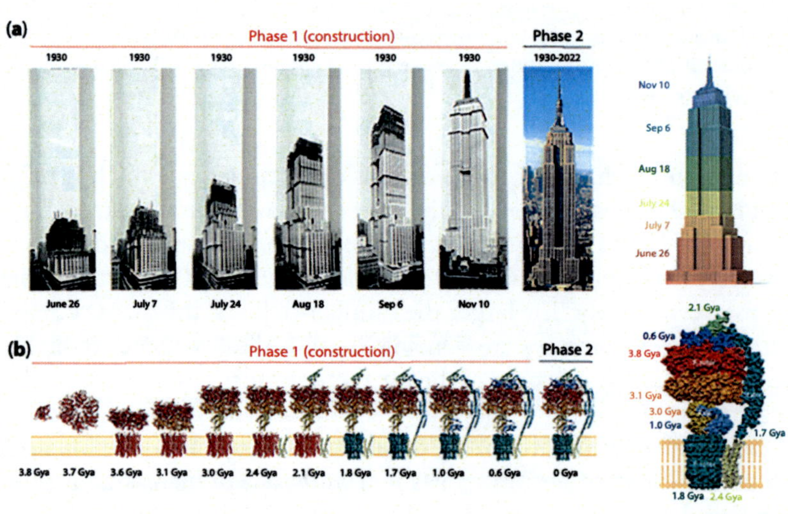

Figure 10.2 Two phases in the history of an engineering object and a biomolecular complex, a construction-unification Phase 1 and a diversification Phase 2. (a) Evolution of the Empire State Building recorded by public domain photographic evidence. (b) The most likely and parsimonious scenario of the origin and evolution of t he F_1/F_0 ATP synthase complex (see details in [10.17]). In the right of the two panels, schematic representations of the building or protein complex map the time of origin of the different parts by tracing them onto the modern structures and coloring them with different color shades for quick identification.

the energy efficiency of the functional ensemble. Phase 1 involves the construction of the entire ensemble while Phase 2 delimits the branching diversification of the complex in different organism lineages.

Thus, the evolution of both an engineering and a biomolecular object follow the biphasic model of module emergence and illustrate how entities must be first built (unified) before they can fully diversify by adapting to various environmental contexts. Remarkably, this "double tale" is carefully recounted in Empedocles' Strasbourg papyrus [10.43], beginning with the introductory thesis [*P. Strasb. Gr. Inv. 1665–6,* lines 233–235]:

> A double tale I'll tell. At one time one thing grew to be just one from many, at another many grew from one to be apart. Double the birth of mortal things, and double their demise.

There is an important philosophical difficulty that arises from the evolutionary analogy presented in Figure 10.2. The engineered object is biological in the sense that it represents a product of human intelligence. It serves a purpose in the quest to satisfy a number of human goals — economic, societal, architectural and aesthetic to name a few. The Empire State Building was one of many high-rise buildings erected to accommodate more population per square foot area (i.e., increasing floor area ratio) in Manhattan, a trend that started at the beginning of the 1900s as a consequence of rapid population growth (reviewed in [10.12]). High-rise buildings rapidly accreted with increasing building heights. Buildings also diversified in Phase 2 of the biphasic model and generated the complex and beautiful skyline of New York City. In contrast, there seems to be no teleological (i.e., goal-driven) factor involved in the evolutionary assembly of the protein complex responsible for ion pumping and energy management of the cell. When asking what drives the molecular growth of this complex, I claim that such a statement must be carefully reexamined since there is an important teleology hidden in molecular communication that I try to dissect.

I conclude this section by discussing a crucial aspect of evolutionary growth—the principle of increasing granularity. This concept refers to the idea that the parts at lower levels of a system become more cohesive and interconnected as the system evolves and new layers of organization are added. This principle was first proposed by Simon [10.64] decades ago, who noted: "Each of the parts of a nearly-decomposable system has strong internal links among its sub-parts, but the several top-level parts are bound together with each other only by comparatively weak linkages." Only recently has this principle been tested in an evolving molecular system representing cellular metabolism [10.45]. The results uncovered evolutionary trends in modularity, power law scaling, and small-world statistical properties that support Simon's proposal. In particular, the analysis revealed that the evolutionary constraints acting on metabolic network structure are stronger at the lower "enzyme" level than at the higher "metabolic subnetwork" level. The culprit for this phenomenon appears to be sets of enzymes with significant network centralities, which keep the network structure tightly knit along time series describing metabolic network evolution. The implication of this finding is significant: as systems evolve, the evolutionary constraints acting at the lower levels of organization increase over time, resulting in collectives of parts that are more cohesive and interlinked.

10.3 Communication

Communication is a pervasive phenomenon in living systems [10.72] [10.73]. Without communication biological structure is not possible; and life as we know it would have no meaning and would not exist. To begin with, all molecular and cellular makeup is established through interactions that involve interpretations of signs. Here, signs (signals) are semiotic entities that communicate meaning and are related to objects through interpretation. This relationship can be considered causal. Peirce's semiosis, for example, is multilevel and embodies constraints of efficient, formal and final causation [10.54]. In fact, semiosis in biological multi-level organization is affected by how the levels constrain processes that occur at other levels through determinative relationships. For example, when an aminoacyl-tRNA synthetase (aaRS) enzyme aminoacylates a specific tRNA molecule by attaching its cognate amino acid to its 5' end, it recognizes (interprets) a message (a discrete unit of communication) embedded in chemical "words" of a structural alphabet that is present in the molecular structure of both the tRNA molecule and the amino acid moiety. This molecular alphabet is responsible for delimiting another molecular alphabet operating downstream, the genetic code, under the auspices of the ribosome. This genetic alphabet is in turn constrained by a complex network of interactions that unfolds between the many molecules that facilitate the process of translation, including mRNA, rRNA and a host of translation factors and other molecules in interaction with the proteins they encode, the ultimate communication agents. The interactions of these agents are part of a world of carefully curated causal links that materialize as an integrated and highly optimized functional system created in the course of billions of years of evolution.

In the example, the aaRS enzyme acts as sign, object, and interpretant, embodying Peirce's triad [10.50]. The aminoacylation capacity acts as a sign of the enzyme because its "signifying element" is the enzymatic activity that is encoded in its catalytic domain module. This module attaches a specific amino acid to its cognate tRNA. Without that functionality, the aaRS loses its central communication ability, its signification, its meaning. The aaRS enzyme besides being a sign is also an object, a 3-dimensional atomic structure that must exist for signification to occur. This "enzymatic representation" (representamen) or "causal connection" of the aminoacylation functionality embodies the catalytic domain module with its active site and all of its functionalities, a constraint necessary for signification and therefore for communication. The initial sign-object dyad, conceptually initiated by a molecular function (aminoacylation), now requires a third element that is capable of linking the first two, the interpretant. This link now subjects the sign-object dyad to context and capabilities, formalizing meaning. Finally, the aaRS enzyme is also an interpretant because it mediates (translates, decodes) the physical connection between the sign and the object, i.e., the aminoacylation reaction and its enzymatic representation. This triadic relation now fully conveys meaning trough a message embedded in the enzymatic representation of the molecular module. More specifically, and following Peirce's logic of "existential relation" made explicit with graphs [10.51], CP 4.347, the sign-object-interpretant triad can be linearized by conceptual order into "*what may be*" (aminoacylation), "*what characterizes it*" (catalytic domain module), and "*what it means*" (amino acid charging of tRNA necessary for the genetic code to be fulfilled). The progression defines a temporal communication sequence in which a function encodes a message in a molecular module that is then decoded back for interpretation. Note that the linearized triad follows Peirce's more elaborate conception of Firstness, Secondness and Thirdness: "First is the conception of being or existing independent of anything else. Second is the conception of being relative to, the conception of reaction with, something else. Third is the conception of mediation, whereby a first and a second are brought into relation" [10.50], CP 6.32. Here, Firstness embodies potentialities endowed by external effects (the existence of many possible functions, including aminoacylation) that poise the rise of efficient causal events and more material Secondness (structural domain modules capable of hosting the aminoacylation function). Thirdness solidifies the initial dyad by endowing semiotic regularities and constraints over the multilevel system (specificities added to the aminoacylation function and domain module that will support the development and stability of the genetic code).

Shannon's Theory of Communication, which lays down the foundation for the entire field of information theory [10.60], can be understood as a biphasic communication system embodied in a linearized triad form. The emergence of this biphasic communication sequence can be explained with the biphasic model of module emergence: a communication module must first be constructed before it can be interpreted. The original communication system that Shannon envisioned had five components (Figure 10.3a): (i) An information *source* (sender) capable of producing a message; (ii) A *transmitter* that transforms the message into a signal; (iii) A *channel* that transfers the signal and experience contextual effects from the environment; (iv) A *receiver* that transforms the signal back into the message; and (v) An information *sink* (destination) that ultimately receives the message. The communication system may require a computing machine to encode information into signals. Shannon was originally concerned about the information source of an engineering type producing messages of sequence of letters as in telegraph or teletype systems, a function of time $f(t)$ or time and space $f(x,y,z,t)$ in various combinations as seen in radio, telephone, and television, and functions of time interacting with each other as in sound transmission [10.15]. All these examples involve message transmission unfolding relatively quickly, especially because in engineering the goal was to transfer the information in a minimum amount of time. In sharp contrast, information in biology unfolds at different timescales and occurs between entities at different levels of biological organization. While communication of enzymes and their substrates, protein complexes and their cognate helper molecules and cofactors, ribosomes and their translation components, and transcriptomes modulating gene expression may occur at timescales of seconds to days, deeper communication channels exist that occur at developmental and evolutionary timescales spanning days to Gy. This dichotomy can be illustrated with the ribosome, the most central molecular complex of the cell.

The ribosome is comparable to a universal Turing machine with two tapes [10.15], a theoretical machine capable of computing any computable function (Figure 10.3b). Von Neumann referred to this Turing machine as a "universal constructor," an algorithm in the form of an "automaton" capable of self-replication and universal computation [10.70]. A universal Turing machine has the unique ability of reading tapes of other Turing machines and computing what these other machines can do, with computation speed proportional to the number of tapes it reads [10.31]. In other words, the machine can simulate any Turing machine on arbitrary input in the form of a "stored-program" computer in the same way as any modern computer can emulate other computers by using an "emulator" program. Superficially, however, the ribosome is like a standard Turing machine, a *finite state machine*,

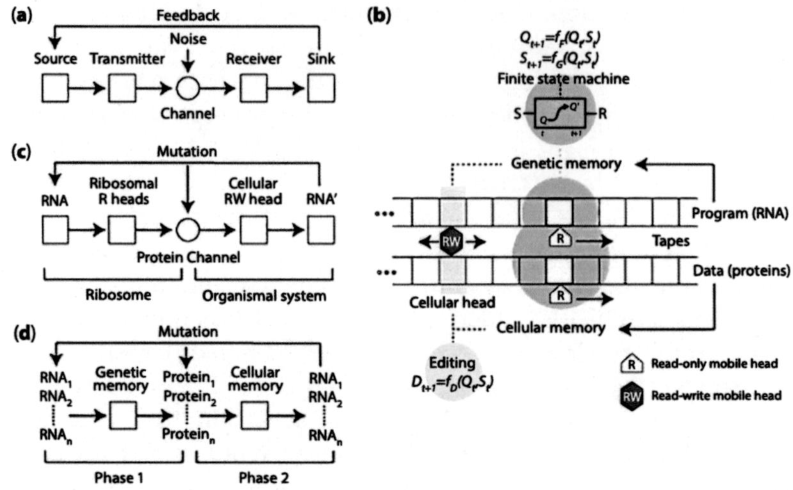

Figure 10.3 The ribosome as a biphasic communication system: (a) Shannon's model of communication describes the encoding and transmission of messages, which are then decoded and received. Messages are affected by environmental noise during transmission through a channel. Norbert Weiner extended the original theory by integrating the sender-receiver (source-sink) of information through feedback (i.e., the receiver acts by "responding" to the message). (b) The ribosome is like a universal Turing machine with two tapes with placeholders for symbols of two alphabets. One tape stores the "program" for reading machines (nucleic acid memory embedded in RNA) and the other stores the data of machines (protein memory embedded in polypeptides). The ribosomal machine operates recursively by moving Read-only (R) mobile (turnstile) heads, which move both tapes towards the left during biosynthesis. The Read-Write (RW) "cellular head" has editing capabilities that operate at different timescales and in multiple cells by sieving mutations through selective optimization. The head can move left or write in the tape according to function D. This effectively edits symbols embedded in nucleic acid and protein sequences. The ribosomal 2-tape system acts as a finite state machine (a black box). It receives a stimulus (input symbols S), which then changes the internal state of the machine (from state Q to Q') generally through a non-linear response R. The cellular RW head turns the finite state machine into a universal Turing machine. (c) The ribosome is a communication system. RNA information is processed by the finite state machine into proteins, making up a "protein channel," which must integrate into the communication of the cell. The cellular head tests the protein messages for appropriate integration and modifies the RNA message through mutation at DNA level, which is then selected at the organism level. This constitutes a cellular RW head, which decodes the protein message with functions associated with organism persistence. (d) The ribosomal communication system evolves through a "double tale" of genetic and cellular memories, one encoding protein modules and the other diversifying them for better integration into the communication system of the cell.

but atypical in that it controls two mobile heads operating on two infinite tapes, one storing a program (the nucleic acid tape) and the other storing the data of an object it constructs (the polypeptide tape). Tapes are strings of placeholders for an alphabet of symbols, which typically are subjected to three operations: "read" a symbol in the placeholder, "write" (or overwrite) a symbol, or "move the head" (right or left) to gain access to another placeholder in the tape. The ribosome, however, is atypical in respect that it can only read tapes. It cannot edit them as true universal machines do, acting instead as a simple "double turnstile" operation. Previously, we proposed that such computational limitation of non-universality is deceiving for the trivial reason that the ribosome is not an isolated molecular complex but a central subsystem tightly integrated with the rest of the cell [10.15]. While ribosomal finite state machines do not have visible editing capabilities and simply convert one type of sequence into another by reading the information from the RNA "tape" and translating that information into the protein "tape," the ribosome as communication system can gain read-write (RW) capabilities by selective production of proteins that better integrate into the communication network of the cell. Organism lineages that harbor "nucleic acid" programs facilitating appropriate protein functions become enriched in evolving populations and more persistent. In other words, the RW "cellular" mobile head of the Turing machine tests protein messages that best integrate into the cellular communication system, decoding best protein messages by modifying RNA messages through mutations at the DNA level. Remarkably, the cell-integrated ribosomal Turing machine complies with Shannon's communication model (Figure 10.3c). RNA information is processed by the ribosomal finite state machine into proteins, making up a "protein channel" of communication. This channel is tested for integration into other cellular functional networks through the RW cellular head operating at the organism level. Feedback occurs when computations result in a modified RNA, cycling information back from sink to source via mutations in the DNA. The communication system is biphasic and can be explained by the biphasic model of emergence (Figure 10.3d). The information source and the ribosomal transmitter generate protein modules in Phase 1 (construction) at timescales from seconds to days. These modules are transmitted through a cellular channel, in which proteins are affected by both environments: internal and external to the cell (e.g., crowding, external signals, mutations). This "noise" induces diversity, which is then selected through cellular computation at the organism level in Phase 2 (diversification). This occurs at timescales spanning from days to Gy-s of evolution depending on organisms and channels. Thus, the ribosomal communication system evolves through a "double tale" of genetic and cellular memories that generate protein modules capable of integrating into the communication system of the cell. This is done by diversifying modules at higher levels of organization.

So, what are the messages that living organisms produce? What are the meanings they convey? Meaning is entrenched in widely different forms of information, but also in matter and energy. For example, the genome, transcriptome, or proteome of cells embeds information either in the individual molecule types or in their combinations [10.22]. The ribosomal communication system illustrated in Figure 10.3 involves tapes, strings of symbols that *per se* provide little information about how the communication system operates. The same can be said about the genetic code that translates one set of symbols into another. Despite advances and a multitude of hypotheses, we are still struggling to understand how and why the genetic code was constructed. While it is likely that answers lie in the structure and function of proteins, there is no theory that explains the structural makeup of proteins, not to mention that of RNA. The latest release 2.08 of the SCOPe database [10.23], a gold standard of protein classification, catalogues 5,084 families of structural domains into 2,067 superfamilies holding 1,257-fold structures, each one representing a unique architectural and functional design. The number of superfamilies and fold structures approximates the diversity of the world of proteins, as very few of them are expected to be newly discovered. The meaning of this vocabulary of 3-dimensional atomic structures and functions has not been dissected, including its syntax, semantics or pragmatics. However, it is plausible that organisms also use other "languages" channeling information, matter and energy at higher and lower levels of organization. Indeed, fungi communicate with other fungi through oscillations of extracellular electrical potential that propagate along the mycelium network [10.1].

The distributions of lengths of trains of electrical spikes in the mycelium are similar to the distributions of word lengths in human languages. These trains of spikes form a lexicon of about 50 words (with a core of 15–20 words), and get combined into "sentences" with Liz-Zempel complexity, algorithmic complexity, and logical depth that varied in different fungal species. Thus, different communication "dialects" appear to exist in fungi. It turns out that electrical spiking activity is common in protozoans, fungi and plants, and in neuronal cells. In fact, spike activities of neuronal communication appear to embody a "language of the brain" in higher organisms [10.53]. Note that this Morse code-like communication of electrical potential is ultimately enabled by facilitated diffusion and transport, which are cellular processes mediated by the protein machinery of ion channels and pumps. At lower complexity levels, chemical reactions can be considered communication devices, powerful molecular recognition machines that behave as computing automata. One remarkable example is the oscillatory chemistry of the Belousov–Zhabotinsky reaction, which can be turned into a programmable chemical Turing machine capable of context-sensitive symbol processing [10.27]. The challenge is to understand how all of these different languages and computations are integrated with each other and how to uncover elementary messages, their meaning and rationale.

Shannon's unidirectional model (with or without feedback) cannot portray the information flow through the incredibly complicated network of channels within the cell. Instead, a directed graph describing the flow of information and matter-energy through biological networks makes much more sense. For example, many proteins of the ribosome-enabled protein channel of Figure 10.3d are themselves finite state machines or Turing machines responsible for enzymatic activities that process matter and energy. This fact links the communication system of ribosomes with communication in metabolic networks. For example, the F_1/F_0 ATP synthase complex of Figure 10.2b mediates the flow of energy and matter through the cell by linking the F_0 rotor in the membrane, which is driven by ion gradients (analog information), with the extramembranous F_1 rotor, which is driven by ATP hydrolysis and transmits digital information associated with energy. The complex therefore modulates different channels of information contributing to both analog and digital signaling.

Chomsky's hierarchy of formal languages describes sets of terminal symbols (alphabets), which are often presented as combinations of strings organized by rules [10.24] [10.25]. Here, grammars are rewritings defined by a tuple (N,Σ,P,S), where *nonterminal* (N) and *terminal* (Σ) symbols are arranged according to *production rules* (P) and a *sentence symbol* (S) that starts the algorithmic implementation of language. It turns out that there are four classes of formal languages and associated grammars in nested arrangement from simple to increasingly complex computations: (i) Regular, (ii) Context-free, (iii) Context-sensitive, and (iv) Unrestricted. These classes are driven by finite state machines, pushdown automatons, linear bounded automatons, and Turing machines, respectively. Biomolecular machines depend on each and every one of these grammars for effective communication. The regular grammars of metabolic enzymes are embedded in macromolecular turnstiles driven by the stoichiometry of substrates. Their biosynthesis is dependent on the unrestricted language of the ribosome. The context-free grammars of functional RNA (e.g., rRNA, tRNA) require the activity of RNA polymerase protein, which is also synthesized by the ribosome. Context-sensitive grammars of the non-ribosomal peptide synthetase (NRPS) assembly lines are known to synthetize over 500 small peptides (these are analogous to Turing machines with linearly bounded tapes). Both functional RNA and NRPSs rely on ribosomes for production of associated proteins, making them dependent on unrestricted languages. Thus, a number of computational machines in the form of molecular complexes containing protein, RNA, or both, are vehicles of communication in biology. All of them are organized around the ribosomal communication system and make up a vast computational grid of communication channels.

The recent development of a universal Turing machine capable of learning is particularly significant because it embeds a network architecture (a recurrent neural network) that can be trained to, for example, learn simple algorithms, answer synthetic questions that emulate reasoning, complete a moving blocks puzzle with changing goals, or learn to ride the London Underground rapid transit system [10.28] [10.29]. This Neural Turing Machine (NTM) architecture [10.28] and the Differentiable Neural Computer (DNC) [10.29], which is its more advanced successor, have a neural network "controller" that interacts with the outside world via inputs and outputs. The novelty, however, is that the controller also interacts with a memory matrix via "blurry" RW operations that behave as multiple read and write heads of a Turing machine. Interactions are computationally differentiable and can therefore be trained by gradient descent. Blurriness is implemented by differentiable attention mechanisms that constrain memory interactions. The hybrid computing system was able to learn how to operate on data structures, adapt to new predictive distributions, and learn complex task such as finding shortest paths or missing links in random graphs. Its power rests on both its "working memory," a concept important for psychology and neuroscience, and its network structure capable of recursive and parallel processing, a concept important for cognitive science and linguistics. I contend that the NTM and DNC implementations mimic the workings of the immensely more powerful computational grids of communication channels embedded in cellular machinery, which reroute information and matter-energy flows in recursive and parallel fashion through the entire communication system and use the cellular RW head of the ribosome as "working memory" for learning operations.

10.4 Learning Demands Communication

The standard teleological definition of learning is the acquisition of understanding, knowledge, behaviors, skills, values, attitudes and preferences. From a systems biology perspective, however, a more appropriate definition of learning is a system's use of information obtained during one interaction with its environment to improve performance in future interactions [10.10]. In computer science, improving the performance of database management systems, adaptive control systems or concept formation systems involve altering assertions, parameters or structures, respectively. Improving biological systems involves altering assertions, parameters and structures that impinge on fitness [10.22]. Note that biological systems are adaptive, storing and diffusing information at different timescales as they become structured. I will elaborate on this concept and describe how fitness is ultimately related to persistence.

Communication in biological systems involves machines that process information, matter and energy. One amazing feature of these machine "agents" is that they can learn, adapt, organize, repair themselves, and even construct other machines. The universal Turing machine model of a ribosome indicates that this macromolecular complex can produce most component parts of the cell (including its own) and use its cellular RW head editing capabilities to perfect them in the course of evolution. In computer science, three approaches drive construction of computational machines that can adapt and learn: (i) adaptive control, (ii) pattern recognition, and (iii) artificial intelligence [10.10]. All of these approaches involve computational intelligence, i.e., the design of *intelligent agents* that act in their own environments to perform tasks [10.52]. Agent activity is adaptively adjusted to circumstances and goals and is flexible regarding internal and external cues. In other words, intelligent agents are capable of learning from experience. They are also robust to disturbance and are mindful of budgetary constraints of the system. Symbols in abstract and material embodiments are at the center of intelligence. The physical symbol system of Newell and Simon [10.49] uses manipulation (processing) to combine symbols (e.g., physical patterns) into expressions (structures) and change existing expressions to produce new expressions. Newell and Simon considered this system necessary and sufficient to generate intelligent action. Thus, biological and computational machines that use and edit signs during communication can be expected to act as intelligent agents to become functional, robust, sustainable, and self-supporting.

I suggest that intelligent agents can be natural (i.e., systems found in nature) or artificial (i.e., systems constructed by humans and other organisms). Artificial intelligence is synthetic in that it is a synthesized product of a natural intelligence system. For example, a city is an artificial intelligence system that can learn, adapt, organize and repair itself. I have illustrated its capabilities with biphasic patterns of growth of high-rise buildings in Manhattan (Figure 10.2a). Human population growth stimulated the emergence of new technologies of building construction in order to accommodate more people per square footage. The example shows adaptive control but also generation of novelties (e.g., steel beams, rivets, elevators) that would not only "organize" construction but also "repair" failing structures (e.g., the crash of a B-25 bomber into the structure, the effect of a hurricane). The city, however, is an artificial intelligence system constructed by a natural intelligence system (i.e., humans). Similarly, an analog or digital computer or a temperature-regulating ant mound are examples of artificial engineered intelligence synthesized by life. My point is that all "intelligence" can be considered a natural extension of life, with origins appearing very early and at the molecular and cellular level. It is therefore important to explore the very foundational systems that were formed in early evolution to attempt dissecting the learning phenomena.

10.5 Biological Vocabularies and Their Evolution

We recently proposed that the *Lexicon* ⇒ *Syntaxis* ⇒ *Semantics* ⇒ *Pragmatics* progression in natural languages corresponded to the *Module* ⇒ *Structure* ⇒ *Function* ⇒ *Fitness* paradigm in biology [10.13]. The proposal, modified from Searls [10.57], is based on analogies between linguistics and molecular biology, i.e., the transfer of meanings (metaphors) from one symbolic representation of an abstraction to another with the goal of uncovering cognitive similarities.

Lexicons are message "wordbooks," catalogs of elementary units of meaning (lexemes). Genomic biology and bioinformatics have been uncovering lexicons in proteins and nucleic acids of different types. These lexicons are made of molecular modules, structural, functional and evolutionary units that in the course of evolution spread in time and space in different molecular contexts. I described above a lexicon of structural domains of proteins that is comprised of about 5,000 molecular "words." Our phylogenomic analyses have shown the origin of these words unfolded during 3.8 Gy of protein history [10.18] [10.19]. For example, we have recently traced the origin and evolution of domains and their combinations in proteins with evolving networks [10.5]. Figure 10.4 illustrates the result of the study. Domains often combine in evolution to form multidomain proteins (Figure 10.4a). The time of origin of individual domains in single-domain proteins or their combinations in multidomain proteins can be determined by generating a phylogenomic tree of domains and domain combinations reconstructed from a proteomic census (Figure 10.4b). The leaves of this tree are proteins and the phylogenetic characters used to construct the phylogenies are proteomes. The features being studied are SCOPe-database superfamily assignments based on genomic sequences processed by the highly effective hidden Markov models (HMMs) of

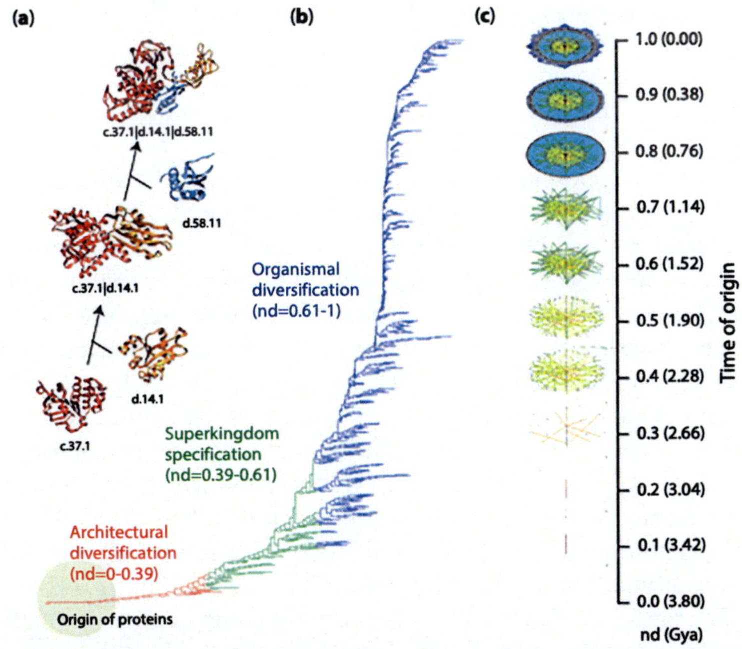

Figure 10.4 An evolving network describing the organization of structural domains in proteins: (a) Phylogenomic analyses allow construction of timesteps describing how domains combine with other domains in proteome evolution. The rationale is illustrated with 3 domains that are ubiquitous in dehydratase enzymes and elongation factors. Domains in three-dimensional atomic models visualized in ribbon format with Chimera are identified in protein "architectures" with different colors and are described using *ccs* domain identifiers defined at superfamily level of SCOPe classification. (b) A most-parsimonious phylogenomic tree of protein architectures defined at SCOPe superfamily level shows the three epochs of the protein world and the approximate span of time of origins of domains in node distance (nd) given in a scale from 0-1, with one being the present. (c) Snapshots of networks of protein domain organization of the CX type [10.5] taken at regular evolutionary intervals are described in radial format as they evolve along a timeline of 3.8 billion years (Gy) of history. Nodes of the networks represent architectures and color-coded arcs represent donor-acceptor recruitments and the flow of time from ancient to recent nodes. The time of fold emergence (Gya) is predicted from node distance (nd).

structural recognition. The reconstructed tree is rooted and highly unbalanced allowing to establish a time of origin (age) for each leaf (protein architecture) by counting nodes from the base of the tree. A molecular clock of folds places this age within the geological record. Times of origin describe a sequence of protein architecture accumulation which can be transferred to a network of protein architectures connected by links that represent the presence of the same domains (Figure 10.4c). A "big bang" of domain rearrangements occurred late in protein evolution visualized by the massive increase of network connectivity. This connectivity of nodes in the network embedded a syntax of domain combination responsible for the structure of proteins, which we showed was evolving in time. This syntax can be made explicit with grammar dependencies, for example, with the use of catenae driven by evolutionary governors [10.13].

Each domain "word" is also composed of structural parts that are highly conserved in evolution, the protein loops. Loops are regions of functional importance that connect standard elements of secondary structure and often change the direction of the polypeptide chain. These molecular parts resemble morphemes, the smallest grammatical units of language that stand alone or combine with other morphemes to form words. Analysis of predicted structures uncovered the existence of 7,078 loop prototypes in the proteomes of 8,127 organisms belonging to superkingdoms Archaea, Bacteria and Eukarya and the viral supergroup [10.7]. These prototypes, which were defined by bracing secondary structures and structural geometries of the loops, were divided into two main subsets. A subset of 5,125 "non-modular" loop prototypes mapped uniquely to individual domains. These loops failed to be recruited by other domains, and consequently, their times of origin were that of their associated domains. In sharp contrast, a subset of 1,937 "modular" loop prototypes mapped to more than two domains, with times of origin spanning the timeline of domain history. These prototypes acted as true modular units, engaging in a combinatorial process of recruitment, i.e., the reuse of gradually accumulating molecular innovations to perform related functions in different molecular and cellular contexts. [10.13]. Note that recruitment of innovations is evolutionarily advantageous because it enhances functional novelties associated with the protein structure. Using phylogenomic methods, we traced the origin and evolution of loop prototypes onto evolving networks and uncovered remarkable patterns of structural diversification [10.6] [10.7]. Thus, a fractal-like multitude of lexicons are embedded into each other in protein structure and accumulate in a piecemeal way driven by the evolutionary force of recruitment.

Semantics describes the meaning of signs and their combinations at different hierarchical levels of the message. For example, the lexical semantics of natural languages explores the meaning of linguistic expressions such as words and sub-words (the syntax-semantic interface) in specific context, while conceptual semantics explores the meaning of words as general concepts. Semantic networks such as those of the WordNet® lexical database (https://wordnet.princeton.edu) seek conceptual semantic relationships that highlight synonymy, hyponymy, and hypernymy, i.e., sameness, subordination, and superordination in meaning, respectively. Similarly, lexical semantic relationships that standardize the biological functions of genes in the Gene Ontology (GO) database are organized as hierarchical networks (directed acyclic graphs, DAGs), in which child GO terms connect to parent terms and lower DAG levels describe more specialized functions [10.4]. The vocabulary of GO terms describes biological processes (*bp*), molecular functions (*mf*) and cellular components (*cc*) of organisms and viruses. GO *bp* terms define events (objects and properties of agents unfolding in time), *mf* terms define activities (actions of agents), and *cc* terms define components (parts and places). The collective of GO terms of an organism represents a "functionome," the functional meaning of an organism. We have been exploring the evolution of functionomes, determining the time of origins of individual GO terms defined at different levels of the DAG [10.35] [10.36] [10.37]. For example, Figure 10.5 illustrates our analysis with a tree of GO *mf* terms defined at level 2 of its DAG. The tree describes the evolution of the molecular activities of the cells with definitions at less specialized level. Remarkably, the basal splits of the tree support the early evolutionary origin of metabolic activities through binding and catalysis, which is in line with the evolutionary analysis of biocatalytic mechanisms [10.48] and many other lines of evidence. A recent review discusses some aspects of these types of semantic analyses, including the use of "word clouds" arranged in temporal order to explore functional evolution [10.13].

Finally, pragmatics (Greek, πραγματικό ς, fit for action) defines a context-dependent relation between the sign and a sign user according to rules governing communication. The triangle of effective communication, for example, places main goals of communication at the triangle's three vertices, *Attraction*, *Persuasion* and *Relevance* [10.13]. The neo-Gricean inspired triangle concept is based on the "organon model" of Bühler [10.11] and the six "constitutive factors" of Jakobson [10.34], which ultimately rely on Shannon's communication model. The triangle embodies trade-off solutions that are necessary when a trait cannot increase without the decrease of another trait. Trade-offs maximize some measure of communication performance and involve three pairs of conflicting functions. First, navigation toward the *Persuasion* vertex of the triangle requires the sender of the message to influence the receiver through an emotive–conative trade-off relationship. Second, navigation toward

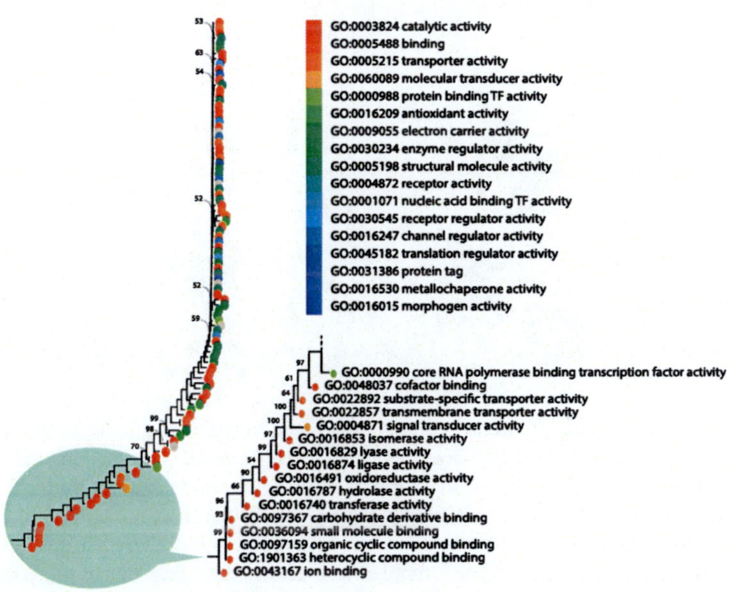

Figure 10.5 Evolution of molecular functions: A phylogenomic tree of GO terms of molecular functions defined at level 2 of the directed acyclic graph is shown, with leaves portrayed as colored dots according to level 1 associated terms described in the color key. The phylogenetic relationships at the base of the phylogeny are made explicit in a subtree and highlight the ancestral nature of binding and metabolic functions. Number on branches indicate bootstrap support values exceeding a 50% threshold.

the *Relevance* vertex demands the context of the message be both impactful and faithful via trade-offs between referential and rhetorical functions. Third, navigation towards the *Attraction* vertex requires attracting attention via trade-offs between relational and reflexive functions. A similar triangle for molecular biocommunication can be proposed, in which the communication system navigates a trade-off space with best fitness solutions manifesting in Pareto fronts, boundaries in multidimensional performance spaces with mathematically elaborated geometries [10.62].

Figure 10.6a illustrates the rationale of transforming the triangle of effective communication in conversation into a triangle of molecular communication driven by molecular persistence, and then into a generic triangle of persistence that describes any system that may exist in biology. If we take ribosome as an example, a triangle of effective molecular communication for this molecular ensemble transforms under a matter-energy-information-based model of optimization, the Persuasion, Relevance and Attraction vertices into Budget, Specificity and Mechanism vertices, respectively. The Persuasion vertex seeks to economize communication by minimizing effort of both sender and receiver through brevity, order and other minimalistic principles. Similarly, ribosome seeks to optimize the GTP-mediated

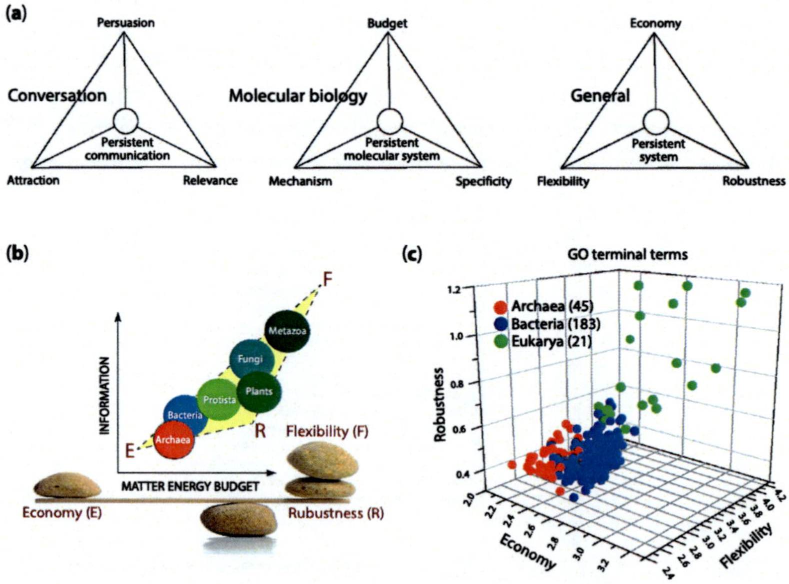

Figure 10.6 Exploring the pragmatics of biomolecular communication with triangles of persistence: (a) The persuasion-relevance-attraction triangle of conversation explains the trade-off solutions that maximize the persistence of communication (left), while the budget-specificity-mechanism triangle of molecular persistence maximizes persistence of a molecular system in biology (center). The "general" economy-robustness-flexibility triangle of persistence explains the trade-off solutions for any biological system, subsuming all previous alternatives. (b) Trade-off patterns for all complements of structural domains in proteomes that impact the proteomic vocabulary of the six organism kingdoms, generating clear patterns of distribution in a general triangle of persistence. (c) The impact of the persistence triangle on the functionome vocabulary of superkingdoms. The three-dimensional scatter plot evaluates the persistence strategies of organisms by analyzing the presence and frequency of their corresponding terminal GO terms. The axes on the graph are logarithmically scaled, and the number in parentheses represent the number of proteomes analyzed. (Data from Nasir *et al.* [10.46])

Table 10.1 The three drivers of persistence.

Economy	The matter and energy budget needed to maintain the functional mechanisms of a system.
Flexibility	The structural and functional mechanisms responding to internal and external communication of the biological system that require processing of information and enable adaptation to environmental change.
Robustness	Mechanisms that use information to maintain the structure and function of a system despite external influences, system malfunction, damage and change.

energetic budget of peptide synthesis by controlling its chemical budget through economy of "catalyst" function. The Relevance vertex seeks persistence of meaning in the presence of varied syntactic and pragmatic contexts through robustness (relevance) of context and message of the conversation. Similarly, the ribosome seeks to maximize "gatekeeping" discrimination by maintaining genetic code specificities and cellular memory, mostly via specificity of aaRS enzymes and translation factors that maximize robustness. Finally, the Attraction vertex requires a sender-receiver conduit that follows a faithful but flexible code of communication that is able to dissect meaning within the space of possible syntactic and pragmatic contexts. Similarly, the ribosome seeks to optimize its molecular mechanism, its dynamic "turnstile" of tape recognition and movement, which fosters molecular flexibility. Thus, conversation and molecular communication in biology are isomorphic and can be best described in general terms by a "triangle of persistence" with Economy, Robustness and Flexibility vertices. See Table 10.1 for definitions.

Yafremava *et al.* [10.74] carefully elaborated almost a decade ago a general framework supporting the "triangle of persistence." The framework, which was inspired by von Uexküll's organism-centric view of the environment and Miller's general theory of living systems has been put to the test experimentally by, for example, exploring the lexicon of domains in proteomes of the six organismal kingdoms of life (illustrated in Figure 10.6b). Similarly, an analysis at semantic level of terminal GO terms of molecular functions revealed that the "triangle of persistence" could dissect the functionomes of superkingdoms in three clearly defined groups (Figure 10.6c), showing three different strategies are operating in the cellular world [10.46]. Note, however, that the lexicon, syntaxis, semantics and pragmatics of languages are not isolated levels of communication complexity. In fact, they are all related to each other through integration in a hierarchy of semiotic experiences.

10.6 Persistence, the Evolutionary Master

"When I use a word," Humpty Dumpty said in a rather scornful tone,
"it means just what I choose it to mean – neither more or less."
"The question is," said Alice, "whether you can make words mean so many things."
"The question is," said Humpty Dumpty, "which is to be master – that's all."

–*Through the Looking Glass* by Lewis Carroll

Lewis Carroll, the deeply subversive pragmatist, makes it clear: context is central to meaning. His work is especially relevant for those interested in "everything you ever wanted to know about language but were afraid to ask" [10.39]. But why "afraid"? Well, because context is both a "master" and a challenging proposition for the impact that languages have on reality (e.g., politics, society, economy, cultural identities). Just take for example the negative influence of "fake news" and other forms of misinformation. In the same way, context is a master in biology when it impacts biological reality, including *L'influence des circonstances* and its effects on deep molecular communication. There is no doubt that there are many possible contextual landscapes operating in evolutionary biology, including physical, structural, cellular and functional environments. So, which of all of these is a "master"? Life carries hundreds of definitions but there is one central theme that overrides them all: *persistence.*

Colloquially, persistence is a continued or prolonged existence. Philosophically, persistence is maintaining identity in time and space. The continuance of a structural feature and the succession of entities (lineage) through time illustrate the problem within a temporal framework. A gene, for example, represents a unit of inheritance that is maintained within the confines of a lineage. A gene in itself is a string of nucleotide residues in a nucleic acid molecule. It often encodes a protein that carries information in a string of amino acid residues. When sequences of genes or encoded proteins come from evolutionarily closely related organisms, larger portions of these sequences can be aligned as compared to sequences from distantly related organisms. If sequences are sampled from increasingly distant organisms, then correspondences become increasingly weak, up to a point where they become meaningless. These properties of alignments illustrate the problem of persistence and the difficulty of maintaining identity in the presence of accumulating mutations and stochasticity. Mathematically, modeling persistence with Markov chains explains how time erodes information that exists in the present [10.65] [10.66]. Erosion occurs exponentially and is only delayed when drift or balancing selection are at play.

Experimentally, we can track mutation accumulation in real time in viral outbreaks and pandemics, such as in the case of the fast-evolving COVID-19 virus [10.20] [10.68] [10.69]. Mutations accumulate in bursts in regions of functional significance, including the N-terminal domain and the receptor-binding domain of the spike protein that modulate seasonal behavior and immunogenicity of SARS-CoV-2, respectively, and often optimize protein flexibility by modulating intrinsic disorder. The mutational landscape shows continual recruitment from mutational bursts, which result in network-like patterns in the variants of concern (VOCs) that are continuously arising [10.20]. This makes clear that change occurs in clumps. This clumping ultimately gives rise to hierarchical relationships of modules, which in VOCs take the form of mutant constellations. Note that only events that have occurred relatively recently can be tracked in real time. If changes in protein or nucleic acid sequences unfold in a distant past, then only phylogenetic methodology (with all its might) can help us reconstruct the past.

Persistence is a manifestation of existence, a conjectured state of things. This fact has been the subject of deep philosophical debate for decades [10.41]. Two schools of thought, perdurantism and endurantism agree that perduring systems (things) are "persistents." However, both schools disagree on how things persist. Perdurantists believe that things have both "spatial parts" in different places and "temporal parts" at different times. They are four-dimensional persisting "occurrents," worms stretching through time and space.

Endurantists reject the existence of temporal parts. Instead, they believe things are wholly present at every moment of their existence, i.e., things only exist in the present. These "continuants" are therefore three-dimensional persisting entities that "endure." New views of the debate support perdurantism, viewing instead endurantism as the denial of perdurantism [10.71]. The four-dimensional view of perdurantists provide a historical and predictive view that is tenseless (eternal). That tenseless view allows studying processes instead of mechanisms demanding explanation of time. The view is tightly aligned with modern bioinformatic implementations of evolutionary biology. The "evolving conduit" view of Figure 10.1a is historical and informative. It explains change by describing occurrents (lineages) as collections of temporal parts and using mereological arguments of parthood, ordering relationships, and temporary intrinsics to solve the difficulties of the philosophical "problem of change" (see [10.19] for discussion). Parts and their properties can "transmit" change [10.33]. Furthermore, identity can be made strict or loose depending on context, as has been made explicit for biology.

If persistence is the evolutionary "master" (contextually speaking), how does the "triangle of persistence" [10.74] embody fitness functions? When factors, functions, and principles of biological languages find optimal solutions that strike a balance in the trade-off space of economy, flexibility and robustness, that balance applies to the entity set of occurrents that represents the system (the four-dimensional worms). Each temporal part of a worm makes a point in a temporal cloud of the trade-off space. This cloud, for example, could represent trade-off solutions for lineages of organisms in superkingdoms (Figure 10.6b). Here, the context of interactions with internal and external organismal environments condition the strategies being adopted throughout the lifetime of the organismal occurrents, past, present and future. An axis of matter-energy costs attracts the cloud towards the Economy vertex by, for example, modulating the "reuse" of protein and nucleic acid component parts or the metabolic production of matter-energy in the cell. Conversely, an axis of redundancy costs attracts the cloud towards the Robustness vertex by modulating the maintenance of identity within a stochastic background (the "stasis" of the system). Finally, an axis of novelty costs attracts the cloud towards the Flexibility vertex by modulating the generation of novelties important for function and fitness. The circles of Figure 10.6a and b are meant to illustrate clouds describing collectives of occurrents. Trade-offs in the triangle manifest constraints in matter-energy and information flows unfolding in spacetime. These constraints "canalize" the functioning and evolution of the system. Because systems are occurrents, constraints act as historical memories imparted by the emergent layers of biological organization, i.e., by the effects of modularity and hierarchy.

Operationally, the "triangle of persistence" maps matter-energy costs of a system onto a budgetary "economy" axis and information associated with context onto both an "umwelt" axis of "flexibility" and a "gap" axis of "robustness" [10.74]. Here, umwelt ("the world around us") corresponds to the total number of signs (information) that are processed by the system, gap corresponds to signs that are not perceived or processed but that are indirectly associated with the robustness of component parts of the system. Both umwelt and gap, the collective of which is "scope," involve all signs reflecting behavioral, physiological, epigenetic, and genetic experiences triggered by interactions of the system with suprasystems (the environment). These semiotic entities communicate meaning that is causally related to objects they represent. For example, terminal GO terms of molecular functions encoded in the genomes of 249 organisms belonging to the three superkingdoms of life [10.46] describe the semantics of their biological functions. This semantic landscape maps to axes of economy, redundancy and flexibility, separating the functionomes of the three superkingdoms into three separate clouds of a 3D scatter plot (Figure 10.6c). The cloud of the archaeal superkingdom unfolded maximum economy but least flexibility and robustness. The bacterial cloud was similarly clustered but exhibited higher levels of flexibility. In sharp contrast, the eukaryotic cloud occupied a large volume in the trade-off space of persistence, exchanging economy for the greatest levels of flexibility and robustness. The three metrics describe differential deployment of molecular functions in the three superkingdoms., supporting the significant semantic and pragmatic compression that molds the vocabularies of microbial life [10.13].

10.7 Laws of Language in Biological Communication

If communication and language exist in molecular biology, quantitative linguistic patterns should exhibit themselves as statistical laws of language that are universal, i.e., non-contextual [10.13]. It is important to note that these statistical generalizations are distinct from the rules that govern syntax and semantics, which, while conserved, can be altered by contextual factors. There are three types of language laws: probability distribution laws, functional-type laws, and developmental-type laws [10.2].

The laws of probability distribution make forecasts. For instance, the Zipf's law states that the frequency of a word in a piece of writing is in inverse proportion to its ranking in a frequency chart. This law is a special case of the scale-free distribution commonly found in many biological networks (e.g., metabolism, protein-protein interaction networks). The law explains power law behavior in the occurrence of Pfam domains [10.58] and SCOPe structural domains [10.14] in proteomes. The power law decay values for SCOPe domains averaged ~2 for the proteomes of Bacteria and Archaea and ~1.4 for the proteomes of Eukarya [10.14] and followed the same progression observed in Figure 10.6c for superkingdoms. These decay values were similar to those for the English and Chinese languages, respectively [10.42].

Functional-type laws follow functions that link two types of variables. A typical example is the Menzerath-Altmann law describing how longer syllables contained shorter articulated units of sound [10.3]. The law was summarized by the motto *"The greater the whole, the smaller its constituents,"* and was found present in natural languages, vocal communication outside humans, genomes, proteins, organization of structural domains in proteins, and evolution of RNA structure. When plotted in log-log graphs, we found that the length of structural domains decreased in a linear fashion as the number of domains increased [10.59]. Mathematical elaborations supported parameters consistent with the distribution of superkingdoms in the triangle of persistence of Figure 10.6c. Similarly, we found that in the history of molecular accretion, substructural parts of RNA decreased their size when RNA structures enlarged [10.67]. Thus, a principle of diminishing returns exists in both natural languages and biology.

Finally, developmental-type laws describe processes of development, growth and evolution of languages. In other words, they place languages in the context of time. For example, the Heaps' law describes a scaling relationship between the number of words in a document or corpora and the number of words in a word database [10.32]. Remarkably, the Heaps' law paraphrases how the vocabulary of SCOPe structural domains (defined at superfamily level) scales with the total number of domains in proteomes [10.47]. The scaling relationship had four regimes of sublinear growth, i.e., slowdown of vocabulary growth, each corresponding to the proteomes of viruses, Archaea, Bacteria and Eukarya, in that order. Exponents for viruses approached unity and were not far away from those of the tightest regime of Chinese, Japanese and Korean languages, which have limited dictionary sizes. This regime embodies the ancient kernel-like vocabulary of natural languages. In turn, exponents for Archaea matched those of English text corpora, exponents of Bacteria, those of the second slowdown of Chinese language, and exponents of Eukarya, those of word novelties with extreme evolutionary slowdown.

10.8 Conclusions

The hypothesis that modularity and hierarchy are interlinked and necessary for evolution of biocommunication appears well supported by evolutionary genomics. Here I extend the hypothesis to claim that learning in evolution occurs when modules combine with other modules to generate novel vocabularies in an entangled network of communication clustered around the universal Turing machine of the ribosome. The theoretical framework aligns with Peirce's triadic semiotic system, Shannon's Theory of Communication, Chomsky's hierarchy of formal languages, von Neumann's universal constructor, and the arguments of computational intelligence. It is clear that the makeup of molecular vocabularies retains memories of their histories and can therefore display patterns that reflect those of a 'triangle of persistence' driven by economy, flexibility and robustness. We have shown that the triangle is general in its application to a number of problems in biology, including that of communication. The triangle highlights the central role of persistence as an evolutionary, intelligent and teleological "master" when biological systems navigate the contextual landscape defined by suprasystems and environment. Persistence is evolutionarily deeper than adaptation. It encompasses the entire historical breath of life and all evolutionary constraints, from the very early appearance of molecular and cellular complexity to the late development of computational capabilities in the brain. Persistence is an intelligent master because it drives intelligent agents to learn from previous historical and contextual experiences with the ultimate goal of furthering the life of the biological occurrent. Again, intelligence arises naturally when a system is persistent by unfolding and interlinking biological languages, developing the computational tools of their interpretation (e.g., translation), embodying structural representation by mapping and mismapping them onto concrete biological operators, and ultimately materializing "cognition." I end by noting that persistence provides a goal-driven explanation (*telos*) versus cause-driven explanation of biological phenomena, since it overrides or subsumes all other goals when it pushes the system towards perpetuity. The standard evolutionary view of absence of actors with intentions is therefore overridden by a "canalizing" agency that tends to produce long-lived occurrents.

Acknowledgments

Research was supported by grants from the National Institute of Food and Agriculture (ILLU-802-909 and ILLU-483-625), the Illinois Campus Cluster Program (ICCP), and Blue Waters supercomputing allocations from the National Center for Supercomputing Applications (NCSA). I wish to thank the three reviewers that examined my manuscript and the editor, Alexei Sharov, for valuable comments and suggestions, all of which have improved the manuscript.

Declaration

The author declares no conflict of interest.

References

[10.1] Adamatzky, A. (2022) Language of fungi derived from their electrical spiking activity. *Royal Society Open Science* 9, 211926.

[10.2] Altmann, E.G. and Gerlach, M. (2016) Statistical laws of linguistics. In: *Creativity and universality in language, Lecture Notes in Morphogenesis*, Degli Esposti, M., Altmann, E.G. and Pachet, F. (eds), Springer, Berlin: 7-26.

[10.3] Altmann, G. (1980) Prolegomena to Menzerath's law. *Glottometrika* 2, 1-10.

[10.4] Ashburner, M., Ball, C.A., Blake, J.A., Botstein, D., Butler, H., Cherry, J.M., Davis, A.P., Dolinski, K., Dwight, S.S. and Eppig, J.T. (2000) Gene Ontology: tool for the unification of biology. *Nature Genetics* 25, 25–29.

[10.5] Aziz, M.F. and Caetano-Anollés, G. (2021) Evolution of networks of protein domain organization. *Scientific Reports* 11, 12075.

[10.6] Aziz, M.F., Caetano-Anollés, K. and Caetano-Anollés, G. (2016) The early history and emergence of molecular functions and modular scale-free network behavior. *Scientific Reports* 6, 25058.

[10.7] Aziz, M.F., Mughal, F. and Caetano-Anollés, G. (2023) Tracing the birth of structural domains from loops during protein evolution. *Scientific Reports* 13, 14688.

[10.8] Barbieri, M. (2009) A short history of biosemiotics. *Biosemiotics* 2, 221-245.

[10.9] Bard, J.B.L. (2011) The next evolutionary synthesis: From Lamarck and Darwin to genomic variation and systems biology. *Cell Communication and Signaling* 9, 30.

[10.10] Belzer, J., Holtzman, A.G. and Kent, A. (1978) Models of learning systems. *Encyclopedia of Computer Science and Technology* (Marcel Dekker, New York) 11, 24-61.

[10.11] Bühler, K. (1934) *Sprachtheorie. Die Darstellungsfunktion der Sprache.* Gustav Fisher, Jena.

[10.12] Caetano-Anollés, D., Caetano-Anollés, K. and Caetano-Anollés, G. (2018) Evolution of macromolecular structure: a 'double tale' of biological accretion and diversification. *Science Progress* 101(4), 360-383.

[10.13] Caetano-Anollés, G. (2021) The compressed vocabulary of microbial life. *Frontiers in Microbiology* 12, 655990.

[10.14] Caetano-Anollés, G. and Caetano-Anollés, D. (2003) An evolutionary structured universe of protein architecture. *Genome Research* 13, 1563-1571.

[10.15] Caetano-Anollés, G. and Caetano-Anollés, D. (2015) Computing the origin and evolution of the ribosome from its structure — Uncovering processes of macromolecular accretion benefiting synthetic biology. *Computational and Structural Biotechnology Journal* 13, 427–447.

[10.16] Caetano-Anollés, G. and Janko, R. (2022) The rise of hierarchy and modularity in biological networks explained by Empedocles' double tale ~2,400 years before Darwin and systems biology. *Frontiers in Genetics* 13, 973233.

[10.17] Caetano-Anollés, G. and Seufferheld, M.J. (2013) The coevolutionary roots of biochemistry and cellular organization challenge the RNA world paradigm. *Journal of Molecular Biology and Biotechnology* 23, 152-177.

[10.18] Caetano-Anollés, G., Aziz M.F., Mughal, F., Grater, F., Koç, I., Caetano-Anollés, K. and Caetano-Anollés, D. (2019) Emergence of hierarchical modularity in evolving networks uncovered by phylogenomic analysis. *Evolutionary Bioinformatics* 15, 1176934319872980.

[10.19] Caetano-Anollés, G., Aziz M.F., Mughal, F., Koç, I., Caetano-Anollés, K. and Caetano-Anollés, D. (2022) Recruitment: A problem of entangled temporal parts. *Frontiers in Bioscience* 27(4), 128.

[10.20] Caetano-Anollés, K., Hernandez, N., Mughal, F., Tomaszewski, T. and Caetano-Anollés, G. (2022) The seasonal behavior of COVID-19 and its galectin-like culprit of the viral spike. *Methods in Microbiology* 50, 27-81.

[10.21] Caetano-Anollés, G., Sun, F.J., Wang, M., Yafremava, L.S., Harish, A., Kim, H.S., Knudsen, V., Caetano-Anollés, D. and Mittenthal, J.E. (2008) Origins and evolution of modern biochemistry: Insights from genomes and molecular structure. *Frontiers in Bioscience* 13, 5212-5240.

[10.22] Caetano-Anollés, G., Yafremava, L. and Mittenthal JE (2010) Modularity and dissipation in evolution of macromolecular structures, functions, and networks. In: *Evolutionary bioinformatics and systems biology*, G. Caetano-Anollés, (ed.) Wiley-Blackwell, Hoboken, NJ, USA: 443–449.

[10.23] Chandonia, J.M., Guan, L., Lin, S., Yu, C., Fox, N.K. and Brenner, S.E. (2022) SCOPe: Improvements to the Structural Classification of Proteins—extended database to facilitate variant interpretation and machine learning. *Nucleic Acids Research* 50, D553–559.

[10.24] Chomsky, N. (1956) Three models for the description of language. *IRE Trans Information Theory* 2(3), 113-124.

[10.25] Chomsky, N. (1957) *Syntactic structures.* Mouton, The Hague/Paris.

[10.26] Cottam, R., Ranson, W. and Vounckx, R. (2016) Hierarchy and the nature of information. *Information* 7, 1.

[10.27] Draper, T.C., Dueñas-Díez, M., and Pérez-Mercader, J. (2021) Exploring the symbol processing 'time interval' parametric constraint in a Belousov–Zhabotinsky operated chemical Turing machine. *Royal Society of Chemistry Advances* 11, 23151–23160.

[10.28] Graves, A., Wayne, G., Danihelka, I. (2014) Neural Turing machines. *arXiv* 1410.5401v2.

[10.29] Graves, A., Wayne, G., Reynolds, M., *et al.* (2016) Hybrid computing using a neural network with dynamic external memory. *Nature* 538, 471-475.

[10.30] Hartwell, L.H., Hopfield, J.J., Leibler, S. and Murray, A.W. (1999) From molecular to modular cell biology. *Nature* 401, c47–c52.

[10.31] Hennie, F.C and Stearns, R.E. (1966) Two-tape simulation of multitape Turing machines. *Journal of the ACM* 13(4), 533-546.

[10.32] Heaps, H.S. (1978) *Information retrieval: Computational and theoretical aspects.* Academic Press, New York.

[10.33] Hull, D.L. (1978) A matter of individuality. *Philosophy of Science* 45, 335–360.

[10.34] Jakobson, R. (1960) Linguistics and poetics. In: *Style in language*, Sebeok, T. (ed.), MIT Press Cambridge, MA: 350–377.

[10.35] Kim, K. M. and Caetano-Anollés, G. (2010) Emergence and evolution of modern molecular functions inferred from phylogenomic analysis of ontological data. *Molecular Biology and Evolution* 27, 1710–1733.

[10.36] Kim, K. M., Nasir, A. and Caetano-Anollés, G. (2014) A tree of cellular life inferred from a genomic census of molecular functions. *Journal of Molecular Evolution* 79, 240–262.

[10.37] Koç, I. and Caetano-Anollés, G. (2017) The natural history of molecular functions inferred from an extensive phylogenomic analysis of Gene Ontology data. *PLoS One* 12, e0176129.

[10.38] Kull, K., Deacon, T., Emmeche, C., Hoffmeier, J. and Stjernfelt, F. (2009) Theses on biosemiotics: Prolegomena to a theoretical biology. *Biological Theory* 4(2), 167-173.

[10.39] Lakoff, R.T. (1993) Lewis Carroll: Subversive pragmaticist. *Pragmatics* 3(4), 367-385.

[10.40] Lamarck, J.B. (1809) *Philosophie Zoologique*. Detu, Paris, France.

[10.41] Lewis, D. (1986) *On the plurality of worlds*. Blackwell, Oxford, UK. 1986.

[10.42] Li, S., Lin, R., Bian, C., Ma, Q.D.Y. and Ivanov, P.C. (2016) Model of the dynamic construction process of texts and scaling laws of words organization in language systems. *PLoS One* 11(12), e01688971.

[10.43] Martin, A. and Primavesi, O. (1999) L'Empédocle de Strasbourg. De Gruyter, Berlin.

[10.44] Mittenthal, J.E., Caetano-Anollés, D. and Caetano-Anollés, G. (2012) Biphasic patterns of diversification and the emergence of modules. *Frontiers in Genetics* 3, 147.

[10.45] Mughal, F. and Caetano-Anollés, G. (2019) MANET 3.0: Hierarchy and modularity in evolving metabolic networks. *PLoS ONE* 14(10), e0224201.

[10.46] Nasir, A., Kim, K.M. and Caetano-Anollés, G. (2014) A phylogenomic census of molecular functions identifies modern thermophilic Archaea as the most ancient form of cellular life. *Archaea* 2014, 706468.

[10.47] Nasir, A., Kim, K.M. and Caetano-Anollés, G. (2017) Phylogenetic tracings of proteome size support the gradual accretion of protein structural domains and the early origin of viruses from primordial cells. *Frontiers in Microbiology* 8, 1178.

[10.48] Nath, N., Mitchell, J.B.O. and Caetano-Anollés, G. (2014) The natural history of biocatalytic mechanisms. *PLoS Computational Biology* 10, e1003642.

[10.49] Newell, A. and Simon, H. A. (1976) Computer science as empirical inquiry: Symbols and search. *Communications of the ACM* 19 (3), 113–126.

[10.50] Peirce, C.S. (1931) Principles of Philosophy. In *Collected Papers of Charles Sanders Peirce*. C. Hatshome & P. Weiss (Eds.), Harvard University Press, Cambridge, MA.

[10.51] Peirce, C.S. (1976) *The new elements of mathematics. Volume II, Algebra and Geometry*. Mouton Publishers, The Hague.

[10.52] Poole, D., Mackworth, A. and Goebel, R. (1998) *Computational Intelligence: A Logical Approach*. Oxford University Press, New York.

[10.53] Pruszynski, J.A. and Zylberberg J. (2019) The language of the brain: real-world neural population codes. *Current Opinion in Neurobiology* 58, 30–36.

[10.54] Queiroz, J. and El-Hani, C. (2012) Downward determination in semiotic multi-level systems. *Cybernetics and Human Knowing* 1(2), 123-136.

[10.55] Salthe, S.N. (1985) *Evolving hierarchical systems: their structure and representation*. Columbia University Press, New York.

[10.56] Salthe, S.N. (2012) Hierarchical structures. *Axiomathes* 22, 355-383.

[10.57] Searls, D.B. (2001) Reading the book of life. *Bioinformatics* 17(7), 579-580.

[10.58] Searls, D.B. (2002) The language of genes. *Nature* 420, 211-217.

[10.59] Shahzad, K., Mittenthal, K.E. and Caetano-Anollés, G. (2015) The organization of domains in proteins obeys Menzerath-Altmann's law of language. *BMC Systems Biology* 9, 44.

[10.60] Shannon, C.E. (1948) A mathematical theory of communication. *The Bell System Technical Journal* 27, 379-423 623-656.

[10.61] Sharov, A.A. (2016) Evolutionary biosemiotics and multilevel construction networks. *Biosemiotics* 9. 399-416.

[10.62] Sheftel, H., Shoval, O., Mayo, A. and Alon, U. (2013) The geometry of the Pareto front in biological phenotype space. *Ecology and Evolution* 3, 1471–1483.

[10.63] Simon, H.A. (1962) The architecture of complexity. *Proceedings of the American Philosophical Society* 106, 467–482.

[10.64] Simon, H,A. (1997) *Models of bounded rationality.*: MIT Press, Cambridge, MA.

[10.65] Sober, E. and Steel, M. (2014) Time and knowability in evolutionary processes. *Philosophy of Science* 81, 537-557.

[10.66] Sober, E. and Steel, M. (2015) Similarities as evidence for common ancestry: A likelihood epistemology. *British Journal of Philosophical Science* 68, 617-638.

[10.67] Sun, F.J. and Caetano-Anollés, G. (2021) Menzerath–Altmann's law of syntax in RNA accretion history. *Life* 11, 489.

[10.68] Tomaszewski, T., Gurtler, V., Caetano-Anollés, K. and Caetano-Anollés, G. (2022) The emergence of SARS-CoV-2 variants of concern in Australia by haplotype coalescence reveals a continental link to COVID-19 seasonality. *Methods in Microbiology* 50, 233-268.

[10.69] Tomaszewski, T., Asif, M.A., Caetano-Anollés, K. and Caetano-Anollés, G. (2023) Seasonal effects decouple SARS-CoV-2 haplotypes worldwide. *F1000Research* 12, 267.

[10.70] von Neumann, J. (1966) *The theory of self-reproducing automata.* University of Illinois Press, Urbana, Illinois.

[10.71] Wasserman, R. (2016) Theories of persistence. *Philosophical Studies* 173, 243-250.

[10.72] Witzany, G. (2014) Language and communication as universal requirements for life. In: *Astrobiology: An Evolutionary Approach.* V. Kolb, (ed.) CRC Press, Boca Raton, Florida: 349–370.

[10.73] Witzany, G. (2016) The biocommunication method: on the road to an integrative biology. *Communicative and Integrative Biology* 9, e1164374.

[10.74] Yafremava, L.S., Wielgos, M., Thomas, S., Nasir, A., Wang, M., Mittentha,l J.E. and Caetano-Anollés, G. (2013) A general framework of persistence strategies for biological systems helps explain domains of life. *Frontiers in Genetics* 4, 16.

11

Meaning Relies on Codes but Depends on Agents

Robert Prinz

Rechenkraft.net e.V., Marburg, Germany

Online resources:
https://www.researchgate.net/profile/Robert-Prinz
https://rechenkraft.academia.edu/RobertPrinz

Abstract

This chapter gives an overview of biological codes based on Marcello Barbieri's concept of codes as *mapping between two independent worlds through an adaptor*. Biological codes in this framework are arbitrary connection rules established in evolution through an adaptor that mediates between two other molecules of different types and cannot be explained based solely on physical or chemical laws. All codes in the narrow (Barbierian) sense thus consist of a set of adaptors that map a set of components from one independent world to a set of components from another independent world. Each instance of a code is a concrete realization of such a mapping. In the genetic code, for example, a codon on the messenger RNA is mapped by the corresponding aminoacyl-transfer RNA synthetase to the encoded amino acid (linked to that transfer RNA) that will be included in the growing protein by the cellular translation machinery. The genetic code is only one example of this kind of code and many more are described by Barbieri and others. This review includes two major types of codes: *organic* and *neural*, both of which can be further divided into *structural* and *regulatory* codes. This categorization allows ordering the countless codes published in the literature based on their role and context. The role of codes in meaning-making is based on the views of both, Barbieri, who assumes objective meaning within a code, and Kalevi Kull, who proclaims interpretation as a prerequisite to any meaning-making activity. My critical reflection is that meaning is not located in codes, but can be derived from the action of *codeusers*, i.e., molecular machines like the ribosome, that follow coded rules during their performance. It is proposed that all forms of meaning(-making) rely on a continuity of codes, from molecular to cellular codes, literally from molecules to cognition, and, at the same time, depend on agents as *meaning-makers*.

Keywords: Meaning, code biology, organic codes, neural codes, structural codes, regulatory codes, Marcello Barbieri

Email: robert-prinz@web.de

11.1 Introduction

In a series of books [11.5] [11.6] and seminal papers, Marcello Barbieri develops the concept of semantic biology, organic and neural codes, and their relation to information and meaning [11.7] [11.9] [11.11] [11.12] [11.13] [11.14] [11.16] [11.17] [11.18] [11.19]. Together with other authors, the foundations of the code biology field has been established under the biosemiotics research agenda [11.8] [11.10]; and later developed as a separate discipline that investigates the dynamics of codes in the context of networks, modularity, hierarchy, robustness, and during the evolution of life [11.33] [11.53] [11.67] [11.89] [11.94] [11.95] [11.96] [11.97] [11.95] [11.96] [11.123]. Code biology emphasizes the evolutionary importance of codes and that the steps towards life most probably coincided with the origin of the first codes. Today, the notion of codes is well developed and several authors in addition to Barbieri present plausible scenarios on how codes contributed to major evolutionary transitions [11.67] and how they can be used in defining species [11.97], or interactions between species [11.33]. Also, codes are connected to the concepts of information, meaning, and semiosis [11.7] [11.9] [11.16]. Since its discovery, the idea of genetic code has dominated modern molecular biology, whereas other codes nearly went unnoticed until Marcello Barbieri pointed towards their significance as a fundamental principle of the living world [11.17].

The hallmark of *organic codes* is an arbitrary *mapping* between two independent worlds of molecules realized by a set of *adaptor* molecules. Therefore, codes consist of three components and Barbieri defines them in conjunction with concrete examples as follows:

> "A code has been defined as 'a mapping between the objects of two independent worlds that is implemented by the objects of a third world called adaptors'[11.6].The genetic code, for example, is a mapping between codons and amino acids that is implemented by transfer-RNAs; a signal-transduction code is a mapping between first and second messengers that is implemented by molecular receptors; the Morse code is a mapping between the letters of the alphabet and groups of dots-and-dashes that is implemented by neural circuits according to a set of learned rules" [11.17] p. 6.

> "This definition allows us to make experimental tests because organic codes are relationships between two worlds of organic molecules and are necessarily implemented by a third type of molecules, called *adaptors*, that build a bridge between them. The adaptors are required because there is no necessary link between the two worlds, and a fixed set of adaptors is required in order to guarantee the specificity of the correspondence. The adaptors, in short, are essential in all organic codes. They are the molecular *fingerprints* of the codes, and their presence in a biological process is a sure sign that that process is based on a code" [11.15] p. xii.

> "More precisely, we can prove that an organic code exists, if we have three things: (1) two independent worlds of molecules, (2) a set of adaptors that create a mapping between them, and (3) the demonstration that the mapping is arbitrary because its rules can, at least in principle, be changed" [11.15] p. xii.

The major role of the adaptor is to map one molecular component to the other. Therefore, we can speak of a tripartite relation, as a code always consists of three independent parts or components. We will see in the following sections that many, if not most, biological codes pertain to this tripartite pattern, whereas further definitions of biological codes not relying on adaptors are presented in the literature as acknowledged by Barbieri:

> "The actual situation, however, is more complicated than that, because *codes-without-adaptors* have been reported in biology too. In the 1980s and 90s, Edward Trifonov proposed that the genome carries several codes simultaneously, in addition to the classic genetic code, and gave them the collective name of sequence codes [11.113] [11.114] [11.115]. This conclusion is based on Trifonov's definition that "*a code is any pattern in a sequence which corresponds to one or another specific biological function*" [11.114], p. 424. As in the case of the engineering codes, the sequence codes transform sequences into other sequence that belong to the same alphabet world, and no adaptors are required. In addition to the codes described by Trifonov, furthermore, various other sequence codes have been brought to light. Among them, the transcription regulatory codes [11.109] [11.126], the translation regulation codes [11.63], the operon codes [11.24] [11.100], the riboswitch codes [11.85] [11.117] [11.124] and the histidine kinase code [11.22]." [11.17] p. 6.

Barbieri's notion of "*codes-without-adaptors*" is further developed by Heng and Heng who claim that there may be codes supported not by concrete molecules as adaptors, but rather by processes that are connected to other processes or events through time and space [11.58]. If such mappings of processes to events can be deduced from arbitrary rules, then indeed one could speak of "adaptorless" codes. Similarly, Paredes *et al.* present a network-view on codes in which any mapping between components of a network that are functionally or informationally related to each other in a rule-based manner constitute a code [11.89].

Also, many mathematical code models may not properly reflect Barbieri's code definition (personal communication). A growing interest in codes inevitably yields even further and wider definitions of biological codes than those initially proposed by Barbieri in regard to the role of adaptors. Stephen Cowley draws on the idea of wide cognition and presents the concept of *wide coding* where, e.g., in the Morse coding, the human operator is considered an adaptor and "*a human code intermediary*" [11.36]. Based on this notion, we can widen the concept of "adaptor" to comprise entire organisms or species that mediate the interaction of, e.g., symbioses in a rule-based manner. The number of codes in a strict *Barbierian* sense and the *non-Barbierian* codes to be discovered in the future is therefore expected to grow substantially.

While many publications on adaptorless codes remain abstract in their definition of codes and often lack concrete examples, countless publications during the past two decades support Barbieri and other proponents of codes as being based on adaptors and follow a tripartite mapping pattern (see following sections). The core idea of a tripartite, rule-based, arbitrary mapping of molecular worlds is thus strengthened through the emergence of more evidence without being softened by alternative code concepts. A long list of published codes of all types, i.e., Barbierian and non-Barbierian codes, is available as an online resource from the website of the International Society of Code Biology [11.93].

11.2 Organic Codes and Cellular Core Functionalities

Life builds on many essential processes, like anabolism, catabolism, inheritance, signal processing, and others. Without them, organisms can hardly survive. They can therefore be referred to as *core processes* and *core functionalities*. The underlying codes are called "structural codes" as they are crucial in establishing the basic structure of organisms, anatomically, and functionally. Examples of well-described codes from the literature are presented below.

The *genetic code* is probably among the oldest and most conserved codes we know today. This code maps a nucleic acid tripled (codon) encoded in an mRNA to a specific amino acid attached to a tRNA by aminoacyl synthetase. The tRNA (with aminoacyl synthetase) acts as an adaptor through codon pairing with its anti-codon portion and thereby allows the protein translation machinery to repeatedly follow this arbitrary pairing rule that leads to regular protein structures based on a sequence of nucleic acid triplets [11.38] [11.78]. Protein translation cannot be explained by deterministic physics, as Barbieri claims [11.18], and until today, no convincing model based on thermodynamic or other physical principles is proposed that could explain the origin of the genetic code. Zolyan goes as far as to compare the genetic code to semiosis comprising vocabulary and grammar [11.136]. The genetic code has a superior status compared to other codes, as without it, the protein world as we know it could not have a heritable basis and existence. The *splicing code* establishes rules that allow a cell to make combinatorial use of large coding sequences which can be shuffled and assembled from pre-mRNAs, when so-called introns or non-coding sequences are excised and the remaining coding sequences (i.e., exons) form a single transcript ready for protein translation [11.6] [11.37] [11.48] [11.77]. In this way, proteins can be produced that are constituted of distinct functional modules depending on the actual requirements of certain cell types, or on temporal or local context [11.4]. Barbieri stated that most probably, multiple, overlapping codes are involved in what is also called alternative splicing [11.15].

The *DNA repair code* allows a cell to efficiently restore damaged DNA, or chromosomes. Proper condition and functioning of chromosomes and DNA are important not only for inheritance, but also for maintenance of somatic processes. Double strand breaks of DNA can be repaired when the involved molecular machines follow certain molecular rules, including response of histone variants that recognize the double strand break and recruit as adaptor other checkpoint proteins to initiate the repair process [11.39]. Posttranslational modifications like phosphorylation, or biotinylation of involved histones play an important role in coordinating the response and outcome, forming sub-codes as they alter the behavior of involved adaptors [11.56].

The *HOX codes* are involved in pattern formation of the developing embryo [11.21], p. 353-356, and are referred to by Barbieri as "*codes of the body plans*" [11.15]. The family of hox genes encode transcription factors proteins that coordinate the expression of a variety of genes in a combinatorial manner [11.62].The adaptors in this code are the hox transcription factors that relate an input in the form of small molecules (like retinoic acid) to gene products they regulate, and which ultimately determine the identity of anatomical segments in the developing embryo and the overall body axes [11.57]. The *signaling codes* comprise defined rules, based on which cells can transmit signals to each other and translate messages from the outside of the cell membrane to meaningful messages within the cell [11.6]

[11.27] [11.43]. Classic cell biology calls the external signal molecules *first messengers*. Their binding to membrane-bound protein receptors triggers the release of the *second messengers* inside the cell that in turn activate or deactivate other molecular targets, e.g., in the nucleus [11.69] [11.75]. Barbieri refers to codes based on first messengers as signal transduction codes and to those involving second messengers and transcription factors as signal integration codes [11.15], p. 43 ff.

Many codes are based on posttranslational protein modifications. The *sugar code* relates patterns of glycans (polysaccharides) that are attached to certain residues of proteins on the cell surface, e.g., to distinct recognition processes of the immune system mediated by lectin receptors as adaptors [11.50] [11.51] [11.52] [11.61]. The *ubiquitin code* is a system of labels that allows the cell to mark and recognize proteins for degradation [11.129] and proteolysis [11.64]. Ubiquitin is attached to certain protein residues, and is recognized later by other proteins of degradation pathways. Thus, ubiquitin plays the role of an adaptor. Ubiquitylation is widespread in signal transduction and autophagy pathways [11.68]. The so-called *zip codes* determine the recognition of cargo like mRNAs, proteins, or vesicles, and locations to which it must be transported within the cell [11.23] [11.31] [11.41] [11.80] [11.82] [11.88]. The *exit codes* determine protein export from the endoplasmic reticulum [11.99]. In all such transport codes, a distinct adaptor molecule binds to structural RNA or protein motifs; and on the other hand, connects to transport proteins. The *tubulin code* is based on the posttranslational tubulin modifications that function as adaptors when they determine what kind of motor proteins bind to the tubulin strands [11.72] [11.121] [11.128] [11.133].

11.3 Codes Control Other Codes

Life not only depends on core processes like metabolism and inheritance, but must also adjust such processes to ecological context. Those control or feedback processes turned out to be based on codes as well. We can therefore talk of regulatory codes that control or fine-tune core processes in a context-dependent manner. Many biological regulatory systems are described as codes in the literature and fit the definition of regulatory codes.

The *microRNA codes* are part of a cell's posttranscriptional regulatory network and constitute one instance of controlling the production of protein. In principle, an adaptor molecule (or machinery) uses a short RNA strand that pairs to a complementary non-coding region of a messenger RNA [11.103] or within its coding region [11.47] and initiates the degradation of the mRNA. A related code is the *RNA interference code* [11.101]. The *transcriptional codes* allow combinatorial regulation of gene expression through regulation of transcription factor binding to gene promoters [11.55]. Such "condition-enabled" transcription factors like the Msn2 transcription factor mediate the binding of other transcription factors to promoters depending on environmental context, and therefore function as adaptors; in other words, they can directly link the world from outside a cell, to gene transcription by interfering with transcription factors. The *histone code* refers to histone modifications regulating the accessibility of DNA through chromatin remodeling [11.35] [11.60] [11.110] [11.118] [11.119]. Histones are adapters that integrate a variety of signals in terms of methylation or acetylation and therefore can silence or activate readout of larger DNA regions [11.34] [11.60]. Although the nature of the histone code is still debated, the histone code fulfills all necessary code criteria [11.15] [11.65]. Histone modifications, which are inherited from one to the next generation, are referred to as *epigenetic codes*, i.e., the chromatin-mediated silencing or activation of gene transcription patterns over generations follow distinct rules [11.46] [11.119].

The *chaperone codes* refer to the selective change of chaperones' activity in assisting the folding and unfolding of proteins [11.32]. Chaperones' activity can be modified through phosphorylation, acetylation, methylation, glycosylation, SUMOylation, ubiquitination, and others [11.3] [11.32] [11.84]. Following such modifications, the chaperones may alter their activity or target client proteins. This means that the basic functionality of one type of protein can be used in many different situations, instead of "inventing" new chaperones for every possible circumstance. The *membrane code* is the change of membrane fluidity or curvature in a rule-based manner depending on the composition and three-dimensional conformation of proteins [11.127] or lipids [11.49]. The many causes that are able to change the membrane composition (world 1) can be considered as an independent world connected to a second world by membrane-bound signaling proteins functioning as adaptors, i.e., mediating to the triggered effectors within the cell (world 2) [11.127] p. 478.

The *phosphorylation codes* are widespread codes based on posttranslational phosphorylation patterns of proteins. Attachment of phosphate groups to certain adaptor molecules can subsequently be "read out" by effectors and trigger the recruitment and activity of, e.g., kinases [11.86] [11.105]. Therefore, phosphorylation patterns can establish rules that regulate the activity of proteins in a context-dependent manner. Such posttranslational modifications can be commonly found in carboxy-terminal protein domains like in the case of RNA Polymerase II, where the literature refers to them as *CTD code* [11.28] [11.40] [11.42] [11.76].

The *karyotype code* forms a non-Barbierian code in that it is proclaimed an adaptorless code that supposedly connects the three-dimensional arrangements of entire chromosomes in space and time to control recombination, chromosome inheritance, and gene expression in general [11.58]. In other words, a chromosome scaffold serves as an adaptor between effectors of chromatin structure and chromosome segregation and other outputs. This may play an important role in plants that show a high variability of (poly)ploidy and therefore determines the identity of that species.

11.4 Coding the Mind: Neural Codes Enable Perception and Cognition

Barbieri differentiates between organic and neural codes, the latter being specific to perceptive and cognitive processes [11.14] [11.15] [11.19]. We can, however, also find organic codes that are pertinent only to neurons or glia cells (e.g., [11.81] [11.107] [11.116]). Cortical oscillation patterns can emerge from synchronous firing activity of multiple neurons, which is sometimes referred to as *population codes* (e.g., [11.29] [11.106] [11.120]). The *neural codes* are highly dependent on such "*synchrony codes*" [11.108]. For both, the neural organic codes and the neural codes related to perception and cognition, structural and regulatory codes are described in the literature. Regulation in the neurosciences is often presented as *modulation, adaptation, plasticity*, or *crosstalk*, independent of whether the concerned process is structural or regulatory in the sense as described for organic codes in the previous section.

The *cadherin code* determines the assembly of neurons mediated in a rule-based manner by cadherins [11.98] [11.102]. The pattern of cadherin expression contributes to neuronal identity and enables the assembly of many neurons into higher order nuclei of the brain [11.90]. Cadherins, in this context, serve as adaptors mediating the recognition of self vs. non-self among the multiple cells interacting to form the nuclei through distinct cell-cell adherence or self-avoidance (repulsion) processes [11.30] [11.79]. Actually, the cadherins are involved in the formation of synapses between neurons [11.132]. The *neurexin code* bridges the pre-synaptic to post-synaptic worlds of different neurons and allows for proper synapse formation [11.112]. The *neurotransmitter codes* regulate the kinetics of neurotransmitter release from vesicles into the synaptic cleft [11.59]. In detail, the fusion of the vesical membrane with the pre-synaptic membrane is modulated by adaptors that mediate slow and fast vesicle fusion pathways leading to complex release dynamics [11.59]. Such fine-tuning of neurotransmitter release based on codes appears a common mechanism [11.25] [11.26].

The *olfactory code* underlies the ability of organisms to distinguish the perception of (combinatorial) odors and the coupled response based on cognitive signal integration, like avoidance of bad smelling food. Olfaction is based on specific receptors that after odor binding trigger the axon firing of the receptor neuron, thus mapping *discrete* odorant molecules via the nervous system to cortical information processing [11.54]. The translation from receptor-type binding event to neuron firing pattern as well as cortical signal integration involves multiple coded processes [11.54] [11.131]. The *pheromone code* connects a volatile odor to sexual identity. Pheromones in insects are sexual attractants that are emitted by one individual and perceived by the opposite-sex individual, which determines a decision on initiating or continuing mating behavior [11.83].

The *auditory codes* connect objects emitting sounds to organisms that perceive soundwaves through the sense organ (e.g., ears) following cortical integration of location and other information encoded or inferred from the sound pattern [11.87] [11.91] [11.125]. The pathway from ears to brain involves a complex series of codes, starting with the translation of soundwaves into the bending of cilia in the ear that triggers neuron firing that encodes the initial signal into a transmissible form. Signal integration in the auditory cortex can involve a series of codes that relate the sounds to behavioral responses, based, for example, on the recognition of comates [11.92]. Ecologically, sound is a major communication channel in many animal species, and it is especially important for members of species to unambiguously hear their conspecifics even in a background of sounds produced by other animals. Biological niches often include soundscapes [11.70] of overlaying voices and signals from different species that may constitute unique *acoustic codes* [11.44] [11.45] [11.74].

The *visual codes* underlie visual perception and cortical integration of visually perceived objects [11.120] [11.134] [11.20] [11.104] [11.111] [11.135]. As all forms of neural (perceptive) coding, visual coding involves several staggered codes across levels of hierarchy from the light capturing neuronal cells of the retina to cortical integration. It is beyond the scope of this chapter to present the details of those codes. Basically, the visual codes relate color, shape, texture, motion, and related characteristics to learned objects, and behavioral responses towards the respective objects, like grasping, fleeing, and so forth. Attention is among the important neural activities allowing proper response to moving objects and the underlying attention codes may be considered as a regulatory neural code [11.106].

The central nervous system as an information processing system turned out to be an endless source of new codes [11.71]. Lyall *et al.* accurately describe what neural codes do; namely, they enable the "...linear discrimination of arbitrary subsets of sensory input combinations, for learning complex classes and rules" [11.71], p. 13. The brain is the master code-generator, which most probably can support some yet unrecognized mechanisms of learning and memory. The *archetype codes* are ancient behaviors encoded in human brains that become overt only in situations of ambiguity, where people elicit actions that could be attributed to very ancient response patterns inherited over generations [11.73]. Actually, a certain situation is mapped to a specific behavior, so the adaptor in this context must be located in the cortical circuitry that relates the "memorized" input situation to the "memorized" output behavior.

11.5 Meaning *in* or *from* Codes?

The many biological codes summarized in the sections above show the relevance of codes for biological processes. Whereas these processes undoubtedly have an objective meaning for the organism, depending on the outcome or end product of the respective processes, Barbieri proposes that codes in themselves carry meaning. He develops the concept of *organic meaning* residing *in* organic codes as follows:

> "By the same token, the rule of the genetic code that a group of three nucleotides (a codon) corresponds to an amino acid is equivalent to saying that amino acid is the *organic meaning* of that codon. Anywhere there is a code, be it in the mental or in the organic world, there is meaning. We can say, therefore, that *meaning is an entity which is related to another entity by a code*, and that organic meaning exists whenever an organic code exists [11.1] [11.2] [11.6]." [11.15], p. 26.

In other words, meaning is in his view literally an object, in which a certain molecule from world 1 is the meaning of the corresponding other molecule from the independent world 2. With this transformation from mapping to meaning, codes become carriers of meaning in their own right. It remains, however, highly debatable if meaning resides within codes, if a code with its two components and adaptor in its entirety is meaning, or if meaning is derived from multiple codes [11.66]. Considering the genetic code and the process of translation, it is actually the aminoacyl-tRNA synthetases that "...establish the rules of the genetic code" [11.38] p. 440, by attaching the correct amino acids to the proper tRNAs, and ultimately the ribosome as a *codeuser* making a *meaningful* product, i.e., a functional protein. A codeuser is thus a molecular machine that uses the components of a code and follows the respective code to create a "code product." The amino acid linked to the respective tRNA adaptor does not mean anything to the codon, nor does the codon mean anything to the respective amino acid. It obviously requires an external frame of reference outside the concrete instance of a code to recognize the meaning of a code as a whole. Meaning thus does not reside within a single component of a code alone, but may be derived from code(s) when one considers the output of a coded process as meaning.

The codon as a part of the messenger RNA is a product of the RNA polymerase, the amino acid is a metabolic product, and its binding to the specific tRNA is a product of the aminoacyl-tRNA synthetase. All three components are mere biochemical artifacts. As their respective relation or mapping amongst each other is not determined from within, but from outside, as Barbieri would call it; they consequently cannot bear meaning from within. Codes however are parts of molecular and cognitive processes and the outcomes of those processes bear potential meaning in that all have certain consequences in the context they are expressed. For example, proteins being expressed by codeusers do something, and even if they would not do anything, they have a distinct three-dimensional structure and surface charge which will interact with other molecules and affect their distribution and so forth.

Box 1 Definition of terms

Artifacts: Molecules, biomacromolecules, or assemblies of these that come into place only through the action of molecular machines that follow rules, i.e., depend on coded relations at least at one step during their assembly or generation, can be considered artifacts. The existence of artifacts cannot be predicted solely from physical or chemical laws, yet natural laws set the boundary conditions within artifacts are made.

Codemaker: Biomacromolecules, molecular machines, or any entity that establishes the correspondence between adaptor and the components it maps. In the genetic code, the aminoacyl-tRNA-synthetases are codemakers as they load the amino acids to the corresponding tRNAs (adaptors in the genetic code), and thereby determine the mapping the tRNA (adaptor) performs between codon and amino acid.

Codeuser: A codeuser is any biomacromolecule, molecular machine, or entity that uses adaptors and the components mapped by the adaptors to produce a molecular or mental artifact. There is no adaptorless codeuser in a narrow, Barbierian interpretation.

Interpretation: Interpretation is a *call-to-action*, i.e., any interpretation is not only followed by a decision, but actually caused by the need for a decision. Interpretation is based on multiple meanings and arbitrary rules, and driven by goals, e.g., the need for adaptation to changing conditions. Interpretation is thus an *act* of relating the *status quo* to the status *to be*.

Interpreter: Any entity that uses arbitrary rules to act upon the state or product of multiple meaning makers, i.e., meanings. The interpreter becomes the first instance of subjectivity in that from its objective inputs (meanings), decisions are derived (evidenced by output), but not determined.

Meaning: The state of something (including artifacts), material or mental, that is (potentially) used to make a decision based on arbitrary rules. The presence, absence, or modified form of molecules, or certain memories, can become meaning (to an interpreter). Meaning is therefore the objective input of an interpreter.

Meaning maker: Any entity that draws on artifacts, and optionally also on non-artifacts, to generate a derived molecular or mental state that serves as input (meaning) for an interpreter to act upon.

11.6 Meaning *with* or *without* Interpretation?

Kalevi Kull argues that semiosis requires agency and that "...the core aspect of agency is interpretation" [11.66], p. 139. A code in his line of argument is necessary but not sufficient for meaning-making. Kull further makes learning and choice based on interpretation a prerequisite to the emergence of new codes through variations of existing codes:

> "A single code does not include a source of its own change. The code-maker preserves a code; it does not actively change it, because it is not an agent. A new or a changed code is a result of learning or choice, or interpretation. However, this does not imply that the code itself carries meaning. Meaning as a phenomenon appears only due to the choice-like (also called abductive) element in semiosis, i.e., in interpretation. Codes are commonly products of interpretation, or semiosis." [11.66] p. 139.

The introduction of choice and interpretation by Kull is deeply rooted in a Peircean-grounded biosemiotics account of the topic. If not copying or coding errors, then Barbieri most probably would account chance and changed constraints as a cause for new codes, but not choice and interpretation at the cellular level. Nevertheless, Barbieri simply bridges from his code-biology grounded terminology to a biosemiotic terminology when referring to semiosis as "the production of signs" and applying it to the genetic code, e.g.,

> "In biology, the presence of semiosis at the molecular level is documented by the existence of the genetic code in all cells, and this implies that *molecular (or organic) semiosis is produced by coding.*
>
> In the humanities, the dominant view is the Peircean concept that semiosis is always an interpretive process, and this implies that *semiosis is produced by interpretation.*" [11.15] p. 151.
>
> "That is what divided animals from single cells, and that is why there are at least two distinct types of semiosis in living systems: one that is based on coding and decoding (*organic semiosis*) and one that is based on interpretation (*animal semiosis*)." [11.15] p. 164.

Barbieri proclaims two types of semiosis, one accounting for organic meaning based on coding and decoding, and the other covering "classical" semiosis based on interpretation. He then turns around the argument in that not meaning requires interpretation but that interpretation requires two forms of meaning:

> "There is, in short, a real difference between the semiosis of single cells and that of animals. A semiosis based on coding and decoding requires only one type of meaning (sense) whereas a semiosis based on interpretation requires two distinct types of meaning (sense and reference)." [11.15] p. 165.

In his "Code Model of Semiosis," Barbieri states "A sign, to start with, is always linked to a meaning." and "Sign and meaning, in other words, cannot be taken apart because they are the two sides of the same coin" [11.15], p. 164.

Barbieri further states

> "A semiotic system, therefore, is not any combination of two distinct worlds. It is *a combination of two worlds between which there is no necessary link*, and this has an extraordinary consequence. It implies that a bridge between the two worlds can be established only by *conventional* rules, i.e., by the rules of a *code*. This is what qualifies the semiotic systems, what makes them different from everything else: *a semiotic system is a system*

> *made of two independent worlds that are connected by the conventional rules of a code.* A semiotic system, in conclusion, is necessarily made of at least *three* distinct entities: *signs, meanings* and *code*." [11.15], p. 164.

The relation between meaning and sign does not come into existence without an agent enabling the respective mapping. Barbieri calls those agents *codemakers*, independently of the level of hierarchy where they occur:

> "Signs, meanings and conventions, however, do not come into existence of their own. There is always an 'agent' that produces them, and that agent can be referred to as a codemaker because it is always the making of a code that gives origin to semiosis. In the case of culture, for example, the *codemaker* is the human brain, since it is the brain that produces the mental objects that we call signs and meanings and the conventions that link them together. We come in this way to a general conclusion that can be referred to as 'the Code model of semiosis': *a semiotic system is a set of signs, meanings and code that are all produced by the same agent, i.e., by the same codemaker*." [11.15], p. 30.

> "We come in this way to a general conclusion that can be referred to as 'the code model of semiosis': a semiotic system is a triad of signs, meanings and code that are all produced by the same agent, i.e., by the same codemaker." [11.9], p. 26.

Barbieri uses the concept of agents to explain the existence of codes and it appears controversial if a codemaker can be considered an agent or not. Especially when we think of the aminoacyl-tRNA synthetase—a single enzyme, that actually is a codemaker in the sense of Barbieri. A codemaker also generates organic meaning when we take Barbieri literally. The recognition process of the aminoacyl-tRNA synthetase is in his line of argumentation not an act of interpretation, and therefore meaning within the code is independent of interpretation as Barbieri argues. We therefore see two positions here: the first locates meaning within codes uncoupled from interpretation at the molecular/cellular level as presented by Barbieri, and the second calls for interpretation as a precondition for meaning-making as argued by Kull.

11.7 Critical Conclusion

Codes are a hallmark of the living world and organisms highly rely on codes for core processes of life. Codemakers establish codes and are called agents by Barbieri. As, according to Barbieri, codes contain meaning, codemakers are therefore also meaning-makers, yet never called so by Barbieri. To assign the role of an agent to a single molecule that can at the same time generate organic meaning, is an ambitious approach because consequent application of Barbieri's notions of codes comes down to this: Codes are rules and there are codes (rules) that control or regulate (rule) other rules (codes), all enabled by codemakers as agents whose byproduct is objective meaning residing in codes. Recurrency of codes through codemakers as agents is thus the single sufficient ingredient necessary to explain organic meaning when we take Barbieri literally.

Yet, many biologists would assign agency to complex entities like cells, microorganisms, or multicellular organisms, and indeed we may ask if we can find single codes outside a cell, or more concretely, what keeps codes and codemakers alive? Whole cells and organisms turn out as carriers of codemakers and codes, and with their (macro) behavior constrain and influence their action and inheritance, briefly put. It is thus ultimately the behavior, more generally known as the agency, of *code carriers* that is key to understand how codes are handled and propagated by living systems. It is to those code carriers, to which codes mean something, if at all, and one should follow the route towards carriers in defining concrete agents.

When agents prepare the ground for codes, what is their motivation and intention? This provocative and anthropocentric question can translate to selective or competitive pressure, and to survival and adaptation, respectively. Survival and adaptation are not evaluated at the level of organic codes but definitely comprise multiple codes across hierarchies and will *require* their integration and therefore subjective meaning. The outputs of coded processes are meaningful to the underlying system in terms of persistence, propagation, and survival — necessitating at the same time choices between response options. Meaning reveals two sides here: it is objective through the artifacts or outputs of coded processes, and subjective through the consequences of their readout (or interpretation) by its carrier.

When codes contain meaning, they are self-referential in that no external point of reference is needed to objectify this meaning. Agency in a non-Barbierian sense appears as the only external point of reference that would qualify as meaning-harvesting instance. This again means an instance outside or beyond a code is necessary to make meaning that may be harvested by further instances or entities. But to harvest could implicate to have something seeded and grown before. This assumption has two important implications, (*first*) *chance* may have played a role in the origin of the first codes, whereas its importance substantially decreased with every new code, and (*second*) *agents* emerged as the first instance of meaning-making feeding on the products or artifacts of coded processes, accountable for the main part of evolutionary change. It is out of the scope of this chapter to discuss the minimal requirements for agency, and if they are already met by viruses, bacteriophages, or similar entities. There is evidence enough to support a view that viruses act like (minimal) agents spreading and rearranging genomes, and therefore significantly contribute to or at least grow the potential fields for new codes [11.122] [11.130]. Most probably, agents comprise multiple codes, and this is for two reasons: When two codes interact, there should be an "objective" external reference point of the same kind, and that would be a third code. Objective meaning can thus arise in a literal *coordination* system of codes. Meaning is associated with a *consequence* that always happens when a code or rule is followed. Interpretation is a *call-to-action* that necessitates *choice* between options, based on readout of meaning(s). Without choice, it was a mere mechanism. In more complex agents, codes act in a physical space, on the one hand, and yet rely on "memorized" events or information from the past during the progression from one to the next generation, on the other hand. Interpretation thus requires at least two kinds of codes, those accounting for event or process evaluation in space and others relating the past outcomes of such code action to the present, i.e., codes "in time." In a hypothetic minimal agent, the former codes most probably would be structural in nature, and the latter codes regulatory.

In conclusion, meaning may be grounded in a set of codes involved in the production of artifacts as outputs of the processes they are involved in. Codemakers establish codes, and codeusers produce artifacts according to rules of a code. There may be meaning-makers that act in certain ways upon the outputs of coded processes and interpreters that integrate the states of meaning-makers. It remains not so much a matter at which stage agency comes into play, but what constitutes an agent based on its activity. However, it becomes clear that meaning relies on codes but depends on the agents implementing them.

Acknowledgments

I want to thank Marcello Barbieri for his foundational work in code biology and the kind collaboration during compilation of the code list that he edited for publication on the website of the International Society of Code Biology. The contributions by Kalevi Kull to evolutionary aspects of code biology, and more specifically to agency and interpretation, set important cornerstones that helped me recognize the limits of code-based explanations not only during preparation of this publication. The work of Günther Witzany gave me confidence in that the molecular foundation of code biology and agency is valid to the core when explaining the origin of the first molecular agents.

Declaration

I declare no conflict of interest.

References

[11.1] Artmann, S. (2007) Computing codes versus interpreting life: two alternative ways of synthesizing biological knowledge through semiotics. In: *Introduction to biosemiotics: the new biological synthesis.* M. Barbieri, (ed.) Springer, Dordrecht.

[11.2] Artmann, S. (2009) Basic Semiosis as Code-Based Control. *Biosemiotics* 2(1), 31-38.

[11.3] Backe, S.J., Sager, R.A., Woodford, M.R., Makedon, A.M. and Mollapour, M. (2020) Post-translational modifications of Hsp90 and translating the chaperone code. *J Biol Chem* 295(32), 11099-11117.

[11.4] Baralle, M. and Baralle, F.E. (2018) The splicing code. *Biosystems* 164, 39-48.

[11.5] Barbieri, M. (1985) *The Semantic Theory of Evolution.* Harwood Academic Publishers; Reissued 2020 by Routledge.

[11.6] Barbieri, M. (2003) *The organic codes. An introduction to semantic biology.* Cambridge University Press.

[11.7] Barbieri, M. (2006) Life and semiosis: The real nature of information and meaning. *Semiotica* 2006(158), 233-254.

[11.8] Barbieri, M. (ed.) (2007) *Introduction to Biosemiotics. The New Biological Synthesis.* Springer, Dordrecht.

[11.9] Barbieri, M. (2008) The code model of semiosis: The first steps toward a scientific biosemiotics. *The American Journal of Semiotics* 24(1-3), 23-37.

[11.10] Barbieri, M. (ed.) (2008) *The Codes of Life - The Rules of Macroevolution.* Springer, Dordrecht.

[11.11] Barbieri, M. (2008) The Mechanisms of Evolution. Natural Selection and Natrual Conventions. In: *The Codes of Life. The Rules of Macroevolution.* M. Barbieri, (ed.) Springer, Dordrecht: 15-35.

[11.12] Barbieri, M. (2012) Code Biology – A New Science of Life. *Biosemiotics* 5(3), 411-437.

[11.13] Barbieri, M. (2012) Codepoiesis – the Deep Logic of Life. *Biosemiotics* 5(3), 297-299.

[11.14] Barbieri, M. (2013) Organic Codes and the Natural History of Mind. In: *Origins of Mind*. Springer, Dordrecht: 21-52.
[11.15] Barbieri, M. (2015) *Code Biology - A New Science of Life*. Springer, Dordrecht.
[11.16] Barbieri, M. (2016) What is information? *Philos Trans A Math Phys Eng Sci* 374(2063), #20150060.
[11.17] Barbieri, M. (2018) What is code biology? *Biosystems* 164, 1-10.
[11.18] Barbieri, M. (2019) Evolution of the genetic code: The ambiguity-reduction theory. *Biosystems* 185, #104024.
[11.19] Barbieri, M. (2019) A general model on the origin of biological codes. *Biosystems* 181, 11-19.
[11.20] Barnes, L., Goddard, E. and Woolgar, A. (2022) Neural Coding of Visual Objects Rapidly Reconfigures to Reflect Subtrial Shifts in Attentional Focus. *J Cogn Neurosci* 34(5), 806-822.
[11.21] Barresi, M. and Gilbert, S. (2020) *Developmental Biology*. Sinauer, Oxford University Press.
[11.22] Bilwes, A.M., Alex, L.A., Crane, B.R. and Simon, M.I. (1999) Structure of CheA, a Signal-Transducing Histidine Kinase. *Cell* 96(1), 131-141.
[11.23] Bisio, H., Chaabene, R.B., Sabitzki, R., Maco, B., Marq, J.B., Gilberger, T.W., Spielmann, T. and Soldati-Favre, D. (2020) The ZIP Code of Vesicle Trafficking in Apicomplexa: SEC1/Munc18 and SNARE Proteins. *mBio* 11(5), #e02092-20.
[11.24] Blumenthal, T. (2004) Operons in eukaryotes. *Briefings in Functional Genomics* 3(3), 199-211.
[11.25] Boudkkazi, S., Carlier, E., Ankri, N., Caillard, O., Giraud, P., Fronzaroli-Molinieres, L. and Debanne, D. (2007) Release-dependent variations in synaptic latency: a putative code for short- and long-term synaptic dynamics. *Neuron* 56(6), 1048-1060.
[11.26] Boudkkazi, S., Fronzaroli-Molinieres, L. and Debanne, D. (2011) Presynaptic action potential waveform determines cortical synaptic latency. *J Physiol* 589(Pt 5), 1117-1131.
[11.27] Bruni, L.E. (2007) Cellular semiotics and signal transduction. In: *Introduction to Biosemiotics: The New Biological Synthesis*. M. Barbieri, (ed.) Springer, Dordrecht: 365-408.
[11.28] Buratowski, S. (2003) The CTD code. *Nat Struct Biol* 10(9), 679-680.
[11.29] Campbell, R.A., Honegger, K.S., Qin, H., Li, W., Demir, E. and Turner, G.C. (2013) Imaging a population code for odor identity in the Drosophila mushroom body. *J Neurosci* 33(25), 10568-10581.
[11.30] Canzio, D. and Maniatis, T. (2019) The generation of a protocadherin cell-surface recognition code for neural circuit assembly. *Curr Opin Neurobiol* 59, 213-220.
[11.31] Chaudhuri, A., Das, S. and Das, B. (2020) Localization elements and zip codes in the intracellular transport and localization of messenger RNAs in Saccharomyces cerevisiae. *WIREs RNA* 11(4), #e1591.
[11.32] Cloutier, P. and Coulombe, B. (2013) Regulation of molecular chaperones through post-translational modifications: decrypting the chaperone code. *Biochim Biophys Acta* 1829(5), 443-454.
[11.33] Cohen, I.R. and Marron, A. (2023) Evolution is driven by natural autoencoding: reframing species, interaction codes, cooperation and sexual reproduction. *Proceedings of the Royal Society B: Biological Sciences* 290(1994), #20222409.
[11.34] Cosgrove, M.S. (2012) Writers and readers: deconvoluting the harmonic complexity of the histone code. *Nat Struct Mol Biol* 19(8), 739-740.
[11.35] Cosgrove, M.S. and Wolberger, C. (2005) How does the histone code work? *Biochem Cell Biol* 83(4), 468-476.
[11.36] Cowley, S.J. (2019) Wide coding: Tetris, Morse and, perhaps, language. *Biosystems* 185, #104025.

[11.37] Craig, N.L., Green, R., Storz, G., Greider, C., Wolberger, C. and Cohen-Fix, O. (2021) RNA Splicing. In: *Molecular Biology - Principles of Genome Function*. Oxford University Press, Oxford: 398-410.

[11.38] Craig, N.L., Green, R., Storz, G., Greider, C., Wolberger, C. and Cohen-Fix, O. (2021) Translation. In: *Molecular Biology - Principles of Genome Function*. Oxford University Press, Oxford: 434-497.

[11.39] DeMicco, A. and Bassing, C.H. (2010) Deciphering the DNA damage histone code. *Cell Cycle* 9(19), #3845.

[11.40] Dieci, G. (2021) Removing quote marks from the RNA polymerase II CTD 'code'. *Biosystems* 207, #104468.

[11.41] Drepper, C. and Sendtner, M. (2011) A new postal code for dendritic mRNA transport in neurons. *EMBO Rep* 12(7), 614-616.

[11.42] Eick, D. and Geyer, M. (2013) The RNA polymerase II carboxy-terminal domain (CTD) code. *Chem Rev* 113(11), 8456-8490.

[11.43] Faria, M. (2008) Signal Transduction Codes and Cell Fate. In: *The Codes of Life: The Rules of Macroevolution*. M. Barbieri and J. Hoffmeyer, (eds.) Springer, Dordrecht: 265-283.

[11.44] Farina, A. (2018) Ecoacoustic codes and ecological complexity. *Biosystems* 164, 147-154.

[11.45] Farina, A. and Pieretti, N. (2014) Acoustic Codes in Action in a Soundscape Context. *Biosemiotics* 7(2), 321-328.

[11.46] Fatemi, M. and Wade, P.A. (2006) MBD family proteins: reading the epigenetic code. *J Cell Sci* 119(Pt 15), 3033-3037.

[11.47] Forman, J.J. and Coller, H.A. (2010) The code within the code: microRNAs target coding regions. *Cell Cycle* 9(8), 1533-1541.

[11.48] Fu, X.D. (2004) Towards a splicing code. *Cell* 119(6), 736-738.

[11.49] Futerman, A.H. and Schuldiner, M. (2010) The plasma membrane code. *Nature Chemical Biology* 6(7), 487-488.

[11.50] Gabius, H.-J. (2008) Biological Information Transfer Beyond the Genetic Code: The Sugar Code. In: *The Codes of Life: The Rules of Macroevolution*. M. Barbieri and J. Hoffmeyer, (eds.) Springer, Dordrecht: 223-246.

[11.51] Gabius, H.J. (2000) Biological information transfer beyond the genetic code: the sugar code. *Naturwissenschaften* 87(3), 108-121.

[11.52] Gabius, H.J., Cudic, M., Diercks, T., Kaltner, H., Kopitz, J., Mayo, K.H., Murphy, P.V., Oscarson, S., Roy, R., Schedlbauer, A., Toegel, S. and Romero, A. (2022) What is the Sugar Code? *Chembiochem* 23(13), #e202100327.

[11.53] Görlich, D. and Dittrich, P. (2013) Molecular codes in biological and chemical reaction networks. *PLoS One* 8(1), #e54694.

[11.54] Grabe, V. and Sachse, S. (2018) Fundamental principles of the olfactory code. *Biosystems* 164, 94-101.

[11.55] Harbison, C.T., Gordon, D.B., Lee, T.I., Rinaldi, N.J., Macisaac, K.D., Danford, T.W., Hannett, N.M., Tagne, J.B., Reynolds, D.B., Yoo, J., Jennings, E.G., Zeitlinger, J., Pokholok, D.K., Kellis, M., Rolfe, P.A., Takusagawa, K.T., Lander, E.S., Gifford, D.K., Fraenkel, E. and Young, R.A. (2004) Transcriptional regulatory code of a eukaryotic genome. *Nature* 431(7004), 99-104.

[11.56] Hassa, P.O. and Hottiger, M.O. (2005) An epigenetic code for DNA damage repair pathways? *Biochem Cell Biol* 83(3), 270-285.

[11.57] He, S., Del Viso, F., Chen, C.Y., Ikmi, A., Kroesen, A.E. and Gibson, M.C. (2018) An axial Hox code controls tissue segmentation and body patterning in Nematostella vectensis. *Science* 361(6409), 1377-1380.

[11.58] Heng, J. and Heng, H.H. (2021) Karyotype coding: The creation and maintenance of system information for complexity and biodiversity. *Biosystems* 208, #104476.

[11.59] Hu, Z., Tong, X.J. and Kaplan, J.M. (2013) UNC-13L, UNC-13S, and Tomosyn form a protein code for fast and slow neurotransmitter release in Caenorhabditis elegans. *Elife* 2, #e00967.

[11.60] Jenuwein, T. and Allis, C.D. (2001) Translating the histone code. *Science* 293(5532), 1074-1080.

[11.61] Kaltner, H., Abad-Rodríguez, J., Corfield, A.P., Kopitz, J. and Gabius, H.J. (2019) The sugar code: letters and vocabulary, writers, editors and readers and biosignificance of functional glycan-lectin pairing. *Biochem J* 476(18), 2623-2655.

[11.62] Kessel, M. and Gruss, P. (1991) Homeotic transformations of murine vertebrae and concomitant alteration of Hox codes induced by retinoic acid. *Cell* 67(1), 89-104.

[11.63] Kindler, S., Wang, H., Richter, D. and Tiedge, H. (2005) RNA transport and local control of translation. *Annual Review of Cell and Developmental Biology* 21(1), 223-245.

[11.64] Komander, D. and Rape, M. (2012) The ubiquitin code. *Annu Rev Biochem* 81, 203-229.

[11.65] Kühn, S. and Hofmeyr, J.-H.S. (2014) Is the "Histone Code" an Organic Code? *Biosemiotics* 7(2), 203-222.

[11.66] Kull, K. (2020) Codes: Necessary, but not Sufficient for Meaning-Making. *Constructivist Foundations* 15(2), 137-139.

[11.67] Kun, Á. (2021) The major evolutionary transitions and codes of life. *Biosystems* 210, #104548.

[11.68] Kwon, Y.T. and Ciechanover, A. (2017) The Ubiquitin Code in the Ubiquitin-Proteasome System and Autophagy. *Trends Biochem Sci* 42(11), 873-886.

[11.69] Lodish, H., Berk, A., Kaiser, C.A., Krieger, M., Bretscher, A., Ploegh, H., Martin, K.C., Yaffe, M. and Amon, A. (2021) *Molecular Cell Biology*. W.H. Freeman and Company, New York.

[11.70] Louder, M.I.M., Lawson, S., Lynch, K.S., Balakrishnan, C.N. and Hauber, M.E. (2019) Neural mechanisms of auditory species recognition in birds. *Biol Rev Camb Philos Soc* 94(5), 1619-1635.

[11.71] Lyall, E.H., Mossing, D.P., Pluta, S.R., Chu, Y.W., Dudai, A. and Adesnik, H. (2021) Synthesis of a comprehensive population code for contextual features in the awake sensory cortex. *eLife* 10, #e62687.

[11.72] Magiera, M.M. (2022) The tubulin code: Empowering microtubules. *Semin Cell Dev Biol.*

[11.73] Major, J.C. (2021) Archetypes and code biology. *Biosystems* 208, #104501.

[11.74] Malavasi, R., Kull, K. and Farina, A. (2014) The Acoustic Codes: How Animal Sign Processes Create Sound-Topes and Consortia via Conflict Avoidance. *Biosemiotics* 7(1), 89-95.

[11.75] Marks, F., Klingmuller, U. and Muller-Decker, K. (2017) *Cellular Signal Processing: An Introduction to the Molecular Mechanisms of Signal Transduction*. Garland Science, Taylor & Francis Group, LLC.

[11.76] Materne, P., Anandhakumar, J., Migeot, V., Soriano, I., Yague-Sanz, C., Hidalgo, E., Mignion, C., Quintales, L., Antequera, F. and Hermand, D. (2015) Promoter nucleosome dynamics regulated by signalling through the CTD code. *Elife* 4, #e09008.

[11.77] Matlin, A.J., Clark, F. and Smith, C.W.J. (2005) Understanding alternative splicing: towards a cellular code. *Nature Reviews Molecular Cell Biology* 6(5), 386-398.

[11.78] Meneely, P., Hoang, R.D., Okeke, I.N. and Heston, K. (2017) Translation: From Nucleic Acids to Amino Acids. In: *Genetics - Genes, Genomes, and Evolution*. Oxford University Press, Oxford: 517-554.

[11.79] Mountoufaris, G., Canzio, D., Nwakeze, C.L., Chen, W.V. and Maniatis, T. (2018) Writing, Reading, and Translating the Clustered Protocadherin Cell Surface Recognition Code for Neural Circuit Assembly. *Annu Rev Cell Dev Biol* 34, 471-493.

[11.80] Muslimov, I.A., Patel, M.V., Rose, A. and Tiedge, H. (2011) Spatial code recognition in neuronal RNA targeting: role of RNA-hnRNP A2 interactions. *J Cell Biol* 194(3), 441-457.

[11.81] Ngai, J., Chess, A., Dowling, M.M., Necles, N., Macagno, E.R. and Axel, R. (1993) Coding of olfactory information: topography of odorant receptor expression in the catfish olfactory epithelium. *Cell* 72(5), 667-680.

[11.82] Nie, C., Wang, H., Wang, R., Ginsburg, D. and Chen, X.W. (2018) Dimeric sorting code for concentrative cargo selection by the COPII coat. *Proc Natl Acad Sci U S A* 115(14), E3155-E3162.

[11.83] Nieberding, C.M., Fischer, K., Saastamoinen, M., Allen, C.E., Wallin, E.A., Hedenström, E. and Brakefield, P.M. (2012) Cracking the olfactory code of a butterfly: the scent of ageing. *Ecol Lett* 15(5), 415-424.

[11.84] Nitika, Porter, C.M., Truman, A.W. and Truttmann, M.C. (2020) Post-translational modifications of Hsp70 family proteins: Expanding the chaperone code. *J Biol Chem* 295(31), 10689-10708.

[11.85] Nudler, E. and Mironov, A.S. (2004) The riboswitch control of bacterial metabolism. *Trends in Biochemical Sciences* 29(1), 11-17.

[11.86] Örd, M., Möll, K., Agerova, A., Kivi, R., Faustova, I., Venta, R., Valk, E. and Loog, M. (2019) Multisite phosphorylation code of CDK. *Nature Structural & Molecular Biology* 26(7), 649-658.

[11.87] Oxenham, A.J. (2018) How We Hear: The Perception and Neural Coding of Sound. *Annu Rev Psychol* 69, 27-50.

[11.88] Paolillo, M., Comincini, S. and Schinelli, S. (2021) Fostering "Education": Do Extracellular Vesicles Exploit Their Own Delivery Code? *Cells* 10(7), 1741.

[11.89] Paredes, O., Morales, J.A., Mendizabal, A.P. and Romo-Vázquez, R. (2021) Metacode: One code to rule them all. *Biosystems* 208, #104486.

[11.90] Pearson, Caroline A., Butler, Samantha J. and Novitch, Bennett G. (2014) Neuronal Organization: Unsticking the Cadherin Code. *Current Biology* 24(23), R1127-R1129.

[11.91] Plack, C.J., Barker, D. and Hall, D.A. (2014) Pitch coding and pitch processing in the human brain. *Hear Res* 307, 53-64.

[11.92] Plakke, B., Diltz, M.D. and Romanski, L.M. (2013) Coding of vocalizations by single neurons in ventrolateral prefrontal cortex. *Hear Res* 305, 135-143.

[11.93] Prinz, R. (2022) Code Biology Database – A List of Biological Codes. International Society of Code Biology (http://www.codebiology.org/database.pdf).

[11.94] Prinz, R. (2022) The modularity codes. *Biosystems* 219, #104735.

[11.95] Prinz, R. (2022) A simple measure for biocomplexity. *Biosystems* 217, #104670.

[11.96] Prinz, R. (2023) Nothing in evolution makes sense except in the light of code biology. *Biosystems* 229, #104907.

[11.97] Prosdocimi, F. and de Farias, S.T. (2021) Life and living beings under the perspective of organic macrocodes. *Biosystems* 206, #104445.

[11.98] Redies, C. and Takeichi, M. (1996) Cadherins in the developing central nervous system: an adhesive code for segmental and functional subdivisions. *Dev Biol* 180(2), 413-423.

[11.99] Roy, G., Chalfin, E.M., Saxena, A. and Wang, X. (2010) Interplay between ER exit code and domain conformation in CFTR misprocessing and rescue. *Mol Biol Cell* 21(4), 597-609.

[11.100] Salgado, H., Moreno-Hagelsieb, G., Smith, T.F. and Collado-Vides, J. (2000) Operons in *Escherichia coli*: Genomic analyses and predictions. *Proceedings of the National Academy of Sciences* 97(12), 6652-6657.

[11.101] Schorr, A.L. and Mangone, M. (2021) miRNA-Based Regulation of Alternative RNA Splicing in Metazoans. *Int J Mol Sci* 22(21), 11618.

[11.102] Shapiro, L. and Colman, D.R. (1999) The diversity of cadherins and implications for a synaptic adhesive code in the CNS. *Neuron* 23(3), 427-430.

[11.103] Siomi, H. and Siomi, M.C. (2009) On the road to reading the RNA-interference code. *Nature* 457(7228), 396-404.
[11.104] Skyberg, R., Tanabe, S., Chen, H. and Cang, J. (2022) Coarse-to-fine processing drives the efficient coding of natural scenes in mouse visual cortex. *Cell Rep* 38(13), #110606.
[11.105] Sluchanko, N.N. (2020) Reading the phosphorylation code: binding of the 14-3-3 protein to multivalent client phosphoproteins. *Biochem J* 477(7), 1219-1225.
[11.106] Snyder, A.C., Yu, B.M. and Smith, M.A. (2021) A Stable Population Code for Attention in Prefrontal Cortex Leads a Dynamic Attention Code in Visual Cortex. *J Neurosci* 41(44), 9163-9176.
[11.107] Spehr, M. and Leinders-Zufall, T. (2005) One neuron-multiple receptors: increased complexity in olfactory coding? *Sci STKE* 2005(285), pe25.
[11.108] Stanley, G.B. (2013) Reading and writing the neural code. *Nat Neurosci* 16(3), 259-263.
[11.109] Stergachis, A.B., Haugen, E., Shafer, A., Fu, W., Vernot, B., Reynolds, A., Raubitschek, A., Ziegler, S., LeProust, E.M., Akey, J.M. and Stamatoyannopoulos, J.A. (2013) Exonic Transcription Factor Binding Directs Codon Choice and Affects Protein Evolution. *Science* 342(6164), 1367-1372.
[11.110] Strahl, B.D. and Allis, C.D. (2000) The language of covalent histone modifications. *Nature* 403(6765), 41-45.
[11.111] Stringer, C., Michaelos, M., Tsyboulski, D., Lindo, S.E. and Pachitariu, M. (2021) High-precision coding in visual cortex. *Cell* 184(10), 2767-2778.e2715.
[11.112] Südhof, T.C. (2017) Synaptic Neurexin Complexes: A Molecular Code for the Logic of Neural Circuits. *Cell* 171(4), 745-769.
[11.113] Trifonov, E.N. (1989) The multiple codes of nucleotide sequences. *Bulletin of Mathematical Biology* 51(4), 417-432.
[11.114] Trifonov, E.N. (1996) Interfering contexts of regulatory sequence elements. *Bioinformatics* 12(5), 423-429.
[11.115] Trifonov, E.N. (1999) Elucidating sequence codes: Three codes for evolution. *Annals of the New York Academy of Sciences* 870(1), 330-338.
[11.116] Trotier, D. (1994) Intensity coding in olfactory receptor cells. *Semin Cell Biol* 5(1), 47-54.
[11.117] Tucker, B.J. and Breaker, R.R. (2005) Riboswitches as versatile gene control elements. *Current Opinion in Structural Biology* 15(3), 342-348.
[11.118] Turner, B.M. (2002) Cellular memory and the histone code. *Cell* 111(3), 285-291.
[11.119] Turner, B.M. (2007) Defining an epigenetic code. *Nature Cell Biology* 9(1), 2-6.
[11.120] Turner, M.H., Krieger, A., Pang, M.M. and Clandinin, T.R. (2022) Visual and motor signatures of locomotion dynamically shape a population code for feature detection in *Drosophila*. *Elife* 11, #e82587.
[11.121] Verhey, K.J. and Gaertig, J. (2007) The tubulin code. *Cell Cycle* 6(17), 2152-2160.
[11.122] Villarreal, L.P. (2005) *Viruses and the Evolution of Life*. ASM Press.
[11.123] Vinicius, L. (2010) The new "Chain of Being": Hierarchical evolution and biological complexity. In: *Modular Evolution: How Natural Selection Produces Biological Complexity*. Cambridge University Press: 186-207.
[11.124] Vitreschak, A.G., Rodionov, D.A., Mironov, A.A. and Gelfand, M.S. (2004) Riboswitches: the oldest mechanism for the regulation of gene expression? *Trends in Genetics* 20(1), 44-50.
[11.125] Wang, X. (2018) Cortical Coding of Auditory Features. *Annu Rev Neurosci* 41, 527-552.
[11.126] Weatheritt, R.J. and Babu, M.M. (2013) The Hidden Codes That Shape Protein Evolution. *Science* 342(6164), 1325-1326.
[11.127] Wells, A. (2019) Breaking the Cybernetic Code: Understanding and Treating the Human Metacognitive Control System to Enhance Mental Health. *Front Psychol* 10, #2621.

[11.128] Wethekam, L.C. and Moore, J.K. (2022) Microtubule cytoskeleton: Revealing new readers of the tubulin code. *Current Biology* 32(18), R960-R962.

[11.129] Williamson, A., Werner, A. and Rape, M. (2013) The Colossus of ubiquitylation: decrypting a cellular code. *Mol Cell* 49(4), 591-600.

[11.130] Witzany, G. (2009) Noncoding RNAs: Persistent Viral Agents as Modular Tools for Cellular Needs. *Annals of the New York Academy of Sciences* 1178(1), 244-267.

[11.131] Wolfe, J.M., Kluender, K.R., Levi, D., Bartoshuk, L., Herz, R., Klatzky, R. and Merfeld, D. (2020) *Sensation and Perception*. Oxford University Press.

[11.132] Wu, Q. and Jia, Z. (2021) Wiring the Brain by Clustered Protocadherin Neural Codes. *Neurosci Bull* 37(1), 117-131.

[11.133] Yu, I., Garnham, C.P. and Roll-Mecak, A. (2015) Writing and Reading the Tubulin Code. *J Biol Chem* 290(28), 17163-17172.

[11.134] Zhang, Y., Huang, R., Nörenberg, W. and Arrenberg, A.B. (2022) A robust receptive field code for optic flow detection and decomposition during self-motion. *Curr Biol* 32(11), 2505-2516.e8.

[11.135] Zheng, Y., Jia, S., Yu, Z., Liu, J.K. and Huang, T. (2021) Unraveling neural coding of dynamic natural visual scenes via convolutional recurrent neural networks. *Patterns (N Y)* 2(10), #100350.

[11.136] Zolyan, S. (2022) From matter to form: the evolution of the genetic code as semio-poiesis. *Semiotica* 2022(245), 17-61.

12

Evolutionary Growth of Meanings in the Relational Universe of Intercommunicating Agents

Abir U. Igamberdiev

Department of Biology, Memorial University of Newfoundland, St John's, NL, Canada

Online resources:
https://www.mun.ca/faculty/aigamberdiev/
https://scholar.google.ca/citations?user=NEeEsGcAAAAJ&hl=en
https://www.researchgate.net/profile/Abir-Igamberdiev

Abstract

Meaning appears in the relational universe through the interaction of individual occasions that become agents at higher levels of organization. In the autopoietic living systems, the meaningful information is associated with maximization of system power and efficiency in the course of evolution. It is realized via the network of codes that span different levels of organization. The whole process of generating novel coding statements that substantiate organizational complexification leads to the expansion of the system that incorporates the elements external to it for supporting newly generated complex structures. During this complexifying process, the meanings are assigned to previously unproven statements. The processing of meaningful information refers to the achievement of resilient self-maintaining configurations acting as the attractors resistant to external and internal perturbations. In these configurations, the evolving system achieves the condition of maximization of its power via synergistic effects. The subdivision of the system into the controlling "self" and the controlled "body" is the precondition of its self-sustainable operation, and of the growth of meanings in the system plus the environment that it signifies, which appears as a part of the growing biosemiosphere formed by the organisms inhabiting it.

Keywords: Attractor, autopoiesis, biosemiosphere, goal-directedness, kinetic perfection, ontolon

12.1 Introduction: Agency and Meaning

Meanings correspond to a relationship between signs and the objects they signify and refer to the information that semiotic systems hold [12.52]. In other words, meaningful

Email: igamberdiev@mun.ca

information is reality in potential form [12.9], and the growing of semiotic system corresponds to the increase of meaningful information in it. Meaning, having epistemological nature, is supported and established through a dynamic ontological process. As soon as living beings originate, meaning appears in the processes establishing their survival and reproduction. Living beings arise as complex agents that communicate and thus generate meanings in the relational semiotic interaction between them. This expanding process underlies the whole biological organization and its continuing evolution [12.62] [12.63].

The agency has a structure that becomes a result of the epistemic cut, which separates energy-degenerate, rate-independent symbols from the rate-dependent dynamics of construction that they control [12.58] [12.59]. The idea of agency and of the functional nature of meaning originating in the communication of agents introduces the dynamic aspect of semiosis. The dynamic representation of meaning was originally introduced in the concept of *Funktionkreis* of Jakob von Uexküll [12.67], which depicts the semiotic interconnection between low-energy recognition and high-energy work following this recognition. Such structure represents the meaningful information dynamically and can be viewed in any biological function [12.3]. In other words, it is characterized by the structure, whose fundamental principles are based on the essential distinction between the informational and energetic constituents of the integral system.

Philosophically, Aristotle formulated this distinction and, in fact, he was the first who developed the concept of agency corresponding to living system [12.63]. In the most detailed form, this concept is expressed in his tractate "*De Anima*" [12.9]. Meaning in the action of agent refers to the Aristotelian final cause, which, in his views, retrocausally directs the development of the system shaped by the material, the efficient, and the formal causes. The unique dynamic organization of the individual agent refers to the "closure to efficient causation" [12.60], while the final cause corresponds to the idea of teleonomy in biological development and evolution [12.16]. In the Darwinian framework, it corresponds to the survival of the fittest, whose basic criterion could be the increase of system efficiency according to the maximum power principle [12.51]. The concept of maximum power can be defined as the maximum rate of useful energy transformation [12.57], and matches the establishment of the thermodynamic stable non-equilibrium state [12.4] [12.5].

Although the problem of agency and meaning refers to living beings and their communication, i.e., to the biosemiosphere, the origin of semiosis arises to the basic principles of organization of the physical world, i.e., to the percolation between different levels of physical reality from the basic quantum phenomena to the upper levels [12.14]. In order to comprehend the evolutionary growth of meanings in living matter, we turn in the next section to the physical background of the agential nature. We substantiate the fundamental principles of reality that make possible the appearance of agents and thus origination of the self-growing process of biological evolution.

12.2 Physical Background of the Agential Nature

Living systems act as agents. The proto-agents having reduced representation capacity [12.64] also denote primitive biological systems. Their origin corresponds to the appearance of the epistemic cut, prior to which we cannot talk about agents and their communication [12.59] [12.68]. However, in the pure physical reality, if we do not accept absolute space and time, we assume the existence of individual measuring units acting as observers. The relational paradigm was introduced to physics as a dependence of the spatiotemporal structure on the observer. The observers appear as living subjects only at the highest level of organization, while at the physical level they materialize as the dynamic units of complex reality, i.e., the units of dynamic change. It has been suggested to define these units as *ontolons* to emphasize their profoundly ontological rather than epistemological nature [12.9] [12.38].

In fact, Aristotle attempted to formulate physics as the studies of beings that arise on their own, with regard to their being [12.25], and thus introduced the relational principle in it. In the frame of the inertia principle of Galileo, who claimed the equality of all inertial systems, it became the basis of the scientific exploration of reality. Although the Aristotelian physics was not successful from the standpoint of modern science, as it did not formulate explicitly the reference frame for the physical world, in the original physical concept of Aristotle we can find very important relational statements that require further development. In particular, Aristotle distinguished between the two types of time: the time, which is measured, and the time by which we measure, which corresponds to the separation of an agent and of the objective reality measured by the agent [12.36]. In the early 18th century, Leibniz viewed the world as a multiverse of individual units (monads) that perform a kind of internal "computation," corresponding to the physical idea of measurement. The spatial order appears as a consequence of the communicatory arrangement between monads [12.49]. In the early 20th century, Whitehead suggested to view the world as a set of occasions of experience (momentary monads) engaged in the temporal/causal succession. This opens a possibility of the interpretation of physics in the framework of causal sets [12.11].

The relational framework of physics based on the events of individual occasions is not yet developed; however, there are a few approaches which represent the first steps for formulating this agenda. The standpoint of quantum measurement could be at the center of this approach. In the original interpretations of quantum mechanics, such as the Copenhagen, the multi-world, as well as later developed consistent histories interpretation, all have essential limitations related to the absence of analysis of the structure of measuring event causally shaped by an agent. A step forward to finding this structure became the concept of objective reduction of Hameroff and Penrose [12.24]. The measuring occasions introduced by Penrose upon reaching the curvature of the value of Planck's mass, represent individual units that possess the property of singular events in the sense of Whitehead. The main problem in this interpretation appears to be the absence of visual evidence that the reduction takes place upon reaching the value of Planck's mass of the space curvature. The gravitational self-energy that characterizes the degree of space-time separation specified by the superpositioned mass is related to the time of superposition before the quantum reduction in accordance with the Heisenberg's uncertainty ratio energy-time [12.24]. However, this may not be necessarily the gravitational energy but it can refer to the energy of other fundamental forces. The transactional interpretation of quantum mechanics introduced by Cramer [12.17] and developed by Kastner [12.43] brings forth a different type of structure that shapes the event of quantum measurement. According to this interpretation, the measurement event results from the interference of the forward and the backward (in time) "handshaking" waves that constitute a transaction. In the transactional interpretation, "the future can affect the past only very indirectly, by offering possibilities for transactions" [12.17], p. 669.

The transaction is framed in correspondence with the values of Planck's units; however, this is not straightforwardly the imposition of the Planck's mass curvature. While the relation of the individual transaction to the Planck's mass has not been clarified, as Hameroff and Penrose [12.24] initially suggested, the dynamical process of transaction has the minimum action value and the maximum value of transmission through the physical space-time. It also can have different degrees of certainty of the final output, which is defined in accordance with the Heisenberg's uncertainty ratio energy-time [12.8] [12.31]. If the transaction process is held for a prolonged time of measurement (transaction), then the dissipation of energy is low, and the process of quantum measurement has a precise output. This property of transaction has an essential consequence for the origin and function of living systems since it reduces uncertainty and provides a possibility of the efficient operation of macromolecular structures. In fact, the complex structure of the biomacromolecules operating as quantum measurement devices aims to reduce energy dissipation and provide efficient realization of individual processes, which includes enzyme catalysis, membrane transport, or signal transduction [12.33] [12.61]. It becomes a physical basis of semiosis, as discussed by Howard Pattee [12.58], and became the central assumption in the semiotic interpretation of quantum mechanics by Christiansen [12.13].

The representation of physics on the basis of intercommunicating ontolons refers to the pattern of the physical world as a universe of individual occasions or actions that form the relational space-time. These occasions represent the cases of single quantum measurement events or transactions, in which efficient (in the Aristotelian sense) cause cannot be defined externally but has certain physical limitations. The efficient causes always precede the effects that shape regular events in our world. The irreversible transactions define the unidirectional flow of time in the real world, while the basis of transaction is atemporal and occurs in the "implicate order" of the potential whole in the sense of David Bohm [12.7].

The spatiotemporal configuration of the world is determined by the coordinate systems choice made by observers within the Universe [12.36]. This principle, describing the world's relational arrangement as a set of systems with the internal causation, represents a generalized principle of relativity. A particular consequence of this principle is the fact that the absolute maximum value of the speed of signal is related not to the absolute time-space but to the agent (or, in the physical reality, to the individual occasion event). Another consequence is represented by a standpoint that the causation is possible only if there exists a materialized analogy of the infinitesimal acquiring the finite value in order to preserve the ability to measure everything that happens in the world. This "atomization" in the sense of ancient Greek philosophy frees the world from the immobility logically postulated by Parmenides and proved by Zeno's aporias (paradoxes). Thus, the internal causation event is shaped by the fundamental constants of the speed of light and of the minimum action as originally suggested by Liberman [12.50].

The physical space of occasions appears as objective structure because it possesses additional universal characteristics such as the gravitational constant as well as the constants that describe the additional compact dimensions, which includes the elementary charge and the electric constant (unified in the fine structure constant) for the electromagnetic interaction. To this "objective" world, a unified subset of the laws common to all observers is applied, which is expressed in the formal logic having a substantially epistemic nature. However, there is a necessity to refer to the world consisting of multiple acting ontological units (ontolons) operating and communicating as observers [12.9]. These units appear as conscious subjects only at the highest level of cognitive organization. At the level of the physical world, they represent "occasions" in the sense of Whitehead, which correspond to individual quantum measurement transactions [12.17] and constitute the relational universe [12.36]. The origin of the "universe of transactions" can be substantiated very generally within the framework of the dialectical logic of Plato's dialogue *Parmenides*. Plato demonstrates there that the existing one can appear in the reality as many. The vast multiplicity of primary elements that include both the potential and the actual constituents becomes the consequence of actualization process in the real world.

Ontolons operating as agents produce the actions that actualize information and have the minimum price (the Planck's constant); they can communicate by exchanging signals that have the maximum speed (the speed of light), and they generate the field that defines their interrelation (the gravitational field and other fields). Ontolons resemble Descartes' vortexes forming an interconnected network with the Universe. However, the concept of vortexes was rather purely mechanical, while ontolons represent the individual beings with the internal determinism, which is framed in such fundamental properties as the absolute speed of signal (relative to ontolon and not to the absolute space) and the minimum price for sending this signal (the Planck's constant) [12.9]. The observed complex structure of the physical world can be a consequence of the individual actions shaped by the limits of the establishment of three spatial dimensions of the gravitational field and of several additional compact dimensions corresponding to other fundamental forces [12.6].

12.3 Meaningful Information in Autopoietic Systems

The biological unity (cell, organism) as an agency is unique in abstracting and holding its self-identity, and, compared to pure physical bodies, biological systems support their organization with the use of the material act of measurement as a form of abstraction [12.53]. Thus, the meaningful information as a reality in its potential form [12.9] in a living system, aims to hold the system's self-identity and thus support their autopoietic nature, i.e., their internal capacity to reproduce and organize themselves [12.70].

The spatiotemporal organization of biosystems is fundamentally shaped by their dynamic structure, which, in turn, is grounded on their metabolic organization. The efficient energy autonomy of living systems from local, high-energy potentials is achieved via their internal metabolic structure by sensitivity to non-local low-energy fluxes. Jakob von Uexküll [12.67] placed this idea in the center of the theoretical basis of biology and thus substantiated the semiotic nature of living organization. Uexküll introduced the elementary biosemiotic structure defined as a *Funktionkreis*, which captures the basic structural interaction between the living organism and its environment. This idea was further advanced in the dynamical theory of information processing of James Barham [12.3] who suggested that every act of recognition involves a low-energy interaction between a component of a non-linear system and an environmental input signal that causes the component to undergo a state of transition. It also became the basis of the concept of the epistemic cut of Howard Pattee [12.59]. In the transition from low energy, non-local input to a high-energy response of the system, an exchange of signals loaded by meaning takes place. This transition is mediated by the system of codes of biological system [12.1]. There are many different codes in all biological systems spanning from low to high levels of organization [12.1]. In addition to the symbolic and digitally coded genetic structures such as DNA and natural language, biological systems possess the analog-coded structures that are based on indexical reference and also represent a semiotic mode in their dynamics [12.28]. The latter structures bear meanings in the whole system, which are transduced through the shape and communication of coding molecules and structures, e.g., in the signal transduction through a topologically channeled cascade of reactions mediated by individual enzymes and membrane proteins.

Thus, biological dynamics include both the low-energy and the high-energy processes separated through the epistemic cut. To function successfully, the biological organism should efficiently maximize its power via synergistic effects to perform work in an efficient way [12.51]. According to Bauer [12.4] [12.5], a capacity for the increase in external work is the main characteristic feature of the evolutionary process. Biological systems, during self-organization, maximize power intake and energy transformation for the goal of reinforcing production and efficiency [12.56]. Thus, in the autopoietic systems, the meaningful information is associated with the maximization of power, and the informational reality in its potential form represented by the systems of codes supporting the whole system, is actualized efficiently into the processes and structures performing work efficiently. This is the basis of the meaningful operation of living systems. The development of biological systems continuously transforms their organization and energy flow structure in a way that the useful energy transformation becomes maximized via the constrained release of energy that delays the production of entropy [12.44]. This assumes that energy in biological evolution can be analyzed via such economic criteria as productivity, efficiency, and the costs and benefits ("profitability") of various mechanisms for capturing and utilizing energy to build biomass and do work [12.15].

The meaningful operation of biological systems toward the most efficient utilization of energy becomes possible due to the effective control by their internal subsystems efficiently shielded from energy flows. This represents an internal epistemic cut which complements the recognition of external low stimuli and is reached in highly ordered coherent states within the system [12.29] [12.33] [12.54]. The function of nucleic acids that store the information and thus maintain the field of meanings, representing the potential for development and adaptation of a particular biological system, makes possible the persistence and reproduction of biological organization as an invariant entity. In fact, in livings systems, the *Funktionkreis* is implemented internally via the system of reflexive loops generating the network of biological codes. This becomes possible in the frames of the structure called *chemoton* [12.21], which exerts a similarity with the (M,R) system suggested by Rosen [12.60]. In this metabolically closed structure, which may represent the basic feature of life, its elements fulfill the role of "double duty," serving both as energetic and as informational components that are equilibrated through the balance of energy and information fluxes. The meaning of information held by the system corresponds to the most optimal and efficient way of energy utilization in order to achieve system optimal performance.

The network of biological macromolecules holds the internal ordered potential state shielded from thermal fluctuations [12.29] [12.33]. Its most coherent part corresponds to perceptive actions and conscious activity. It governs the rest of the body by perceiving the whole system, sending the commands to it, and ensuring that the organism operates with maximum power and efficiency. These coherent internal states are experienced by the system as qualia. Understanding this would lead us to approaching the "hard problem of consciousness" [12.12], which actually refers not only to consciousness but rather to any information-bearing systems, e.g., to any self-perception of the observer. The original idea of Dubrovsky [12.18], who formulated it explicitly in the 1960s, that the informational processes in the brain are inseparable from the qualia, allows us to assume the existence of the protophenomenal properties in any natural observer such as living cell, organism, or even in any transactional measurement event in the physical world.

12.4 Generation of New Meanings in the Evolutionary Process

While autopoiesis denotes self-maintenance of the system via its internal regulation and the support of its stable non-equilibrium state, the complexification of organization in evolution corresponds to the transformation beyond the limits of autopoietic system. Marcello Barbieri [12.1] defined the basis of this transformation as *codepoiesis*. The codes underlying biological organization possess a capacity of growing up through the attainment of indefinite statements within them by meaning [12.37]. This intrinsic capacity triggers those evolutionary phenomena that are accompanied by the complexification of organization (defined as the macroevolutionary events). While most evolutionary changes take place within one level of hierarchy and do not require major transformations (represent the microevolutionary phenomena), evolutionary process can also lead to the complexification of biological organization corresponding to progressive evolution and thus to the appearance and growth of new meanings.

The whole process of generating novel coding statements that substantiate organizational complexification leads to the expansion of the system that often incorporates the elements that originally were external to it. As a result, newly generated complex structures arise. During this complexifying process, new meanings are assigned to the previously unproven and thus previously "meaningless" statements. The set of meanings in the biological semiotic system has its limits, but it also has a property of overcoming these limits, and this corresponds to evolution. According to Gödel's incompleteness theorem, any sufficiently powerful consistent logical system has true statements, which cannot be proven within its framework. In relation to biological systems, this assumes that the foundation of these statements exists outside the formal language of the system, being imposed through the creative process of evolutionary transformation. Through the transformative process, the elements of formal system acquire the meanings through the procedure defined as Gödel numbering. In this procedure, new semiotic relations appear that can attain interpretations in the formalized language. Codepoiesis appears as an intrinsic property of biological systems through the mapping of living system as a whole into the finite set of its molecular structures [12.30] [12.31] [12.37].

There are multiple ways of codepoiesis, which all involve unique novel significations by using new codes or rearranging the old ones. This realization appears as irreversible reduction of the fundamental uncertainty appearing in the self-referential living process. Evolutionary complexification denotes a metasystem transition [12.66] that results in the increase of external work by the system based on the division of labor between its components. The principle of metasystem transition determines the emergence of novel structures during evolution of complex systems. It is based on the appearance of new interpretations within the system of internal codes of complex systems and on the generation of new coding systems (codepoiesis). The principle of metasystem transition transforms cyclic dynamics of autopoietic systems into a spiral spatiotemporal pattern that results from memorization of new information in the reflexive loops that incorporate externality [12.9] [12.42].

New traits in evolution originate as "exaptations," which appear as the features available for useful cooptation by descendants [12.22] [12.23]. The term exaptation was suggested to define the process of adaptation of a trait for a purpose other than what the trait was evolved for. It is considered as more precise as compared to the term "pre-adaptation," which assumes a direct anticipation of new meanings in the course of evolution. The exaptation configurations become empowered by meaning in the timeline of evolution. Originally, they do not hold a definite functional role but they can gain it in the course of temporal development. The subdivision of the system into the controlling "self" and the controlled "body" through the epistemic cut is a precondition of the complexification process. The growth of meanings in the system that is incorporated in its environment takes place in the expanding biosemiosphere formed by the organisms inhabiting our planet [12.52].

The growth of meanings in evolution corresponds to the increase of kinetic perfection of biosystem, which occurs in the rates of metabolic processes, in the spatial movement of organisms, in their mechanical construction, hydro- and aerodynamic characteristics [12.65]. The kinetic perfection as a part of the power maximization of the system, determines the choice of a unique efficient, and thus meaningful, trajectory in the multidimensional space of biological evolution. The kinetically meaningful transformation of thermodynamic characteristics of living systems results in their continuous complexification [12.69]. At earlier stages of evolution, the achievement of kinetic perfection is expressed in the refinement of the mechanisms of replication, transcription and translation toward their higher reliability. At later stages, overcoming the diffusion limitations of kinetic processes becomes important via the active maintenance of ionic gradients in cells supporting vectorized fluxes of energy and matter, the development of transport systems in multicellular organisms, morphogenesis including the morphological progress with major alterations in topology of organisms, and the mechanisms of active motion in space. The latter is controlled by a continuous perfectioning of the nervous system, whose operation is apparently based on holding the long-range coherent states [12.41] [12.45].

The principle of maximum useful energy flow transformation, by substantiating the kinetic perfection of biological systems as their evolutionary attractor, determines the durability of their non-equilibrium state and thus their competitiveness in the process of adaptation to environment. The hypercyclic structure that underlies these properties is based on the active search of the system towards fitting the slow non-equilibrium fluxes and the fast equilibration processes in metabolic structure [12.39]. In fact, the equilibria of coenzyme nucleotides and substrates established in cells generate simple rules that provide optimal conditions for the non-equilibrium fluxes of major metabolic processes [12.40]. The equilibrium "buffering" enzymatic reactions serve as control gates for the non-equilibrium fluxes through the "engine" enzymes, providing the directed diffusion of substrates to their active sites and establishing the balance of the fluxes of load and consumption of metabolic components. The search of best "buffering" configurations for "engine" enzymes represents an optimized strategy in the developmental trajectories. The evolutionary optimization is achieved under the coordinated operation of the buffering and the engine enzymes that determine the establishment of stable and organizationally invariant non-equilibrium states that organize the fluxes of energy spatially and temporally by controlling the rates of major metabolic fluxes that follow thermodynamically and kinetically defined computational principles.

The genetic system is formed from the total pool of nucleoside triphosphates, and DNA is metabolically isolated from the major metabolic pool of nucleotides because it uses deoxynucleotides whose synthesis is separately controlled. Nucleotides serve as metabolites (free coenzymes) of enzymes, while their association with nucleic acids generates matrices for the reproduction of enzymes themselves. Polymerization of co-substrates into nucleic acids generates a self-referential set of arrows for the set of catalysts, resulting in the appearance of the digital information of the genetic code forming the internal programmable structure of biosystems and representing the meaningful potential for realization of multiple biological functions [12.31]. To understand the metabolic aspect of the organizational invariance parameter is important for drawing the equations linking RNA biosynthesis with the availability of nucleoside triphosphates and cofactors like magnesium and potassium, whose concentrations are established upon equilibrium of nucleoside triphosphates. Even in a very general form, these equations would reflect the chemoton [12.21] or the (M,R) structure [12.60] and could be described and analyzed thermodynamically by determining the actual rates of each process and its regulation. A functional component and not a material part as such is replicated in this structure. From this point of view, the distinction between genotype and phenotype is dynamic and flexible, so it is not useful to try defining what is primary and what is secondary. The evolutionary process is grounded on the reestablishment of the parameter of the organizational invariance rather than on the independent and casual changes in the genotype [12.24].

The semiotic pattern of living system opens the possibilities for potentially infinite evolutionary unfolding of codepoiesis. Active combinatorial process of self-modification of information corresponds to the growing of meanings in the continuous language game, which preconditions any evolutionary transformation. The operational features that include negative feedbacks, feedforward links, equifinality, and flexibility toward external influences can be viewed as final causes (in the Aristotelian sense) in this evolutionary development. They represent the far-from-equilibrium meaningful attractors of a dynamical system achieved with the participation of particular mechanisms preventing possible deviations in development from various external perturbations [12.10] [12.26] [12.27].

For reaching a new meaningful attractor by a system, natural selection not necessarily acts as the driving force of evolution, but rather the attractor itself becomes a retrocausal reason for natural selection [12.16]. Biological entities as anticipatory systems exploit the meaning-based retrocausality as their basis of evolutionary change in which stable non-equilibrium states appear as natural attractors achieved and continuously reestablished in the course of evolution. For the appearance of new genetic changes, the mobile genetic pool of all organisms serves as a source of horizontal gene transfer [12.55], and this genetic pool represents a part of biosemiosphere common for all living beings. However, biosemiosphere is not restricted only to genetic elements, but it comprises all symbolic elements that include digital-, index- and analog-coded structures that operate in the semiotic dynamical mode. All these structures attain symbolic functions in the whole systems that are entered into the processes of meaningful interaction and intercommunication.

12.5 The Growth of Meanings in the Evolutionary Transition from Biological to Social Systems

The large-scale expansion of meanings accompanies the evolutionary transition from biological to social systems. This transition is characterized by the origination of the structure of the subject (conscious agent) incorporating the internal image of external world [12.32]. Such structure is formed via the meaningful referral of the subject (self-conscious agent) to its symbolic image. This referral gains a power of rationally describing the external world in the frames of the semiotic structure of human language. Sigmund Freud metaphorically described the fundamental reflexive structure of social behavior within the framework of the Oedipus complex that represented in his concept a type of logical pattern establishing the connection between consciousness and the outside world, including the interconnections between different conscious subjects with the possibility to substitute others in the course of reflection [12.20]. Jacques Lacan [12.46], who suggested a structuralist interpretation of the Oedipus complex, viewed the basic subject structure as a flow of the signifier and the signified, characterized by the continuous acquirement of meanings in the course of this process. The structure of conscious subject necessarily incorporates the image of total world into the dynamical process of acquiring knowledge. This became a starting point of social evolution initially determining the origin of religion and arts, and later of philosophy and natural sciences [12.35].

The structure of conscious agent can be represented using the algebra of simple relations. The model suggested by Vladimir Lefebvre [12.48] introduces fundamental opposition of the two basic complementary types of reflection, which produce the opposite models of behavior and the contrasting organizations of societies. This involves the interactions not only between the societies but also within the same society, triggering the changes in its homeostasis and dynamics. The invention of new modes of thinking and putting novel technologies into practice continuously alter the probability patterns of reflexive choices, and result in the direct changes in the preference of reflexive modes of behavior, which leads to further expansion of meanings in the course of development of the human language. Throughout the history of human civilization, the expansion of meanings resulted in the development and spreading of new technologies determining the formation and splitting of large empires, and in the alterations of consecutive phases of social prospering and decline.

The growth of meanings in social systems is based on the development of language. The origin of language can be compared to the origin of life in the way that it represents the appearance of the system of arbitrary rules that interprets the world through the generation of meanings [12.2]. The semantic activity associated with language assigns meanings to the elements of the spoken or written language in the process of interpretation. Thus, the growth of language can be considered as a vast expansion of abstracting capacity inherent in all living beings [12.53]. Language continuously generates new meanings as the choices for alternative realizations of the modes of action in exploring real world. The abstracting capacity arises from the set of relations generated by the reflective capacity of the self-conscious agent. This structure generates a broad field of new meanings that appear in the transformation of external reality to fit it for the needs of the transforming agents.

As a result, the potential of conscious reflexive agents to generate new meanings exceeds in many orders the corresponding capacity of biological systems.

Emile Durkheim [12.19] defined social progress as a development of the increased differentiation of properties and functions in the social organism. This differentiation is grounded in the intercommunication of conflicting reflexive systems. The memorization of new knowledge introduces linearity in the cyclic dynamics of civilizations and fixes the meanings to transfer them to other people and to new generations, which constitutes the progress of social development. The fundamental part of social progress based on the natural science and advancement of technology is cumulative contrary to the ethical knowledge, which is not cumulative and generally needs to be recaptured in the meaningful reflexive process.

The fundamental reflexive choice of intercommunicating conscious agents appears as a source of social energy that transforms into the actual work in the course of social dynamics. This energy can be constructive or destructive but, in both cases, it possesses the power for introducing social change. At the current historical stage, we face the process in which the interaction and dialogue between different reflexive types generating opposite meanings turns to the internal dynamical process occurring within the global civilization. This process may finally open into a new dialogue between civilizations at the cosmic scale [12.47], which will expand the evolutionary growth of meanings in the realm of intercommunicating agents to cosmic levels and thus extend the semiosphere to the scale of the Universe.

12.6 Conclusion

Evolution occurs as the emergence of functional information in natural agents [12.54]. This corresponds to an expansion of reality in the potential form, represented by the meaningful information in living systems [12.33]. Any evolutionary change obtains evolutionary value only if it gets an interpretation through which it gives rise to a meaningful function. Complexification in the evolutionary process corresponds to the generation of new meanings and thus to the construction of new ecological niches when the previous are occupied. At higher levels of organization, evolutionary innovations emerge as niche constructions, behavioral choices, and finally the phenomenon of cognition. The subdivision of the system into the controlling "self" and the controlled "body" is the precondition of the growth of meanings in the system plus the environment that it signifies, which appears as a part of the growing biosemiosphere formed by the organisms inhabiting it. The evolutionary growth of meanings acquires major expansion with the emergence of consciousness that incorporates the image of the whole world into the dynamical process of knowledge acquisition and creates the conditions for development of the global civilization.

Declaration

The author declares no conflict of interest.

References

[12.1] Barbieri, M. (2015) *Code Biology. A New Science of Life*. Springer, Cham.

[12.2] Barbieri, M. (2020) The semantic theory of language. *Biosystems* 190, #104100.

[12.3] Barham, J. (1996) A dynamical model of the meaning of information. *Biosystems* 38(2-3), 235-241.

[12.4] Bauer, E. (1920) Die Definition des Lebewesens auf Grund seiner thermodynamischen Eigenschaften und die daraus folgenden biologischen Grundprinzipien. *Naturwissenschaften* 8(18), 338–340.

[12.5] Bauer, E. (1935) *Theoretical Biology*. Republished in 1982. Akadémiai Kiadó, Budapest.

[12.6] Blumenhagen, R., Körs, B., Lüst, D. and Stieberger, S. (2007) Four-dimensional string compactifications with D-branes, orientifolds and fluxes. *Physics Reports* 445(1-6), 1–193.

[12.7] Bohm, D. (1980) *Wholeness and the Implicate Order*. Routledge, London.

[12.8] Braginsky, V.B. and Khalili, F.Y. (1996) Quantum non-demolition measurements: the route from toys to tools. *Reviews of Modern Physics* 68(1), 1–11.

[12.9] Brenner, J.E. and Igamberdiev, A.U. (2021) *Philosophy in Reality. A New Book of Changes.* Springer, Cham.

[12.10] Busseniers, E., Veloz, T. and Heylighen, F. (2021) Goal directedness, chemical organizations, and cybernetic mechanisms. *Entropy (Basel)* 23(8), #1039.

[12.11] Carlson, C.R. (2017) *The Mind-Body Problem and Its Solution*. LitFire Publishing, Atlanta, Georgia, 2nd.

[12.12] Chalmers, D.J. (1996) *The Conscious Mind: In Search of a Fundamental Theory*. Oxford University Press, Oxford.

[12.13] Christiansen, P.V. (1985) *The Semiotics of Quantum Non-Locality* (Tekster fra IMFUFA [ISSN 0106-6242], text number 93). University of Roskilde, Roskilde.

[12.14] Conrad, M. (1996) Cross-scale information processing in evolution, development and intelligence. *Biosystems* 38, 97–109.

[12.15] Corning, P.A. (2020) 'Thermoeconomics': Time to move beyond the second law. *Progress in Biophysics and Molecular Biology* 158, 57–65.

[12.16] Corning, P.A. (2022) A systems theory of biological evolution. *Biosystems* 213, #104630.

[12.17] Cramer, J.G. (1986) The transactional interpretation of quantum mechanics. *Review of Modern Physics* 58 (3), 647-687.

[12.18] Dubrovsky, D.I. (2019) The hard problem of consciousness. Theoretical solution of its main questions. *AIMS Neuroscience* 6(2), 85–103.

[12.19] Durkheim, E. (1893) The Division of Labor in Society. Republished in 1997. Free Press, New York.

[12.20] Freud, S. (1917) *Introductory Lectures on Psychoanalysis*. Republished in 1989. Liveright, New York.

[12.21] Gánti, T. (2003) *The Principles of Life*. E. Száthmary, J. Griesemer (eds.). Oxford University Press, Oxford.

[12.22] Gould, S.J. (2010) Of coiled oysters and big brains: how to rescue the terminology of heterochrony, now gone astray. *Evolution & Development* 2(5), 241–248. https://doi.org/10.1046/j.1525-142x.2000.00067.x

[12.23] Gould, S.J. and Vrba, E.S. (1982) Exaptation – a missing term in the science of form. *Paleobiology* 8(1), 4–15.

[12.24] Hameroff, S. and Penrose, R. (2014) Consciousness in the universe: a review of the 'Orch OR' theory. *Physics of Life Reviews* 11(1), 39-78.

[12.25] Heidegger, M. (1955-1956) *The Principle of Reason*. Republished 1991. Indiana University Press, Bloomington.

[12.26] Heylighen, F. (2014) Cybernetic principles of aging and rejuvenation: the buffering-challenging strategy for life extension. *Current Aging Science* 7(1), 60-75.

[12.27] Heylighen, F. (2021) Entanglement, symmetry breaking and collapse: correspondences between quantum and self-organizing dynamics. *Foundations of Science* 1-29.

[12.28] Hoffmeyer, J. (2000) Code-duality and the epistemic cut. *Annals of New York Academy of Sciences* 901, 175-186.

[12.29] Igamberdiev, A.U. (1993) Quantum mechanical properties of biosystems: a framework for complexity, structural stability, and transformations. *Biosystems* 31(1), 65-73.

[12.30] Igamberdiev, A.U. (1998) Time, reflectivity and information processing in living systems: a sketch for the unified information paradigm in biology. *Biosystems* 46(1-2), 95–101.

[12.31] Igamberdiev, A.U. (1999) Foundations of metabolic organization: Coherence as a basis of computational properties in metabolic networks. *Biosystems* 50(1), 1-16.

[12.32] Igamberdiev, A.U. (1999) Semiosis and reflectivity in life and consciousness. *Semiotica* 123(3-4), 231-246.

[12.33] Igamberdiev, A.U. (2007) Physical limits of computation and emergence of life. *Biosystems* 90, 340–349.

[12.34] Igamberdiev, A.U. (2014) Time rescaling and pattern formation in biological evolution. *Biosystems* 123, 19–26.

[12.35] Igamberdiev, A.U. (2017) Evolutionary transition from biological to social systems via generation of reflexive models of externality. *Progress in Biophysics and Molecular Biology* 131, 336-347.

[12.36] Igamberdiev, A.U. (2018) Time and life in the relational universe: Prolegomena to an integral paradigm of natural philosophy. *Philosophies* 3, 30.

[12.37] Igamberdiev, A.U. (2021) The drawbridge of nature: evolutionary complexification as a generation and novel interpretation of coding systems. *Biosystems* 207, #104454.

[12.38] Igamberdiev, A.U. and Brenner, J.E. (2021) Mathematics in biological reality: The emergence of natural computation in living systems. *Biosystems* 204, #104395.

[12.39] Igamberdiev, A.U. and Kleczkowski, L.A. (2009) Metabolic systems maintain stable non-equilibrium via thermodynamic buffering. *Bioessays* 31(10), 1091–1099.

[12.40] Igamberdiev, A.U. and Kleczkowski, L.A. (2019) Thermodynamic buffering, stable non-equilibrium and establishment of the computable structure of plant metabolism. *Progress in Biophysics and Molecular Biology* 146, 23–36.

[12.41] Igamberdiev, A.U. and Shklovskiy-Kordi, N.E. (2017) The quantum basis of spatiotemporality in perception and consciousness. *Progress in Biophysics and Molecular Biology* 130, 15-25.

[12.42] Iurato, G. and Khrennikov, A.Y. (2020) Entropy, externality and human evolution. Biosystems 191-192, #104130.

[12.43] Kastner, R.E. (2012) The possibilist transactional interpretation and relativity. *Foundations of Physics* 42(8), 1094–1113.

[12.44] Kauffman, S. (2020) Answering Schrödinger's "What Is Life?" *Entropy (Basel)* 22(8), #815.

[12.45] Khrennikov, A. (2022) Order stability via Fröhlich condensation in bio, eco, and social systems: The quantum-like approach. *Biosystems* 212, #104593.

[12.46] Lacan, J. (1970) Ecrits. Republished in 2006. W.W. Norton and Co, New York.

[12.47] Lefebvre, V.A. (1997) *The Cosmic Subject*. Russian Academy of Sciences Institute of Psychology Press, Moscow.

[12.48] Lefebvre, V.A. (2017) Theoretical modeling of the subject: western and eastern types of human reflexion. *Progress in Biophysics and Molecular Biology* 131, 325-335.

[12.49] Leibniz, G.W. (1714) *Monadology and Other Philosophical Essays*. Republished in 1965. Bobbs-Merrill, Indianapolis, USA.

[12.50] Liberman, E.A. (1979) Analog-digital molecular cell computer. *Biosystems* 11(2-3), 111–124.
[12.51] Lotka, A.J. (1922) Contribution to the energetics of evolution. *Proceedings of the National Academy of Sciences of the United States of America* 8(6), 147–151.
[12.52] Lotman, Yu.M. (1990) *Universe of the Mind: A Semiotic Theory of Culture*. Indiana University Press, Bloomington.
[12.53] Matsuno, K. (2014) Self-identities and durability of biosystems via their abstracting capacity. *Biosystems* 120, 31-34.
[12.54] Matsuno, K. and Paton, R.C. (2000) Is there a biology of quantum information? *Biosystems* 55(1-3), 39-46.
[12.55] Mikhailovsky, G. and Gordon, R. (2020) Shuffling type of biological evolution based on horizontal gene transfer and the biosphere gene pool hypothesis. *Biosystems* 193-194, #104131.
[12.56] Odum, H.T. (1995) Self-organization and maximum empower. In: *Maximum Power: The Ideas and Applications of H.T. Odum*. C.A.S. Hall (ed.). Colorado University Press, Colorado.
[12.57] Odum, H.T. and Pinkerton, R.C. (1955) Time's speed regulator: The optimum efficiency for maximum output in physical and biological systems. *American Scientist* 43, 331–343.
[12.58] Pattee, H.H. (1967) Quantum mechanics, heredity and the origin of life. *Journal of Theoretical Biology* 17(3), 410-420.
[12.59] Pattee, H.H. (2001) The physics of symbols: bridging the epistemic cut. *Biosystems* 60(1-3), 5-21.
[12.60] Rosen, R. (1991) *Life itself: A comprehensive inquiry into the nature, origin, and fabrication of life*. Columbia University Press, NY, USA.
[12.61] Rosen, R. (1996) Biology and the measurement problem. *Computers & Chemistry* 20(1), 95–100.
[12.62] Sharov, A.A. (2016) Evolution of natural agents: preservation, advance, and emergence of functional information. *Biosemiotics* 9(1), 103-129.
[12.63] Sharov, A.A. and Tønnessen, M. (2022) *Semiotic Agency. Science beyond Mechanism*. Springer, Cham.
[12.64] Sharov, A.A. and Vehkavaara, T. (2015) Protosemiosis: agency with reduced representation capacity. *Biosemiotics* 8(1), 103-123.
[12.65] Shnoll, S.E. (1979) *Physico-Chemical Factors of Biological Evolution*. Nauka, Moscow.
[12.66] Turchin, V. (1977) *The Phenomenon of Science. A Cybernetic Approach to Human Evolution*. Columbia University Press, NY, USA.
[12.67] Uexküll, J. von (1909). *Umwelt und Innenwelt der Tiere*. Springer, Berlin.
[12.68] Umerez, J. (2001) Howard Pattee's theoretical biology – a radical epistemological stance to approach life, evolution and complexity. *Biosystems* 60(1-3), 159-177.
[12.69] Vanchurin, V., Wolf, Y.I., Koonin, E.V. and Katsnelson, M.I. (2022). Thermodynamics of evolution and the origin of life. *Proceedings of the National Academy of Sciences* 119, #e2120042119.
[12.70] Varela, F.G., Maturana, H.R. and Uribe, R. (1974) Autopoiesis: the organization of living systems, its characterization and a model. *Current Modern Biology* 5, 187–196.

Part III

MEANINGS IN ORGANISM BEHAVIOR AND COGNITION

13
The Sentient Cell

Arthur S. Reber[1*], František Baluška[2] and William B. Miller, Jr.[3]

[1]*Department of Psychology, University of British Columbia, Vancouver, BC, Canada*
[2]*Institute of Cellular and Molecular Botany, University of Bonn, Bonn, Germany*
[3]*Bioverse Foundation, Paradise Valley, Arizona, USA*

Online resources:
Arthur S. Reber:
https://en.wikipedia.org/wiki/Arthur_S._Reber
https://psych.ubc.ca/profile/arthur-reber/
https://www.researchgate.net/profile/Arthur-Reber
František Baluška:
unb15e@uni-bonn.de
https://www.thethirdwayofevolution.com/people/view/frantisek-baluska
https://www.plantbehavior.org/contact/
William B. Miller, Jr.:
https://humansandnature.org/william-b-miller-jr/
http://www.themicrocosmwithin.com/evolutionary-biologist-william-b-miller-jr

Abstract
The premise of this volume is that there is a continuum of semiotic agents across biology, extending from the first individually competent cells to their colonial aggregations and then further becoming functioning tissue ecologies, multicellular eukaryotic organisms, and holobionic species. This same agency propels collective human thought and our organizations, including nations or political movements. Our contribution explains this causal chain by focusing on the conscious cell. In recent publications, we have argued and presented empirical evidence for sentience/cognition in all organisms, from unicellular prokaryotes to *Homo sapiens*. Put simply, life and sentience are coterminous. All living organisms, all biological entities, feel, learn, remember, choose, react, and communicate. All are biosemiotic agents. Therefore, all biological and evolutionary development is directly related to intelligent cells and their suite of sentient, problem-solving behaviors. This perspective has been codified as the Cellular Basis of Consciousness (CBC), and its particulars are presented in this chapter. Several critical ramifications derive from within the parameters of cellular sentience. Notably, cellular consciousness exists as an organized whole such that even cells have meaningful experiences

**Corresponding author*: areber@pointroberts.net

at their scale that include semantic elements. Cellular consciousness requires a competent interactive external membrane, separating the internal milieu of the cell from the external environment. That perceptive membrane must link to retrievable and deployable memory to enable cellular learning and effective problem-solving. This complex receptive apparatus supporting cellular sentience is termed the senome, representing the pivotal architecture connecting the cellular interior to the external environment. In CBC, all aspects of the cell coordinate to support sentience, and all of its internal architecture, including genes, are its tools. Because any information a cell has about its external environment must travel across media to interact with the cell and transit an intervening membrane, all environmental information has been degraded compared to its source. This inescapable informational ambiguity means that cells must measure that information to gauge its validity. Consequently, all the information that any cell has is self-generated in its interior. Consequently, cells create their own reality. So do all other living agents. CBC asserts that all living creatures are conscious, including plants. On the other hand, non-biological agents cannot be conscious since they are not living. Creating consciousness means creating life, which, to date, we are unable to do.

Keywords: Cellular sentience, biological information, biosemiotics, plant cognition, anesthetics, senome

13.1 Introduction

In this chapter we:

a. Provide an abridged overview of the operative theory, the Cellular Basis of Consciousness (CBC),[1] outline its implications, and show how it helps resolve a number of sticky philosophical problems that have bedeviled the field of consciousness studies.
b. Explore various biomolecular candidates for the underlying mechanism(s) responsible for cellular sentience, arguing that these are best viewed as whole-cell phenomena — any part of the cell could be its epicenter.
c. Critique the notion that individual cell sentience is gene-centric.
d. Examine the concept of biological information and how it differs from the type of data derived from the inanimate realm — the concept of how an observer — participant achieves meaning is pertinent.
e. Argue against the possibility of non-biological entities becoming conscious agents, in which the "bio-" part of "biosemiotics" is an essential feature.
f. Examine the role of anesthetics, outline their biomolecular functions, and their role in mitigating negatively valued, subjective experiences.
g. Outline the basic principles of evolutionary biology that support the proposition that the CBC applies to plants — along with relevant empirical evidence.
h. Sum it all up.

---

[1] Apologies and thanks to Lynn Margulis, whose seminal 2001 paper titled "The Conscious Cell" was one of the inspirations for the development of the CBC theory of the origins of minds.

13.2 The Issues

13.2.1 The Cellular Basis of Consciousness (CBC) Model

The CBC approach is based on a simple presumption: life and sentience are coterminous. All living forms have a large number of functions and processes that are associated with having internal, subjectively valued experiences. They include sensing and perceiving, learning, forming memories, ascertaining the value of events, including anticipation, making decisions, and communicating. There is ample evidence for each function in unicellular species.

(i) Sensory and perceptual mechanisms: Cell biology has long been based on the presumption that unicellular organisms make appropriate adjustments in how and where they move based on sensing the positive and/or negative features of specific elements encountered [13.1] [13.18] [13.26] [13.30]. Prokaryote cells, bacteria and archaea, are richly supplied with receptors for a host of stimuli, including temperature, light, osmolarity, pH levels, touch, diurnal cycles, magnetism, and, through quorum sensing, each other [13.57] [13.69]. Even the simplest species like *E. coli* that lack membrane-bound organelles, have upwards of 10k chemoreceptors and each has multiple binding sites. The cell's experiences are manifestly perceived as meaningful in that systematic locomotions toward and away from stimuli are matched with the positive or negative nature of the stimuli and their impact on the well-being of the cell. In short, these events are *perceived* in the standard sense of that term, signaling that the cell has fully expressed internal, subjective experiences of them. In the terminology of biosemiotics, the signs have semantic elements, meaning, and significance for the organism.

(ii) Learning: That unicellular species learn has been a staple of cell biology for some time [13.26] [13.67]. Individual cells learn and, in fact, learn rather complex tasks, including navigating complex pathways. Boisseau *et al.* [13.14] reported that the slime mold *Physarum polycephalum learns* to navigate a bridge between two food sources, avoiding locations where caffeine and quinine had been placed. And several laboratories have reported simple anticipatory learning pattern in bacteria based on sequences of alternating stimuli. Other researchers [13.45] alternated the nutrients maltose and lactose and found that *E. coli* learn to "foresee" (their term) the next nutrient by making anticipatory shifts in metabolic function.

(iii) Memory: Bacterial behaviors are based on memory, specifically memory of events of the recent past. Without such an ability to detect change, there would be no direction, no focus to cellular life. If a cell did not recall that the low temperature of the environment it just moved out of was uncomfortable it would not show systematic locomotion toward a more acceptable environment. Uncomfortable in this context relates to relative states of preference that microbes clearly exhibit. For example, nearly all infectious disease medical diagnosis relies on known microbial preferences to predict a likely pathogen for a presenting disease pattern. Furthermore, the choices behind bacterial surface colonization are complex and non-random, relating to intercellular interactions, relative population densities, surface textures, and memory [13.26]. As Lyon [13.35] put it, arguing for cognitive functioning in bacteria, "Without memory, present circumstances have no context; the detection of change is impossible." It is memory of the past that allows for actions oriented on what is happening now and in the near future. Indeed, all the examples of *learning* discussed above are based on the establishment of memories of events. Interestingly, the first reports of memory in the unicellular realm were from Macnab and Koshland in the early 1970s [13.36] but the effects were presented as "memory-like" in places and the term *memory* itself was in scare quotes (e.g., "memory"). There was a reluctance to view the data as being evidence of sentience in unicellular species.

(iv) Decision-making: Bacteria make decisions constantly. All these decisions involve a complex interaction among genes whose outputs seem stochastic at the individual cell level but are deterministic at the level of populations where individual cells differentiate in a variety of phenotypes, each with its own particular attributes [13.13]. When in collectives such as biofilms or microbial mats, the entire colony will make decisions based upon stressors that are present and the information about their circumstances is communicated to others. Cells that are experiencing nutrient deficits secrete molecules expressing their condition. Other cells in the colony respond by reducing cell division and nutrient uptake. When balance is restored, a different electrical signal is emitted and the colony's functions return to normal [13.12]. These kinds of information exchanges, accompanied by context-relevant decisions, also take place between colonies, a process dubbed "time sharing," as each colony takes its turn at utilizing the available nutrients [13.49]. As noted elsewhere [13.50], these kinds of adaptive and cooperative behaviors fulfill the standard definition of *altruism* in that some cells put themselves at risk in response to information about the stressors other cells are experiencing. In a review of these and other similar behaviors in bacteria, biophysicist Jané Kondev referred to them as a "kind of free will" [13.29].

(v) Summary of the CBC: The model is developed in more detail in several places [13.6], [13.64], [13.19] [13.20] [13.21] [13.50] [13.54] but the basic principle is straightforward. All life is based on cells, sentient cells. All existing and extinct species evolved from these original cells and all forms of sentience, cognition, consciousness, with its referential, biosemiotic elements, did as well. The former proposition is a fundamental truth of evolutionary biology and accepted by virtually all scientists and knowledgeable laypersons. The latter is still regarded by many as a radical idea, a bridge too far. The more commonly accepted "standard" model is that consciousness first appeared with the evolution of nervous systems [13.19] [13.20] and likely in the early Cambrian [13.23] [13.25]. It is true that cognitive functions changed when neurons emerged and nervous systems evolved but this is very different from claiming that they did not exist before. Many

> biological events had similarly dramatic impact on life, photosynthesis, an immune system, warm-blood metabolism, homoeothermism, flight — but they did not cause consciousness or minds to emerge. Mental life has been an essential feature of life since the origins of living, reproducing species. Each of these (and other) major evolutionary events simply modified the manner in which sentience was expressed.

The standard model is also saddled with a problem that is rarely acknowledged: the *emergentist's dilemma*. Assuming that consciousness, sentience, didn't appear until later in evolution brings an obligation to identify the underlying biomolecular mechanism(s) responsible, to explicate why a planet filled with nonsentient species was suddenly populated with ones that were conscious. Elsewhere [13.19] [13.20] [13.54] we offered several candidate mechanisms, all of which are linked with cell membrane biomechanics. In the next section we review these.

13.2.2 Biomolecular Candidates for Sentience

During the first two billion years after Earth settled into being a livable planet, all life was unicellular. The most ancient species were primitive, small prokaryotic organisms which evolved into the presently living archaea and bacteria. As noted above, in order to deal with extremely unstable and challenging environmental conditions, the very first cells needed to be able to respond properly and adapt effectively, which required being able to learn, establish stable memories of experiences and make decisions about appropriate actions. The key was the evolution of limiting membranes which could pick up information about environmental parameters using sensory proteins and effectively control the intracellular space using prototypes of the current ion channels and transporters. In addition, the very first cells needed both sensors capable of measuring extracellular space and ones responsible for the intracellular homeostasis of diverse biological parameters, including pH levels, patterns of signaling and feedback pathways, and nucleic acid-based protein synthesis.

Excitable plasma membranes not only controlled fluxes of energy, matter, and metabolites but also governed tight co-ordination of internal structures and processes with the external circumstances. The division of a cell into this kind of inside-outside dichotomy played the central role in the establishment of cellular sentience by defining a subjective inside and distinguishing that state from the insentient exterior environment [13.5] [13.10]. Only cells with these traits survived and embarked on biological evolution.

A closer look at intracellular cytoplasmic[2] space reveals that it is extremely complex and densely filled with biological macromolecules situated within a gel-like cytomatrix [13.34] which, based on numerous charged ions and reactive molecules, generates a bio-electromagnetic field-like space [13.5]. In order to appreciate this complexity and its role in establishing cellular sentience, we characterize this sensory-order-based bioelectric field as a *senome* that is proposed to be upstream of both the genome, and the epigenome in handling the sensory information for cognitive-like control of cellular sentience, behavior,

[2] The cytoplasm is the gelatinous material made up of water, ions, and reactive molecules that fills the inside of a cell. There are also other encapsulated organelles, such as mitochondria and the nucleus (in eukaryotic cells), that are in the cytoplasm, but they are enclosed in their own membranes.

and communication [13.4]. Consequently, the senome represents a pivotal gateway between the cellular interior and its external environment, integrating the sensory reception of information with the cellular epigenome and genome to maintain crucial homeorhetic balance [13.4]. The plasma membrane is a vital part of an entire senomic apparatus, linking all receptive aspects of the cell with its interior tools to enable the accurate internal measurement of information. Accordingly, there are three biomolecular sources of senomic fields: (a) the limiting plasma membrane and the extra- and intra-cellular vesicles derived from it, (b) the intracellular macromolecules and, (c) the cytoskeletal polymers [13.20] [13.21] [13.54].

It will help to understand our approach here if we take a short "aside" on the notion of a *senome*. Following Darwin's insights, the role of natural selection in evolution and the importance of genes became accepted dogma among biologists. Thomas Huxley, famously dubbed "Darwin's bulldog," argued that focusing on the genome would prove to be problematic. He recommended that the cell periphery-based systems (representing the *epigenome* and the *senome*) be included as fundamental to understanding evolution. At the time, the epigenome was not considered to be a factor, largely because evolutionists had rejected Lamarckian mechanisms. However, recent research has shown that it plays an important role in evolution and epigenetics is now considered part of the standard model. The senome, however, hasn't played a role in the thinking of most evolutionary biologists. We argue that this is a mistake and have, in several papers [13.20] [13.43], made an effort to resurrect the concept, show how important its role is and, of course, to reveal once again how prescient Huxley was.

In eukaryotic cells,[3] organelles like mitochondria, plastids, and the nucleus are of endosymbiotic origin and, as noted, have their own membranes [13.2]. From the senomic point of view, these organelles generate their own senomic fields that integrate with the senome of the host cell. We have proposed an emphatic coupling of cellular senomes [13.20] that generates supracellular N-space episenomic fields [13.43] [13.44]. These complex interactions could well explain unitary and conscious awareness of environment and sense of self. They are also the functions that are suppressed when we are exposed to anesthetics — importantly for our overall CBC model, all cells and all symbiotic organelles are sensitive to anesthetics. From an evolutionary point of view, it strongly suggests that all cells have subjective, discriminating experiences. Why else would a species be sensitive to anesthetics? We will discuss this important topic below.

One way to think of these complex biomolecular functions of individual cells is to take note of the title of this volume: these are the foundations of meaning. This is where biosemiotics begins — with the establishment of internal, valenced feelings and a conscious awareness of both the internal functions of cellular self and a sensitivity to the events and objects in the external world. In short, sentience.

[3] The emergence of the first eukaryote was almost certainly an endosymbiotic event in which a prokaryote was incorporated by a larger prokaryote. The cells in all animals and plants are eukaryotic. While the standard approach to the first eukaryotes is to treat them as unicellular, they are, as we note below, multicellular organisms.

13.2.3 Biological Information

The exchange of "information" among cells is not, in our view, sufficient to presume that there are semiotic elements. Information itself does not carry knowing. Biological information, on the other hand, does and is an essential feature of all organisms. Here we take a look at how it differs from the type of data derived from the inanimate realm; the concept of how an observer/participant achieves meaning is pertinent.

One easily overlooked aspect of biological information is its categorical distinction from the conceptual framework of information used by engineers or physicists. Typically, and even among biologists, information is considered according to the familiar nomenclature of *Shannon Information* which underpins communication through bits and bytes, or algorithmic information as a cornerstone of computational programming. Each is an attempt to optimize efficiency, reproducibility, and precision in information transfer [13.33]. However, in living systems, information has a very different profile. Biological information, as opposed to computer data, is inherently ambiguous [13.40] [13.41] [13.60].

The sources of imprecision that constitutively afflict the quality of biological information have been previously reviewed in depth [13.41] [13.42] [13.43]. The uncertainty stems from multiple sources. For example, among cells, information is typically a broadcast and often, neither the sender nor receiver of the information is readily apparent. Furthermore, any information a cell receives from the environment must cross interstitial spaces between cells where biomolecular degradations can occur or bioenergetic fields might be distorted. Obviously, any communication from a source that involves travel among cells is associated with unavoidable time delays.

Secondly, all cells are self-referential. Consequently, each cell represents a distinct observer – participant when evaluating any environmental cue. Therefore, any observer/participant might achieve significantly different interpretant – object meaning from any environmental stimulus. Accordingly, there are embedded, unavoidable variances among self-referential observers of the same object. For example, an object might be an environmental cue to two observers, but as they cannot be in identical places, each will observe a different facet of that object, which might provide a differential meaning to one or the other in a process that has been described as *antipodal* information [13.62].

The most important reason for the ambiguity of information is that cells have an external plasma membrane. All the information that a cell can have about the external environment must travel across it. As already emphasized, the plasma membrane is a vital component of the cellular senomic architecture that links across every cell [13.4]. Informational reverberations from the plasma membrane convey across the cellular cytoplasm and its bioactive constituents, encompassing cellular organelles, the cytoskeleton, and its nuclear compartment. All of these are part of the critical cellular measuring capacity within its cell-wide information management system [13.42] [13.43].

There is an unobvious entailment from this absolute requirement. All the information that any cell has is self-generated; it is *infoautopoietic* [13.43]. There is no absolute outer reality for any cell. Anything it can glean about its external milieu is a product of its individual internal self-interpretation and self-representation. Since the cell-wide interpretation of the meaning of that self-generated cellular interpretation/representation is the product of a complex set of physical connections, some degree of noise is inevitable. Nonetheless, the purpose of self-produced information is quite clear. That information is directed to sustaining the self-identity of the cell, which equates with its state of homeorhetic equipoise, the tendency of a system to return to a trajectory or path that it was on when disturbed [13.39] [13.41].

This conditional reality carries another crucial ramification. Homeorhetic equipoise cellular state of flux. Cells are negentropic zones compared to their external environment [13.61]. In biological terms, the interior of the cell is relatively "negentropic" with respect to its external environment, corresponding to its more ordered interior status. However, in a self-referential framework, any self-identified state of homeorhetic equipoise equates with a cellular state of preference [13.41] [13.43]. Consequently, cellular sentience, consciousness is *knowing* that the information the cell has at its disposal is imperfect but must still be used to sustain homeorhetic preferential equipoise. It follows that self-referential cellular consciousness is directly related to the cellular measurement of ambiguous information. This specific feature of cellular life has been experimentally confirmed. Individual cells will demonstrate highly variable responses if exposed to identical concentrations of epidermal growth factor since intracellular signaling mechanisms respond differently to the same concentration of growth factor from cell to cell [13.31]. Unquestionably, cells are not automatons as many assume. When dealing with unicellular organisms, the standard approach is to treat them as insentient and governed by functions programmed into their genes. Philosophers [13.22] and biological scientists [13.25] [13.23] specifically defend this position which, from our analysis here, is clearly wrong.

From the foregoing, it is clear that any environmental cue that impacts the senome and is appraised as something other than environmental noise is a felt experience. The logic is straightforward. All cellular information is ambiguous by definition. Due to those embedded uncertainties, it must be internally measured through a senomic-centered, reverberating cell-wide process [13.10]. These connections necessarily comprise cellular sentience, i.e., self-awareness of its environmental status. All of these actions that support cellular homeorhetic equipoise as the product of self-referential measurement are, in effect, predictions to sustain crucial self-identity [13.42]. To achieve the goal of sustaining a cellular state of homeorhetic preference, cells must measure information for both quantitative and qualitative features in order to make those useful predictions. Thus, all cellular measurements are cell-based predictions to determine whether its scant energy resources should be expended in cell-cell communication or bioactive expression [13.42]. Unquestionably though, predictions are themselves forms of experiences since they are inherently subjective. All cellular information is an experience as its full sentience, at its scale. We experience it in our distinctive aggregated cellular form at our human scale. Simply put, a theory of conscious mind is a cell theory.

13.2.4 Cell Sentience is not Gene-Centric

When it is appropriately understood that all biology centers on the intelligent measuring cell, it is fruitful to explore the role of genes, as they have represented the *idée fixe* of evolutionary biology as articulated in the Modern Synthesis. However, not all agree that genes should be considered the primary drivers of biological and evolutionary development. Others insist that issues of information flow should take precedence over genes in any modern evolutionary narrative [13.37] [13.38].

In considering that alternative viewpoint, the ambiguity of information is directly relevant. Since all biological information is uncertain, cells cope through collective measurement, which is why multicellularity as biofilms or holobionts[4] are the dominant forms of planetary life. The collective measurement of information is simply the proverbial "wisdom of crowds" expressed in cellular form. Collective measurement improves information validity as effective and actionable [13.41]. The collective appraisal of environmental information carries greater measuring value which translates into enhanced prediction through suppression of surprise and correlates with minimizing the variations of free energy [13.31]. The free energy concept relates to the critical issue of the flow of energy in self-organizing systems. Ordered states require the conservation of energy that can be devoted to work. In biological terms, that energy resourcing shares equivalences with enhanced cellular prediction and a minimization of surprises that would require unanticipated energy outputs. These latter concepts are one way in which consciousness has been approached through thermodynamics. Simply put, better predictions mean fewer wasted resources.

Accordingly, collective measurement of information is the glue that binds collaborative, multicellular life. Notably, this collaborative tendency represents a hallmark feature of the origin of life and, thus, by inference, is an embedded attribute of sentient self-reference. Single cells do not leave traces. The stromatolites and evidence of microbial mats that are used to estimate the origin of life as at least 3.7 billion years ago are evidence of collective cellular action [13.46]. So, the collective measurement of environmental information, linked to cellular preferences, is the stimulus for multicellular life and the niche constructions that resulted in biofilms for billions of years and eukaryotic holobionts for some 600 million years. Cells seamlessly maintain these complex networks through coordinated problem-solving, which requires conjoint predictions of the meaning of environmental stresses. As noted above, the cell's plasma membrane is the gateway for all information to the cell as the outer edge of its complex senomic apparatus. From this, several conclusions can be drawn. The components of consciousness were instantiated in the cellular form co-incident with the origin of life which is constitutively dependent on cell-cell collaborations, and those collaborations necessitate a competent plasma membrane.

All biofilms and holobionts are comprised of aggregated tissue ecologies that consist of varied types of sub-specialized individual cells, differentiated eukaryotic cells, and a coincident microbiome, the latter made up of multiple species with separate genotypes and phenotypes [13.40]. Nonetheless, they coordinate seamlessly and engage in collaborative natural cellular engineering to produce needed metabolic substrates and obtain and distribute nutritional substances. Yet, each participating cell is required to measure its own information internally.

Consequently, the requirement for a common informational platform is unavoidable since there must be some type of common referencing system to permit concordant measurement. That common measuring platform has been described in detail as an N-space Episenome [13.10] [13.44], a common platform that serves as a compartmentalization of outer N-space, which we view as the cell's entire universe of potential informational cues that might require assessment by any organism. In effect, the N-space Episenome represents an informational boundary architecture that permits any individual cell to self-organize its informational inputs. Accordingly, an N-space Episenome, as an informational

[4] A holobiont is a collection of organisms living together and forming a distinct ecological unit. The individual parts are *bionts* and the totality of the genome is known as the *hologenome*. All animals, including humans, and plants are holobionts.

architecture, is wholly integrated with physical cells, particularly its obligate plasma membrane as its bioactive interface with the external environment [13.44].

Accordingly, a conscious cell requires a circumscribed attachment to information space-time (the summation of all possible environmental cues as its N-space Episenome) reinforced physically by its sensitive, intelligent plasma membrane. Notably, and often overlooked, the plasma membrane is a crucial aspect of the retrievable and deployable memory of the cell upon which its survival and conscious self-reference depends. Indeed, all aspects of the cell are included in this memory storage system as the crucial element of the information management system of all cells. From these requisites, the place of genes is clarified. Genes are part of the retrievable and deployable memory system of cells.

Experiments have demonstrated that the genome does not template morphogenesis as previously presumed [13.44] [13.63]. Furthermore, the earlier view of the genome as a sacrosanct repository of heritable memory has been superseded. Instead, the genome is now assessed as a flexible and responsive participant in cellular problem-solving, actively and continuously modulating cellular activities and its responses to stress as a read-write informational architecture [13.56].

One of the reasons why the genome was considered the primary focus of biology was Crick's renowned Central Dogma where the flow of information was from DNA to RNA to proteins. It could never be the reverse. Contemporary biology now contradicts that viewpoint [13.16]. Among the primary reasons for this is the power of epigenetics to modulate our genomes. Furthermore, the actual genetic complement of cells is much more robust and fluid than had been previously understood. Our working genetic complement includes a vast array of micro-RNAs, circular intracytoplasmic DNAs, transposable elements, and various other small genetic elements that constitute an active genetic mobilome [13.66]. Furthermore, all multicellular eukaryotes are holobionts with a huge panoply of obligate microbial life that contributes to our development, metabolism, and physiology and whose genes number in the millions [13.40]. Consequently, contemporary research confirms that genes are tools of cells and not the reverse as long presumed [13.42].

It is apparent that all the components of consciousness were instantiated in cellular form some 3.7 billion or more years ago and have continued unabated to the present. Thus, it is not coincidental that the cellular plasma membrane has been continuously reproduced in an uninterrupted stream from the first competent cell forward to today. The plasma membrane is the vital intercessory between the external environment of the cell and its inner self-referential state that is entirely dependent on the measuring assessment of information that links to prediction and grants survival.

In order to make sense of these complexities, some scientists have argued that a governing agency is necessary. Walker *et al.* [13.65] suggested the requirement of a *control kernel* connecting the information to the causal mechanisms of the network. Within the CBC model, that control kernel is readily identified. It is cellular consciousness since self-reference is self-organization [13.42]. All biology and evolutionary development emanate from that sentient ground state, and our crucial genome is among its essential tools.

13.2.5 Non-Biological Entities Cannot Become Conscious Agents

The stepping off place here is simple, we maintain that the "bio-" part of "biosemiotics" is an essential feature, that non-biological entities cannot become conscious. There is an approach in philosophy, in particular the philosophy of mind, that assumes that mental life is, at its core, computational. If, as Hilary Putnam mused back in the 1960s, an artificial entity, a computer, were to carry out all the operations that a human brain did it would itself have consciousness, cognitive states, a mental life. Putnam referred to this proposition as "machine-state functionalism" or "hardware independent functionalism" though, as a number of variations on the key argument were put forward, they all became clustered under the umbrella term "functionalism" — and that's how we'll treat them here. The core argument is that it is what a system does that counts, and these operations have causal roles to play in its mental life no matter what the nature or form of the entity is. Functionalism was widely embraced and versions have been put forward by philosophers such as Jerry Fodor and Daniel Dennett and cognitive scientists such as Douglas Hofstadter and Giulio Tononi. Interestingly, while functionalism became accepted and promoted by many, Putnam himself later changed his mind and argued against his original proposal.

It has also been the target of vigorous criticism, most effectively, in our view, by philosopher John Searle who pointed out that functionalism only deals with syntax and not semantics. His famous (some would say *in*famous) Chinese Room thought experiment outlined why this is the case [13.55]. Searle described a situation where a mono-lingual English speaker is in a sealed room with an "In" slot and an "Out" slot for exchanging information. Outside the room are several monolingual Chinese speakers who ask questions of the person in the room which are, of course, written in Chinese. The person in the room has a series of code books that explain how to respond to each of these questions and how to put the answers in Chinese — which are dutifully sent through the "Out" slot. The Chinese speakers, upon seeing the accuracy with which their queries are replied to, assume that the person inside the room is fluent in Chinese. But, as Searle noted, there's no understanding in the room, no knowledge of Chinese. All the appropriate calculations are being carried out but there's no mental state that captures the semantics of Chinese. Lacking semantics, semiotics, there can be no meaning, no intentionality, no *aboutness*. In short, none of the classic features of cognition are present. Searle's argument has been the subject of intense debate over the forty-plus years since it was put forward, much of it critical and by defenders of functionalism. From our perspective, the core argument, that one cannot get semantics from syntax, is correct whether or not the logic of the Chinese Room holds up.

Antonio Damasio also criticizes functionalism but takes a different tack [13.19]. He argues that an artificial entity that relied on computational algorithms wouldn't have feelings or emotions and, without this affective element, wouldn't have anything approximating the internal states of entities based on biomolecular processes. Consciousness, in his view, is an embodied experience, not one tied to the computational operations of brains. Tying subjectivity to cerebral functions would be analogous to simply having a recipe for cooking a meal. It would describe the food accurately but have nothing associated with the experience of eating it. There would be no feelings, no experience and, of course, no consciousness or sentience associated with it. There would be syntax (the recipe) but no semantics (consuming the meal, tasting the food).[5]

---

[5] Damasio, we note, is also a critic of the CBC [13.20] claiming that the many behaviors of prokaryotes discussed above are all merely "intelligent but unminded competences." Elsewhere [13.51] Damasio's position is reviewed and criticized in more detail than we provided above.

From our perspective, the difficulties with functionalism are derived from a strategic error. The efforts almost always begin with human consciousness and focus on efforts to identify the essential features of the most complex and evolved forms of mentation. We see this approach as misguided. The one taken in the CBC theoretical framework is the classic: "reverse engineer" the issues — begin at the beginning. Identify the essential biomolecular processes that operate in the simplest species, the unicellular prokaryotes, then unpack the many complex processes, characteristics, and functions that evolved on this biomolecular platform. Elsewhere [13.9] [13.10] we present an analysis of the first critical step here—how the first eukaryotes came to have a unified subjective experience. It's not always appreciated, but a unicellular eukaryote is essentially a multicellular organism (having cells within a cell). The critical biomolecular problem the first such species solved was to have a unified sense of self while each of the cells that make it up have their own sentience. Keep in mind that this was not a simple process. It took several billion years to accomplish it.

13.2.6 The Role of Anesthetics

All cells are sensitive to anesthetics [13.28] [13.59]. This simple fact is among the most compelling bits of evidence to support the core assumption of the CBC. It also raises the obvious question asked above: "What is the point of reacting to anesthetics if there are no feelings, representations, no valenced experiences?" The biomechanisms required to react to the administration of anesthetics and to recover appropriately when they are removed use cellular resources. A basic principle in evolutionary biology is that is it generally adaptive to reserve resources for essential and important functions. In this section we will provide a largely non-technical overview of the molecular functions that modulate the manner in which anesthetics function within cells and why they operate to basically neutralize the normal states of consciousness and block sentient self-experience.

Recently, the plasma membrane was reported to represent a primary target to anesthetics both in plant and animal/human cells. As several studies showed, under anesthesia the plasma membrane loses its excitability properties, specificity for ion transport, and becomes incapable of supporting proper endocytosis and action potentials [13.11] [13.28] [13.57] [13.70]. The implication of these findings is that the plasma membrane plays a crucial role in cellular sentience. The fact that ordered lipids are targets of anesthetics is relevant for well-known observations that anesthesia is reversed with an increasing external pressure on cells [13.64], which also increases the ordering of lipids in the plasma membrane [13.68].

Other evidence from anesthetics-based experiments comes from studies on the cytoskeleton of eukaryotic cells. Actin[6] filaments and microtubules provide large intracellular hydrophilic surface areas associated with the so-called ordered (or exclusion zone) water[7] [13.17]. A variety of anesthetics have been shown to affect hydrophilic surfaces, such as the cytoskeleton and the plasma membrane, in a concentration-dependent manner. Kundacina *et al.* [13.32] presented evidence that strongly supports the role of intracellular surfaces and interfaces of biological macromolecules in generating senome-based cellular sentience.

Matrix-like networks of charged ions and reactive molecules permeate the cellular interior and are organized in code-like patterns which are ancient, preceding the genetic code

[6] Actin is a family of proteins found in virtually all eukaryote cells that form microfilaments in the cytoskeleton.

[7] A zone in which particles of a colloidal suspension in water are repelled from hydrophilic surfaces.

in cellular evolution [13.27]. These codes are closely linked with the cellular clocks which are also sensitive to anesthetics [13.8]. Cognitive circadian clocks are inherent to every cell and they co-assemble into a supracellular organismal clock in multicellular organisms in a manner similar to that which governs how the sentience of individual cells co-assembles into unitary conscious experience in humans and other multicellular organisms.

Put in simple terms, virtually every aspect of the biological functions of cells is vulnerable to anesthetics while no non-biological entities are. We view this phenomenon as compelling evidence of sentience in all life forms.

13.2.7 Basic Principles of Evolutionary Biology and the Question of Floral Sentience

In their highly regarded 1995 work, *The Major Transitions in Evolution*, Smith and Szathmáry categorize those signal transitions that energized our evolutionary narrative [13.58]. At the biomolecular level, these include switching from replicating molecules to molecular networks, transforming independent genes to chromosomes, and complex linkages between RNA, DNA, and proteins. Further evolution depended on the crucial endosymbiotic transition from prokaryotes to eukaryotes with indwelling organelles, the drive to sexual reproduction, and multicellularity. Similarly, at the human cultural level, a dominating transition has been from pre-linguistic societies to human language with its rich semiotic competencies. Notably, Szathmáry and Smith emphasize that all of these are grounded within recurrent themes of the cellular division of labor, collective encounters with environmental stresses, and reiterating levels of multicellular ecological complexity. Moreover, as they further emphasize, all these actions are dependent on some type of central control. The CBC comports quite comfortably with this expansive and articulate narrative since we assert that this controlling agency is cognition invested in each individual cell.

Accordingly, their identification of "dynamics" as the most systematic biological factor can now be better addressed as the dynamic flow of self-referential information which distinctly drives multicellularity. Reiteratively, our human language is expressly devoted to that flow, thereby permitting our contingent responses to environmental cues. Of course, it must be so since we are cellular beings and deploy those cellular competencies in our human expressions.

Smith and Szathmáry emphasize collaboration by placing it into "fraternal" terms. We agree and further assert that this relational fundament of biology can now be identified as elemental cellular cognition/sentience established approximately four billion years ago. Consequently, the essential collaborations that characterize all biology and the critical divisions of labor that exemplify multicellular life follow through as an unbroken chain to encompass human cooperation, sociality, and language as expressions of shared human trust and norms. All adhere to structural correspondences that permit the cellular self-referential appraisal of information and its attendant communication that undergirds all planetary life.

The model we're defending here is based on some very simple, noncontroversial principles of evolutionary biology. When a function, form, or trait emerges and is adaptive in the classic Darwinian sense, it is maintained and becomes a part of both the species' genome and its senome. It then becomes the foundation for other forms, traits, and functions to evolve from it. Highly adaptive traits are rarely, if ever, lost. Since life and sentience are inextricably linked, all forms of mental expression, all of the vast varieties of sentience and consciousness across the evolutionary tree can be traced to the *ur*-sentience of prokaryotes. The primate mind, avian cognition, cephalopod mental representations, plant sentience are all varied tokens of an underlying cognitive system that's been an inherent component of all life forms.

Within this larger theoretical framework, the issue of plant sentience arises. If, as noted, the jettisoning of an adaptive function or trait is so rare, it would be remarkable if sentience, which lies at the very core of early life forms, were to have been discarded when the first plants evolved. The issue of cognitive functions in the world of flora is, of course, vigorously debated with passionate defenses going back to Darwin's book *The Power of Movement in Plants* (written with his son Francis) [13.21], to more recent reports from researchers such as Monica Gagliano [13.24], Paco Calvo and Natalie Lawrence [13.15], and one of us [13.3]. Much of this research is carried out under the rubric, *plant neurobiology*. Reber, in his 2019 book *The First Minds* [13.50], took an agnostic stance on plant cognition — since modified on the grounds that jettisoning sentience would have been an evolutionary dead-end and, given how essential it is for survival, wildly unlikely.

Acknowledging cognitive functions in plants carries a number of practical and ethical issues [13.48] and are far from resolved. A full treatment of this topic isn't possible here but we note that many people become vegetarians and vegans specifically over ethical issues associated with consuming animal food products. The issue will be of more than passing interest to them.

13.3 The Last Word

We'll end by identifying the primary reason the CBC framework is not uniformly recognized — despite our efforts here and elsewhere [13.50] [13.6] [13.7] [13.52] [13.10]. It stems, as we noted earlier, from the way in which the issues of sentience and consciousness have historically been approached. The explorations, for obvious reasons, virtually always begin with human consciousness, looking at *Homo sapiens*' minds. We are conscious. We know we are conscious. We assume that other humans, who look like us, act like us, are similarly self-aware. In short, we overcome what philosophers of mind have identified as the "first person problem." We become generous; we endow our fellow humans with the same internal, mental states that we know we have. But when it comes to other species, ones that resemble us (e.g., other primates) and ones that certainly do not (octopuses, oak trees), things get a bit tricky.

The efforts to resolve this issue have produced two branches of research, one that has a neuroscience slant, the other a behavioral one. In the former, the search is to identify species whose brains have the neural centers and pathways that are deemed to be ones that are similar to or homologues of those responsible for cognitive functions in humans. In the latter, the exploration is for species that display cognitive behaviors, ones that appear to lie on a continuum with those of humans.

We have no problems with either of these research strategies. We applaud the many intriguing findings that have come from them, ones that have broadened our view of the rich panoply of the extraordinary manifestations of life on the evolutionary tree. But they're not going to solve the enigma of subjective experiences — and they're not going to identify the point in evolution where consciousness emerged. As we've noted elsewhere [13.50]

[13.53] [13.54], the results of these explorations have produced an intriguing (and amusing) pattern of outcomes. Researchers who operate within this framework invariably conclude that the first true expression of consciousness occurred in the species that have been the focus of their research. Those who focus on cuttlefish and octopuses identify cephalopods as the first conscious species; avian experts typically site corvids (and occasionally parrots); entomologists put the emergent moment at the evolution of the first insects; cell biologists maintain, as does our CBC, that the first unicellular prokaryotes had a primitive form of cognition. In short, you will find consciousness, sentience where you look for it for the simplest of reasons — it's always been there, an inherent element of all life forms.

References

[13.1] Adler, J. (1969) Chemoreceptors in bacteria, *Science* 166, 1588–1597.

[13.2] Baluška, F. and Lyons, S. (2021) Archaeal origins of eukaryotic cell and nucleus. *Biosystems* 203: #104375.

[13.3] Baluška, F. and Mancuso, S. (2018) Plant cognition and behavior: From environmental awareness to synaptic circuits navigating root apices. In: *Memory and Learning in Plants.* F. Baluška, M. Gagliano, & G. Witzany (Eds.), Springer, Cham, SWZ: 51-78.

[13.4] Baluška, F. and Miller, W. B. Jr. (2018) Senomic view of the cell: Senome versus genome. *Communicative & Integrative Biology* 11, 1-9.

[13.5] Baluška, F. Miller, W. B. and Reber, A. S. (2021) Biomolecular basis of cellular consciousness via subcellular nano-Brains. *International Journal of Molecular Sciences* 22, #2545.

[13.6] Baluška, F. and Reber, A. S. (2019) Sentience and consciousness in single cells: How the first minds emerged in unicellular species. *BioEssays* Mar;41(3):e#1800229.

[13.7] Baluška, F. and Reber, A. S. (2020) The biomolecular basis for plant and animal sentience: senomic and ephaptic principles of cellular consciousness. *Journal of Consciousness Studies* 28, 34-59.

[13.8] Baluška, F. and Reber, A. S. (2021) CBC-Clock Theory of life - integration of cellular circadian clocks and cellular sentience is essential for cognitive basis of life. *BioEssays* 23, 43, e2100121.

[13.9] Baluška, F. and Reber, A. S. (in press) Cellular and Organismal Agency – Not Based on Genes: A Comment on Baverstock. *Progress in Biophysics and Molecular Biology* 167, 161-162.

[13.10] Baluška, F., Reber, A. S. and Miller, W. B. Jr. (2022) Cellular sentience as the primary source of biological order and evolution. *Biosystems* 218, #104694.

[13.11] Baluška, F. and Yokawa, K. (2021) Anaesthetics and plants: from sensory systems to cognition-based adaptive behaviour. *Protoplasma* 258, 449-454.

[13.12] Beagle, S. D. and Lockless, S. W. (2015) Microbiology: Electrical signaling goes bacterial. *Nature* 527, 44-45

[13.13] Ben-Jacob, E., Lu, M., Schultz, D. and Onuchic, J. N. (2014) The physics of bacterial decision making. *Frontiers in Cellular and Infection Microbiology* 4:#154.

[13.14] Boisseau, R. P., Vogel, D. & Dussutour, A. (2016) Habituation in non-neural organisms: Evidence from slime moulds. *Proceedings of the Royal Society B: Biological Sciences*, 283, #20160446. http://dx.doi.org/10.1098/rspb.2016.0446.

[13.15] Calvo, P., and Lawrence, N. (2022) *Planta Sapiens: Unmasking Plant Intelligence.* Little Brown, Boston, MA.

[13.16] Chandramouly, G., Zhao, J., McDevitt, S., Rusanov, T., Hoang, T., Borisonnik, N., ... and Pomerantz, R. T. (2021) Polθ reverse transcribes RNA and promotes RNA-templated DNA repair. *Science Advances 7*(24), e#abf1771.

[13.17] Chierici, F., Dogariu, A. and Tuszynski, J. A. (2022. Computational investigation of the ordered water system around microtubules: implications for protein interactions. *Frontiers in Molecular Biosciences* 9, 3e#884043.

[13.18] Clayton, R. K. (1959) Phototaxis of purple bacteria. *Encyclopedia of Plant Physiologie* (W. Ruhland, Ed.), Vol. XVII, 371–387. Springer-Verlag; Berlin/Heidelberg

[13.19] Damasio, A. (2021) *Feeling and Knowing: Making Minds Conscious*. Pantheon, New York.

[13.20] Damasio, A. (2022) Perspectives on the cognitive unconscious. In *The Cognitive Unconscious: The First Half-Century*. A. S. Reber & R. Allen (Eds.), OUP Press, New York: 366-372.

[13.21] Darwin, C. and Darwin, F. (1898) *The power of movement in plants*. Appleton.

[13.22] Dennett, D. C. (2017) *From bacteria to Bach and back: The evolution of minds*. Norton, New York.

[13.23] Feinberg, T. E., and Mallatt, J. (2018) *Consciousness demystified*. MIT Press, Cambridge, MA.

[13.24] Gagliano, M. (2018) *Thus spoke the plant: A remarkable journey of groundbreaking scientific discoveries and personal encounters with plants*. North Atlantic Books, Berkeley, CA.

[13.25] Ginsburg, S. and Jablonka, E. (2019) *The evolution of the sensitive soul: Learning and the origins of consciousness.*: MIT Press, Cambridge, MA.

[13.26] Jennings, H. S. (1906) *Behavior of the Lower Organisms*. Republished by Indiana University Press, Bloomington, IN, 1962.

[13.27] Jones, D. and Sies, H. (2015) The redox code. *Antioxidants and Redox Signaling* 23, 734-746.

[13.28] Kelz, M. B. and Mashour, G. A. (2019) The biology of general anesthesia from Paramecium to primate. *Current Biology* 29, R1199-R1210.

[13.29] Kondev, J. (2014) Bacterial decision-making: Stochastic fluctuations in the expression of genes allow a cell to exhibit a kind of free will. *Physics Today*, 67, 31-37.

[13.30] Koshland, D. E. (1977) Sensory Response in Bacteria. In *Advances in Neurochemistry*. B. W. Agranoff, M. H. Aprison (Eds.), Vol. 2. Springer, Boston, MA. 5

[13.31] Kramer, B. A., Sarabia del Castillo, J., and Pelkmans, L. (2022) Multimodal perception links cellular state to decision-making in single cells. *Science* 377, 642-648.

[13.32] Kundacina, N., Shi, M., and Pollack, G. H. (2016) Effect of local and general anesthetics on interfacial water. *PLoS ONE* 11, e#0152127.

[13.33] Lombardi, O., Fortin, S. & Vanni, L. (2015) A pluralist view about information. *Philosophy of Science* 82, 1248-1259.

[13.34] Luby-Phelps, K. (2013) The physical chemistry of cytoplasm and its influence on cell function: an update. *Molecular Biology of the Cell* 24, 2593-2596.

[13.35] Lyon, P. (2015) The cognitive cell: Bacterial behavior reconsidered. *Frontiers in Microbiology* 6, 264.

[13.36] Macnab, R. M. and Koshland, D. E. (1972) The gradient-sensing mechanism in bacterial chemotaxis. *Proceedings of the National Academy of Science* 69, 2505-2512.

[13.37] Marijuán, P. C., Navarro, J., and del Moral, R. (2015) How the living is in the world: an inquiry into the informational choreographies of life. *Progress in Biophysics and Molecular Biology* 119, 469-480.

[13.38] Marijuán, P. C., and Navarro, J. (2022) The biological information flow: From cell theory to a new evolutionary synthesis. *Biosystems* #104631.

[13.39] Miller, W. B. Jr. (2013) *The microcosm within: Evolution and extinction in the hologenome*. Universal-Publishers, Boca Raton, FL.

[13.40] Miller, W. B. Jr. (2016) Cognition, information fields and hologenomic entanglement: Evolution in light and shadow. *Biology* 5, 21.

[13.41] Miller, W. B. Jr. (2018) Biological information systems: Evolution as cognition-based information management. *Progress in Biophysics and Molecular Biology* 134, 1-26.

[13.42] Miller, W. B. Jr., Torday, J. S., and Baluška, F. (2019) Biological evolution as the defense of self. *Progress in Biophysics and Molecular Biology* 142, 54-74.

[13.43] Miller, W. B. Jr., Torday, J. S. and Baluška, F. (2020a) The N-space Episenome unifies cellular information space-time within cognition-based evolution. *Progress in Biophysics and Molecular Biology* 150, 112-139.

[13.44] Miller, W. B. Jr., Baluška, F. and Torday, J. S. (2020b) Cellular senomic measurements in Cognition-Based Evolution. *Progress in Biophysics and Molecular Biology* 156, 20-33.

[13.45] Mitchell, A., Romano, G. H., Groisman, B., Yona, A. Dekel, E., Kupiec, M., Dahan, O., and Pilpel, Y. (2009) Adaptive prediction of environmental changes by microorganisms. *Nature* 460, 220-224.

[13.46] Papineau, D., She, Z., Dodd, M. S., Iacoviello, F., Slack, J. F., Hauri, E., ... and Little, C. T. (2022) Metabolically diverse primordial microbial communities in Earth's oldest seafloor-hydrothermal jasper. *Science Advances* 8(15), e#abm2296.

[13.47] Pavlovič, A., Jakšová, J., Kučerová, Z., Špundová, M., Rác, M., Roudnický, P. and Mithöfer, A. (2022) Diethyl ether anesthesia induces transient cytosolic [Ca2+]] increase, heat shock proteins, and heat stress tolerance of photosystem II in Arabidopsis. *Frontiers in Plant Science* 13, #995001.

[13.48] Pelizzon, A. and Gagliano, M. (2015)The sentience of plants: Animal rights and rights of nature intersecting? *Australian Animal Protection Law Journal* 11, 5-13.

[13.49] Prindle, A., Liu, J., Asally, M., Ly, S., Garcia-Ojalvo, J., and Süel, G. (2015) Ion channels enable electrical communication in bacterial communities. *Nature* 527, 75-76.

[13.50] Reber, A. S. (2019) *The First Minds: Caterpillars, 'Karyotes, and Consciousness.* Oxford University Press, Oxford, UK.

[13.51] Reber, A. S. (2022) Implicit learning: Background, history, and theory. In *The Cognitive Unconscious: The First Half-Century*. A. S. Reber and R. A7len (Eds.), Oxford University Press, Oxford UK: 3-21.

[13.52] Reber, A. S. and Baluška, F. (2020) Cognition in some surprising places. *Biochemical and Biophysical Research Communications*, 564, 150-157

[13.53] Reber, A. S. and Baluška, F. Where minds began: A friendly amendment to Joseph LeDoux's *The Deep History of Ourselves*. *Philosophical Psychology*, (2023) 36, 745-755.

[13.54] Reber, A. S., Baluška, F. and Miller, W. (2023) *The Sentient Cell: The Cellular Foundations of Consciousness*. Oxford University Press, Oxford, UK.

[13.55] Searle, J. (1980) Minds, brains, and programs. *Behavioral and Brain Sciences* 3, 417-424.

[13.56] Shapiro, J. A. (2017) Biological action in Read–Write genome evolution. *Interface Focus* 7(5), #20160115.

[13.57] Shimuzu, T. S., Le Novère, N., Levin, M.D., Beavil, A.J., Sutton, B.J. and Bray, D. (2000) Molecular model of a lattice of signaling proteins involved in bacterial chemotaxis. *Nature Cell Biology* 2, 792-796.

[13.58] Smith, J. M. and Szathmáry, E., (1995) *The Major Transitions in Evolution*. Oxford University Press, Oxford, UK.

[13.59] Sonner, J. A. (2008) A hypothesis on the origin and evolution of the response to inhaled anesthetics. *Anesthesia & Analgesia* 107, 849-854.

[13.60] Torday, J. S. and Miller, W. B., Jr. (2017) The resolution of ambiguity as the basis for life: A cellular bridge between Western reductionism and Eastern holism. *Progress in Biophysics and Molecular Biology* 131, 288-297.

[13.61] Torday, J. S. and Miller, W. B., Jr., (2020) *Cellular-Molecular mechanisms in epigenetic evolutionary biology*. Springer Nature: Cham, Switzerland.

[13.62] Tozzi, A. and Peters, J. F. (2017) What is it like to be "the same"? *Progress in Biophysics and Molecular Biology* 133, 30–35

[13.63] Tung, A., and Levin, M. (2020) Extra-genomic instructive influences in morphogenesis: A review of external signals that regulate growth and form. *Developmental biology 461*(1), 1-12.

[13.64] Vadakkan, K. I. (2015) A pressure-reversible cellular mechanism of general anesthetics capable of altering a possible mechanism for consciousness. *Springerplus* 4, 485.

[13.65] Walker, S. I., Kim, H., & Davies, P.C. (2016) The informational architecture of the cell. *Transactions of the Royal Society A: Philosophy* 374, #20150057.

[13.66] Witzany, G. (2018) Evolution of Genetic Information without Error Replication. In: *Theoretical Information Studies. Information in the World.* World Scientific Series in Information Studies: London, UK, 11, 295–320,

[13.67] Wood D. C. (1992) Learning and adaptive plasticity in unicellular organisms. In: *Encyclopedia of learning and memory.* L. R. Squire (ed.), Macmillan, New York: 623-624.

[13.68] Worcester, D. L. and Weinrich, M. (2015) Hydrostatic pressure promotes domain formation in model lipid raft membranes. *Journal of Physical Chemistry Letters* 6, 4417-4421.

[13.69] Yi, T.-M., Huang, Y., Simon, M.I. and Doyle, J. (2000) Robust perfect adaptation in bacterial chemotaxis through integral feedback control *PNAS* 25;97(9):4649-53.

[13.70] Yokawa, K., Kagenishi, T., Pavlovič, A., Gall, S., Weiland, M., Mancuso, S. and Baluška, F. (2018) Anaesthetics stop diverse plant organ movements, affect endocytic vesicle recycling and ROS homeostasis, and block action potentials in Venus flytraps. *Annals of Botany* 122, 747-756.

14

A Hypothesis about How Bacterial Cells Sustain and Change Their Lives in Response to Various Signals

Vic Norris[1] and Alexei A. Sharov[2*]

[1]*Bacterial Communication and Anti-infectious Strategies, University of Rouen, Mont Saint Aignan, France*
[2]*Elixirgen Scientific, Baltimore, Maryland, USA*

Online resources:
Vic Norris:
https://www.researchgate.net/profile/Vic-Norris
https://loop.frontiersin.org/people/177422/overview

Abstract

Bacterial cells are composite agents whose functions are mediated by assemblies of molecules and macromolecules that are termed hyperstructures. Most hyperstructures are cell subagents characterized by goal-directed contextual activity and a certain level of autonomy. Because cells live in highly dynamic and diverse environments and solve varying internal and external problems, they need to switch effectively between multiple modes of living on demand. Such adaptive plasticity of phenotypes requires sensing of various components of the environment and internal state of the cell, evaluating the relative importance of either preserving the existing state or switching to the possible alternative phenotypes, construction of a functionally competent next state, and preserving the overall background identity of the system capable of restoring each of the possible phenotypes. Bacterial cells implement their plasticity responses via a network of chemical and physical signals that emerged in evolution by natural selection. The proposed model of *competitive coherence* describes how such a network can assemble and function through two levels of selection among components: certain molecules are selected by signaling of either preceding or co-occurring molecules to form hyperstructures, and certain hyperstructures are selected to produce a phenotype that is consistent with the environment and the needs of the cell. This model is consistent with the exploration-stabilization principle in evolution and learning and it helps to elucidate the "language" of hyperstructures that supports competitive coherence. These ideas are illustrated in particular with respect to the bacterial cell cycle, which comprises distinct phases of DNA replication, DNA segregation, and subsequent cell division.

**Corresponding author*: sharov@comcast.net

Keywords: Bacterial evolution, adaptation, hyperstructure, competitive coherence, biosemiotics, molecular signaling, now and next processes

14.1 Introduction

Bacteria such as *Escherichia coli* are among the most intensely studied living systems. Despite this, much remains to be understood about their behavior. Such living systems can be modeled as networks of interconnected elements or nodes, such as genes, proteins, or RNA molecules. In a pioneering investigation, Kauffman highlighted a fundamental problem that confronts cells containing many interacting constituents, namely, how such cells manage to obtain the reproducible phenotypes needed for performing living functions when a "hyper-astronomical" number of combinations of these constituents is apparently available [14.44]. The solution we favor is to consider that out of all a cell's elements, only a few are selected to regulate cell's behavior or phenotype at any given time [14.72].

A bacterial cell can be characterized as an *agency* — an autonomous unit with "a capacity for acting purposefully and using signs to make informed choices" [14.100]. In biology, such purposeful activities are known as functions or habits; they include sensing, resource consumption, metabolism, reproduction, and locomotion. Contemporary biological understanding of agency rejects supernatural entities such as homunculus or soul, and assumes that coordination of all activities is achieved by self-organization and interaction of lower-level components, such as subagents, tools, scaffolds, and resources [14.100]. This assumption does not mean that cell functions are reduced to the activity of its smaller components, because when these constituents are isolated they lack many of these functions. Most cell activities rely on the coordinated action of a large number of components that are organized into a highly integrated whole. A better description is that cell activity becomes represented by a small subset of its constituents, which perform higher-level regulatory functions (e.g., the centrioles in mitosis in eukaryotic cells). Such organization where higher-level functions can be performed by a subset of subagents is known as a heterarchy [14.59].

Functional interactions between cell components are always associated with certain long-term goals such as increasing the amount of nutrients, making new subagents (e.g., ribosomes) or scaffolds (e.g., cytoskeleton filaments), or duplicating genomic DNA as a preparation for cell division. Such interactions are different from non-functional random events (e.g., Brownian movement) because they are mediated by mechanisms that were encoded genetically and epigenetically, and selectively retained through millions of generations. Such meaningful interactions can be understood better by using the language of *semiotics*, a theory of signs, signification, and meaning [14.81]. But what exactly are signs? According to Charles Peirce, "A sign . . . is something which stands to somebody for something in some respect or capacity" [14.85], EP2.228. This means that a sign, besides its physically real form, has a connection to something else (e.g., cause, goal, or other kind of meaning) for an interpreting agent. According to Peirce, a sign is a triadic relation between a sign vehicle (representamen), which represents an object, and the result of interpretation, which he called an interpretant [14.85], EP2.478. For example, smoke is a sign vehicle of fire, which is the object and cause, and the interpretant could be "danger." But the interpretant is more than just a result of interpretation; it is also something that existed before as a part of knowledge that *enabled interpretation*. In other words, an interpretant is a tool for interpretation that was called "ground of the representamen" [14.84], CP2.228, or "context" [14.17], p. 46.

Peirce proposed three classes of signs: iconic signs (e.g., portraits), indexical signs (e.g., symptoms of a disease), and symbolic signs (e.g., words). Put differently, ". . . the relation of sign to object may be based on a possible resemblance, an actual reaction, or a habitual connection. Insofar as a sign is related to its object by virtue of a possible resemblance, it functions iconically; insofar as this relationship depends on an actual reaction or set of such reactions, it functions indexically; and insofar as this relationship results from a habitual connection, it functions symbolically" [14.16] (cited from [14.42]). An important kind of sign for bacteria is a signal which is produced by an organism (or its subagent) and "mechanically or conventionally triggers some reaction on the part of the receiver" [14.96], p. 121. We term "cues" those signs that come directly from the environment, and that have not been constructed by agents (unlike signals).

A wide variety of cues and signals are known to be used by bacteria and probably many are still to be discovered. These include molecules, macromolecules, and some inorganic ions (e.g., calcium). In the case of proteins, signals may take the form of post-translational modifications, such as phosphorylation, methylation, acetylation, deamidation of asparagines, succinylation, glycosylation, and oxidation of methionines [14.54]. One of the goals of post-translational modifications is to change the location of a protein so that it either joins or leaves a specific protein complex or causes the assembly, disassembly, fusion or fission of protein complexes. Phosphorylation of proteins, which is one of the best understood types of modification, is mostly done by protein kinases that phosphorylate proteins on the amino acids serine, threonine, tyrosine and histidine. Two-component signal transduction systems operate to allow bacteria to sense and respond to many environments. In these systems, a sensor histidine kinase catalyses its autophosphorylation and then transfers the phosphoryl group to a response regulator, which may then change gene expression [14.50].

Signals used for communication between cells include alarmones (such as (p)ppGpp, Ap(4)A, ZTP, cAMP and cGMP), the energy charge (ATP+1/2ADP) to (ATP+ADP+AMP), various lipids (e.g., fatty acids), inorganic ions (such as calcium), amino acids (such as leucine), polyamines, polyphosphate, small RNAs, short peptides, and the well-characterized transcription factors and sigma factors.

Chemical communication between bacterial cells can take the form of quorum-sensing molecules such as auto-inducing peptides and acyl-homoserine lactones [14.106]. Bacterial cells need not even be alive to communicate: dead cells release a "necrosignal" that activates antibiotic survival pathways in bacterial swarms [14.8]. Although communication can be achieved via simple diffusion, it can also be applied through protein appendages such as sex pili or nanotubes [14.21]. In this context, the methylation of specific sequences in chromosomal DNA by host enzymes can help the cell distinguish this DNA from incoming, foreign DNA brought by bacteriophages (bacterial viruses) in which these sequences are unmethylated; this discrimination allows the cell to only destroy the latter parasitic DNA. The CRISPR-Cas nuclease system is another way by which cells can destroy such incoming DNA, but bacteriophages can protect themselves from such nucleases by building a proteinaceous shield around their genome [14.60]. Communication between bacteria can also take physical forms as in the case of electromagnetic and sonic communication [14.56]

[14.75] [14.87]; it has been proposed that such sonic communication depends on pH variations induced by pulses propagating along membranes [14.24].

Communication also occurs between bacteria and eukaryotes [14.18]. For example, *E. coli* not only performs chemotaxis in response to sugars (see below) but also in response to signaling molecules in the gut that include melatonin and spermidine (as repellents) as well as to dopamine and norepinephrine (as both attractants and repellents depending on concentrations) [14.53]. Not surprisingly, stress hormones can therefore act as cues to initiate major changes in bacterial physiology [14.27] as when *Streptococcus pneumoniae* changes from a commensal to a parasitic state [14.2]; whilst, in the other direction, bacteria may affect host behavior [14.12] [14.77] as when an enzyme, glyceraldehyde-3-phosphate dehydrogenase, is on the surface of bacterial cells or is released from cells into the surrounding medium where it is proposed to play a role in virulence [14.28].

An important environmental cue is temperature with 37 °C informing pathogenic bacteria, such as *Yersinia pseudotuberculosis*, that are inside a eukaryotic host. This sensing can take the form of "RNA thermometers," in which the structure of some mRNAs adopts a conformation that allows these mRNAs to be translated into proteins according to the temperature [14.93].

In this chapter we discuss a hypothesis about how bacterial cells sustain and change their phenotypes in response to various cues and signals. First, we define functional components of cells composed of molecules and macromolecules as *hyperstructures*. And second, we develop a model of *competitive coherence* that explains functional selection at two levels: (a) at the first level, certain molecules are selected to form hyperstructures by signaling of either preceding or co-occurring molecules, and (b) at the second level, certain hyperstructures are selected to produce a phenotype that is consistent with the environment and the needs of the cell. Finally, we apply this model to various phenotype transitions in bacterial cells, including the phases of the cell cycle.

14.2 The Hypothesis

Bacteria are hierarchically structured, and the most important functional roles are performed by what we have termed hyperstructures, which are assemblies of molecules and macromolecules that function as subagents. The existence of this intermediate level of organization in a hierarchical system helps to reduce the background noise from the level below (i.e., in molecules) and hence facilitates the emergence of coherent behavior at the level above [14.51] [14.91]. This is just one of the advantages of hyperstructures, which are many and varied; for references see [14.48].

We have also proposed that, at the lower level, hyperstructures self-organize by competitive coherence within a subset of molecules and macromolecules, and, at the higher level, the bacterial cell itself results from competitive coherence via selecting a subset of active hyperstructures. In other words, the operation of competitive coherence regulates both the level of hyperstructures and the level of the phenotype of the cell. The subset of selected elements with regulatory functions must be coherent, which means that the elements in this subset must work together and not against one another. However, highly coherent systems have a tendency to exhibit super-stability, which prevents the organism from switching to alternative functional states in cases of environmental change or internal misbalance. Thus, in order to survive and reproduce, cells have to combine coherence with a capacity to switch to alternative modes of living either under stressful conditions or to support episodic functions, such as cell division. Different phenotypes (modes of living) are partially incompatible; for example, a bacterium cannot simultaneously adopt both a growth strategy appropriate for good conditions and a survival strategy appropriate for extreme environmental conditions. Thus, phenotype changes imply a competition between some cell components that support different modes of living. Competition between components at the time of phenotype change does not exclude coherence between components at the previous and the next time points, when one behavior-regulating subset of elements continues the work started by the previous subset. Such organization of the signaling network within a cell that supports both coherence of components within a single phenotype and competition between components that support conflicting phenotypes is termed "competitive coherence" [14.72].

Competitive coherence is a solution to the problem of how a living system can behave in both (1) a coherent way over time so as to maintain historical continuity and (2) a coherent way at a particular time that makes sense in terms of both internal and external (environmental) conditions [14.68] [14.72]. At a particular level of organization, competitive coherence describes the way that a key subset of constituents—the *Active* subset—is selected to determine the behavior of the system at that level. This selection is the result of a competition for inclusion in the *Active* subset between elements with connections that confer coherence over time (i.e., continuity) and connections that confer internal and external coherence at the present time. In the case of a bacterium, competitive coherence describes the way that (1) a key subset of its constituent macromolecules—the *Active* subset at the first level—is selected to determine the existence of a hyperstructure (see Subsection 14.3.3), and (2) a key subset of hyperstructures—the *Active* subset at the second level—is selected to determine the behavior of the cell, that is, the phenotype (see Subsection 14.3.4). This selection is the result of a competition for inclusion in the active subset between elements with connections that confer coherence over time (i.e., continuity) and connections that confer internal and external coherence at the present time.

We have explored how the concept of competitive coherence can be used in a neural network to model the phenotype of the cell at any one time as determined by a subset of macromolecules. In particular, each element (e.g., macromolecule) is modeled as a unit with an address and two fields that carry addresses of other elements that either tend to be active simultaneously with that element (the *Now* field), or tend to become active in the following time step (the *Next* field) (Figure 14.1). The addresses in these fields are interpreted as preferences to activate components either at the same time step (*Now*) or at the immediately following time step (*Next*). Some versions of this model also included weights (frequencies) for each connection between components [14.68].

Only a small proportion of elements is active at a time, and the set of elements to become active at the next time set is determined as follows. First, the total counts of *Next* preferences of all active elements are used as a score to select candidate elements, but only the top candidate with the highest score becomes the first active element in the following time step. Then, the *Now* preferences of this first element are used to calculate scores and two scores are compared: the second *Next* score and the highest *Now* score. If the *Next* score is higher, then the corresponding element becomes the second active element in the new time step. Because there are already two elements in the new time step, their *Now* preference scores are combined and compared to the following (i.e., third) *Next* score. If the *Now*

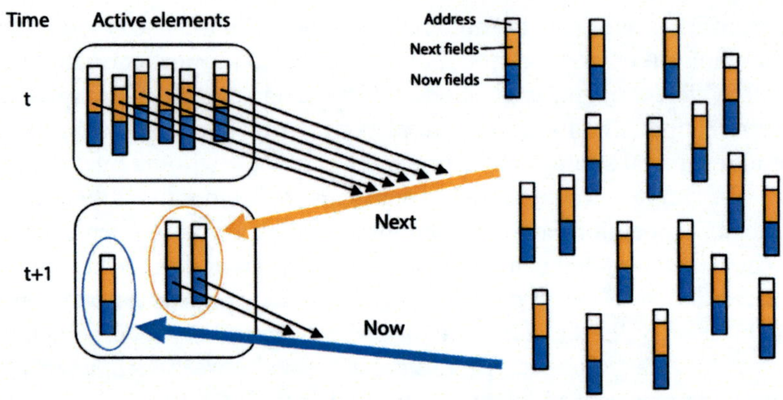

Figure 14.1 Competitive coherence model implemented as a simple neural network with *Now* and *Next* competing preferences for selecting elements to become active in the following time step.

score appears higher, then the corresponding element is added to the *Active* subset. This procedure is continued till the *Active* subset of elements is filled. Because of the competition between *Next* and *Now* processes (based on the highest score), the combined process is called "competitive coherence." After rewarding and punishing procedures were added to the model, the system was able to generate new output patterns that were rewarded during the training session [14.72].

The notion of competitive coherence is closely related to the principle of exploration-stabilization in evolution, development, and learning proposed by Ginsburg and Jablonka [14.29] [14.30]:

> "Exploration-stabilization processes are manifest at all levels of biological organization and are based on a common principle — the generation of variations from which only a subset is eventually stabilized (selected). Examples are the selection of genetic mutations in populations; selective stabilization of biochemical networks within a cell; developmental selection processes that occur during ontogeny in plants and animals and lead to homeorhesis [...]; stabilization of exploratory motor behaviours. In all cases, variations that confer benefits, or, more generally, that lead to a set-point (an attractor state) are stabilized/selected. As the nervous system evolved, new levels of developmental selection were introduced: in addition to selection among neurons during embryogenesis, differential stabilization of synaptic connections takes place." [14.30], p. 3.

The quote contains "the generation of variations from which only a subset is eventually stabilized (selected)," and competitive coherence describes how this stabilization/selection occurs. Exploration strategies can range from mostly random (within physiological limits) to more regulated and organized patterns of variation, as assumed in the competitive coherence principle. The main advantage of organized variation is that it is less likely to fail and more likely to provide benefits in terms of viability and adaptability in unpredictable conditions. Bateson [14.5], p. 173, distinguished between "simple learning" and "learning to learn," which is aimed at developing more successful learning strategies. The latter strategy is described as "[...] the acquisition of that class of abstract habits of thought with which this paper is concerned; that the states of mind which we call 'free will,' instrumental thinking, dominance, passivity, etc., are acquired by a process which we may equate with 'learning to learn.'"[1] In light of these studies we can say that competitive coherence is a result of "learning to learn" within the evolutionary domain, which provided an infrastructure that facilitates both adaptive plasticity and natural selection.

To model competitive coherence, we present a living system as a network of interactions between elements that goes through a trajectory of states, each characterized by a subset of active elements that regulates the system's behavior. In the case of bacteria, a cell goes through a trajectory of behaviors or phenotypes. This trajectory comprises a series of states with different properties and compositions; and each state is the result of a particular combination of active hyperstructures.

Some of the elements in the competitive coherence model of bacterial signaling correspond to cues and signals. To be effective, bacterial signals need to be powerful, cheap to produce, and easy to degrade. And signals and cues can also be effectors, which are molecules or macromolecules that alter processes like enzyme-catalysed reactions. When molecular-level signals and cues play effector roles, they usually represent lower-level metabolic functions, whereas final effectors can perform organism-level action or production. A signal can be powerful if it is derived from the fundamental constituents of a hyperstructure because this allows it to reflect the state of that hyperstructure (e.g., whether that hyperstructure is functioning or not). In other words, a hyperstructure has the potential advantage of releasing its constituents so that they can be reused by other hyperstructures and/or act as signals. No new synthesis or degradation is needed. Hence, organization at the level of hyperstructural signaling is flexible, economical and rapid. This is why enzymes can act as signals going from the disassembly of one hyperstructure to nucleate the assembly of another hyperstructure.

Competitive coherence operates at the level of composition of both the hyperstructure and the cell. At the hyperstructure level, signaling takes the form of *Now* processes and *Next* processes. With *Now* processes, the importance (or intensity) of the signals leading to hyperstructure assembly is likely to increase as the hyperstructure is assembled (i.e., the signal is amplified), whereas with *Next* processes, this importance decreases. In this approach, the same constituents of a hyperstructure can play both functional roles (i.e., be effectors, where an effector is a molecule or macromolecule that alters a process like an enzyme-catalyzed reaction) and signaling roles. Indeed, a functional constituent of a hyperstructure, like a macromolecule, can be considered both as a *Now* signal that helps promote the assembly of the hyperstructure and as *Next* signal when that same macromolecule is modified during the disassembly of the hyperstructure that helps promote the assembly of the successor hyperstructure.

At the level of the composition of the cell itself, signaling again takes the form of *Now* processes and *Next* processes. Here, the signals take on a more global nature since they must lead to coherent sequences of coherent subsets of hyperstructures. This approach embraces both cell-to-cell signals and environmental cues. At both the hyperstructure level and the cell level, a signal or cue can be characterized by its relative contribution to *Now* and *Next* processes, used as coordinates in a two-dimensional space. However, the extent to which the values of these coordinates is attributed to a particular signal depends on the scale of the analysis: at the lower level of the selection of an individual hyperstructure, the signal or cue

[1] Bateson also called it "deuterolearning" [14.5], p. 174.

might be classed as mainly *Next*; whilst, at the higher level of an individual cell, the same signal or cue might be classed as mainly *Now*.

14.3 The Evidence

14.3.1 Hyperstructures

Bacterial cells include numerous functional hyperstructures that can be roughly classified into cytoskeletal filaments, enzymatic structures, microcompartments, membrane-associated structures, appendages, DNA-containing structures, phase separation condensates, and others (Table 14.1). Earlier ideas that bacteria lack cytoskeletal filaments have been disproved, and now many bacterial cytoskeleton proteins have been described [14.102]. Enzymatic structures include helices of cold shock proteins and so-called regrowth-delay bodies that are both important for survival [14.38] [14.111]. Hyperstructures that include membranes support a large variety of functions from sensing to production of energy-rich resources (oxidative phosphorylation and photosynthesis), whereas DNA-containing complexes are involved in such functions as regulation of gene expression, DNA repair, replication, and segregation.

Hyperstructures differ in their relation to the flow of energy or nutrients. Non-equilibrium (NE) hyperstructures actively use energy and nutrients to support active growth and

Table 14.1 Examples of hyperstructures in bacterial cells.

Types of hyperstructures	Examples
Cytoskeletal filaments	Actin-like filaments, tubulin-like filaments, intermediate-like filaments
Structures made from enzymes	CTP synthase, RNA degradosome, elongation factor (thermal unstable), helices of cold shock proteins, regrowth-delay body
Microcompartments	Carboxysome, metabolosome, acidocalcisome, polyphosphate granule
Membrane-associated structures	Chemoreceptor arrays, lipid rafts (based on flotillin or on cardiolipin), RecA bundles, E1 clusters, segrazones, phosphorylation complexes, and photosystem complexes (e.g., phycobilisome)
External appendages	Flagellae, pili, fimbria, nanotubes
DNA-containing structures	Bacterial "nucleolus," foci of NAPs, DNA replication complex, DNA segregation complex, topoisomerase complex, RNA polymerase complex, metabolic enzyme complexes, associations of genes, nascent RNA, and nascent proteins
Phase separation condensates	Clusters of RNA polymerase and NusA, stress granules
Other	Bacteriophage factory, magnetosome

reproduction of cells. Many of them support transcription and translation processes and consume ATP and GTP. Other, equilibrium (E)—strictly, quasi-equilibrium—hyperstructures do not require energy and resources after their construction, and they function mostly to promote survival and protect sensitive molecules from degradation. Typical examples are the Nucleoid-Associated Protein (NAP) hyperstructures or the co-crystals of DNA-Dps formed in starved cells of *E. coli* [14.52].

14.3.2 Competitive Coherence at the Hyperstructure Level: Selection of Molecules and Macromolecules

Individual hyperstructures contribute to different processes within the competitive coherence system that supports both growth and reproduction in favorable conditions and survival in stressful conditions [14.68] [14.72]. Strong *Now* connections are particularly important for E hyperstructures that are crucial for survival of bacteria in harsh conditions. Thus, the E hyperstructures that form in such conditions are integrated by interactions only between the constituents of these structures [14.78]. This is because it is important for some bacteria in the population, like persisters, to become insensitive to growth opportunities so that they can survive unpredictable disasters [14.4]. Once the E hyperstructure (appropriate for survival) has formed, there should be few *Next* connections to NE hyperstructures (appropriate for growth). In other words, the strength of the *Next* connections from NE elements *into* an E hyperstructure is much greater than the strength of the *Next* connections *out* from an E hyperstructure to NE elements; moreover, the strength of the *Now* and *Next* connections from an individual E element to other E elements is much greater than the strength of *Now* and *Next* connections from an individual E element to NE elements.

Since the majority of *Next* connections from an E hyperstructure are to E elements, going from an E to an NE state (e.g., from a non-growing to a growing state) should be difficult and may even entail abandoning the E material completely, as in the germination of a spore where much material is abandoned. One way to achieve the transition from E to NE states may be for much more weight to be given to *Now* connections and to the environmental cues so that the new *Now* connections that correspond to the environmental cue dominate the selection for the *Active* subset (rather than this selection being dominated by the E outgoing *Next* connections to E elements). To preserve the heterogeneity of the population, this E to NE transition should only happen for some of the time in some of the cells; this might be the case if the life of a bacterium at equilibrium were to correspond to a state cycle with some elements in just one of the states being responsive to an environmental *Now*.

A hyperstructure results from a selection of molecules and macromolecules locally by cues and signals. This local action can be done by transcription factors (activators and repressors of transcription) of which there are over 200 in *E. coli* [14.92]. It seems likely that many of the genes under the control of these factors form part of hyperstructures in which they are transcribed into mRNA, this mRNA then translated and the protein products then staying within the hyperstructure [14.64]. The *lac* operon, when expressed, is a paradigm for such NE hyperstructures. A sufficiently high concentration of the substrate, lactose (the actual inducer is allolactose, an isomer of lactose formed when lactose enters the cell), induces expression of the operon by binding to the LacI protein and thereby relieving the repression caused by LacI binding to the operators in and near the genes. This results in the formation of a hyperstructure comprising the *lac* genes dynamically attached to tens of nascent mRNAs and to hundreds of ribosomes and the nascent enzymes, one of which is a membrane permease that may help create a membrane structure as part of the hyperstructure [14.47].

A very different E hyperstructure is formed when the *lac* operon is repressed in the absence of an inducer. There are around 10 copies per cell of LacI, which binds in an on-off fashion to the operator O1 and to two auxiliary operators, O2 and O3, which are close to O1; this dynamic binding to the operators plus their proximity to one another increases the local concentration of LacI at these operators to effectively block transcription [14.65]. Lactose can therefore be considered as both a nutrient and a cue (indeed, the induction by allolactose can be mimicked by a non-metabolizable analogue Isopropyl-β-D-thiogalactopyranoside) with the LacI protein acting as a lactose sensor. A similar but opposite argument can be made for glucose in this context since *the* lac operon is also repressed in the presence of glucose with a protein, CAP, acting as the sensor of cAMP (the concentration of which is inversely related to that of glucose).

The above examples show how a certain sort of cue like lactose can bring together the constituents of a hyperstructure to cause its assembly or disassembly. Moreover, in the progressivity of repression with increasing local concentration, LacI and the operators of the *lac* operon shows some of the characteristics of a *Now* connection.

14.3.3 Competitive Coherence at the Level of the Cell: Selection of Hyperstructures

The phenotype of the cell can be interpreted in terms of sets of hyperstructures selected *globally* by *Active* cell components that perform regulatory functions. In bacteria, such global action can be done by so-called "sigma factors," which are molecules that are responsible for the specificity of promoter DNA binding of bacterial RNA polymerases and control of transcription [14.22]. A *regulon* is a group of genes under the control of a particular regulator, such as a sigma factor, that are organized into more than one operon; they may be distributed throughout the genome and help regulate certain aspects of cell phenotype [14.13]. The seven sigma factors in *E. coli* are sigma70 or sigmaD (housekeeping sigma) and the "alternative" sigmas: sigma32 (heat shock response), sigmaF (flagellar synthesis), sigmaS (adaptation to starvation and stationary phase), sigmaN/sigma54 (nitrogen metabolism), sigmaFecI (iron metabolism), and sigmaE (response to extracytoplasmic stress).

Global selection of hyperstructures can also be affected by cues and signals in the form of Simple, Universal Molecules or Inorganic ions [14.80]. These ions include calcium, which is generally at a much higher concentration outside the bacterial cell than inside it. This permits cells to use a flux of calcium down the gradient as a powerful signal [14.20]. Low calcium in the environment can therefore be an important cue because, for example, it may be interpreted to mean that the bacterium is inside a eukaryote and should produce virulence factors [14.31]. Calcium can alter the structure of membranes and water, can condense onto DNA, and can activate a variety of different enzymes, including proteases. This has led us to speculate that calcium could constitute a reset signal allowing the cell to proceed from one *Active* subset of hyperstructures to the next [14.74].

Global selection can also take the form of structural regulation by NAPs and their hyperstructures can both compact DNA and regulate its transcription, thereby protecting the chromosome or promoting/suppressing expression of regions of the chromosome. The involvement of NAPs in cellular processes besides transcription in *E. coli* includes their roles in DNA replication, recombination, and DNA repair and the SOS response (see below). Going from one *Active* subset of hyperstructures to another could involve post-translational modifications of NAPs like H-NS (Histone-like Nucleoid-Structuring protein) and HU (Histone-like protein, once called factor U) so as to alter their self-association as well as their association with other macromolecules [14.19]. Indeed, it has been proposed that a transition can occur from a DNA bundle to a filament, depending both on the ration of HU to DNA and on environmental factors, and that this would alter the structure of the nucleoid [14.33] [14.88].

Another way to globally select hyperstructures is by a gradient of negative supercoiling along the replication origin-to-terminus axis of the bacterial chromosome. This gradient can be created by RNA polymerases and topisomerases depending on the energy provided by the environment, and would be interpreted by the sensitivity of different promoters of genes to supercoiling at different positions in the chromosome, thereby coordinating the transcription and selection of hyperstructures with environmental cues [14.66] [14.103].

14.3.4 Cell Cycle Hyperstructures and Signaling

The bacterial cell cycle involves transitions between different states (Figure 14.2). The starting point is a growing cell characterized by the activity of metabolic hyperstructures that expand together with the cell size. At some point, the cell initiates the sequence of steps resulting in replication of the chromosome, segregation of daughter chromosomes, and

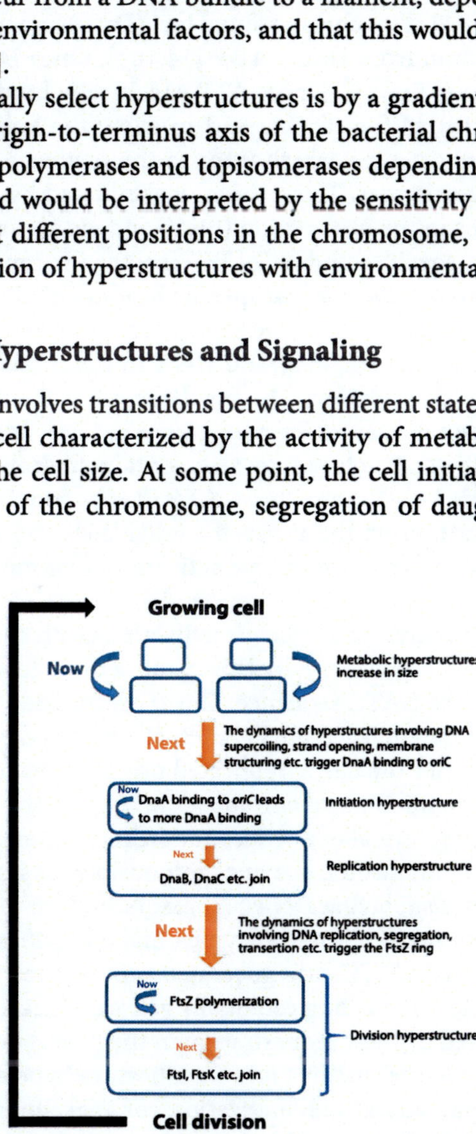

Figure 14.2 Control of the bacterial cell cycle by hyperstructures and competitive coherence: The boxes represent hyperstructures. The large font of Now (blue) and Next (orange) represents these processes acting during transition between hyperstructures, and the small font represents them acting within hyperstructures. The non-specialist should be aware that this figure is highly speculative and simplified, and likely would require an update as more experimental data becomes available.

then, cell division. In addition, we have proposed that within the period that the chromosome is being replicated, the replication hyperstructure can pause in different places as part of a mechanism to select different combinations of hyperstructures, and hence different states [14.76].

The state of the DnaA protein is considered to be the key signal for the initiation of chromosome replication in *E. coli* [14.43]. The binding of over twenty copies of DnaA to sites in the origin of replication of *E. coli*, *oriC*, is one of the earliest steps in the assembly of a hyperstructure that initiates DNA replication, also known as initiation or pre-replication complex [14.32] [14.41]. Two chromosomal loci (DnaA-reactivating sequences) promote DnaA-ATP production from DnaA-ADP [14.104]. Once DnaA in its ATP or ADP form has bound all R-boxes in *oriC*, DnaA-ATP specifically binds to other sequences in *oriC* to trigger the opening of the strands and to allow the helicase, DnaB, and DnaC proteins to join the hyperstructure, which leads to the further separation of the strands and recruitment of more enzymes. This process is facilitated by: DNA supercoiling; the transcription of two genes in the *oriC* region (*mioC* and *gidA*); and the binding of NAPs, which includes sequence-specific binding of FIS (Factor for Inversion Stimulation) and IHF (Integration Host Factor) and the less specific binding of HU and H-NS. The resulting initiation complex/hyperstructure contains several different sorts of protein, RNA and *oriC*. Specific proteins bind single-strand DNA in the initiation hyperstructure and protect it from enzymatic degradation. The involvement of the membrane in the initiation of replication has been discussed for a long time [14.14], and it is tempting to speculate that the initiation hyperstructure may also contain cardiolipin, given, for example, that the conversion of DnaA-ADP into DnaA-ATP *in vitro* requires acidic phospholipids in the bilayer in a fluid state. From the above, it could also be argued that DnaA has characteristics of *Now* signals with its progressive activation, oligomerization and binding to different sites.

Newly replicated DNA is transiently hemimethylated (with the old strand methylated and the new strand still to be methylated). This hemimethylated state of the DNA constitutes a signal if it contains the GATC sequence. This is because SeqA, an oligomeric protein, binds preferentially to hemimethylated GATC sites, of which there are over 19,000 in the *E. coli* chromosome, to sequester the newly replicated origin of replication and other regions so as to prevent multiple reinitiations of chromosome replication. The pattern of clustering of these sites has led to the proposal of a sequestration/replication hyperstructure based on SeqA that would comprise not just the enzymes responsible for synthesizing DNA but also the enzymes responsible for supplying DNA precursors, and for DNA repair and recombination, the genes encoding many of these enzymes, and a specific region of the membrane [14.73]. The hemimethylated GATC sites generated progressively by the initiation hyperstructure might be considered as acting essentially as *Next* signals given that their appearance in one state is followed by their sequestration in the next state.

This initiation hyperstructure matures into a replication hyperstructure that contains a DnaE–DnaE or PolC–DnaE strand polymerization complex along with the clamp loader, primase, helicase, and single-stranded DNA binding proteins. Additional polymerases are needed to help with the recombination and repair of DNA as shown by the presence of RarA in the hyperstructure. Ribonucleoside diphosphate reductase, which catalyzes the synthesis of the dNTPs (deoxynucleotide triphosphates; DNA building blocks), is also present. The dynamic nature of the replication hyperstructure is consistent with its disruption that occurs when the replication fork encounters barriers before reaching the terminus, or upon reaching the terminus.

There is another aspect to signaling in the context of replication. DNA synthesis is coupled to nutrient richness and growth rate in a metabolic control over replication. One way this occurs is via (p)ppGpp [14.10] [14.23], and another is via central carbon metabolism, which comprises around thirty enzymes that transform nutrients into the molecules and energy used for the production of the biomass (including monomers for macromolecules such as DNA). For example, in *Bacillus subtilis*: (1) subunits of pyruvate dehydrogenase (PdhC) and related enzymes bind the origin of replication region, the helicase (DnaC) and the primase (DnaG), and/or inhibit the initiation of replication; (2) mutations in the genes encoding the enzymes of central carbon metabolism suppress the defects in the initiation and/or the elongation steps of DNA replication in mutants of DnaC, DnaG, and the lagging-strand polymerase, DnaE; (3) mutations in the genes encoding glyceraldehyde 3-phosphate dehydrogenase (GapA) perturb the metabolic control of replication; (4) pyruvate kinase (PykA) can both inhibit initiation and stimulate elongation via proposed interactions with DnaC, DnaG, and DnaE as modulated perhaps by phosphorylation [14.37]. The connections between DNA replication and central carbon metabolism in *E. coli* include the promotion of DnaA to its active DnaA-ATP form and its binding to oriC by cAMP (a regulator of this part of metabolism), the suppression of defects of a DnaA mutant by changes in pyruvate and acetate metabolism and the inhibition of DnaA conversion to DnaA-ATP and its binding to *oriC* by acetylation of DnaA (with acetyl-CoA and acetyl-phosphate as donors) [14.37]. DnaA clearly has a role as an integrative signal in the transition from the non-replicating to the replicating state with actions typical of both AND and OR gates.

Cell division in *E. coli* also involves the assembly of a hyperstructure. This assembly entails the tubulin-like, tread-milling protein FtsZ forming a dynamic ring-like structure (Z ring) at the nascent division site [14.110] where it is joined by FtsA and other proteins that help anchor the Z ring to the cytoplasmic membrane; the hyperstructure may also contain the heat shock protein GroEL. The hyperstructure matures with the subsequent arrival of proteins such as FtsK, a DNA translocase, FtsI, a transpeptidase, and AmiC, an amidase. Phospholipids are also associated with the hyperstructure and in *B. subtilis* most of the phospholipid synthases are located in the membrane part of the hyperstructure, which is enriched in cardiolipin and phosphatidylethanolamine.

A variety of signals are required for the formation and operation of the division hyperstructure. At the earliest known stage, membrane domains appear to define where the division hyperstructure will assemble [14.26]. These domains can be created by the coupling of transcription, translation and insertion of proteins into the membrane and hence their dynamics essentially acts as a *Next* signal to promote division [14.25]. Successful cell division requires the division hyperstructure to act between the segregated chromosomes. Here the signaling is performed in part by the DNA-binding proteins SlmA in *E. coli* [14.89] and Noc in *B. subtilis* [14.1], which inhibit FtsZ assembly over the nucleoids. Signaling systems exist to prevent division when DNA has been damaged. These include the SOS system [14.55] that, when induced by DNA damage, produces the SulA protein (along with forty other proteins) to interfere with FtsZ. The induction of the SOS system is equivalent to (1) an OR gate, given that DNA damage can have many causes and (2) a *Now* signal, given that the environmental cue forces an immediate change in the cellular state. Phosphorylation can also be implicated in the coupling between DNA damage and cell division as in *Deinococcus radiodurans*, where DNA damage leads to the phosphorylation of FtsA and FtsZ by a serine/threonine kinase [14.57]. In *E. coli*, the Min system confines the division hyperstructure to the middle of the cell. This system comprises MinC, which inhibits FtsZ, and MinD and MinE, which generate pole-to-pole oscillations to create a concentration of MinC that is lowest at midcell [14.86]. The size of the cell of *E. coli* and *B. subtilis* depends on the richness of the growth medium: in nutrient-rich media, the high intracellular concentration of UDP-glucose stimulates UgtP to interact with FtsZ in the division hyperstructure and so delay division and increase cell size, whereas in nutrient-poor media, the production of UgtP is low and the protein is sequestered away from the division hyperstructure, which can then generate smaller cells [14.15]. A similar system based on OpgH operates in *E. coli*. At this cellular level, UDP-glucose is primarily acting as a *Now* signal, but it also appears to have another role as a global signal that acts on the stationary phase transcription factor, which regulates gene expression if carbon supply is limited [14.9]. Another signal that couples the nutrient status of the medium with cell division is pyruvate, which causes relocation of one of the subunits of pyruvate dehydrogenase to the chromosome and, as proposed, promotes the assembly of FtsZ [14.63]. Finally, the membrane-associated, actin-like protein, MreB, can undergo calcium-induced structural changes, and it has been proposed that these changes may be important during cell division [14.105].

14.4 Using Semiotics to Explain Cues and Signals in Bacteria

14.4.1 Types of Semiosis, Signs and Effectors

Semiotics originated as a part of philosophy and linguistics to explain human language and other forms of human communication. However, soon it became clear that animals also use signs for thinking and communication. Biologist Jakob von Uexküll was the first to systematically study the generation of meaning in living organisms, and these studies became the cornerstone of his theoretical biology and theory of meaning [14.107] [14.108]. He argued that meanings emerge from biological functions and habits. In particular, animals construct internal models of their environment where objects are used as signs (meaning-carriers) associated with living functions, and thus, become functional tokens such as food, enemy, or shelter. These meaningful components are functionally linked to the organism and together constitute an *Umwelt*, which is a subjective representation of the organism's environment. "[t]he umwelt is the ecological niche as the animal itself apprehends it," explained Jesper Hoffmeyer [14.34], p. 54. Perceptions and actions of animals are thus integrated into "functional circles" that work as cybernetic feedback loops. Von Uexküll [14.108], p. 34, wrote: "Even the simple blink-reflex, caused by the eye being approached by a foreign body, does not consist of a mere sequence of physical causes and effects, but of a simplified functional circle, beginning with perception and ending with effect."

The theory of von Uexküll was adopted and popularized by semiotician Thomas Sebeok, who worked on the project of zoosemiotics, which is the application of semiotic ideas to animals [14.95]. Eventually Sebeok realized that semiotic analysis is applicable to all living organisms, and thus to how *biosemiotics* emerged as a transdisciplinary study that integrates biology and semiotics [14.3]. Because organisms of different evolutionary lineages vary in their capacity to create and interpret signs, one of the important tasks of biosemiotics was delineation of semiotic thresholds (or threshold zones) that separate different levels of types of semiosis. These studies developed in parallel with the theories of cybernetic complexity [14.45] and major evolutionary transitions in biology [14.58]. Biosemiotic studies emphasized the importance of different sign processes in such transitions [14.36]. To separate semiotic systems from non-semiotic, Kull [14.49] used the functional cycle of Uexküll as a basic model of semiosis. Then he proposed to distinguish between vegetative, animal, and cultural levels of semiosis, ordered by increasing complexity. He argued that these three levels of semiosis can be associated with three types of signs defined by Peirce: icons, indexes, and symbols, respectively.

However, the notion that complexity increases from bacteria to humans remains debatable. According to author #1 of this chapter, the cytoplasm of bacteria per unit volume or weight is just as complex as that of eukaryotes, as indicated in the above section on hyperstructures [14.70]. A specific measure of cellular complexity—or rather *hypercomplexity*—could be derived from the combination of different levels of organization and competitive coherence. In this proposal, hypercomplexity could be quantified via a set of general parameters $\Sigma_{Now}C_{i,j}$ and $\Sigma_{Next}C_{i,j}$ at a given level of organization where $C_{i,j}$ is the strength of the connectivity (e.g., number of *Now* and *Next* links where these are of equal strength) from constituent *i* to constituent *j* at a particular level [14.71] [14.72]; there are variants of this quantification that take into account whether the members of the *Active* subsets have equilibrium or non-equilibrium characteristics [14.71]. Author #1 argues that the bacterial system of plasmids, transposons, bacteriophages, mutators and horizontal gene transfer vastly outstrips the efficiency and adaptability of eukaryotic systems.

In contrast, author #2 assumes that semiotic complexity needs to be measured as the number of non-redundant functions per agency at the dominating level (e.g., organism or holobiont) [14.100], where functions are counted in all included subagents (e.g., cells and hyperstructures). Although we still do not know many functions a cell has, one way to evaluate functional complexity at the genome level is via the total length of non-redundant functional genome segments, which shows an increasing trend (from 4.6- to 15.3-fold per 1 billion years) in the macro-evolution of eukaryotes [14.97]. Author #2 has argued that prokaryote lineages have a much slower rate of genome complexity increase, which is 1.9- and 2.5-fold per 1 billion years in *Archaea* and *Eubacteria*, respectively (ibid.), which is an indicator of evolutionary constraints that cannot support faster growth of complexity.

According to author #2, besides quantitative measures of complexity there is a qualitative difference between primitive semiosis (protosemiosis) in prokaryotes and more advanced semiosis (eusemiosis) in eukaryotes. He wrote: "In protosemiosis, agents associate signs directly with actions without considering objects, whereas in eusemiosis, agents associate signs with objects and only then possibly with actions. ... Eusemiosis is qualitatively different from protosemiosis because it cannot be reduced to a small number of specific signaling pathways." [14.101], p. 103. However, protosemiosis and eusemiosis include similar classes of signs. In particular, "Proto-signs can be classified into proto-icons that signal via single specific interaction, proto-indexes that combine several functions, and proto-symbols that are processed by a universal subagent equipped with a set of heritable adapters" (ibid.). In this classification, "Examples of proto-icons include both simple molecular bindings (e.g., binding of cAMP to a CAP receptor) and complex multi-component bindings (e.g., complementary binding of nucleic acids in DNA replication, transcription, and miRNA binding)" [14.98], p.1062. The idea is that an icon could correspond to the cellular object that binds a molecule; an important qualification would be that the origin of this molecule be environmental because it is the icon that adapts to the molecule rather than the reverse (the landscape does not adapt to the painting). In Sharov's classification, "an adaptor molecule is a proto-index because it has two binding surfaces for interaction with two different kinds of molecules, which are physically connected. Proto-indexes often form chains of signal transduction which transfer signals detected at the cell membrane to the nucleus" (ibid.). In bacteria, the proto-index class would therefore include the signals of the sensor kinase and response regulator, which transduce environmental cues so as to regulate gene expression. Finally,

> Proto-symbols are conventional sign relations established for vertical and/or horizontal communication. For example, DNA plays the role of a proto-symbol in a sequence of processes that start from mRNA synthesis on the DNA template and ending with protein synthesis on the mRNA template. Protein synthesis follows the rule of the genetic code which is a heritable "natural convention" based on a set of molecular adapters (tRNAs and acyl-transferases). [14.98], p. 1062.

Another example of a proto-symbol is natural antibiotic that can be used by a cell not as a weapon but as a warning (or invitation) to other cells; it is analogous to a man waving an axe as a threat to another person, where the axe is also not a weapon but a symbol.

In the hypothesis about bacterial semiotics presented here, there is an important difference between signs and effectors. One distinction that might be made is in terms of the number of qualitatively different steps between the action of the sign or effector and the final action. It could be argued that an effector is close to—or indeed constitutes—the final action, whilst a sign is further away from this final action. This can be illustrated by the program simulating competitive coherence, where there can be several different *Active* subsets between an input and an output: a sign could be considered as corresponding to an element in an early *Active* subset with its *Next* connections leading to the final *Active* subset, which contains the output element; this output element—and arguably the other elements in the final *Active* subset connected to the output element via the *Now* connections—would then be effectors. All that being said, having made this distinction, the connecting elements in bacteria can have characteristics of both pure signs and pure effectors just as they can have both *Now* and *Next* connections (see below).

Consider the case of chemotaxis in *E. coli*, which entails sensing the concentration of an attractant or a repellent, calculating the gradient of this concentration[2] and, as a consequence, rotating the flagella so as to either continue in a straight line (by "running") or adopt a new direction (by "tumbling"). This sensing can operate successfully over five orders of magnitude with high sensitivity at low concentrations but low sensitivity at high concentrations, a remarkable achievement due to the sensitivity of the receptors depending on whether or not the receptors are grouped into extensive hyperstructures within which they can communicate with one another [14.11]. This receptor hyperstructure is perhaps a proto-icon, given that an icon is a sign that would possess the character which renders it significant even if its object were no longer to exist, whilst the signal transduction downstream of the binding of an attractant to the hyperstructure might qualify it as a proto-index (moreover,

[2] According to author #2, protosemiotic agents do not calculate the gradient and even do not "know" that such gradient exists. Instead, they follow an innate sequence of actions that includes habituation and sensitization.

it would lose the character which makes it a sign if its object—such as an attractant—were removed). The movement part of chemotaxis is provided by flagella, which might therefore be seen as effectors. A flagellum, however, can also be seen as a sensor of humidity [14.109] or, to put it differently, the same hyperstructure can be both effector and sign.

According to Colapietro [14.16] (cited from [14.42]), signs signify in multiple ways: "The fact that a sign functions in one of these ways (as an icon, index or symbol) does not preclude it from functioning in one or both of the other ways; in fact, the most perfect signs function in all three of these ways." For example, semiotic methods can help in communication by helping people learn languages via the use of emoticons and colors, which involve several senses [14.82].

A molecule, macromolecule or other type of potential information is more likely to be used as a cue or signal by bacteria if it is already influential in determining the phenotype. It is not therefore surprising that many of the simple, universal molecules and inorganic ions that play important roles in modern cells were probably important as signals at the origins of life. Multiple sign function occurs when, for example, a metabolic enzyme has a role in addition to the one normally ascribed to it [14.6], as when the metabolic enzyme, pyruvate kinase, interacts physically (presumably to signal metabolic conditions) with some of the classical DNA replication enzymes to modulate their activity [14.37]. In eukaryotes, other glycolytic enzymes can also act as DNA binding proteins [14.90]. A possible general interpretation of the multi-function of some enzymes would be in the context of a change in the dynamics of hyperstructures with the disassembly of a hyperstructure freeing a constituent enzyme as a signal that changes the behavior of other hyperstructures.

14.4.2 Hierarchy, Agency, and Their Features

To be classed as semiosis, it has been proposed that "code-based mapping needs to be embedded in a functional agent and support the functions of this larger system. Thus, semiosis requires interaction between systems of two (or more) hierarchical levels, which can be taken as an important clue for reconstructing the origin of life" [14.98]. The hyperstructure concept is relevant here not only because hyperstructures constitute a level between the molecule and the bacterial cell but also because hyperstructures are the descendants of the composomes from which life possibly originated in the scenario of a prebiotic ecology [14.39]. Multi-level functioning is evidenced by chemotaxis, which operates at the level of the cell in addition to that of the hyperstructure such that the calculation that leads to running or tumbling also depends on whether the bacterium is alone or in a swarm [14.83].

Agents can be defined as living or non-living systems with goal-directed programmed behavior. All share the ability to perform functions for the purpose of reaching certain goals. These functions are encoded and controlled by a set of signs [14.98]. Agents have a hierarchical structure[3] and their subagents can be nested forming what Hoffmeyer termed "boundaries within boundaries" [14.35]. But what exactly is a boundary or frontier? A very general definition would be that a frontier is a discontinuity in the connectivity of a system

[3] Some agents (e.g., a ribosome or thermostat) may seem isolated and lack hierarchical structure if taken within a short time interval. However, such agents are always produced by parental agents at earlier times, and this production-relation is equivalent to a hierarchy. Additional connections of these agents may include communication to their parental agents (or their descendents) or niche construction.

[14.69]. It is this discontinuity that permits a thing to have a distinct behavior and identity (e.g., the discontinuity in connectivity that separates a person from the chair on which the person is sitting). Growth can reduce the degree of connectivity of a system and therefore threaten its identity; restoring connectivity by division may be one of the factors driving the bacterial cell cycle [14.69]. In competitive coherence, groups of elements (whether hyperstructures or macromolecules) can be found in the *Active* subset; such groups can be identified by the fact that the elements within a group are more tightly connected to one another than they are to elements outside the group.

Kauffman has compared the autonomous agents that characterize living systems to Carnot engines, in which constraints are essential to exploiting a flow of energy [14.46]. From the point of view of competitive coherence and hyperstructures, a bacterial cell contains both E and NE hyperstructures which, in principle, show different patterns and types of connectivity. One question then is whether the signaling in bacteria differs according to the equilibrium or non-equilibrium characteristics of the hyperstructures involved. For example, it might be argued that going from a growing to a non-growing state involves *Now* signals that become increasingly strong as the new state develops. It is conceivable that quorum-sensing molecules could be put into this class as well.

The semiotic theory of Peirce is based on three basic categories, one of which, Firstness, is taken to include chaos, potentiality, pure feeling and potential qualia. The nature of *qualia alias* subjective experiences, remains a profound mystery. It may be therefore relevant in the context of semiotics that the operation of competitive coherence to generate *active* states in bacteria, brains and other living systems can be considered as inseparable from *qualia* [14.70].

14.5 Discussion

At the level of the cell, competitive coherence describes how a subset of all possible hyperstructures is selected in the process of creating a meaningful phenotype that is coherent both with the previous state of the cell and with its nascent state and environment. This process entails sensing, regulating and coordinating. It therefore entails integrating a wealth of signals and environmental cues. The bacterial cell cycle, which comprises the replication and segregation of DNA and subsequent cell division, is an example of such integration with hyperstructure-related signals varying from alarmones (such as cAMP and (p)ppGpp) to the fluid state of the membrane (via the DnaA protein), the structure of the nucleoid (via NAPs and the SlmA protein) and the dynamics of protein assemblies (such as those formed by FtsZ and its partners).

Competitive coherence has been used as the basis for a learning neural-net program. The advantage of doing this is that the value (or lack of value) of its concepts can be tested according to whether they facilitate or hinder the program learning a task. This has been done in the contexts of biology [14.72] and economics [14.79]. Learning depends on the program having elements with *Now* and *Next* connections, which affect the composition of the nascent state of the system and of its future state, respectively. Learning also depends on elements that correspond to environmental inputs and system outputs and, we would argue, signals. Learning takes the form of changing the connections by changing the addresses of elements held in the *Now* and *Next* fields of an element; each field of an element could therefore be represented by a vector of n-dimensions, where n is the total number of elements in the program but, more simply, in the versions of the program where the size of the fields can vary, each field could be represented by a point on a single dimension that corresponds to its size. In the latter case, the implication for bacteria is that each cue and signal would exist in a two-dimensional space with *Now* and *Next* coordinates.

Insofar as an element is in an *Active* subset that contains an output element, that element could be considered as both a signal and as an effector. This is the position adopted in the bacterial hypothesis presented here: signals and cues can also be effectors. One advantage of this is that the power of a signaling molecule/macromolecule is increased if it also has a function, as in the case of a nutrient or an enzyme, which often have signaling roles.

It appears that there are significant differences in the cues and signals that influence the transition from NE to E hyperstructures and *vice versa*. For example, shutting down systems in stressful conditions avoids the dangers of growth (when bacterial cells are vulnerable, for example, to antibiotics). If the duration of the stress is unpredictable, the safest course of action for most of the population would be to remain in a non-growing state. Hence, we might expect the E hyperstructures that largely constitute a non-growing cell to make only a few, relatively weak, connections to the NE hyperstructures needed for growth. In terms of signals, the strength of the *Next* signals emitted by an E hyperstructure to promote selection of an NE element should be less than the signals needed to promote continued selection of an E element.

This study demonstrates the value of the exploration-stabilization basis of all learning-like processes that include natural selection, habituation, and sensitization in bacteria. The main advantage of the concept of competitive coherence is in the integration of *Now* and *Next* processes and signaling: the former supports stability of phenotypes and functions, whereas the second supports active exploration and growth. Additionally, competitive coherence implies preservation of the basic cell identity despite transitions between phenotypes and functions, which makes transitions cohesive.

However, there is no consensus on whether bacterial cells have a capacity to learn. Ginsburg and Jablonka define learning as

> a process leading to an experience-dependent behavioral response of a system. It requires that: (i) A sensory stimulus that originates either from the activities of the system or from the external world leads to a change in the internal state of the system. (ii) A memory trace of this change is stored; storage requires active stabilization and involves valence mechanisms of positive or negative reinforcement. (iii) Future interactions with the stimulus or associated stimuli lead to a change in the threshold of the behavioral response [14.40].

This definition seems to fit cases of habituation and sensitization in bacteria because a sensory stimulus changes the internal state of the system (e.g., methylation of a chemoreceptor protein). The change is stored; and future interaction with the stimulus changes the threshold of the behavioral response. The only questionable feature is "active stabilization" of information storage; but it is possible at least in principle because bio-molecular mechanisms often include repair functions. The whole process of memory storage and retrieval is heritable and automated, and thus, author #2 has reservations for calling it "learning" since bacterial cells "learn" only those signals for which they already have specialized devices.

Thus, author #2 prefers to classify habituation and sensitization in bacteria as "adaptive plasticity" rather than "learning" [14.99].

This discussion is also relevant to the question of whether bacteria have Pavlovian conditioning. Many biologists assume that bacteria are not capable of conditioning [14.29], but some scientists have attempted to rebuff this notion. For example, Mitchell *et al.* [14.62] showed that *E. coli* can learn to anticipate a carbon source change after exposure to another stimulus. However, in contrast to Pavlovian conditioning, which develops within one and the same animal, bacterial "conditioning" requires hundreds of generations. One conclusion could be that this phenomenon is an example of associative natural selection rather than associative learning.

The proposed model is compatible with the notion of practopoiesis [14.67], which defines hierarchy levels in adaptive systems: (T_0) one-level of self-organization, (T_1) two-level system capable of regulating self-organization, (T_2) three-level system with supervision at the highest level and plasticity of regulation, and (T_3) second-order supervision capable of context-dependent regulation and abduction (i.e., reverse engineering). In the versions of competitive coherence model simulated so far, there has only been one level of regulation (i.e., T_1) operating via the change of status (active or non-active) of functional units at the lower level that are available as resources. The hypothesis does, however, allow the creation of additional levels as proposed in [14.70]; for example, each Active subset at the first level could be treated as an object in its own right and competitive coherence could then act at a second level on these objects to determine their membership within an Active subset at the second level. This transition between levels would constitute a "traverse" in the language of practopoesis. Bacteria are capable of context-dependent regulation, and therefore they qualify to be called T_3 practopoietic systems. But they belong to the lowest category within T_3 systems because individual bacterial cells are not capable of abduction, although evolving populations of bacteria can acquire new structures and activity algorithms through natural selection, which is a learning-like and abduction-like process.

Although this paper is focused on bacteria, the principle of competitive coherence is fully applicable to eukaryotic organisms, multicellularity, and multi-organism agents. Norris *et al.* [14.72], p. 1, wrote:

> The network coherence idea turned out to be scale-free and to address one of the most important problems that confronts bacteria, eukaryotic cells, collections of cells (including brains), and even social organisations. This problem is how a system can behave in (1) a coherent way over time so as to maintain historical continuity and (2) a coherent way at a particular time that makes sense in terms of both internal and environmental conditions.

Thus, the principle of competitive coherence is relevant to any adaptive agency (including subagents) striving to keep the balance between self-preservation and change. It is reminiscent of the philosophy of Simondon who assumed that the identity of living systems is always in the making and is never completed [14.94]. Agents not only learn how to preserve their identity (sf. Darwin's principle of survival and reproduction), but also learn how to change: in what direction, to what extent, and how to preserve coherence and viability despite change, which constitutes robustness (ct. Bergson's [14.7] idea of creative evolution).

In our model, these two aspects of learning are represented by the *Now* and *Next* components. Finally, the principle of competitive coherence assumes that the capacity of coherent change is rooted in the diversity of subagents (e.g., hyperstructures) and their communication. Our hypothesis assumes that active subagents establish *Now* and *Next* relations with other components of the system and participate in a kind of "voting" to select the future state of the whole agent. This is reminiscent of *the society of mind* hypothesis [14.61], where it is assumed that mind has a collective nature, and its functional components (subagents) communicate with each other and compete for influence.

Declaration

All authors declare no conflict of interest

Acknowledgment

We thank anonymous reviewers for their helpful comments and suggestions.

References

[14.1] Adams, D.W., Wu, L.J. and Errington, J. (2015) Nucleoid occlusion protein Noc recruits DNA to the bacterial cell membrane. *EMBO J* 34(4), 491-501.

[14.2] Alghofaili, F., Najmuldeen, H., Kareem, B.O., Shlla, B., Fernandes, V.E., Danielsen, M., Ketley, J.M., Freestone, P. and Yesilkaya, H. (2021) Host stress signals stimulate pneumococcal transition from colonization to dissemination into the lungs. *mBio* 12(6), #e0256921.

[14.3] Anderson, M., Deely, J., Krampen, M., Ransdell, J., Sebeok, T.A. and von Uexküll, T. (1984) A semiotic perspective on the sciences: Steps toward a new paradigm. *Semiotica* 52(1/2), 7-47.

[14.4] Balaban, N.Q., Merrin, J., Chait, R., Kowalik, L. and Leibler, S. (2004) Bacterial persistence as a phenotypic switch. *Science* 305(5690), 1622-1625.

[14.5] Bateson, G. (1987 [1972]) *Steps to an ecology of mind. Collected essays in anthropology, psychiatry, evolution, and epistemology*. Jason Aronson, Northvale, NJ.

[14.6] Beaufay, F., Coppine, J. and Hallez, R. (2021) When the metabolism meets the cell cycle in bacteria. *Curr Opin Microbiol* 60, 104-113.

[14.7] Bergson, H. (1998 [1911]) *Creative evolution*. Dover, New York.

[14.8] Bhattacharyya, S., Walker, D.M. and Harshey, R.M. (2020) Dead cells release a 'necrosignal' that activates antibiotic survival pathways in bacterial swarms. *Nature Communications* 11, #4157, 4151-4159.

[14.9] Bohringer, J., Fischer, D., Mosler, G. and Hengge-Aronis, R. (1995) UDP-glucose is a potential intracellular signal molecule in the control of expression of sigma S and sigma S-dependent genes in *Escherichia coli*. *Journal of Bacteriology* 177(2), 413-422.

[14.10] Boutte, C.C., Henry, J.T. and Crosson, S. (2012) ppGpp and polyphosphate modulate cell cycle progression in Caulobacter crescentus. *J Bacteriol* 194(1), 28-35.

[14.11] Bray, D., Levin, M.D. and Morton-Firth, C.J. (1998) Receptor clustering as a cellular mechanism to control sensitivity. *Nature* 393(6680), 85-88.

[14.12] Buffington, S.A., Dooling, S.W., Sgritta, M., Noecker, C., Murillo, O.D., Felice, D.F., Turnbaugh, P.J. and Costa-Mattioli, M. (2021) Dissecting the contribution of host genetics and the microbiome in complex behaviors. *Cell* 184(7), 1740-1756.

[14.13] Burgess, R.R. (2021) What is in the black box? The discovery of the sigma factor and the subunit structure of *E. coli* RNA polymerase. *J Biol Chem* 297(5), #101310.

[14.14] Castuma, C.E., Crooke, E. and Kornberg, A. (1993) Fluid membranes with acidic domains activate DnaA, the initiator protein of replication in *Escherichia coli. Journal of Biological Chemistry* 268(33), 24665-24668.

[14.15] Chien, A.C., Hill, N.S. and Levin, P.A. (2012) Cell size control in bacteria. *Curr Biol* 22(9), R340-349.

[14.16] Colapietro, V.M. (1996) The ground of semiosis . In: *Peirce's Doctrine of Signs: Theory, Applications, and Connections*. V.M. Colapietro and T.M. Olsjewsky, (eds.) Mouton de Gruyter, BerlIn 129-140.

[14.17] Deely, J. (1992) Semiotics and biosemiotics: Are sign-science and life-science coextensive? In: *Biosemiotics. The semiotic web 1991*. T.A. Sebeok and J. Umiker-Sebeok, (eds.) Mouton de Gruyter, New York: 45-75.

[14.18] Dicks, L.M.T. (2022) How does quorum sensing of intestinal bacteria affect our health and mental status? *Microorganisms* 10(10), #1969.

[14.19] Dilweg, I.W. and Dame, R.T. (2018) Post-translational modification of nucleoid-associated proteins: an extra layer of functional modulation in bacteria? *Biochem Soc Trans* 46(5), 1381-1392.

[14.20] Dominguez, D.C., Guragain, M. and Patrauchan, M. (2015) Calcium binding proteins and calcium signaling in prokaryotes. *Cell Calcium* 57(3), 151-165.

[14.21] Dubey, G.P. and Ben-Yehuda, S. (2011) Intercellular nanotubes mediate bacterial communication. *Cell* 144, 290-600.

[14.22] Feklistov, A., Sharon, B.D., Darst, S.A. and Gross, C.A. (2014) Bacterial sigma factors: a historical, structural, and genomic perspective. *Annu Rev Microbiol* 68, 357-376.

[14.23] Fernandez-Coll, L., Maciag-Dorszynska, M., Tailor, K., Vadia, S., Levin, P.A., Szalewska-Palasz, A. and Cashel, M. (2020) The absence of (p)ppGpp renders initiation of *Escherichia coli* chromosomal DNA synthesis independent of growth rates. *mBio* 11(2).

[14.24] Fichtl, B., Shrivastava, S. and Schneider, M.F. (2016) Protons at the speed of sound: Predicting specific biological signaling from physics. *Sci Rep* 6, #22874.

[14.25] Fishov, I. and Norris, V. (2012) Membrane heterogeneity created by transertion is a global regulator in bacteria. *Curr Opin Microbiol* 15(6), 724-730.

[14.26] Fishov, I. and Woldringh, C.L. (1999) Visualization of membrane domains in *Escherichia coli. Molecular Microbiology* 32(6), 1166-1172.

[14.27] Freestone, P.P., Sandrini, S.M., Haigh, R.D. and Lyte, M. (2008) Microbial endocrinology: how stress influences susceptibility to infection. *Trends Microbiol* 16(2), 55-64.

[14.28] Gao, J.Y., Ye, C.L., Zhu, L.L., Tian, Z.Y. and Yang, Z.B. (2014) A homolog of glyceraldehyde-3-phosphate dehydrogenase from Riemerella anatipestifer is an extracellular protein and exhibits biological activity. *J Zhejiang Univ Sci B* 15(9), 776-787.

[14.29] Ginsburg, S. and Jablonka, E. (2019) *The evolution of the sensitive soul. Learning and the origin of consciousness*. MIT Press, Cambridge, London.

[14.30] Ginsburg, S. and Jablonka, E. (2021) Evolutionary transitions in learning and cognition. *Philosophical Transactions of the Royal Society B* 376, #20190766.

[14.31] Goldberg, J.B., Crisan, C.V. and Luu, J.M. (2022) *Pseudomonas aeruginosa* antivirulence strategies: Targeting the type III secretion system. *Adv Exp Med Biol* 1386, 257-280.

[14.32] Grimwade, J.E., Rozgaja, T.A., Gupta, R., Dyson, K., Rao, P. and Leonard, A.C. (2018) Origin recognition is the predominant role for DnaA-ATP in initiation of chromosome replication. *Nucleic Acids Res* 46(12), 6140-6151.

[14.33] Hammel, M., Amlanjyoti, D., Reyes, F.E., Chen, J.H., Parpana, R., Tang, H.Y., Larabell, C.A., Tainer, J.A. and Adhya, S. (2016) HU multimerization shift controls nucleoid compaction. *Sci Adv* 2(7), #e1600650.

[14.34] Hoffmeyer, J. (1996) *Signs of meaning in the universe. The natural history of signification.* Indiana University Press, Bloomington.

[14.35] Hoffmeyer, J. (2008) *Biosemiotics: An examination into the signs of life and the life of signs* University of Scranton Press, Scranton, PA.

[14.36] Hoffmeyer, J. and Stjernfelt, F. (2016) The great chain of semiosis. Investigating the steps in the evolution of semiotic competence. *Biosemiotics* 9(1), 7-29.

[14.37] Horemans, S., Pitoulias, M., Holland, A., Pateau, E., Lechaplais, C., Ekaterina, D., Perret, A., Soultanas, P. and Janniere, L. (2022) Pyruvate kinase, a metabolic sensor powering glycolysis, drives the metabolic control of DNA replication. *BMC Biol* 20(1), 87.

[14.38] Horn, G., Hofweber, R., Kremer, W. and Kalbitzer, H.R. (2007) Structure and function of bacterial cold shock proteins. *Cell Mol Life Sci* 64(12), 1457-1470.

[14.39] Hunding, A., Kepes, F., Lancet, D., Minsky, A., Norris, V., Raine, D., Sriram, K. and Root-Bernstein, R. (2006) Compositional complementarity and prebiotic ecology in the origin of life. *Bioessays* 28(4), 399-412.

[14.40] Jablonka, E. and Ginsburg, S. (2022) Learning and the evolution of conscious agents. *Biosemiotics* 15(3), 401-437.

[14.41] Jameson, K.H. and Wilkinson, A.J. (2017) Control of initiation of DNA replication in *Bacillus subtilis* and *Escherichia coli. Genes (Basel)* 8(1).

[14.42] Ji, S. (2017) Neo-semiotics: Introducing zeroness into Peircean semiotics may bridge the knowable and the unknowable. *Progress in Biophysics and Molecular Biology* 131, 387-401.

[14.43] Katayama, T., Kasho, K. and Kawakami, H. (2018) The DnaA Cycle in *Escherichia coli*: Activation, function and inactivation of the initiator protein. *Frontiers in Microbiology* 8, 2496.

[14.44] Kauffman, S. (1996) *At home in the Universe, the search for the laws of complexity*. Penguin, London.

[14.45] Kauffman, S.A. (1993) *The origins of order: Self-organization and selection in evolution.* Oxford University Press, New York.

[14.46] Kauffman, S.A. (2000) *Investigations*. Oxford Univ. Press, Oxford, New York.

[14.47] Kennell, D. and Riezman, H. (1977) Transcription and translation initiation frequencies of the *Escherichia coli* lac operon. *Journal of Molecular Biology* 114(1), 1-21.

[14.48] Konto-Ghiorghi, Y. and Norris, V. (2021) Hypothesis: nucleoid-associated proteins segregate with a parental DNA strand to generate coherent phenotypic diversity. *Theory Biosci* 140(1), 17-25.

[14.49] Kull, K. (2009) Vegetative, animal, and cultural semiosis: The semiotic threshold zones. *Cognitive Semiotics* 4, 8-27.

[14.50] Laub, M.T. and Goulian, M. (2007) Specificity in two-component signal transduction pathways. *Annual Review of Genetics* 41, 121-145.

[14.51] Lemke, J.L. (2000) Across the scales of time: Artifacts, activities, and meanings in ecosocial systems. *Mind Culture(*4, 273-290.

[14.52] Loiko, N., Danilova, Y., Moiseenko, A., Kovalenko, V., Tereshkina, K., Tutukina, M., El-Registan, G., Sokolova, O. and Krupyanskii, Y. (2020) Morphological peculiarities of the DNA-protein complexes in starved *Escherichia coli* cells. *PLoS One* 15(10), #e0231562.

[14.53] Lopes, J.G. and Sourjik, V. (2018) Chemotaxis of *Escherichia coli* to major hormones and polyamines present in human gut. *The ISME Journal* 12(11), 2736-2747.

[14.54] Macek, B., Forchhammer, K., Hardouin, J., Weber-Ban, E., Grangeasse, C. and Mijakovic, I. (2019) Protein post-translational modifications in bacteria. *Nat Rev Microbiol* 17(11), 651-664.

[14.55] Maslowska, K.H., Makiela-Dzbenska, K. and Fijalkowska, I.J. (2019) The SOS system: A complex and tightly regulated response to DNA damage. *Environ Mol Mutagen* 60(4), 368-384.

[14.56] Matsuhashi, M., Pankrushina, A.N., Takeuchi, S., Ohshima, H., Miyoi, H., Endoh, K., Murayama, K., Watanabe, H., Endo, S., Tobi, M., Mano, Y., Hyodo, M., Kobayashi, T., Kaneko, T., Otani, S., Yoshimura, S., Harata, A. and Sawada, T. (1998) Production of sound waves by bacterial cells and the response of bacterial cells to sound. *J Gen Appl Microbiol* 44(1), 49-55.

[14.57] Maurya, G.K., Modi, K., Banerjee, M., Chaudhary, R., Rajpurohit, Y.S. and Misra, H.S. (2018) Phosphorylation of FtsZ and FtsA by a DNA damage-responsive Ser/Thr protein kinase affects their functional interactions in *Deinococcus radiodurans*. *mSphere* 3(4), #e00325-00318.

[14.58] Maynard Smith, J. and Szathmáry, E. (1995) *The major transitions in evolution*. W.H. Freeman Spektrum, Oxford, New York.

[14.59] McCulloch, W. (1945) A heterarchy of values determined by the topology of nervous nets. *Bulletin Mathematical Biophysics* 7, 89–93.

[14.60] Mendoza, S.D., Nieweglowska, E.S., Govindarajan, S., Leon, L.M., Berry, J.D., Tiwari, A., Chaikeeratisak, V., Pogliano, J., Agard, D.A. and Bondy-Denomy, J. (2020) A bacteriophage nucleus-like compartment shields DNA from CRISPR nucleases. *Nature* 577, 244-248.

[14.61] Minsky, M. (1986) *The society of mind*. Simon and Schuster, New York.

[14.62] Mitchell, A., Romano, G.H., Groisman, B., Yona, A., Dekel, E., Kupiec, M., Dahan, O. and Pilpel, Y. (2009) Adaptive prediction of environmental changes by microorganisms. *Nature* 460(7252), 220-224.

[14.63] Monahan, L.G., Hajduk, I.V., Blaber, S.P., Charles, I.G. and Harry, E.J. (2014) Coordinating bacterial cell division with nutrient availability: a role for glycolysis. *mBio* 5(3), e00935-14.

[14.64] Montero Llopis, P., Jackson, A.F., Sliusarenko, O., Surovtsev, I., Heinritz, J., Emonet, T. and Jacobs-Wagner, C. (2010) Spatial organization of the flow of genetic information in bacteria. *Nature* 466(7302), 77-81.

[14.65] Muller-Hill, B. (1998) The function of auxiliary operators. *Mol Microbiol* 29(1), 13-18.

[14.66] Muskhelishvili, G., Forquet, R., Reverchon, S., Meyer, S. and Nasser, W. (2019) Coherent domains of transcription coordinate gene expression during bacterial growth and adaptation. *Microorganisms* 7(12), #694.

[14.67] Nikolić, D. (2015) Practopoiesis: Or how life fosters a mind. *Journal of Theoretical Biology* 373, 40–61.

[14.68] Norris, V. (1998) Modelling *Escherichia coli*. The concept of competitive coherence. *Comptes Rendus de l'Académie des Sciences - Series III* 321(9), 777-787.

[14.69] Norris, V. (2015) Why do bacteria divide? *Front Microbiol* 6, 322.

[14.70] Norris, V. (2021) Competitive coherence generates qualia in bacteria and other living systems. *Biology (Basel)* 10(10).

[14.71] Norris, V., Cabin, A. and Zemirline, A. (2005) Hypercomplexity. *Acta Biotheor* 53(4), 313-330.

[14.72] Norris, V., Engel, M. and Demarty, M. (2012) Modelling biological systems with competitive coherence. *Advances in Artificial Neural Systems* 2012(703878), 1-20.

[14.73] Norris, V., Fralick, J. and Danchin, A. (2000) A SeqA hyperstructure and its interactions direct the replication and sequestration of DNA. *Mol Microbiol* 37(4), 696-702.

[14.74] Norris, V., Grant, S., Freestone, P., Canvin, J., Sheikh, F.N., Toth, I., Trinei, M., Modha, K. and Norman, R.I. (1996) Calcium signalling in bacteria. *J Bacteriol* 178(13), 3677-3682.

[14.75] Norris, V. and Hyland, G.J. (1997) Do bacteria "sing"? *Molecular Microbiology* 24, 879-880.

[14.76] Norris, V., Kepes, F., Amar, P., Koch, I. and Janniere, L. (2017) Hypothesis: Local variations in the speed of individual DNA replication forks determine the phenotype of daughter cells. *Medical Research Archives* 5, 1-18.
[14.77] Norris, V., Molina, F. and Gewirtz, A.T. (2012) Hypothesis: bacteria control host appetites. *J Bacteriol* 195(3), 411-416.
[14.78] Norris, V., Nana, G.G. and Audinot, J.N. (2014) New approaches to the problem of generating coherent, reproducible phenotypes. *Theory Biosci* 133(1), 47-61.
[14.79] Norris, V., Norris, L. and Wong, W.-K. (2015) The positive feedback advantages of combining buying and investing. *Theoretical Economics Letters* 5, 659-669.
[14.80] Norris, V., Reusch, R.N., Igarashi, K. and Root-Bernstein, R. (2014) Molecular complementarity between simple, universal molecules and ions limited phenotype space in the precursors of cells. *Biol Direct* 10, 28.
[14.81] Nöth, W. (1990 [1985]) *Handbook of Semiotics*. Indiana University Press, Bloomington, IN.
[14.82] Ollerhead, S. (2019) Teaching across semiotic modes with multilingual learners: Translanguaging in an Australian classroom. *Language and Education* 33, 106-122.
[14.83] Partridge, J.D., Nhu, N.T.Q., Dufour, Y.S. and Harshey, R.M. (2019) *Escherichia coli* remodels the chemotaxis pathway for swarming. *mBio* 10(2).
[14.84] Peirce, C.S. (1931–1958) *Collected Papers (eds. Hartshorne and Weiss; Burks)*. Harvard University Press, Cambridge, MA.
[14.85] Peirce, C.S. (1998 [1893–1913]) *The essential Peirce: Selected philosophical writings*. Indiana University Press, Bloomington.
[14.86] Raskin, D.M. and de Boer, P.A. (1999) Rapid pole-to-pole oscillation of a protein required for directing division to the middle of *Escherichia coli*. *Proceedings of the National Academy of Sciences of the U.S.A.* 96(9), 4971-4976.
[14.87] Reguera, G. (2011) When microbial conversations get physical. *Trends Microbiol* 19(3), 105-113.
[14.88] Remesh, S.G., Verma, S.C., Chen, J.H., Ekman, A.A., Larabell, C.A., Adhya, S. and Hammel, M. (2020) Nucleoid remodeling during environmental adaptation is regulated by HU-dependent DNA bundling. *Nat Commun* 11(1), 2905.
[14.89] Robles-Ramos, M.A., Margolin, W., Sobrinos-Sanguino, M., Alfonso, C., Rivas, G., Monterroso, B. and Zorrilla, S. (2020) The Nucleoid Occlusion Protein SlmA Binds to Lipid Membranes. *mBio* 11(5).
[14.90] Ronai, Z. (1993) Glycolytic enzymes as DNA binding proteins. *Int J Biochem* 25(7), 1073-1076.
[14.91] Salthe, S.N. (1985) *Evolving hierarchical systems*. Columbia University Press, New York.
[14.92] Santos-Zavaleta, A., Salgado, H., Gama-Castro, S., Sanchez-Perez, M., Gomez-Romero, L., Ledezma-Tejeida, D., Garcia-Sotelo, J.S., Alquicira-Hernandez, K., Muniz-Rascado, L.J., Pena-Loredo, P., Ishida-Gutierrez, C., Velazquez-Ramirez, D.A., Del Moral-Chavez, V., Bonavides-Martinez, C., Mendez-Cruz, C.F., Galagan, J. and Collado-Vides, J. (2019) RegulonDB v 10.5: tackling challenges to unify classic and high throughput knowledge of gene regulation in *E. coli* K-12. *Nucleic Acids Res* 47(D1), D212-D220.
[14.93] Scheller, D., Twittenhoff, C., Becker, F., Holler, M. and Narberhaus, F. (2021) OmpA, a common virulence factor, is under RNA thermometer control in *Yersinia pseudotuberculosis*. *Front Microbiol* 12, 687260.
[14.94] Scott, D. (2014) *Gilbert Simondon's psychic and collective individuation: A critical introduction and guide*. Edinburgh University Press, Edinburgh.
[14.95] Sebeok, T.A. (1972) *Perspectives in zoosemiotics*. Mouton, The Hague.
[14.96] Sebeok, T.A. (1976) *Contributions to the doctrine of signs*. Indiana University Press, Bloomington, IN.

[14.97] Sharov, A.A. (2006) Genome increase as a clock for the origin and evolution of life. *Biology Direct* 1, 17.

[14.98] Sharov, A.A. (2010) Functional information: Towards synthesis of biosemiotics and cybernetics. *Entropy* 12(5), 1050-1070.

[14.99] Sharov, A.A. (2022) Consciousness and learning from the biosemiotic perspective. *Biosemiotics* 15(3), 483-490.

[14.100] Sharov, A.A. and Tønnessen, M. (2021) *Semiotic agency. Science beyond mechanism.* Springer, Dordrecht. 25.

[14.101] Sharov, A.A. and Vehkavaara, T. (2015) Protosemiosis: Agency with reduced representation capacity. *Biosemiotics* 8(1), 103-123.

[14.102] Shih, Y.L. and Rothfield, L. (2006) The bacterial cytoskeleton. *Microbiol Mol Biol Rev* 70(3), 729-754.

[14.103] Sobetzko, P., Travers, A. and Muskhelishvili, G. (2012) Gene order and chromosome dynamics coordinate spatiotemporal gene expression during the bacterial growth cycle. *Proc Natl Acad Sci U S A* 109(2), E42-50.

[14.104] Sugiyama, R., Kasho, K., Miyoshi, K., Ozaki, S., Kagawa, W., Kurumizaka, H. and Katayama, T. (2019) A novel mode of DnaA-DnaA interaction promotes ADP dissociation for reactivation of replication initiation activity. *Nucleic Acids Res* 47(21), 11209-11224.

[14.105] Szatmari, D., Sarkany, P., Kocsis, B., Nagy, T., Miseta, A., Barko, S., Longauer, B., Robinson, R.C. and Nyitrai, M. (2020) Intracellular ion concentrations and cation-dependent remodelling of bacterial MreB assemblies. *Sci Rep* 10(1), 12002.

[14.106] Uhlig, F. and Hyland, N.P. (2022) Making sense of quorum sensing at the intestinal mucosal interface. *Cells* 11(11), 1734.

[14.107] von Uexküll, J. (1926) *Theoretical biology*. Harcourt, Brace & Co., New York.

[14.108] von Uexküll, J. (1982 [1940]) The theory of meaning. *Semiotica* 42(1), 25-82.

[14.109] Wang, Q., Suzuki, A., Mariconda, S., Porwollik, S. and Harshey, R.M. (2005) Sensing wetness: a new role for the bacterial flagellum. *EMBO J* 24(11), 2034-2042.

[14.110] Whitley, K.D., Jukes, C., Tregidgo, N., Karinou, E., Almada, P., Cesbron, Y., Henriques, R., Dekker, C. and Holden, S. (2021) FtsZ treadmilling is essential for Z-ring condensation and septal constriction initiation in Bacillus subtilis cell division. *Nat Commun* 12(1), 2448.

[14.111] Yu, J., Liu, Y., Yin, H. and Chang, Z. (2019) Regrowth-delay body as a bacterial subcellular structure marking multidrug-tolerant persisters. *Cell Discov* 5, 8.

15

Self-Reinforcing Cycles and Mistakes: The Emergence of Subjective Meaning

Victoria N. Alexander

Dactyl Foundation, New York, New York, USA
IPAK-EDU, Pittsburgh, Pennsylvania, USA

Online resources:
https://dactylfoundation.org/?page_id=1494
https://www.researchgate.net/profile/Victoria-Alexander-4

Abstract

In this chapter, I investigate how subjectivity arises in living systems, both as a consequence of memory—the experience-dependent modification of signal pathways or internal sensing infrastructure—that allows responses to anticipate future conditions, *and* as a result of interpretative responses to the semiosic qualities of the individual steps within a signal pathway. Thus, there are two levels of subjectivity. I argue that understanding certain kinds of responses as *mis*interpretations is necessary to understanding how truly novel, and thus truly subjective, behavior can emerge. I argue that meaning emerges when the physical *qualities* of signs in pathways (arbitrary connection, relative proximity, relative similarity) constrain the internal processes of living systems such that outcomes tend to maintain that system. I explore this hypothesis using examples of the experimental manipulation of organisms that trigger a *mistaken* subjective response, which misuses existing semiosic infrastructure: (1) an anti-fertility vaccine causes an autoimmune response with a vaccine conjugate acting as an adapter transferring the meaning (to the immune system) of tetanus toxoid to human chorionic gonadotropin; (2) classically conditioned slime mold preferentially moves toward cold as an artificial index of food; and (3) angiotensin II receptors bind with the pharmaceutical-industry-designed molecular mimics known as angiotensin-receptor blockers. I specifically use aneural models of learning and memory to stress that it is the semiosic nature of biological processes, not the unique nature of the central nervous system, that allows for intelligence, meaning and subjectivity to emerge. Finally, I note that out of the somewhat plastic local sign readings of individual molecules in a signal pathway, oscillating chemical reactions emerge. These cyclical processes function as further tools for responding to external conditions.

Keywords: Relative objectivity, interpretation and misinterpretation, aneural learning, relative similarity, relative proximity, arbitrary connection, directionality, originality

Email: alexander@dactyl.org

15.1 Introduction

During the fall in upstate New York, farmers sometimes make corn mazes by cutting out intricate pathways through a cornfield with a center space as the goal. Some of the maze runners never find the center, and it always takes longer to get out than one expects. When my husband and I passed one such maze, on a late November afternoon on a lonely rural road, we decided to try it, even though the attraction was officially closed for the season.

The maze was large, about four rectangular acres. On the south side was the unmanned entrance and a separate exit, which suggested to me that the center would have an entrance and an exit that would prevent the maze runners from going back the way they had come. After diving in between the towering rows of eight-foot-high corn, we looked up at the overcast sky, which didn't give us much information about the position of the sun. We did take note of the electrical poles along the east end, which helpfully provided orientation. Looking down, we noticed a gradient on the pathway. Some of the branching paths were somewhat grassy and others had been trampled to hardened mud. We paused to make sense of this and decided that only about half of the maze runners would select the wrong path and everyone would have to take the correct path eventually, but, of course, the wrong path would be trampled twice by each investigator. Then we imagined that the children who had found the center by the correct path might run back to help their lost friends. Thus, the path should be less and less grassy as we got nearer to the center. Trusting this shaky logic, we preferentially followed the less grassy tracks. We were misled by our rule into a few dead-ends, but eventually we made it to the circular center where a pumpkin-head scarecrow grinned.

As I had guessed, on the other side of the circle was an exit pathway. As the sun was now starting to set, we decided we would try to get out as quickly as possible, instead of willfully risking getting lost for the fun of it. My husband said that he was going to follow his sense of direction with the electrical poles as his guide. I said that I was going to mindlessly follow a rule that I had learned (I don't remember where), to put my right hand on the wall of the corn and to go forward keeping my hand on the wall no matter what. This would ensure that I would not inadvertently retrace my steps. The first few turns or so my rule comported well enough with my husband's "sense," but there came a fork in the path where he laughed when I went right. He went left. I followed my rule to a few dead-ends, although I cheated and turned without actually going all the way in. Eventually, I met my husband who was coming in my direction. He turned around and followed me out of the maze.

Slime mold, *Physarum polycephalum,* is a brainless multi-nucleated single-cell organism that is well known for being able to find the most efficient path out of a maze [15.1] [15.34]. Motivated by a goal—such as a 2% glucose solution in a little hollow in the center of a petri dish—slime mold can follow the glucose gradient that is diffusing away from the source, and it can remember which paths turned out to be dead-ends because it leaves traces behind marking where it went and retreated. The activity of slime mold may be considered an example of minimal cognition, displaying chemical memory and adaptable searching behaviors [15.12] [15.15] [15.34]. According to Ginsburg and Jablonka [15.23], p. 2, "learning" is a process involving "sensor–effector coupling, attribution of valence (intrinsic reinforcement), storage and recall involving a change in response threshold (so the system can be said to "anticipate" the effects of learned stimuli)." In this chapter, learning something "new," that goes beyond genetically-determined couplings, will be important.

Even though slime mold does not have a brain, it behaves meaningfully insofar as it adapts its behavior and uses tools/rules in order to pursue its goal of survival. But is *P. polycephalum* capable of creativity? Can it teach itself a new rule or invent a new tool for finding food? Can slime mold be fooled? Can slime mold harness and entrain its own internal noise to respond to a new external regularity? In other words, can slime mold learn to repurpose its own genetically/chemically determined behaviors? Are all living systems capable of such learning and, if so, by what mechanisms?

Comparing the performances of humans and slime mold solving a maze risks the dangers of, on the one hand, anthropomorphizing a single-celled organism, and, on the other hand, reducing the behavior of creative, self-conscious big-brained animals to the chemically dependent, genetically determined behavior of a simple organism. I can avoid both pitfalls by sticking to the principle that livings systems all have some things in common that make them different from non-living systems. I will argue that they both subjectively interpret their environments.

In this chapter exploring how meaning emerges, I start with the assumption that the concept of *meaning* implies a subjective perspective of a living system. Because binary logic may suggest to some that *subjective* meaning is the opposite of *objective*, I counter with the argument that subjectivity is better understood as *relative objectivity*. Subjectivity arises with the formation of internal biosemiosic relations. Subjectivity depends, in part, upon an experience-dependent memory, which Baluška and Levin [15.8] define as a "modification of internal structure, in a stimulus-specific manner that alters the way the system will respond to stimuli in the future as a function of its past." Although the organism's history determines future actions, it must be noted that subjectivity also depends upon the indeterminate way that the living system can use its memory traces. From a biosemiotic perspective, there is no denying the real and significant effects of relative similarity, relative proximity, arbitrary connection in the responses of living systems—from single cells to societies—to sign-vehicles that are associated with useful effects. There are different levels of subjectivity; inherited or habitual and newly formed.

Subjectivity is essentially selfhood, and subjective interpretations about the qualities of sign-vehicles are generally made for the sake of, or rather with the effect of, maintaining the self. We may not be able to point to an actual moment when the emergence of subjectivity takes place because it is a self-reflexive process. As with all teleological phenomena, that is, so-called *future-oriented events*, subjectivity can only be perceived with a backward glance. As I use the term "teleological" here, I do not refer to any Christian or supernatural notions. I am only concerned with meaning and function in living systems. While physicists do not need to consider the "purpose" of this or that phenomenon, biologists quite often ask themselves what this organ, hormone or enzyme is *for* in the body.

To be precise, teleological processes are inherently circular, not forward looking. What happened to work well in the past is selectively retained. If maze runners discover a new tool/rule for avoiding previously explored and abandoned territory, they may reuse and/or pass on that capacity. A capacity is the physical embodiment of past successes and when it is recalled for use, we say, somewhat incorrectly, that the organism is *anticipating* the future. The subjective agent is always like a general fighting previous battles. This is acceptable as long as sufficient aspects of the Umwelt are preserved, such that there is no practical difference between past and future, between recalling and anticipating. But at some point, the general may encounter such an unfamiliar situation that he is forced to try something he has never done before, maybe use a strategy only used for other contexts, not for war. In this way he may stumble upon a new strategy and may even switch his goals.

I will argue that the emergence of biosemiosic meaning depends on the interrelation of distinct evolutionary processes that I call, following the lead of pre-Darwinian teleologists, *directionality* and *originality*. Such distinctions have been glossed over, confused or ignored in the literature since the publication of Darwin's thesis, even though these aspects are key to understanding evolution. In short, I will argue that directional forms in nature arise first and provide the reliable constraints that may be used in ways that are meaningful (self-reinforcing) to the organism. Prior to emergence of life, lifeless matter can self-organize and become *directional*, which seems somewhat lifelike, but it is only the necessary though insufficient conditions for *originality* to emerge.

Originality emerges when new signs are discovered and used. This behavior is brief, and the newly discovered way of negotiating with the environment can become habit and thereafter may be triggered automatically when the organism encounters a similar situation. Thus, originality reverts back to a directional form of behavior, which, though automatic, may be quite complex if the organism has acquired a number of optional directional paths it may choose according to context.

It is worth noting that these directional tendencies are subjective in the sense that they have been acquired through unique experience, either of that individual organism in the case of learning or that of a species in the case of evolution. However, when living systems exhibit directionality, they might as well be automatons designed by natural selection, whose programs have been codified in the genome and are triggered by specific environments at the appropriate times, such that those responses are likely to survive and continue. In this state, the subjectivity of an organism is not all that remarkable. Its acquired tendencies, its semiotic habits, are predictably activated in specific conditions. The act of choosing appropriately between two or more existing tools represents an uninteresting kind of subjectivity. Adding to Kull's [15.31] analysis of the mechanisms underlying choice in organisms, I argue that truly novel behavior can only emerge through *mistaken* choices [15.4].

When the organism takes a new direction, it forges a new path. I call this *misinterpreting*. The use of a negative term for the process of discovery may offend the sensibilities of some readers. I may seem judgmental to characterize behavior according to a normative sense of what organisms should be doing or not, but I do think there is a normative state defined by behavior that worked in the past and allowed the organism to keep going. In the pragmatic sense, there is a behavior that is "normal." The idea of *mis*interpretations or mistakes is not necessarily negative, especially to those who are familiar with the creative process. Many a narrative of inventors and artists testify to the fact that mistakes frequently open the door to new knowledge.

I also want to make a distinction between creative misinterpretation and learning by negative feedback or "trial-and-error," which only helps the agent make gradual improvements on an already existing tool or procedure or just helps the agent find an optimal strategy for reaching its usual, pre-defined goal. With discovery via mistakes, the positive feedback process that I wish to explore here, entirely new tools would be discovered or new goals set.

15.2 Subjective versus Relatively Objective

It is commonly accepted that physical science researchers employ *quantitative* methods; they take measurements, collect and count data points, and formulate equations that model how systems develop. In contrast, humanities researchers employ so-called *qualitative* methods, such as case studies, interviews, observations, and surveys to construct *narratives* about complex systems that are hard to describe quantitatively. The method of physicists is said to be objective and the method of those interested in humanities is said to be subjective. Biologists must work somewhat uncomfortably between these two domains.

Neo-Darwinian evolutionary theorists, wanting to be more like physicists than humanities researchers, try to make the role of an organism's subjective interpretations as insignificant as possible—and also as different as possible from other organisms' subjective interpretations—so that the direction of evolution can be defined statistically from the cumulative effects of discrete and various individual actions. In this way, each singular subjective effect of a single organism acting in its environment is minimized and is rendered statistically insignificant. The only ones that matter are those that are common and add up to something. The idea seems to be that it is best to ignore subjectivity if you want to do real science.

Contrary to this, I have a mind to argue that it is not just that the humanities researchers' subject matter—the behavior of people and societies—is too complicated to quantify easily, so they have fallen back on their own subjective (and therefore imperfect) understandings. As I will try to show, biological processes themselves are subjective and the effects of subjective interpretations—not just those of animals, but also of cells—are significant to the direction of evolution. I argue that subjective experience is coincident with life itself and did not wait upon the arrival of brains.

As I try to dispel the notion that subjective processes can be reduced to that which can be quantified, I will nevertheless try to relate a subjective realism narrative about the kinds of *physical* processes, constraints and relations—involved in the use of symbol, index, and icon signs—that contribute to subjectivity. This argument thus defends more of a realist position, than that of a radical constructivist position.

Although semiotician C. S. Peirce (1839–1914) inspired my use of the terms and concepts, "symbol, index and icon," I use them somewhat differently here because I'm applying them to brainless systems. Like Peirce, I see that all semiosis has three aspects, which I describe as the *sign vehicle*, the *response*, and the *object*, which I consider to be the effect that the response has on the living system's ability to respond to the sign vehicle in that way. I believe that the ultimate object of any semiosic process is reinforcing the conditions that allow that sign reading to recur, not the agent getting to some actual thing. As long as the semiotic habit is reinforced by the response, it does not matter if the reading is true or not.

I believe that self-reinforcing effects, not things or ideas or concepts, shape the interpretations of big-brained animals. Even if the "object" of the word *tree* seems to be some actual tree or the general idea of a tree, the semiosic process depends upon it being reinforced, whether there really is a tree or not and irrespective of whether or not the concept is correct. As long as the interpretive response is habit-confirming, the object is performing its semiosic role (see [15.2]).

In terms of the different types of signs, for me what is important is that a symbol represents arbitrarily (like an encrypted code), an index represents by proximity (like an index finger pointing) and an icon represents by similarity (like a pictograph). Perhaps it is better to speak of the iconic, indexical, or symbolic qualities of signs and sign relations, which can be simultaneously present in any sign. The index that represents by mere proximity (instead of by actual natural connection) has an arbitrary quality to it. An icon that represents by being coincidentally similar in shape also has an arbitrary quality to it. And thus being indexical and iconic may harbor the potential to be symbolic.

15.3 How Symbols Are Used by Brainless Living Systems

A *symbol* can be defined as a conventional sign that arbitrarily represents its object. Using a dictionary one can look up the arbitrary designate "tree" and find it is the symbol for various kinds of large leafy plants with a woody bark. Using an English-Spanish dictionary, one can translate "tree" to *árbol.* Dictionaries are like decryption devices for the arbitrary codes of language. Although the symbol is the most evolved type of sign, I begin with its emergence because most probably life emerged with the first encoded memory and not before.

As authors such as Deacon and Sherman [15.18] and Sharov [15.44] have variously described it, the primordial soup, out of which life emerged, most certainly already had the self-organizing tendencies and the spontaneously-formed structures that seem life-like. But fully semiosic life would not be possible until some kind of physical memory could ensure that lucky self-reinforcing accidents could be retained and improved upon. Code biologist Marcello Barbieri [15.9] argues that the emergence of the genetic code provided that physical memory that could serve as a template for recreating structures. Although it may not be clear how it emerged, suffice it to say, for now, that somehow a genetic code was encrypted and the trace of that process was erased.

That process probably involved indexes, I argue, and I will come back to the question at the end of this section. For the moment, I want to fast-forward to the time when primitive eukaryotes (ancestors of slime mold and humans) have already appeared. Eukaryotes have nuclei in which they can protect their encoded history from the disruptive effects of the wider world. Thus, eukaryotes can accumulate new codes better and faster than their predecessors.

Inside the eukaryote cell, after tRNA is copied from the template of DNA and leaves the nucleus to enter the cytoplasm of the cell, tRNA undergoes a process that can legitimately be called decryption. We may say that tRNA is like the code of a cypher wheel, which the operator can turn so that letters in the alphabet correspond arbitrarily to numbers, for example, A, B, C to 1, 2, 3 or A, B, C to 18, 19, 20. On one end of tRNA is an anticodon with three nucleotides; the other end binds to the corresponding amino acid. There is no natural affinity between anticodons and amino acids; the link is arbitrary, provided by the structure of the tRNA (formed during earlier stereochemical processes) which acts as an *adapter* between materials that do not spontaneously bond together. The adapter is like the rule for translation provided by a dictionary.

Enlarging upon Umberto Eco's observations [15.20], Barbieri [15.9] notes that the tRNA adapters are related by *convention*, not too dissimilarly from the way that words, which arbitrarily represent objects in the world, have evolved as conventions in language over time. Acted on by the ribosome, the decrypted codes (anticodons translated into amino acids) are joined together, according to the way the long strand of mature mRNA has been spliced, and this process produces proteins.

As Barbieri argues, with the appearance of encryption/decryption "machines" and codes—such as the ribosome and tRNA—evolution ceased to be a purely chemical process. Other new codes followed, such as splicing codes, signal transduction codes, actin codes, and cell migration codes. Over two hundred codes have been described in the literature [15.10]. Barbieri emphasizes that new codes do not originate by the copying errors of previous codes, but in the origin of new *rules* for encrypting/decrypting a code. Barbieri notes that Darwin's idea of evolution, as resulting from random variation and selection, cannot account for the encryption process.

Biological codes such as Barbieri describes may certainly be retained by natural selection, but if we want to understand how that process of associating one unrelated thing with another thing could have happened, we need biosemiotics. What sort of natural process could have occurred to bring things together via an adapter?

Evolution may be impossible to observe in process. Generally, we can only recognize new codes after they have been formed and selected. But we do have some examples of artificial biological codes created by researchers by means of an encryption device, which may be compared to Barbieri's concept. Such a new code will not be genetically fixed, but will serve as an example of a code being learned by an individual's system for use during that individual's lifetime or experience.

Some pharmaceutical interventions employ the strategy of encryption to more or less trick the body. Since the 1970s researchers have been developing anti-fertility vaccines. In one case [15.47], the hormone Human Chorionic Gonadotropin (HCG) is linked to a tetanus toxoid by a conjugate. When injected with the product, subjects develop antibody responses to the tetanus *and* the HCG by association. HCG is a necessary component for a successfully developing embryo. If a vaccinated subject becomes pregnant, the embryo will be spontaneously aborted. The conjugate here plays the role of an encryption apparatus. The body learns to mis-associate part of itself (HCG) with a toxic substance. This autoimmune response is a qualitatively different process than the normal process of an antibody response. Thus, we can say that learning has taken place.

With this example, we can confirm that biological systems can learn new encrypted codes, that is, new symbols, arbitrarily relating one thing to another. It seems very strange that meaning (the effect of) one thing can be transferred to another just by physically joining them. But apparently this is so.

15.4 How Indexes Are Used by Brainless Living Systems

Unlike the symbol—which is related to something else in a completely arbitrary manner, forced together, in Barbieri's description, by a physical encrypted connection like an adapter or a rule book—the index is generally a by-product or an effect of whatever it is pointing to. When this is so, an index is a very reliable sign. Sometimes an index can be part of a series of indexes, like stepping stones in a physical process that lead to some goal state, each transduction leading to the next. If you smell smoke (an index) you might look around until you see a plume of smoke (another index), which indicates there is a fire and you might want to evacuate.

When a maze runner follows a gradient to reach the goal, he/she/it is using an index sign. The gradient should become more concentrated as the runner gets closer to the goal, just as smoke is more concentrated the closer you get to the fire. So, a gradient is not like an arbitrary symbol whose meaning one has to have learned beforehand to follow. For example, if my husband and I met a sign at a fork in the path of the corn maze that said, *Ve a la derecha*, as a Spanish speaker, he would have known how to follow that sign's direction, but I probably would not have. That is because *Ve a la derecha* is a conventional symbol. But we both could read the gradient sign without any prior memorization or learning. One can follow a gradient in a pathway without even thinking consciously; when one walks through a field or a forest, one tends to follow the well-beaten path.

However, because an index sign might not be something that is an actual *effect* of whatever it is pointing to, like a diffusing glucose gradient, and indexes can be things that are merely usually *nearby* something that is useful, the reading of indexes can be fallible. For example, a bear in the woods may follow the odor of decaying wood where grubs, which he likes to eat, tend to be. He can follow the odor as an index sign of grubs, but he may not find any in the rotting log. When it turns out that the object (the self-sustaining effect of eating grubs) of the sign-reading does not exist, we can see how responses to signs can be fallible, based on past experiences.

The essence of what makes a sign a sign is its potential to mislead. As Umberto Eco ([15.19]: 7) famously noted, "Semiotics is in principle the discipline studying everything which can be used in order to lie. If something cannot be used to tell a lie, conversely it cannot be used to tell the truth: it cannot in fact be used 'to tell at all.'" Because a sign is not what it represents, the potential to misinterpret a sign inheres in all signs. However, this can be positive if mistakes allow us to find a new pathway that was not available before. Indeed, mistakes may be the only way because how do we grasp that which we do not yet know?

When we learn a new index sign, we may not know if the sign is physically linked to what it represents or is only proximate. I find it useful to think of the formation of a new index sign as a kind of Pavlovian conditioning [15.36], during which *unrelated* but proximate factors in the environment can become associated, and one can become an index sign pointing to the other. The archetypical example of Pavlovian conditioning involves a dog, an animal with a pretty sophisticated brain, but to say that the association between food and Pavlov's dinner bell is a memory in the brain, just begs the question of how memories form linkages in the first place. To avoid such circular reasoning, I offer an example of slime mold learning via Pavlovian conditioning. Since it has neither brain stem or nerve cells, how it learns will depend upon some kind of structural reorganization of its semiotic infrastructure, transitioning from directional semiosis to original semiosis.

In a classical conditioning experiment, Shirakawa *et al.* [15.45] trained *P. polycephalum* to go against its natural (codified) tendency to move toward warmth and instead to move toward cold in search of food. In the experiment, cold-signal and food-signal sensors were simultaneously triggered, activating two separate signal transduction pathways. Per the organism's usual conditions, the food-signal pathway increases its contraction frequency, causing the plasmodium to move; the cold-signal pathway inhibits the reaction that triggers contraction. According to further analysis done by Shirakawa and Sato [15.46], the two signal pathways could become coupled if one pathway catalyzes the other's activation *and* that of an inhibitor. Such interference could lead to the buildup of a by-product that, once present in sufficient amounts, could prevent the inhibition of the contraction. Thus, the cold-signal would indirectly disinhibit its own inhibition function. As noted in [15.5], this mechanism is essentially a self-organizing Turing Switch, which is ubiquitous throughout biological systems [15.26] [15.30] [15.38] [15.48]. As Shirakawa and Sato [15.46] also note, the by-product reserve functions as temporary memory, and the plasmodium could move toward cold even when the food-signal was no longer present. To the particular slime mold trained in the experiment, cold came to *mean* food; it became an index sign for food, just as a sound of a bell had come to mean food to the dog in Pavlov's experiment.

Although the formation of such a Turing Switch mechanism would be a temporary conditioning, this may serve as a model for any kind of associative learning via the entanglement of signal pathways. As with Barbieri's adapter example, here again we find that biological subjectivity emerges from describable *physical constraints*. Thus, we can say that the subjective sign-reading of this particular conditioned slime mold is *relatively objective*. Although cold does not always indicate food, in this special case in the lab, it does.

This would be an artificial index because slime mold cannot expect to find food in cold areas in its natural habitat. The organism is tricked/trained to act as if cold were a true index of food. It is precisely because these researchers were able to train the organism to do something unnatural that they can claim learning did take place through Pavlovian conditioning. As Kalevi Kull remarks in a similar context, "Meaning may appear only in systems in which something sometimes goes wrong" [15.31], p. 617. Here we have caught the process of originality in action.

In a different experiment also involving slime mold, researchers Saigusa *et al.* [15.42] conditioned *P. polycephalum* to move rhythmically by associating a period of time with a cold dry stimulus. In that example as well, the researchers trained the slime mold to do something unnatural. The *arbitrariness* of these associations—cold with food or cold with moving rhythmically—confirms that the conditioning is a semiosic process. It's part of that particular slime mold's subjective experience of the world.

Pavlovian conditioning is a useful mechanism in the natural world—if not in a lab where the researchers can subject the organism to perverse conditions. Index signs usually indicate real physical relationships: where there is smoke, there is probably fire; where there is odor of decaying wood, there are probably grubs; where there is warm wet organic material, there probably is a fungus food source, where the gradient is more concentrated, it is probably closer to the source, and so forth. We can imagine that the slime mold's natural tendency to move away from cold resulted after generations of naturally-imposed Pavlovian conditioning and was finally codified genetically and became the organism's default tendency. Warmth is not food itself, but it is a reliable index of food to *P. polycephalum*. Once codified, behavior is *directional*, insofar as it is automatic and does not require learning by conditioning.

This example of the accidental index, the arbitrary index, the merely nearby index, shows how symbols, which arbitrarily refer to something else, might begin to form. While the new symbol mentioned above, the HCG=toxoid symbol, was intentionally created by *physically binding* two things together that would not normally bind, in the case of Pavlovian conditioning, unrelated things are also associated but by a different method of bringing them into spatial-temporal proximity. In the Shirakawa and Sato [15.46] experiment, when different chemical signal pathways were activated simultaneously, they happened to interact and became linked such that their interaction switched one on/off. Following this logic, we might speculate that if a subject is exposed to environmental toxins at the same time that he is exposed to, say, pollen, he might develop an allergy to pollen. The toxin and pollen would not be physically yoked together, as the two factors are in the anti-fertility vaccine, but it is conceivable to imagine that they could be associated by being caught up in the complicated process of an immune reaction. Nowlin [15.35] lays the groundwork for investigating these possibilities using biosemiotics. The toxins would activate an immune response and somehow the pollen, being nearby like an index, might become associated with the real source of the problem. If the adaptable immune system remembers this, afterward exposure to pollen might trigger a response to it *as if it were* a toxin.

There is some evidence that such a guilty-by-association mechanism or what is called the "bystander activation" mechanism may be a factor in some autoimmune reactions. In a review of the mechanism, Fujinami *et al.* [15.22] explain that bystander activation appears to occur after severe infection when many protein fragments of T-cell-killed infected tissue were released into the extra-cellular matrix, where these tissue proteins were engulfed by macrophages or dendritic cells, which, in turn, presented these self-protein antigens on their surfaces. Surface tags on these immune cells do generally *indicate* the culprit of a serious infection that is in progress, and so naturally, if T cells interact with them, such tags might *indicate* to the T cell that they should treat self-cell tags like non-self tags due to the diseased context in which these self-protein tags are being presented to the T cell.

Later these trained or conditioned cytotoxic memory T cells can become reactivated by cytokines signaling a *new unrelated* infection. These "bystander" autoreactive T cells attack *uninfected* tissue cells; they do not attack cells infected with the new virus. We can suppose that these memory cytotoxic T cells must have learned to confuse infected tissue cells with healthy same tissue cells. This may be a case of indexical semiosis, in which the meaning of one sign vehicle has been transferred to a nearby one.

The fact that an antigen as index might *not* be an actual relevant by-product of a pathogen shows how learning new indexes is fallible and is a subjective experience based on the misinterpretation of the relative proximity of a false index and a true index. In the case of autoimmune disease, the body has left the correct directional path via a mistaken sign. Not all acts of originality turn out well for the organism.

15.5 How Icons Are Used by Brainless Living Systems

Following a well-beaten path toward a goal is to follow an index sign. Have you ever been hiking in the woods and found yourself *off* the designated path, following a gulley or an animal track? In this instance, you followed an icon sign. Gullies and animal tracks *look like* designated foot paths.

In this section, I will consider the ways in which a cell can respond to an icon sign, making generalizations about things or patterns that are similar to each other. For this example, I will turn to signal pathways in humans that are under intense investigations by the pharmaceutical industry, whose objective it often is, as noted already with the case of the anti-fertility vaccine, to trick the body into responding differently. As with both the cases of the anti-fertility vaccine and of the Pavlovian conditioning of slime mold, we are here again interested in the way a living system can be tricked by sign relations. In my analysis of the icon sign, another pharmaceutical intervention will serve to illustrate the emergence of originality, causing a disruption of the directional path. In this case, an icon sign will be used to block a pathway.

For my example I have chosen the renin-angiotensin system (RAS), which modulates blood pressure, because I happened to be familiar with it. During the first months of the SARS-CoV-2 lockdown in 2020, my friend, who has hypertension (i.e., high blood pressure), was worried that she was at higher risk because the virus employs *molecular mimicry*, an icon sign, to use the ACE2 receptor. People with hypertension tend to have insufficient numbers of or improperly functioning ACE2s [15.13] [15.33], which, when functioning properly, facilitates vasodilation, counteracting the vasoconstriction effects of the angiotensin-converting enzyme (ACE). It was a matter of debate at the time whether or not patients with hypertension should continue their medication to lower blood pressure [15.13] [15.40].

In the following, I will describe that overly complicated system, known as the RAS, the details of which will not matter to the argument here so much as the fact that it is a very complicated system with many interacting parts. Triggered by decreased blood flow, the kidneys secrete renin, which converts angiotensinogen, which has been produced by the liver, into angiotensin I, then angiotensin-converting enzyme (ACE) converts angiotensin I into angiotensin II.

Angiotensin II can lead to a number of different reactions relating to blood pressure. 1) It can bind to receptors on blood vessel cells making them contract, increasing blood flow (and presumably switching off renin secretion that started this chain of events). 2) Angiotensin II can also trigger the pituitary to produce a hormone that triggers the kidneys to increase water reabsorption, thereby increasing blood volume and thereby blood pressure. Again, this should switch off renin secretion. 3) Angiotensin II can also promote water reabsorption in the kidneys by affecting the adrenal cortex and aldosterone production, which raises sodium levels and decreases potassium levels. This is the third way the RAS pathway has for shutting itself off.

In addition, there is the angiotensin-converting enzyme 2 (ACE2) pathway that counters the effects of the RAS. ACE2 cleaves angiotensin I and angiotensin II into the inactive peptides Angiotensin (1–9) and Angiotensin (1–7), respectively. Angiotensin (1–9) gets further metabolized into Angiotensin (1–7), which is a *vasodilator*. When ACE2 cleaves angiotensin I, it becomes unavailable to ACE to convert it into angiotensin II. ACE2 can also directly hydrolyze angiotensin II, reducing its levels in plasma. ACE2 leads to multiple ways of reducing blood pressure.

The main point to notice is that the body has multiple, somewhat redundant pathways for maintaining blood pressure and these pathways share molecules and interact with each other in rather complicated ways.

ACE, angiotensin I and II, and ACE2 make up the biological infrastructure, retained by natural selection, which channels a semiotic habit that results in an effect that is self-sustaining for the body, namely by maintaining homeostatic healthy blood pressure. Here again, we have something similar to a Turing Switch mechanism (such as we saw with the slime mold's cold signal and food signal pathways) made possible by the interaction of different signal pathways.

It is significant that there are a number of steps that are involved on the way to the final reinforcing effect. Each step requires the correct interpretation according to its directional orientation, its semiotic habit. In many biological systems, the number of steps is often quite large and complicated. We can say that each step is a means to an end—that is, it exists in the body because it was selected for a function in the course of evolution—which means that each step represents—is an index sign of—the objective it eventually serves. As with any purposeful action, it matters less what actual steps are taken, as long as the objective is satisfied. There may be a number of means to any end. That is, the steps may be somewhat arbitrary. Substitutions can be made for some of the sign-vehicles if similar enough that they work for their purposes.

That means an icon sign can be recruited for substitution. We may say that the RAS and ACE2 signal pathways are overly complicated for functions they serve, whose intermediary steps interact with each other by triggering and/or suppressing each other. Any one part of the complicated machine might be replaced with some other tool (e.g., a protein) that serves the same purpose, cleaving, converting, binding or blocking to achieve the end state of maintaining stable blood pressure. But there are constraints on the kinds of replacements that can be made. For a different molecule to function as a substitute it must *mimic* the shape of the molecule it is replacing. It must be an icon sign.

The flexibility and redundancy of biological systems can be exploited in medical interventions for hypertension. For instance, pharmaceutical-industry-designed angiotensin-receptor blockers (ARBs) *mimic* angiotensin II and bind to and inhibit angiotensin II receptors in the blood vessels, pituitary and adrenal cortex—in scenarios 1, 2 and 3 above—preventing their function to promote arteriolar contraction [15.11].

We can say that interpretation occurs here insofar as ARBs, mimicking shape of angiotensin II, function as an icon sign and bind to its receptors, deceiving them, in a sense. The way that pharmaceuticals work, using molecular mimicry to bind with one of the body's existing proteins, thereby activating/deactivating some process, gives us some insight into how natural evolutionary adaptations might occur. We can imagine that a mutation in the DNA could produce a novel protein that also mimics the shape of angiotensin II and may bind to the same receptors. Thus, those who have such a mutation might have naturally occurring ARBs, and if the production of such protein is triggered by increased blood flow, it may be self-reinforcing. It may be, for all we know, that some individuals with faulty ACE2 may also make novel proteins that work the same way the ARBs do.

To conclude this analysis of the RAS happily, as it turned out, my friend did not need to make the decision to discontinue her medication or not. While debating the question, she contracted a nasty cold, recovered, and afterwards found out that she had had COVID-19. The intricacy of her immune system was sufficient to protect her, even though she was in one of the high-risk categories for the disease.

To recap, in this chapter so far I have presented three examples of artificially-created signs that affected interpretation and led to learning: 1) a pharmaceutical intervention showing how the physical linkage of HCG with a tentus toxoid, or the symbol encryption mechanism, creates (a desired) autoimmunity; 2) slime mold researchers showing how Pavlovian conditioning, or the index mechanism, could make learning an association possible; 3) another pharmaceutical intervention, ARBs, showing how molecular mimicry, or the icon mechanism, could make fooling the body possible. These may be artificially-generated semiosis, but they show how such natural mechanisms, if they exist, could work to create the conditions that might be favored by natural selection. The point is, I am trying to add another layer to Darwinian evolutionary theory. The ways in which adaptations are first tried out must be semiotic. We do not have to imagine that every random change is tested in an infinite trial and error process that gradually shapes organisms over time. That would be very energy intensive. Instead, we can imagine that adaptations are constrained by the degree to which the physical interactions are either *proximate* in space and time (index) and/or are physically *similar* (icon). Biological systems tend to be constrained by indexical and iconic relations because of the nature of biological systems, functioning amid simultaneous inputs and depending on structural affinities in order to interact. It is these semiotic processes, not nucleotide changes, that can be selected or not. Random single-point mutations cannot be the source of significant novelty; the change that matters must be misinterpretation.

15.6 The History of Directionality and Originality as Concepts in Evolutionary Science

An *interpretation* occurs when *different* semiotic habits are triggered by the *same* sign, depending on context, or when *different* signs can all trigger *one* semiotic habit. There are a number of one-to-many and many-to-one signaling pathways in the body that allow the body to be adaptable in the sense of staying the same when things change.

A *misinterpretation* is different. It occurs when a new sign enters and repurposes an existing semiotic habit. If the misinterpretation is lucky for the organism, if, for example, a receptor binds with a novel protein such that it offers an alternative method of maintaining a signal pathway, it can become a new habit, an interpretation (a semiotic act) habituated or fixed by selection.

Living organisms, with brains or without, can detect and act on patterns. Sometimes the pattern may be just coincidental as it is with molecular mimicry or with Pavlovian conditioning. We must say, therefore, that creativity is, in part, the ability to make use of possibly merely coincidental patterns. That coincidence can be similar appearance, as with an icon, or just being associated in space and time, as with an index, or there may be an artificial physical bond between two materials that acts as a connector. Such a sign-reading is not *merely* subjective in the sense of an interpretation imposed upon the process by an external observer. The subjective interpretation occurs in the process itself due to the relational qualities involved in the process.

I have spent so much time illustrating how a symbol is a kind of encryption (related arbitrarity); an index is a kind of association (related proximity); an icon is a kind of generalization (related similarity) because these appear to be the mechanisms by which evolution breaks out of its directional habit (variations on existing genres) to create a novel and original meaning.

When Darwin banished Christian-inspired teleology from science, he threw the baby out with the bathwater so to speak. Secular teleology provides extremely subtle tools for exploring how something truly new and yet meaningful can emerge in nature. In particular, teleology has shown that the concept of purpose entails two distinct kinds of behavior that are important to understanding the evolution of new species. The first is associated with what I have been calling *directionality*, that is the maintenance of biological systems via stabilizing mechanisms that render habituation possible. Directionality results in types, species, genres, and habits. The second is associated with what I have been calling *originality*, or the discovery of wholly new functions. We can say that originality *begins* the process of creating new types, species, genres and habits.

Darwinism, the theory of random variation and selection for reproductive fitness, ignores the possible significance of internal (mis)interpretations, as described here, in the process of evolution. Darwin's theory speaks exclusively to the issue of how originality, in the sense of new functions discovered after random change, is promoted and fixed by natural selection. Darwin and his descendants do not have a theory of biosemiosis, and so, for them, the useful interpretations and misinterpretations of new structures is decided strictly by random chance. Biosemioticians, in contrast, recognize the role of biases, i.e., the significant semiotic constraints on biological systems.

Both directionality and originality emerge from selection processes that are somewhat different from the natural selection for reproductive fitness. With *directional* selection, which favors the robustness of internal formal structures, slight differences are inconsequential, and sameness prevails. Directionality may be associated with *formal* rather than *final* cause. The many-to-one and one-to-many semiosic rules result in a conservation of existing trends and an inherent resistance to change via a kind of inertia. With *original* selection, which favors the *functionality* of internal formal structures vis-a-vis a larger context, slight differences result in significant functional improvements and/or systematic reorganization. The radical changes, such as we have observed when molecular mimicry (accidental icons) or Pavlovian condition (accidental indexes) are behind the emergence of originality. These changes are semiotically constrained, not random.

Among the few teleologists who have specifically noted this dual aspect is Jean Baptiste Lamarck (1744–1829). According to Gould ([15.25], p. 170), Lamarck has been misremembered as being primarily responsible for promoting the "acquired characteristics" theory of evolution; he was merely including what was more or less an accepted notion of his day. Lamarck's actual theory included *two* mechanisms of evolution: one involved an inherent *tendency* toward complexification, and the other involved accidents of history, as organisms dealt with their environments.

English biologist Richard Owen (1804–1892), who is credited with identifying the first "Dinosauria" fossils as extinct species, also noted teleology's dual aspects. As Russell [15.41] notes, Owen's first aspect is "directional" and underpins the stability and similarity or "a vegetative repetition of structure," which is illustrated in the conception of a groundplan or archetype and in the mathematical symmetry of some organisms and even crystals. The other, which Owen called "teleological," spurs change and diversity by shaping an organism according to function. Owen argued that directionality produces similarity of forms across species, while the teleological results in the diversity of organic forms (see [15.41], pp. 102-12).

The lesser-known German anatomist C. B. Reichert (1811–1883) made similar distinctions between "class" and "functional" modifications in the structures of biological organisms (see [15.41], pp. 145-7). In more recent authors, the duality of teleological phenomena has been noted in, for example, Erich Jantsch's [15.27] conformation and novelty and Stanley Salthe's [15.43] distinctions between development and evolution.

Without a conception of directional evolution, it is not possible to understand the significance of a departure from that direction. In their landmark works, Russell [15.41] and

Lenoir [15.32] describe the history of pre-Darwinian evolutionary theorists who were interested in directional processes and sought to understand why certain forms are common in nature, even if they are not necessarily more functional than less common forms. They believed that the direction of evolution was partly predetermined, in the sense of limited, by intrinsic factors not environment. In contrast, as Gould ([15.24], p. 293) observes, neo-Darwinian theorists believe that there is no inherent direction and the only source of direction comes from historical contingencies established by the natural selection of greater reproductive fitness. Countering Gould, Fontana and Buss [15.21] argue that if the metaphorical "tape" of evolution could be rewound and started again, evolution might not depart so drastically from history, and that inherent constraints might be at least as significant as historical contingencies in directing the pathway of evolution.

Another term for directionality is *orthogenesis*, the idea that there are inherent drives that make evolution proceed in more or less a straight line—having tendencies—instead of branching off randomly. Directionality may be a tendency, but that tendency is not toward perfection in any sense. It often just means more and more of the same thing, like bigger and bigger antlers, for instance that, as Reid ([15.39], pp. 267-287, 392) argues, appear to be caused by genetic reduplication, not pressured by reproductive fitness or sexual attractiveness.

Directional tendencies can be functionally neutral. Having the means to sense a change in the environment is to have a directional tool that may be put to use or may not be. Having a means to sense changes in the environment does not guarantee that the organism can respond meaningfully to that change. For example, during development, the map butterfly (*Araschnia levana*) is sensitive to changes in photoperiods [15.29]. Butterflies that emerge in the spring have been exposed to short days and are mostly orange with brown spots. Butterflies that emerge in the summer have been exposed to long days and are dark brown with a white stripe. Extended light periods trigger genes for more ecdysone production, which is a precursor for the dark pigment production pathway.

In this example, the sign-vehicle is the change in the duration of light, the response is the increased production of dark pigment, but there is no known or alleged advantage to this response, no self-reinforcing effect. There is no semiotic object. Thus, I do not consider the different coloration to be a semiosic process. I would guess that this sensitivity to photoperiods is a kind of Gouldian spandrel [15.25]: in other words, at some other point in the insect's development or day-to-day activities, ecdysone production triggered by longer photoperiods probably does serve a function. The fact that directional tools for sensing are available, whether functional or not, opens up the possibility space for a new function to be discovered for that tool. Directionality precedes originality.

15.7 Looping Signal Pathways, Oscillators and Turing Switches as Sensors

To some extent, I have focused on the individual steps in the signal pathways more than their cyclical nature. In this section, I want to pull back and get an aerial view, so to say, of the structure that emerges from individual sign-readings at each step in a signal pathway, which tend to lead to the emergence of spatial/temporal waves or patterns; they are often oscillating chemical reactions that reach a threshold, then recycle.

Different chemical reactions may be affected by changes in temperature, humidity, changes in chemical concentrations in the external or internal environments, etc. When these chemical reactions are affected by such external differences, we can say that the organism has a means of detecting that difference, as we saw with the slime mold cold-signal detection pathway. Importantly, a change in the environment could change the nature of the emergent spatial or temporal pattern produced by the signal pathway. At different phases in the chemical clock, different and corresponding genes may be triggered. Thus, the shape of the spatial pattern and the timing of the periodic patterns may be part of the regulation of the genes and other processes.

Conrad Hal Waddington [15.50] investigated how the process of morphological development might involve standing or traveling chemical waves that differentially trigger genes. Inspired by Waddington, Alan Turing [15.48] posed the idea that reaction-diffusion dynamics, like those that underlie embryogenesis, might also underlie information processing in biological systems more generally as well as in the brain. In 1951, Boris Belousov discovered how chemical clocks operate when the reactions switch back and forth between two regimes. It became clear that such oscillators might be the means by which biological systems respond to changes in the environment, speeding up, slowing down or aligning with an internal oscillating chemical reaction (see [15.51]). In the last several decades, many neuroscience researchers (e.g., [15.17] [15.28] [15.49]) have characterized the activity of the brain in terms of traveling waves that provide differentiation in spatial/temporal patterns that synchronize distant neurons, without creating permanent circuits [15.37]. And to circle back, let me add that, according to Buznikov and Shmukler [15.14] and Baluška and Levin [15.8], the bioelectric mechanisms used originally for morphogenesis, very likely have been extended and used in the far more complex learning and memory functions of the human mind.

The laws of physics and chemistry determine oscillating phases and wave patterns. However, due to their non-linear nature, they are somewhat unpredictable at the local level—we may not be able predict, for instance, precisely which cell will be on or off, participating in the wave or not, at a given moment. Nevertheless, at an emergent level, the patterns may be said to conform to types or, possibly, attractors and are somewhat predictable. Such constraints most certainly underlie what so many early evolutionary theorists noted as the phenomenon of directionality.

Bacigalupi [15.7] explores the very powerful idea that a virtual directional structure can be implied by overlapping chemical oscillators. I have already alluded to this briefly above with regard to the Saigusa *et al.* [15.42] experiment and the entrainment of slime mold's chemical oscillators, which allows it to harness an appropriate internal oscillator to match an external temporal phase. According to Bacigalupi, superposed chemical terrains can constrain, or rather provide, a no-additional-energy-cost avenue for sensing, biasing, and becoming habituated. The salient point for my thesis here is that the potential capacity would not be an existing physical structure. Bacigalupi offers a way of understanding how a sensing tool aligns with relevant noise along the path of least resistance. This is a kind of original discovery but it results in directional behavior.

In Alexander and Grimes [15.6] and Alexander [15.3], I have also observed that semiosis follows the path of least resistance insofar as sign-readings are flexible, which allows a signal pathway, or a behavioral routine, to flow to the first available option that is like enough to work to maintain the pathway, even when it is a misinterpretation. When all the neighboring individual cells in a system are flowing in a similar way, their interactions become somewhat synchronized, as we can observe in the case of starling murmurations, for example. Waddington's [15.50] idea of an epigenetic landscape, metaphorically illustrating the phenomenon of a virtual terrain in biological processes, is worth revisiting with regard to this idea of non-linear directionality. Accordingly, as well, Chomsky's [15.16] notion of "deep structures"—a *pre-existing* algorithm for an alleged universal grammar—might be revised according to Bacigalupi's theory of a virtual or potential structure instead.

But the emergence of spatial wave patterns and chemical clocks from the interpretations and misinterpretations of individual sign readings in a signal pathway is a subject that deserves separate treatment. Like so many authors who have trouble wrapping up, I have already concluded all I meant to say in the previous section and have started to sketch out my next paper here in this section, for this is where those earlier thoughts have led me.

And I did not want to be accused of being guilty of not seeing the forest for the trees. The qualities of semiosic constraints at the local level of sign-reading make the completion of the signal pathway loop reliable, efficient, and robust but also unpredictable and plastic. The underlying algorithm of such processes will necessarily be dynamic and imprecise, changing in order to stay more or less the same, until such time as it is necessary to depart from the usual pathway to find a better way through the maze of existence. With every decision at every fork in the road, a holistic shape begins to emerge that will define the individual's tendencies. The maze runner does not explore every option; and biases executed, (mis) interpretations made, and roads not taken make all the difference[1] in forming the subjective intelligence.

15.8 Conclusions

Although random change and natural selection may be an important factor in evolution, it is not the only one. With Darwinism, organisms are determined by a gene's interaction with the environment and no new pathways are discovered through the unique perspective of the organism's sign systems. One cannot make a coherent argument for real agency in natural processes without 1) a concept of a subjectively-formed directionality—of both individuals and species—based on past unique experiences and 2) a concept of subjectively-formed originality, based on misinterpretations of the directional semiosic processes. In this chapter, I have tried to argue that subjective interpretations and misinterpretations are significant to the direction of evolutionary trajectories and that this subjectivity might be better understood as relative objectivity because it is based on the real physical constraints of similarity, proximity and arbitrary connections.

Declaration

I declare no conflict of interest.

[1] The reader is right to hear echoes of the poem, *The Road Not Taken*, by Robert Frost (1874-1963) here.

References

[15.1] Adamatzky, A. (2012) Slime mold solves maze in one pass, assisted by gradient of chemo-attractants. *IEEE Transactions on NanoBioscience* 11(2), 131-134.

[15.2] Alexander, V.N. (2013) Creativity: Self-referential mistaking, not negating. *Biosemiotics* 6(2), 253-272

[15.3] Alexander, V.N. (2018) Siri fails the Turing test: Computation, biosemiosis and artificial life. *Researches Semiotique* 38(3), 231-249.

[15.4] Alexander, V.N. (2022) When the immune system makes the wrong choice between incompatible options. In: *Tunne loodust! Knowing Nature in the Languages of Biosemiotics*. D. Favareau and E. Velmezova, (eds.) Université de Lausanne, Lausanne: 33-37.

[15.5] Alexander, V.N., Bacigalupi, A.J. and García, Ó.C. (2021) Living systems are smarter bots: Slime mold semiosis versus AI symbol manipulation. *BioSystems* 206 #104430.

[15.6] Alexander, V.N. and Grimes, V.A. (2017) Fluid biosemiotic mechanisms underlie subconscious habits. *Biosemiotics* 10(3), 337-353.

[15.7] Bacigalupi, A.J. (2022) Semiogenesis: A dynamic system approach to agency and structure. *Biosemiotics* 15, 261-284.

[15.8] Baluška, F. and Levin, M. (2016) On having no head: Cognition throughout biological systems. *Frontiers in Psychology* 7(#902), 1-19.

[15.9] Barbieri, M. (2008) *The codes of life: the rules of macroevolution*. Springer, Dordrecht.

[15.10] Barbieri, M. and Pinz, R. (2022) *Code Biology database: A list of biological codes*. Marcello Barbieri's Lab, 22 November 2022

[15.11] Barreras, A. and Gurk-Turner, C. (2003) Angiotensin II receptor blockers. *Baylor University Medical Center Proceedings* 16, 123-126.

[15.12] Bich, L. and Moreno, A. (2015) The role of regulation in the origin and synthetic modeling of minimal cognition *BioSystems* 148, 12-21

[15.13] Bosso, M., A, T.T., Abu-Farha, M., Alanbaei, M., Abubaker, J. and Al-Mulla, F. (2020) The two faces of ACE2: The role of ACE2 receptor and Its polymorphisms in hypertension and COVID-19 *Molecular Therapy - Methods & Clinical Development* 18, 321-327.

[15.14] Buznikov, G.A. and Shmukler, Y.B. (1981) Possible role of "prenervous" neurotransmitters in cellular interactions of rarly rmbryogenesis: A hypothesis. *Neurochemical Research* 6, 55-68.

[15.15] Castro, Ó.G. (2012) Principles of minimal cognition in smart slime molds and social bacteria. *Pensamiento* 67, 787-797.

[15.16] Chomsky, N. (1965) *Aspects of the theory of syntax*. MIT Press, Cambridge.

[15.17] De Assis, L. (2015) Neural binding, consciousness, and mental disorder: Complexity as a common element. *Journal for Neurocognitive Research* 57(3-4), 110-121.

[15.18] Deacon, T. and Sherman, J. (2008) The pattern which connects pleroma to creatura: The autocell bridge from physics to life. In: *A Legacy for Living Systems: Gregory Bateson as Precursor to Biosemiotics*. J. Hoffmeyer, (ed.) Springer, Berlin. 2: 59-76.

[15.19] Eco, U. (1976) *A theory of semiotics*. Indiana University Press, Bloomington, IN.

[15.20] Eco, U. (1984) *Semiotics and the philosophy of language*. Indiana University Press, Bloomington, IN.

[15.21] Fontana, W. and Buss, L.W. (1994) What would be conserved if "the tape were played twice"? *Proceedings of the National Academy of Sciences of the United States of America* 91, 757-761.

[15.22] Fujinami, R.S., Von Herrath, M.G., Christen, U. and L., W.J. (2006) Molecular mimicry, bystander activation, or viral persistence: Infections and autoimmune disease. *Clinical Microbiology Reviews* 19, 80-94.

[15.23] Ginsburg, S. and Jablonka, E. (2021) Evolutionary transitions in learning and cognition. *Phil. Trans. R. Soc. B* 376, #20190766.
[15.24] Gould, S.J. (1989) *Wonderful life: The Burgess shale and the nature of history*. Norton, New York.
[15.25] Gould, S.J. (2002) *The structure of evolutionary theory*. Belknap Press, Cambridge.
[15.26] Hiscock, T. and Megason, S. (2015) Orientation of Turing-like patterns by morphogen gradients and tissue anisotropies. *Cell Systems* 1, 408-416.
[15.27] Jantsch, E. (1982) *The evolutionary vision: Toward a unifying paradigm of physical, biological and sociocultural evolution*. Routeldge.
[15.28] Kelso, J., Bressler, S.L., Buchanan, S., DeGuzman, G.C., Ding, M., Fuchs, A. and Holroydet, T. (1991) A phase transition in human brain and behavior. *Physics Letters* 169(3 Â), 134-144.
[15.29] Koch, P.B. and Bückmann, D. (1987) Hormonal control of seasonal morphs by the timing of ecdysteroid release in *Araschnia levana* L. (*Nymphalidae: Lepidoptera*). *Journal of Insect Physiology* 33(11), 823-829.
[15.30] Kondo, S. (2017) An updated kernel-based Turing model for studying the mechanisms of biological pattern formation. *Journal of Theoretical Biology* 414, 120-127.
[15.31] Kull, K. (2015) Semiosis stems from logical incompatibility in organic nature: Why biophysics does not see meaning, while biosemiotics does. *Progress in Biophysics and Molecular Biology* 119(3), 616-621.
[15.32] Lenoir, T. (1989) *The strategy of life: Teleology and mechanics in nineteenth century German biology*. Chicago University Press, Chicago.
[15.33] Li, J., Wang, X., Chen, J., Zhang, H., Deng, A. and Nakagaki, T. (2000) Association of renin-angiotensin system inhibitors with severity or risk of death in patients with hypertension hospitalized for coronavirus disease 2019 (COVID-19). Infection in Wuhan, China. *JAMA Cardiology* 5(7), 825-830.
[15.34] Nakagaki, T. (2001) Smart behavior of true slime mold in a labyrinth. *Research in Microbiology* 152(9), 767-770.
[15.35] Nowlin, D.M. (2021) The role of biosemiosis and dysfunctional signaling processes in human pathology. In: *Food and Medicine*. Y.H. Hendlin and J. Hope, (eds.) Springer. 22: 155-182.
[15.36] Pavlov, I. (1906) The scientific investigation of the psychical faculties or processes in the higher animals. *Science* 24(620), 613-619.
[15.37] Postle, B. (2006) Memory as an emergent property of the mind and brain. *Neuroscience* 139(1), 23-38.
[15.38] Raspopovic, J., Marcon, L., Russo, L. and Sharpe, J. (2014) Digit patterning is controlled by a Bmp-Sox9-Wnt Turing network modulated by morphogen gradients. *Science* 345, 566-570.
[15.39] Reid, R.G.B. (2007) *Biological emergences: Evolution by natural experiment*. MIT Press, Cambridge.
[15.40] Reynolds, H.R., Adhikari, S., Pulgarin, C., Troxel, A.B., Iturrate, E., Johnson, S.B., Hausvater, A., Newman, J.D., Berger, J.S. and Bangalore, S. (2020) Renin-angiotensin-aldosterone system inhibitors and risk of Covid-19. *New England Journal of Medicine* 382 2441-2448.
[15.41] Russell, E.S. (1916) *Form and function: A contribution to the history of animal morphology*. Chicago University Press, Chicago.
[15.42] Saigusa, T., Tero, A., Nakagaki, T. and Kuramoto, Y. (2008) Amoebae anticipate periodic events. *Physical Review Letters* 100, # 018101.
[15.43] Salthe, S. (1993) *Development and evolution: Complexity and change in biology*. MIT Press, Cambridge Â
[15.44] Sharov, A.A. (2016) Evolutionary biosemiotics and multilevel construction networks. *Biosemiotics* 9(3), 399-416.

[15.45] Shirakawa, T., Gunji, Y.P. and Miyake, Y. (2011) An associative learning experiment using the plasmodium of physarum polycephalum. *Nano Communication Networks* 2, 99-105.
[15.46] Shirakawa, T. and Sato, H. (2013) Construction of a molecular learning network. *Journal of Advanced Computational Intelligence* 17, 913-918.
[15.47] Talwar, G.P., Sharma, N.C., Dubey, S.K., Salahuddin, M., Das, C., Ramakrishnan, S., Kumar, S. and Hingorani, V. (1976) Isoimmunization against human chorionic gonadotropin with conjugates of processed beta-subunit of the hormone and tetanus toxoid. *Proceedings of the National Academy of Science U.S.A.* 73(1), 218-222.
[15.48] Turing, A.M. (1952) The chemical basis of morphogenesis. *Philosophical Transactions of the Royal Society B* 237, 37-72.
[15.49] Uhlhaas, P., Pipa, G., Lima, B., Melloni, L., Neuenschwander, S., Nikolic, D. and Singer, W. (2009) Neural synchrony in cortical networks: History, concept and current Status *Frontiers in Integrative Neuroscience* 3(17), 1-19.
[15.50] Waddington, C.H. (1940) *Organisers and genes*. Cambridge University Press, Cambridge.
[15.51] Winfree, A.J. (1984) The prehistory of the Belousov-Zhabotinsky oscillator. *Journal of Chemical Education* 61(8).

16

On the Energy-Based Limitations of the Information Capacity and Information Processing Rates in the Human Brain

Jack A. Tuszynski[1,2,3]

[1]Department of Mechanical and Aerospace Engineering, Politecnico di Torino, Turin, Italy
[2]Department of Data Science and Engineering, Silesian University of Technology, Gliwice, Poland
[3]Department of Physics, University of Alberta, Edmonton, AB, Canada

Online resources:
https://scholar.google.ca/citations?user=vQh8WdcAAAAJ&hl=en
https://apcari.ca/meet-the-team/jack-tuszynski-research-group/
https://www.researchgate.net/profile/Jack-Tuszynski

Abstract
In this chapter we use simple reasoning based on time, size and energy scales to analyze the possible information processing rates of the human brain based on the energetic cost of encoding a bit of biological information. We use some well-known empirical information about the brain and its constituent neurons as well as sub-neuronal structures to arrive at characteristic information processing rates at all relevant scales. In order to maintain consistent metabolic rates and clocking frequencies for updating information content, we conclude that tubulin is not likely to operate at a quantum level if its functions include information storage and processing. On the other hand, ion channels, if synchronized, may be able to operate at a quantum mechanical regime.

Keywords: Information processing rate, tubulin, ion channels, long-lived quantum excitations, bioenergetics of the brain

16.1 Introduction

The issue of the molecular nature of brain functions related to cognition has been of great interest to neurophysiologists and scientists working in related disciplines. While much is known about the architecture of brain cells and their individual activities in terms of the

Email: jackt@ualberta.ca

action potential propagation, ion channels and ionic current flows, much less is known about such issues as where memories are stored, which molecular mechanisms are involved in information processing and cognitive functions let alone consciousness. Speculations about molecular mechanisms of cognition abound and range from mundane to exotic. Most physical theories of consciousness view the brain as a computational device, with neurons and their synaptic activities acting as fundamental "bit-like" units of information. Action potentials carry signals from the receptors for the five senses to the brain and within the neurons in the brain. Various types of quantum manifestations of the human brain's operations have been discussed [16.45], both in the case of normal physiology and psychopathologies, including neurodegenerative diseases and psychiatric disorders. Attention has been drawn to the possibility that higher cognitive functions operate at a level of quantum physics [16.1]. Experimental determination or even corroboration of such ideas as the Penrose-Hameroff Orch OR theory [16.23], [16.24], [16.25], [16.27] is currently beginning to be achievable in view of the modern experimental methods and instrumentation available to researchers. A step in this direction has been recently made [16.39] but no validation of long-range and long-lived quantum excitations in microtubules has been found yet.

In the absence of strong molecular-level evidence for quantum processes taking place in the human brain, some questions can still be addressed using general principles of physics combined with some of the known information about brain physiology at a level accessible to modern scientific methods. For example, the metabolic rates of human brains have been studied extensively and their energetic demands are well known [16.29]. It is also known how much of the metabolic energy is required to carry out various molecular-level processes such as protein production, motor protein motion, transcription and translation, etc. [16.19]. Most of the brain's functional roles involve motor control, cognition and information storage and processing [16.10]. Additionally, a large percentage of the energy consumed by the brain of warm-blooded animals (including humans) is used to maintain physiological temperature.

Almost 80 years ago, one of the most influential physicists of all time, Erwin Schrödinger, wrote a book entitled *What is Life* [16.62], in which he outlined challenging problems related to the interpretation of biological processes using physical reasoning. Entropy reduction was a major focal point of his book. It is now known that in entropy reduction the generation of biological cells in order to build structure and maintain function is achieved at the cost of metabolic energy consumption and heat dissipation into the environment; hence, no violations of the laws of thermodynamics take place. However, a related problem of information storage and processing by the brain and the energetic cost of such processes is still worth exploring.

The human brain is an immensely complex organ containing approximately 1.5 kg of soft tissue that is composed of highly organized and specialized electrically-excitable cells, mainly neurons and glia. Until recently, it was generally accepted that there are approximately 100 billion neurons, which comprise approximately 50% of the brain cells [16.40]. However, the ratio of neurons to glial cells has been disputed in the literature as it varies significantly between different brain regions, but the average proportion of 1:1 has been firmly established [16.4]. A more recent and a very precise study [16.30] based on isotropic fractionation [16.31] states that the number of neurons varies between 85 and 86 billion. They interact with one another via 10,000 synaptic connections per neuron on average; hence, there are altogether about 1000 trillion synaptic connections which control the central nervous system (CNS), regulate autonomic functions, neuroendocrine function, energy homeostasis, circadian rhythms, cognition and even human behavior [16.56]. Conscious activities strongly correlate with coherent neural oscillations using still largely unknown mechanisms of synchronization. Recently, Kerskens and Perez [16.42] reported that these oscillations are consistent with a quantum mechanism underlying their activity. In particular, they observed NMR signals which were interpreted as possessing characteristics of quantum entanglement correlated with consciousness-related brain functions. Therefore, these authors concluded that brain functions must operate non-classically, i.e., in the quantum régime. However, it is largely unknown how neurons produce synchronization domains over sufficiently long distances and long time periods from a physiological viewpoint, while rapidly switching between coherent and incoherent states that seem to correlate with cognitive functions. Integration of inputs implies merging and consolidation of numerous and diverse information sources. At the level of individual neurons, integration is typically approximated as linear summation of synaptic membrane potentials. However, integration in branching dendrites and soma requires complicated logical operations, amplification of distal inputs, branch point effects, and signaling in dendritic spines. This highly nonlinear behavior, coupled with at least some quantum effects, poses a major challenge to the current state of the art in neuroscience.

According to the dogma of neuroscience established decades ago by the Nobel Prize winning model proposed by Hodgkin and Huxley [16.33], all such factors are represented by the dynamics of transmembrane potentials in terms of the action potentials propagating along neuronal axons. Within this view, the neuron should behave completely algorithmically and deterministically (hence not quantum mechanically), such that for any predetermined inputs, synaptic strengths and firing thresholds, a fixed output in the form of axonal firings is expected to be generated. However, this leaves no place for subjective and non-algorithmic (for example, quantum mechanical) activities and hence free will, which we humans claim to possess. Therefore, representing the human brain as a strictly classical computational device, no matter how complex, always results in treating it as a deterministic machine. Therefore, factors and processes other than inputs, synaptic strengths (measured by coupling coefficients) and the collective dynamics of transmembrane potentials should exist that contribute to the firing decisions. In fact, introducing quantum processes at the level of individual neurons and/or below would resolve the issue of non-algorithmic functioning of the brain due to the inherent non-deterministic nature of quantum physics. Consequently, deeper layers of the structural and functional hierarchy of the brain should be properly investigated at a nanoscale of smaller spatial and temporal dimensions. Therefore, the neuronal cytoskeleton, especially microtubules, actin filaments and motor proteins as well as ion channels, should be the focus of investigations aimed at revealing the brain's fundamental processes and hence bringing neuroscience closer to quantum physics as advocated by Penrose and Hameroff since the late 1990s.

In this chapter, we intend to use simple yet powerful physics-based analysis [16.69] involving metabolic energy consumption, the maximum amount of information processing rates, spatial and temporal scales in conjunction with the order-of-magnitude deduction. Here, we will appeal to the methodology of statistical thermodynamics and quantum physics to analyze information processing in the human brain, based on the energetic cost of encoding information. The analysis is embedded within the following analytic and conceptual framework that employs Shannon's measure of information, Boltzmann's thermodynamic entropy and, to a lesser degree, Planck's quantum of action. This will allow us to conclude with a high level of confidence which subcellular structures, if any, can be possibly involved in information processing at the rates contemplated in various hypotheses, which is of value in the absence of experimental estimates. Our particular focus will be placed on tubulin and microtubules since they have been for over two decades touted as the key to understanding consciousness and cognition by the more exotic proposals. On the other hand, action potentials involving ion channels in neuronal membranes have been known and appreciated by the neurophysiologists since the seminal work of Hodgkin and Huxley [16.33] but somewhat neglected in the context of quantum cognition and consciousness. We will analyze their role at the same level of detail as given to tubulin and microtubules. This analysis will be performer at a classical physics level in order to obtain general order of magnitude estimates. At present, insufficient information is available about quantum biology of the neuron in order to perform a rigorous analysis of this topic at a quantum level, although advances in quantum biology [16.57] are being made indicating real possibilities of measuring quantum effects in other than photosynthetic systems.

16.2 The Neuron: The Brain's Functional Unit

The neuron is an excitable cell that can be viewed as a fundamental functional unit of the brain from an information processing point of view. It represents a highly-structured matrix composed of the membrane, the cytoskeleton, other proteins, including ion channels and membrane receptors, nucleic acids, ions and water molecules, in which these structures are bathed. This complicated intracellular machinery works in concert to produce normal cellular function in addition to its modes of communication within and between the neighboring cells of the brain. The interactions with other cells can be viewed as signal processing and signal propagation, which includes autocrine, paracrine, and endocrine chemical communication, and electrical signals (mediated by ions) traveling across synapses and gap junctions. Signaling processes include binding and activating numerous types of membrane receptors. Cell biology views intracellular signaling and signal processing as taking place by and large in the cytoplasm, which is composed of randomly diffusing ions, proteins and enzymes driven by local forces and agitated by thermal motion. However, this viewpoint ignores the fact that the crowded environment of the cytoplasm includes the cytoskeleton with its highly organized components, namely microtubules (MTs), actin filaments (AFs) also referred to as microfilaments, intermediate filaments, internal membranes and organelles, all of which are mechanically rigid structures with very interesting electrical and mechanical properties. Moreover, the ordered water molecules and ionic concentrations (especially potassium) surrounding them may be important in affecting these properties. In neurons, both in axons and dendrites, the unique architecture of microtubules represents parallel bundles interconnected by microtubule-associated proteins (MAPs) forming structures that resemble semiconducting parallel-processing units [16.26]. Interestingly, the orientation of MTs in axons is parallel while in dendrites, antiparallel, indicating a possible back propagation of signals in the latter structures. Moreover, dendritic MTs make connections with AFs with possible effects on the functioning of the synapse. In summary, the neuron is an extremely crowded and specifically organized environment, far different from the dilute, idealized conditions often envisaged by oversimplified models.

In neurons, which are excitable but non-dividing (except for brain development), large-scale dynamical processes are especially active during neuronal development, cell motility, differentiation, synaptic generation, and synaptic plasticity processes. Active material transport of intracellular material is carried out by motor proteins such as kinesins and dyneins, which bind to MTs or myosins, which bind to AFs. In neurons, motor proteins transport synaptic components such as neurotransmitter molecules along MTs to their final destination guided in this task by MAPs. Structural defects occurring in the cytoskeleton due to genetic mutations or traumatic injuries often lead to neurodegeneration with deleterious consequences for the person's health. For example, in Alzheimer's disease, MAP tau proteins become detached from MTs due to hyperphosphorylation, aggregating into neurofibrillary tangles causing MTs to destabilize and disassemble, leading to neuronal malfunction and eventual cell death [16.72]. Various forms of MT disruption occur in nearly all neurodegenerative diseases, confirming the key role of neuronal cytoskeleton in cognitive functions of the brain. The cytoskeleton's architecture allows for long-range communication both within the neuron and between neurons or other cells, with the active participation of such structures as gap junctions. Ionic and neurotransmitter signals have the ability to alter cellular architecture affecting both its functioning and plasticity. For example, changes in synaptic strength can regulate and modify neural firing patterns, altering such cognitive functions as learning and memory. Importantly, neuronal functioning is unlikely to operate via a command-and-control mechanism since both ion channels and membrane receptors transduce signals bi-directionally between the cell and its exterior. Cell-cell communication using ions and small molecules involves gap junctions.

The cytoskeletal matrix composed of MTs, AFs, intermediate filaments and various accessory proteins organizes structurally and functionally neuronal interiors [16.69]. However, due to their amazing electrical properties with respect to ionic, protonic and electronic conduction (e.g., lossless signal propagation, signal rectification and amplification), it is very likely that they also participate in information processing involving fast electrical and potentially electromagnetic signal propagation across the neuron and possibly between neurons. This communicating network can be synchronized across the brain providing a new paradigm for neuroscience. Furthermore, as hypothesized by Hameroff and Penrose [16.27], it is possible that at least some aspects of synchronization may involve quantum mechanical mechanisms such as entanglement and quantum collapse, which would explain some of the more enigmatic features of consciousness. In this connection, the quantum approaches for the understanding of biological and neural processes have been extensively discussed elsewhere [16.35], [16.36], [16.37], [16.43], [16.44].

16.3 General Power Consumption Considerations

Following the above general discussion about functional aspects of intra-neuronal signal propagation and processing, we now discuss potential limitations placed on the amount of information storage and information processing based on the energy and power limitations provided by human physiology and neuroscience [16.48, 16.49, 16.50]. The human body consists of approximately 3.7×10^{13} cells [16.11] and its power consumption is on the order of 100 W (of course with individual variations and depending on the time of day and the activity pursued by the person at a given time). Therefore, on average each cell consumes approximately 3×10^{-12} W of power (energy per second). Since the human body requires approximately 100 W [16.32] of metabolic power on average (depending on the size of the individual) for its functional demands, this leads to the turnover of its weight in ATP from ADP every day in order to function, which translates into the synthesis of 10^{21} ATP molecules per second. Since each cell has on the order of 10^3 mitochondria [16.12] (a number that strongly depends on the cell type and size), there are approximately 3.7×10^{16} mitochondria in the human body resulting in approximately 3×10^4 ATP production events per mitochondrion per second. This involves a complex set of synchronized biochemical reactions [16.3] called oxidative phosphorylation whose net effect is a conversion of 1 molecule of glucose into 38 molecules ATP. Since each mitochondrion produces 3×10^4 ATP/s and each ATP synthase operates at a rate of 600 ATP molecules/s, we estimate that each mitochondrion has on average 50 ATP synthase enzymes.

Below we specifically discuss the bioenergetics of the brain's energy production in terms of amounts of energy of the numbers of ATP molecules. The brain is the organ with the highest rate of energy consumption, accounting for roughly 20–25% of the total energy demand, or 20–25 W of power on average [16.59]. The vast majority of biochemical energy supply is provided by ATP molecules, each of which gives off on the order of 10 k_BT (where k_B is the Boltzmann constant) of free energy, or 4×10^{-20} J using SI units. Almost all individual processes, such as motor protein motion, enzymatic catalysis, filament polymerization steps, etc., require one or more molecules of ATP per elementary step of each of these processes making the energy of ATP hydrolysis a so-called "quantum of biological energy."

Undoubtedly, the main role of the human brain is to store and process information provided by sensory inputs. A vast percentage of the metabolic energy, approximately 70–80%, is used to maintain the constant body temperature in mammals. Most of the remaining free energy is used in the protein synthesis machinery, i.e., by the ribosomes. Another major energy demand is due to protein repair and removal. It can be therefore safely assumed that less than 5% of the metabolic energy of the brain is utilized for the purpose of information storage (memory) and processing (cognition) demands. We can, therefore, use a conservative estimate of 2.5 W being consumed for these purposes by the human brain. The brain, has on average 10^{11} neurons and at least as many glial cells. Hence, we can assume that each neuron's average power demand is approximately 1.2×10^{-11} W or 12 pW. This is an important number because information encoding and erasure requires a sizeable amount of energy. We will discuss it in a section below. In order to be able to translate power demands into energy transduction by various processes, we need to know the relevant time scales for each of these processes and they can vary by orders of magnitude. On the human sensory perception time scale, Libet *et al.* [16.52] demonstrated that the typical time scale is 500 ms. As we delve deeper into cellular and molecular processes, these time scales become shorter and shorter.

16.4 Energy Cost of Information Processing

Most of the brain's functional roles involve motor control, cognition and information storage and processing. Physics has until recently treated information with some neglect, focusing instead on entropy and its physically measurable manifestation, heat. It is now known that entropy reduction in biological cells takes place at the cost of metabolic energy consumption and dissipation into the environment, which are concepts related to information storage and processing. The cost of such processes is worth exploring. Simple yet powerful analysis involving metabolic energy consumption, the maximum amount of information processing rates, as well as analysis of spatial and time scales will allow us to conclude with a high level of confidence which subcellular structures, if any, can possibly be involved in information processing at the rates contemplated in various hypotheses. The protein tubulin and microtubules have figured prominently for over four decades in this analysis since they have been [16.28] touted as the key to understanding consciousness and cognition by the more exotic proposals. In addition, ion channels need to be analyzed in detail from this point of view due to their role in maintaining cellular homeostasis and in action potential propagation.

It is important here to stress that any form of information, physical or biological, always comes at an energetic cost. Landauer [16.47] first stated that the minimum energetic cost of one bit of information is $\varepsilon = k_B T \ln (2) = 4\times10^{-21}$ J, which basically comes from the thermodynamic formula that F = U-TS and the entropy S being equivalent to negative information, -I. Shannon's formula for information is $I = -k_B T \ln \Omega$ where Ω is the number of equal probability states in the ensemble representing the choices available for information storage at a given step (e.g., the number of cards in a deck from which one is drawn). The value ε is the minimum energetic cost of creating a bit of information. In the context of biological systems, such as the brain's neurons, it is not expected that the minimum of energy cost is actually attained but, rather, the value for ATP (or in a minority of situations, GTP) which is an order of magnitude greater. This is due to biological function demands trumping a simple energy efficiency objective. In order to relate the metabolic energy expenditure to the information processing rate and estimate the maximum value of the latter, the time scale of the predominant "bit switching" processes, $\Delta\tau$, must be estimated. Therefore, for the entire brain we estimate the rate of information processing as

$$\Delta I/\Delta\tau = P / \varepsilon = 2.5 \text{ W}/4\times10^{-21}\text{J} = 6\times10^{20} \text{ bits/s}$$

This is the absolute upper limit estimate on the processing power of the brain within a classical mode of operation. In the absence of relevant quantum mechanical measurements, we can consider this to be a reasonable estimate of the information processing capability of the human brain. This can be very favorably compared to the processing rate of the most powerful modern day supercomputer clusters such as the BlueGene, for which the processing rate is on the order of 10^{16} bits/s but the corresponding power consumption is enormous, on the order of 10 MW. Reassuringly, this can be interpreted as a message of hope for humanity with respect to the highly touted takeover by AI technology.

16.5 The Brain's Time Scale Hierarchies and Information Processing Rates

Among the various sub-cellular intricate structures within the human brain, perhaps the most widely speculated about structural element has been tubulin, which is a constituent protein of microtubules. Tubulin is mid-size (110 kD per dimer) globular protein that forms stable dimers composed of β and α tubulin homologous monomers [16.34]. The processing of information in neurons is likely to involve the synchronization of tubulin operation with motor proteins as well as ion channels. This necessarily rapid synchronization, has been assumed to rely on the existence of the quantum mechanical regime, which is still in need of experimental validation. The quantum mechanical aspect of tubulin operation is discussed, among others, by Matsuno and Paton [16.54], where the authors adopted the existence of long-range quantum coherent states in the tubulin molecules during their functioning. The quantum regime can be analyzed using the model of quantum non-demolition measurements that explain the existence of long-living coherent states based on the timing of conformational relaxations of macromolecular structures. In fact, this can provide a potential solution to the energy problem at different time scales discussed throughout this study due to the "energy–time" uncertainty principle, which imposes the limits on energy dissipation in the prolonged internal quantum measurement of the "non-demolition" mode. However, for most of the discussion in this study, our attention will be focused on strictly classical approaches to the energetic cost of information processing.

Entropy is an additive physical quantity, meaning that the entropy of a set of subsystems is equal to the sum of entropies of each constituent unit. Since information is negative entropy, the same can be said about information. Consequently, the total information stored in the brain is equal to the sum of information contributions stored in each of its constituents, e.g., neurons.

The characteristic time scales of information processing units in the brain are crucial in determining the maximum information processing rates given the limited amount of power consumption of the brain and hence each neuron. Seminal experiments conducted by Libet *et al.* [16.52] determined the pre-processing time (which is roughly equivalent to the reaction time) of the human brain to be approximately 500 ms. Therefore, it can be inferred that for the whole brain, the characteristic time scale, denoted Δt, is on the order of 1 s or less consistent with the frequency of brain waves, f, in the 10–100 Hz range, as recorded by EEG. Therefore, for the whole brain, it appears that the maximum value of the information processing rate is 6×10^{20} bits/s, which is enormous compared to even the largest supercomputers. Within the human brain, the next level of the structural and functional hierarchy involves its functional units, i.e., neurons. There are approximately 10^{11} neurons in the brain, hence assuming all of them being equal, a single neuron is expected to have an information processing rate on the order of 6×10^{9} bits/s or approximately one gigabyte per second as one byte equals 8 bits. The time scale of these processes is largely determined by the neuron's firing rates, commonly occurring on the order of 1 ms, hence with this time scale, the corresponding frequency, f, is in the 1 kHz range and a neuron can hypothetically process as much as one megabyte per ms. This corresponds to a single information package that can be transmitted in a neuron per its single processing event taking place every 1 ms. Assuming that the computational elements within neurons are microtubules (MTs), we can conservatively estimate the number of MTs per neuron since their density has been measured in dendrites as 68 microtubules per μm^2 of cross-sectional area, with the area reaching 10 μm^2 [16.41], and additional microtubules are also present in the soma and the axon. Therefore, the order of magnitude of one thousand (10^3) MTs is a reasonable estimate. This leads to an average processing rate per MT to be on the order of 10^6 bytes/s or one megabyte per second. The corresponding time scales for microtubules are therefore on the order of 1 ms, which is a typical time for a conformational change to occur in a protein. The corresponding frequency, f, is expected to be 1 MHz. Therefore, a single conformational change event can only encode one byte of information. Finally, the lowest level of information processing discussed in the literature is that of the constituent protein, tubulin. For a typical 10 μm-long MT, there are on the order of 10^4 tubulin dimers, hence if a tubulin dimer is the basic computational element (corresponding to a bit of biological information), then its maximum information processing rate is 100 bytes/s. The latter estimate, however, leads to a contradiction with the various claims of tubulin being a biological qubit that processes information on much shorter time scales, namely 1 ns or even shorter. This could only be reconciled with either infrequent information processing events (every 10 million nanoseconds or more) or continuous information encoding of a small fraction of tubulin dimers (one in 10 million or more). Encoding biological information via post-translational modifications of tubulin exists but this process is very slow and not likely utilized for memory formation [16.38]. This also begs the question about which tubulin dimers would be selected for information processing and why. The conclusions reached here indicate the possibility of state changes due to information encoding on the order of tens of milliseconds, which might be mechanistically explained by microtubule coupling to action potentials propagating along the neuron's axon, but most definitely not with the sub-nanosecond electronic transitions and GHz frequency rates that one would expect otherwise. Consequently, if information processing is taking place in the brain, it may do so in a hierarchical manner with neurons forming the second layer and MTs the next one with final steps being made by selected tubulin dimers. In this case, the maximum theoretically possible information processing rates can be: (a) 10^{20} bytes/s per brain, (b) 10^9 bytes/s per neuron, (c) 10^6 bytes/s per microtubule and possibly (with major questions about long-term stability of these events) 10^2 bytes/s per tubulin. These numbers may be a reasonable quantification of biochemical processes occurring in the brain in the classical regime, consistent with various possible biochemical scenarios. For example, as discussed elsewhere, phosphorylation events of tubulin by calmodulin kinase II (CaMKII) enzyme [16.13] could possibly occur on a time scale of tens of milliseconds, as could interactions of MTs with action potential fields due to the ionic waves crossing the axonal membrane with speeds on the order of 1 m/s [16.14]. We discuss the issue of action potential and ion channels within the context of information processing in a section below. However, the underlying mechanisms, both classical and especially quantum, challenge the present-day knowledge about biochemistry and cell biology. Regarding the possibility of quantum mechanisms playing a role in information processing at the level of microtubules or tubulin [16.58] [16.25], such claims inevitably encounter the issue of explaining how such slow processes can be performed in a quantum mechanical regime [16.22], especially because of the ubiquitous presence of thermal decoherence at physiological temperatures [16.66].

The estimates presented above, while very approximate, respect the basic principle of energy conservation together with the energetic cost of processing each bit of information. If the very rapid transition rates associated with quantum states of tubulin [16.55] were to be present, a simple calculation of the corresponding energetic cost of information processing would amount to enormous power consumption in the tens or hundreds of MW, clearly an absurd value in the context of the human brain. Therefore, the speed of information processing is limited by both the amount of energy provided to each neuron and by the time scales on which individual states encoding biological information can change.

16.6 The Ion Channels

The previous analysis requires a minor modification if we apply the same reasoning to ion channels. There are on the order of 10^4 ion channels per neuron [16.71]. Consequently, the maximum information processing rate per ion channel is 10^5 bytes/s, a number that is consistent with the estimated power consumption by the brain. This translates into a time scale of 10 ms for information processing events involving each of the ion channels.

In model channels, such as the bacterial KcsA channel, one K^+ ion crosses the channel per 10-20 ns under physiological conductances of roughly 80–100 pS [16.60]. This allows for a maximum conduction rate of about 10^8 ions/s, estimating the distances between the center of the channel pore and the membrane surface to scale along 5×10^{-9} m (5 nm) and assuming a simple watery-hole and continuum electro-diffusion model of channels [16.2] [16.21], which would provide an average speed of 5×10^{-1} m/s (0.5 m/s) per ion. However, this model does not take into account the ion-channel structure, masses and thermal energies and within filter movement. Many translocation events within the ion selectivity filter region are required by an ion to completely traverse the filter domain [16.6].

More accurately, a single ion takes approximately 5 ps to traverse an ion channel with a length on the order of 4–5 nm giving an average speed of 1000 m/s involved in this process. The corresponding kinetic energy together with the electrostatic potential energy amount to 2×10^{-20} J, closely corresponding to the ATP energy justifying an active transport requirement. It is also known that the ion flow rate per channel is on the order of 10^5 ions/ms giving a clocking time of approximately 10 ns. Therefore, a 5 ps active event is separated by a 2,000 times longer refractory interval of 10 ns. Since the transmembrane potential is on the order of 100 mV, a flow of singly charged ions like sodium or potassium leads to a current on the order of 10 pA. The corresponding frequency range is 2×10^{11} Hz ($f= 1/\tau$). For a single ion transition across the channel, the energy involved in this oscillation is $E=hf\sim 1$ meV, below the thermal energy of k_BT but if the ion channels of a neuron are synchronized (as has been reported in the literature, e.g., [16.65]), then the total energy value of coupled ion channel transitions in a neuron could reach up to 4 eV [16.6], which is well in the quantum range and corresponds to the energy of a photon in the submicrometer-wavelength range (310 nm, which is in the UV range). As the duration of the rather (stable) non-conducting state scales along 10^{-3}–10^{-4} s and within the selectivity filter [16.7]. the translocation time is on the order of 10^{-11}s, we can expect about 10^7 filter state changes during a non-conducting state and about 10^{10} switches per second (10 GHz), providing the fastest information encoding process consistent energetically and physiologically with the limitations discussed above.

We conclude that unlike cytoskeletal proteins, ion channels [16.40] are a much more likely candidate for quantum information processing [16.54] due to both the time and energy scales involved. They are also natural candidates for information processing having been studied in connection to it for decades, albeit using largely classical approaches. Recently, several studies hypothesized that ion channels may operate quantum mechanically under some conditions. Our order-of-magnitude analysis indicates that coupled (synchronized) ion channels [16.53], [16.67], [16.68] in a neuron may collectively generate quantum oscillations.

To further support these conclusions, we may consider the consistency of the frequency and length scales for various levels of the information processing hierarchy in the brain

[16.61]. For collective oscillations, we follow a lattice-like model where the separation between synchronized neighboring entities [16.18] determines the length scale while the inverse of the characteristic time gives the corresponding frequency. The product of the frequency with the length scale (half wavelength) produces the propagation velocity of a collective excitation. Atoms in a protein are separated by angstrom distances and vibrate on the order on femtosecond time scales, giving propagation velocities in the range 10^4–10^5 m/s. Protein-protein distances, such as between tubulins in a microtubule are in the 10^{-8} m range and their characteristic times are in nanoseconds, leading to 10–100 m/s propagation velocities. Microtubule-microtubule interactions (as in an axonal bundle), occur at distances of 100 nm and time scales of fractions of microseconds, giving rise to velocities on the order of m/s such that they can couple with action potential propagation. Ion channels can be separated by distances of tens to hundreds of nm and their characteristic time scale is on the order of 10 ns, giving rise again to propagation velocities between m/s and tens of m/s [16.46].

In terms of a mechanistic explanation of ion channel information propagation, consider a single ion located inside the selectivity filter of an ion channel [16.6] and assume that it encodes 1 bit of information as suggested previously by Bernroider and Roy [16.8] [16.9]. With a state transition from outside to inside the neuron, the corresponding bit switching rate amounts to 10^{10}/s, and further assuming a channel number of 10^4 per cortical neuron, an ensemble of roughly 10^6 neurons engaged in a single perception event (visual or auditory) would perform a maximum of $10^{10}\times10^4\times10^6=10^{20}$ bits/s, which is comparable to the upper limit of the processing rate in the brain as evaluated above. In fact, if this is true of each of the five senses engaged in the communication with the outside environment, then this information processing rate would almost reach the limit of 6×10^{20} bits/s. Therefore, only a fraction of available metabolic energy consumed by the brain would be available for other types of computation on a continuous level.

Finally, neuron-neuron distances (synaptic connections) are in the microns while typical time scales for their interactions are in the millisecond range, leading to mm/s synchronization propagation speeds. The latter number together with Libet's preprocessing time estimate [16.52] may explain the limited size of specialized functional domains in the brain [16.56], also corresponding to the resolution limit of modern fMRI equipment [16.20].

16.7 Discussion

In this study we used simple reasoning based on time, length and energy scales to analyze the possible information processing rates of the human brain based on the energetic cost of encoding a bit of information. Obviously, this approach has some inherent limitations. The forces which integrate cooperative behavior in inanimate matter are short range. The effect of these forces on the storage and transfer of energy and information can be described by the statistical property called thermodynamic entropy, a measure of the number of microstates consistent with a given macrostate. However, the collective behavior of biologically activated macromolecules in biological systems, especially the amino acids that form proteins and the nucleic acids that form DNA, are the result of long-range interactions. The molecular organization that defines the three-dimensional structure of a protein, such as tubulin, for example, is determined by the effect of the instantaneous configuration of the amino acids on the behavior of the other molecules in the protein. These long-range interactions are not consistently incorporated in the formalism of Gibbs and Boltzmann that we relied on in this work and hence cannot encode this cooperativity of the interacting components that define biologically activated macromolecules. A similar situation arises in the study of collective behavior of cells. Cells can be viewed as isothermal chemical machines. They convert the chemical energy provided by the external environment into metabolic energy and biological work, including structure formation, reproduction and cooperative activities of a collective system. The formalism of statistical thermodynamics, and measures such as the Shannon entropy are not best suited to analyze energy transformation in these systems. Therefore, there are major limitations to the methodology employed in this work. On the other hand, concepts such as evolutionary entropy [16.15] [16.16] and quantum metabolism [16.17] have been proposed to study collective behavior in living systems. Future research on the information processing in the brain using such methodologies could explore the importance of including such approaches for the issue of biology information and even consciousness.

16.8 Conclusions

In the present study we used some well-known empirical information about the brain and its constituent neurons and sub-neuronal structures to arrive at characteristic information processing rates at all relevant scales. In order to maintain consistent metabolic rates and clocking frequencies for updating information content, we have concluded that tubulin is not likely to operate at a quantum level if its functions include information storage and processing. On the other hand, ion channels, if synchronized [16.5], may be able to operate in a quantum mechanical regime. We hope that the protracted debate about quantum information processing via tubulin and microtubules [16.28] [16.70] that has continued for close to several decades, especially in regard to the topics of consciousness, can now be directed at a more plausible set of players, namely ion channels. It is entirely possible that no simple answers exist in this highly complex issue and that a number of molecular players perform different functions and information processing may take place in a hybrid scenario where some classical operations are performed by microtubules (e.g., memory encoding) [16.13], while some others (e.g., higher cognitive functions such as abstract reasoning) may involve ion channels operating quantum mechanically [16.64]. We also wish to draw the reader's attention to the issue of signal amplification, which may be critical to the integration of information processing across scales. Pertinent examples of molecular recognition and signal amplification of sensory inputs have been extensively discussed by Loewenstein [16.51].

In a recent study, Bandyopadhyay and his collaborators [16.61] advanced a hypothesis regarding a hierarchical resonant fractal functioning of the human brain. This is a very attractive hypothesis that may explain the presence of different clocking rates at different spatial scales in the brain and the corresponding resonances. Perhaps this can also shed light on the filtering of information, as it becomes coarse grained passing from the molecular to the cellular to the brain domain to the whole organ scales. According to Sahu *et al.* [16.61] the brain has a bandwidth of 10^{30} Hz (from 10^{-15} to 10^{15} Hz). To quote these authors: "The brain hardware has the capacity to simulate a significant amount of events in the universe.

Thus, it is intelligent." Memory can, therefore, be stored in a fractal manner at every level. In Table 16.1 we summarize the relevant time scales and spatial dimension on which the brain and its component structures discussed in this study work.

Based on the discussion presented in this study, we propose a multi-fractal hypothesis since the scaling laws connecting different information processing hierarchies are not identical. Specifically, this is shown in Table 16.2.

This hierarchical organization with different time scales and frequency scales offers a potential solution to the energy problem [16.63]. Namely, infrequent information processing events which are extremely fast, say of a 1 fs duration, occur with a long temporal separation, say on a ms scale. This could include all or almost all tubulin dimers. On the other hand, relatively long events, such as visual perception, may take place almost continuously. As we move from a low-level of the spatio-temporal hierarchy (e.g., from tubulin to microtubules or from microtubules to neurons), a coarse-graining mechanism may be playing a role in filtering signal from noise [16.73], such that very rapid changes in states at the low level of the hierarchy are not perceived by the higher level of the hierarchy. What mechanism or mechanisms are used by these systems to filter signal from noise is not clear but should be the subject of future investigations. The filtering of noise and amplification of meaningful signals is one of the main conditions of the stable operation of living systems. This also brings into focus the question of meaning in the context of biological systems, which are different in this respect from physical systems. Information alone is not sufficient to explain the purpose of living systems' activities and functions. How information

Table 16.1 Hierarchy of information processing elements.

Time (frequency)	Structures	Numbers/Dimension
s	BRAIN	1/10 cm
ms (10 kHz)	NEURONS	10^{11}/100 microns
microsecond (MHz)	MICROTUBULES	10^3/10 microns
fs (10^{15} Hz)	ION CHANNELS	10^4/5 nm
ns (GHz)	TUBULIN	10^5/8 nm

Table 16.2 Hierarchical organization of the information processing elements in the human brain with their numbers, characteristic frequencies (in Hz) and scaling exponents linking the neighboring elements in the hierarchy.

Element	Number N	characteristic frequency f_1	f_N	scaling exponent α: $f_N \sim (f_1)^\alpha$
Neuron-brain	10^{11}	10^4	10^1	¼
MT-neuron	10^3	10^7	10^4	4/7
Ion channel-neuron	10^4	10^8	10^4	½
tubulin-MT	10^4	10^9	10^7	7/9

processing simple biological structures, such as viruses and bacteria, through evolution, adaptation and competitive advantage, resulted in the acquisition of meaning through an ever-increasing complexity of form and function is a topic that exceeds the scope of this chapter but can benefit from the quantitative analyses presented here.

Acknowledgments

I express my gratitude to Prof. Gustav Bernroider for his important comments about ion channels. This work is dedicated to the memory of my colleague, friend and mentor, the late Prof. Massimo Cocchi of the University of Bologna.

Declaration

I declare no conflict of interest.

References

[16.1] Abbott, D., Davies, P.C. and Pati, A.K. (2008) *Quantum aspects of life*. World Scientific.

[16.2] Allen, T.W., Kuyucak, S. and Chung, S.H. (2000) Molecular dynamics estimates of ion diffusion in model hydrophobic and KcsA potassium channels. *Biophysical chemistry, 86*(1), 1-14.

[16.3] Aon, M.A., Cortassa, S. and O'Rourke, B. (2006) The fundamental organization of cardiac mitochondria as a network of coupled oscillators. *Biophysical Journal, 91*(11), 4317-4327.

[16.4] Azevedo, F.A., Carvalho, L.R., Grinberg, L.T., Farfel, J.M., Ferretti, R.E., Leite, R.E., Filho, W.J., Lent, R. and Herculano-Houzel, S. (2009) Equal numbers of neuronal and nonneuronal cells make the human brain an isometrically scaled-up primate brain. *Journal of Comparative Neurology, 513*(5), 532-541.

[16.5] Bennett, M.V. and Zukin, R.S. (2004) Electrical coupling and neuronal synchronization in the mammalian brain. *Neuron, 41*(4), 495-511.

[16.6] Berneche, S. and Roux, B. (2001) Energetics of ion conduction through the K+ channel. *Nature, 414*(6859), 73-77.

[16.7] Bernèche, S. and Roux, B. (2005) A gate in the selectivity filter of potassium channels. *Structure, 13*(4), 591-600.

[16.8] Bernroider, G. and Roy, S. (2004) Quantum-classical correspondence in the brain: scaling, action distances and predictability behind neural signals. *Forma* (Tokyo), *19*(1), 55-68.

[16.9] Bernroider, G. and Roy, S. (2005). Quantum entanglement of K+ ions, multiple channel states, and the role of noise in the brain. In: *Fluctuations and Noise in Biological, Biophysical, and Biomedical Systems III*, 5841, 205-214.

[16.10] Berg, J.M., Tymoczko, J.L. and Stryer, L. (2007) *Biochemistry (Loose-Leaf)*. Macmillan.

[16.11] Bianconi, E., Piovesan, A., Facchin, F., Beraudi, A., Casadei, R., Frabetti, F., Vitale, L., Pelleri, M.C., Tassani, S., Piva, F. and Perez-Amodio, S. (2013) An estimation of the number of cells in the human body. *Annals of human biology, 40*(6), 463-471.

[16.12] Cole, L.W. (2016) The evolution of per-cell organelle number. *Frontiers in cell and developmental biology, 4*, #5.

[16.13] Craddock, T.J., Tuszynski, J.A. and Hameroff, S. (2012) Cytoskeletal signaling: is memory encoded in microtubule lattices by CaMKII phosphorylation?. *PLoS computational biology, 8*(3), #1002421.

[16.14] Crotty, P., Sangrey, T. and Levy, W.B. (2006) Metabolic energy cost of action potential velocity. *Journal of Neurophysiology, 96*(3), 1237-1246.

[16.15] Demetrius, L. (2000) Thermodynamics and evolution. *Journal of theoretical Biology, 206*(1), 1-16.

[16.16] Demetrius, L.A. (2013) Boltzmann, Darwin and directionality theory. *Physics reports, 530*(1), 1-85.

[16.17] Demetrius, L. and Tuszynski, J.A. (2010) Quantum metabolism explains the allometric scaling of metabolic rates. *Journal of the Royal Society Interface, 7*(44), 507-514.

[16.18] Draguhn, A., Traub, R.D., Schmitz, D. and Jefferys, J.G.R. (1998) Electrical coupling underlies high-frequency oscillations in the hippocampus in vitro. *Nature, 394*(6689), 189-192.

[16.19] Daune, M. (1999) *Molecular biophysics: structures in motion.* Oxford University Press.

[16.20] Feinberg, D.A., Vu, A.T. and Beckett, A. (2018) Pushing the limits of ultra-high resolution human brain imaging with SMS-EPI demonstrated for columnar level fMRI. *Neuroimage, 164*, 155-163.

[16.21] Guidoni, L. and Carloni, P. (2002) Potassium permeation through the KcsA channel: a density functional study. *Biochimica et Biophysica Acta - Biomembranes, 1563*(1-2), 1-6.

[16.22] Hagan, S., Hameroff, S.R. and Tuszyński, J.A. (2002) Quantum computation in brain microtubules: Decoherence and biological feasibility. *Physical review E, 65*(6), #61901.

[16.23] Hameroff, S. (1998) Quantum computation in brain microtubules? The Penrose–Hameroff 'Orch OR 'model of consciousness. *Philosophical Transactions of the Royal Society of London. Series A: Mathematical, Physical and Engineering Sciences, 356*(1743), 1869-1896.

[16.24] Hameroff, S.R. (2007) The brain is both neurocomputer and quantum computer. *Cognitive Science, 31*(6), 1035-1045.

[16.25] Hameroff, S. (2012) How quantum brain biology can rescue conscious free will. *Frontiers in integrative neuroscience, 6*, 93.

[16.26] Hameroff, S., Nip, A., Porter, M. and Tuszynski, J. (2002) Conduction pathways in microtubules, biological quantum computation, and consciousness. *Biosystems, 64*(1-3), 149-168.

[16.27] Hameroff, S. and Penrose, R. (1996) Orchestrated reduction of quantum coherence in brain microtubules: A model for consciousness. *Mathematics and computers in simulation, 40*(3-4), 453-480.

[16.28] Hameroff, S.R. and Watt, R.C. (1982) Information processing in microtubules. *Journal of Theoretical Biology, 98*(4), 549-561.

[16.29] Harris, D.A. (2009) *Bioenergetics at a Glance: An Illustrated Introduction.* John Wiley & Sons.

[16.30] Herculano-Houzel, S. (2016) *The human advantage: a new understanding of how our brain became remarkable.* MIT Press.

[16.31] Herculano-Houzel, S. and Lent, R. (2005) Isotropic fractionator: a simple, rapid method for the quantification of total cell and neuron numbers in the brain. *Journal of Neuroscience, 25*(10), 2518-2521.

[16.32] Heusner, A.A. (1985) Body size and energy metabolism. *Annual review of nutrition, 5*(1), 267-293.

[16.33] Hodgkin, A.L. and Huxley, A.F. (1952) A quantitative description of membrane current and its application to conduction and excitation in nerve. *The Journal of physiology, 117*(4), p.500.

[16.34] Howard, J. and Clark, R.L. (2002) Mechanics of motor proteins and the cytoskeleton. *Appl. Mech. Rev., 55*(2), B39-B39.

[16.35] Igamberdiev, A.U. (2004) Quantum computation, non-demolition measurements, and reflective control in living systems. *BioSystems, 77*(1-3), 47-56.

[16.36] Igamberdiev, A.U. (2021) The drawbridge of nature: evolutionary complexification as a generation and novel interpretation of coding systems. *Biosystems, 207*, #04454.

[16.37] Igamberdiev, A.U. and Shklovskiy-Kordi, N.E. (2017) The quantum basis of spatiotemporality in perception and consciousness. *Progress in Biophysics and Molecular Biology, 130*, 15-25.

[16.38] Janke, C. and Kneussel, M. (2010) Tubulin post-translational modifications: encoding functions on the neuronal microtubule cytoskeleton. *Trends in Neurosciences, 33*(8), 362-372.

[16.39] Kalra, A.P., Benny, A., Travis, S.M., Zizzi, E.A., Morales-Sanchez, A., Oblinsky, D.G., Craddock, T.J., Hameroff, S.R., MacIver, M.B., Tuszynski, J.A. and Petry, S. (2023) Electronic Energy Migration in Microtubules. *ACS Central Science*, 9, 3, 352–361.

[16.40] Kandel, E.R., Schwartz, J.H., Jessell, T.M., Siegelbaum, S., Hudspeth, A.J. and Mack, S. eds. (2000) *Principles of neural science*, 4, 1227-1246. New York: McGraw-hill.

[16.40] Karplus, M. and McCammon, J.A. (1983) Dynamics of proteins: elements and function. *Annual reviews of biochemistry*, 52, 263-300.

[16.41] Katrukha, E.A., Jurriens, D., Pastene, D.M.S. and Kapitein, L.C. (2021) Quantitative mapping of dense microtubule arrays in mammalian neurons. *Elife, 10*, #e67925.

[16.42] Kerskens, C.M. and Pérez, D.L. (2022) Experimental indications of non-classical brain functions. *Journal of Physics Communications, 6*(10), #105001.

[16.43] Khrennikov, A. (2021) Quantum-like model for unconscious–conscious interaction and emotional coloring of perceptions and other conscious experiences. *Biosystems, 208*, #104471.

[16.44] Khrennikov, A. (2022) Order stability via Fröhlich condensation in bio, eco, and social systems: The quantum-like approach. *Biosystems, 212*, #104593.

[16.45] Koch, C. and Hepp, K. (2006) Quantum mechanics in the brain. *Nature, 440*(7084), 611-611.

[16.46] Kuyucak, S., Andersen, O.S. and Chung, S.H. (2001. Models of permeation in ion channels. *Reports on Progress in Physics, 64*(11), #1427.

[16.47] Landauer, R. (1996) The physical nature of information. *Physics letters A, 217*(4-5), 188-193.

[16.48] Laughlin, S.B. (2001) Energy as a constraint on the coding and processing of sensory information. *Current opinion in neurobiology, 11*(4), 475-480.

[16.49] Laughlin, S.B., de Ruyter van Steveninck, R.R. and Anderson, J.C. (1998) The metabolic cost of neural information. *Nature neuroscience, 1*(1), 36-41.

[16.50] Laughlin, S.B. and Sejnowski, T.J. (2003) Communication in neuronal networks. *Science, 301*(5641), 1870-1874.

[16.51] Loewenstein, W.R. (1999) *The touchstone of life: molecular information, cell communication, and the foundations of life*. Oxford University Press.

[16.52] Libet, B., Wright, E.W.W., Feinstein, B. and Pearl, D.K. (1979) Subjective referral of the timing for a conscious sensory experience. *Brain, 102*(1), 193-224.

[16.53] Liu, D.S., Astumian, R.D. and Tsong, T.Y. (1990) Activation of Na+ and K+ pumping modes of (Na, K)-ATPase by an oscillating electric field. *Journal of Biological Chemistry, 265*(13), 7260-7267.

[16.54] Matsuno, K. and Paton, R.C. (2000) Is there a biology of quantum information?. *BioSystems, 55*(1-3), 39-46.

[16.55] Mershin, A., Kolomenski, A.A., Schuessler, H.A. and Nanopoulos, D.V. (2004) Tubulin dipole moment, dielectric constant and quantum behavior: computer simulations, experimental results and suggestions. *Biosystems, 77*(1-3), 73-85.

[16.56] Nicholls, J.G., Martin, A.R., Wallace, B.G. and Fuchs, P.A. (2001) *From neuron to brain*. Sunderland, MA: Sinauer Associates.

[16.57] Panitchayangkoon, G., Hayes, D., Fransted, K.A., Caram, J.R., Harel, E., Wen, J., Blankenship, R.E. and Engel, G.S. (2010) Long-lived quantum coherence in photosynthetic complexes at physiological temperature. *Proceedings of the National Academy of Sciences, 107*(29), 12766-12770.

[16.58] Pokorný, J. (2004) Excitation of vibrations in microtubules in living cells. *Bioelectrochemistry, 63*(1-2), 321-326.

[16.59] Raichle, M.E. and Gusnard, D.A. (2002) Appraising the brain's energy budget. *Proceedings of the National Academy of Sciences, 99*(16), 10237-10239.

[16.60] Roux, B. and Schulten, K. (2004) Computational studies of membrane channels. *Structure, 12*(8), 1343-1351.

[16.61] Sahu, S., Ghosh, S., Hirata, K., Fujita, D. and Bandyopadhyay, A. (2013) Multi-level memory-switching properties of a single brain microtubule. *Applied Physics Letters, 102*(12), #123701.

[16.62] Schrödinger, E. (1992) *What is life?: With mind and matter and autobiographical sketches.* Cambridge University Press.

[16.63] Scott, A. (1999) *Stairway to the mind: The controversial new science of consciousness.* Springer Science & Business Media.

[16.64] Summhammer, J., Salari, V. and Bernroider, G. (2012) A quantum-mechanical description of ion motion within the confining potentials of voltage-gated ion channels. *Journal of integrative neuroscience, 11*(02), 123-135.

[16.65] Stiefel, K.M. and Ermentrout, G.B. (2016) Neurons as oscillators. *Journal of neurophysiology, 116*(6), 2950-2960.

[16.66] Tegmark, M. (2000) Importance of quantum decoherence in brain processes. *Physical review E, 61*(4), #4194.

[16.67] Traub, R.D., Draguhn, A., Whittington, M.A., Baldeweg, T., Bibbig, A., Buhl, E.H. and Schmitz, D. (2002) Axonal gap junctions between principal neurons: a novel source of network oscillations, and perhaps epileptogenesis. *Reviews in the neurosciences, 13*(1), 1-30.

[16.68] Tsong, T.Y. (1990) Electrical modulation of membrane proteins: enforced conformational oscillations and biological energy and signal transductions. *Annual Review of Biophysics and Biophysical Chemistry, 19*(1), 83-106.

[16.69] Tuszynski, J.A. (2018) *Molecular and cellular biophysics.* CRC Press.

[16.70] Tuszyński, J.A., Brown, J.A. and Hawrylak, P. (1998) Dielectric polarization, electrical conduction, information processing and quantum computation in microtubules. Are they plausible?. *Philosophical Transactions of the Royal Society of London. Series A: Mathematical, Physical and Engineering Sciences, 356*(1743), 1897-1926.

[16.71] van Steveninck, R.D.R., Anderson, J. and Laughlin, S.B. (1996) The metabolic cost of neural information. *Nature Neuroscience, 1*(1), 36-41.

[16.72] Woolf, N.J., Priel, A., Tuszynski, J.A. (2010) Nanotechnology, Nanostructure, and Nervous System Disorders. *Nanoneuroscience: Structural and Functional Roles of the Neuronal Cytoskeleton in Health and Disease*, 177-226.

[16.73] White, J.A., Rubinstein, J.T. and Kay, A.R. (2000) Channel noise in neurons. *Trends in neurosciences, 23*(3), 131-137.

17

The Peculiar Case of Danger Modeling: Meaning-Generation in Three Dimensions

Hongbing Yu

Department of Languages, Literatures and Cultures, Toronto Metropolitan University, Toronto, Canada

Online resources:
https://www.torontomu.ca/llc/people/faculty/hongbing-yu
https://scholar.google.com/citations?hl=en&user=R-AFcTYAAAAJ
https://orcid.org/0000-0002-1944-1514

Abstract

In the struggle for existence, any known species of organisms must avert danger to ensure survival. To do that, they need to identify—or model—it in ways that are specific to their species and actualized in concrete situations. This is an act of generating meaning that falls perfectly within the purview of biosemiotics. Inspired by the most recent pioneering work by Marcel Danesi on the semiotics of danger, the present chapter uses human danger modeling as a case in point to discuss meaning-generation in the biosemiotic context. The rationale is a three-fold one. Firstly, the study of danger, understood in terms of meaning-generation, serves as a fine example of semiotic analysis, because warning signs are among the most primordial and relevant vehicles for meaning. Secondly, the Sebeokian concept of modeling serves as a useful tool to tackle this problem. The reason is that the concept underscores semiotic agency and promises to be a highly integrative framework for studying meaning-generation. Thirdly, humans have evolved into such complex superorganisms that we not only live by existential modeling and semiotic modeling at the same time but are also deeply entangled in our own semiotic webs while probing into and creating possibilities of new meanings. This threefold rationale underlines the need for a comprehensive approach that considers meaning-generation as actualized in different dimensions. This chapter identifies three such dimensions, namely, the representational, interpretational, and existential dimension. These three dimensions form a system that corresponds roughly to the Peircean triad of semiosis.

Keywords: Danger modeling, meaning-generation, reality, semiosis, semiotic agency

Email: hongbing@torontomu.ca

Two types of people often foresee danger: those who have learnt from experience, very much to their own cost, and the astute, very much to the cost of others.[1]

– Baltasar Gracián (1601-1658)

17.1 Introduction

By any normal standard, there are two fundamental activities that are top priorities for any known species of organism to ensure their survival in the struggle for existence. One is energy preservation, made possible by all forms of sustenance by which organisms manage to live and reproduce, as well as all techniques that the organisms are capable of to maintain autopoiesis. The other is danger aversion. To avert danger, organisms need to be able to identify — or model [17.26] — it in ways that are specific to their species and actualized in concrete situations.

A trait that is common between energy preservation and danger aversion is that both perfectly demonstrate what is known in biosemiotics today as "semiotic agency" [17.27]. In biosemiotic terms, "all living organisms are semiotic agents capable of sensing and information-driven adaptive behavior. Thus, even the smallest creatures such as bacteria, the most abundant organisms on Earth, are agents" [17.27], p. 96. It is the semiotic agential nature of organisms that enables them to achieve the two above-mentioned feats in what Richard Dawkins calls the "greatest show on earth" — the evolution of life [17.9].

These two feats could be so closely intertwined in the fate of an organism that it would be impossible to discuss one of them without presupposing the other, but the present chapter takes interest in the latter feat, with a special focus on our own species, *Homo sapiens*. For one thing, compared with energy preservation, danger aversion often turns out to be a more immediate need (cf. [17.2]), especially in cases of existential danger [17.7] [17.8]. For another, our species have demonstrated superior abilities in causal reasoning compared to other species [17.17], pp. 27-28), enabling us to understand and deduce reasons for both experienced and perceived dangers.

What is the key to the superiority in reasoning? The answer can be found in Marcel Danesi's book *Warning Signs* [17.7], a most recent pioneering work on semiotics of danger, which has inspired the present chapter. According to Danesi, it is the use of a range of symbolic systems that aids in the process of danger aversion, making it possible for humans to represent and even imagine worlds of danger, that is, to create external and internal models of danger. These models are known as representations and mental forms, but we will come back to them in the next section. What we need to note at this point is that, under the biosemiotic lens [17.26] [17.30], danger modeling can be treated as an instance of meaning-generation. This is a trait shared by all organisms, but in the case of *Homo sapiens*, meaning-generation seems to have a multi-dimensional nature. This nature will be the central issue in this chapter.

The rationale behind this study is a three-fold one. Firstly, danger averting, understood in terms of meaning-generation, serves as a fine example of semiotic analysis, because

[1] Gracián, B. (2011) *The Pocket Oracle and Art of Prudence*. Trans. by Jeremy Robbins, Penguin Books.

warning signs are among the most primordial and relevant vehicles for meaning. Secondly, the Sebeokian concept of modeling serves as a useful tool to address this problem, because the concept underscores semiotic agency and promises to be a highly integrative framework. Thirdly, humans have evolved into such complex superorganisms [17.21] that we not only live by existential modeling and semiotic modeling at same time [17.30] but also are deeply entangled in our own semiotic webs while probing into and creating possibilities of new meanings. This three-fold rationale underlines the need for a comprehensive approach that considers meaning-generation as actualized in different dimensions. One such approach, as we will find out later in this chapter, is an updated and integrative concept of modeling [17.30], integrated with Peirce's semiosis.

17.2 Models In, Models Out

How is it possible to couple the concepts of modeling and semiosis? It all comes down to what "model" can mean vis-à-vis "sign," considering that models are end results of modeling, as it has been understood (cf. [17.1] [17.5] [17.14] [17.25] [17.26]), and that semiosis is the irreducibly triadic activity of signs [17.18], CP 5.484, 5.485. One thing for sure is that if we simply understand model as a substitute for sign, signifier, representation, or representamen, and so on, the term would hardly be convincing or necessary. We should go beyond that. For this purpose, we may turn to a working definition of model in Sebeok and Danesi's book *The Forms of Meaning* [17.26], which provides a theoretical framework that treats model as the form of meaning, either a representation or a mental image (also a kind of representation), and considers modeling as the activity of actually producing such forms. See their detailed definition of model below:

> "A model can be defined as a form that has been imagined or made externally (through some physical medium) to stand for an object, event, feeling, etc., known as a referent, or for a class of similar (or related) objects, events, feelings, etc., known as a referential domain. An imagined form can be called, simply, a mental form; a form made externally to stand for a referent can be called an externalized form." [17.26], p. 2.

The authors go on to elaborate:

> "The ability to make models is, actually, a derivative of semiosis, defined simply as the capacity of a species to produce and comprehend the specific types of models it requires for processing and codifying perceptual input in its own way. Semiosis is a capacity of all life forms; representation, on the other hand, is a unique capacity of the human species, which develops during the neonate and childhood periods." [17.26], p. 5.

This seminal conception of model and modeling is a refreshing and relatively comprehensive one, reaching beyond such discrete terms as sign, text, code, and metaphor, which have often created confusion not just in semiotics, but also in many other related fields. However, this conception still deserves further clarification, and perhaps a few "tweaks" as well, to increase its comprehensiveness and effectiveness for today's semiotic analysis.

The reason is that there are some terminological obscurities in the book. For instance, while the authors equate "representation" with model, thus revealing the essence of the latter, they also use the term to denote the capacity to refer to the world in terms of models [17.26], p. 6. In addition, the distinction between models and signifier/signified relations [17.26], p. 8, could be made clearer. But these minor infelicities can be easily tweaked out of the way, and it can be done in the spirit of the book itself. For the present purposes, let's foreground two most promising and time-tested points about modeling that Sebeok implied over the years [17.29] [17.30], especially in *The Forms of Meaning*: the ubiquity and operability (or manipulability) of models.

Sebeok seems to have used the term "model" to mean a few different things [17.30], pp. 643-645, for instance, an ontogenetically functioning system, Umwelt [17.25], and eventually forms of meaning [17.26]. But the ubiquity of models and their operability (or manipulability) have remained two underlying themes in Sebeok's treatment of the term. Especially in *The Forms of Meaning*, the Sebeokian conception of model is so thoroughly model-based that not only are the representational forms of meaning regarded as models, but meanings themselves, equated with the particular concepts that are elicited by specific representational forms, are also models because concepts are mental forms [17.26], p. 8. This conception is little short of a game changer in today's semiotic inquiry, whose central task is to discuss meaning-generation. As Sharov and Tønnessen aptly describe, "because signs are used to transfer meanings, semiotics cannot avoid the discussion of what meanings are and what their relation is to the material world. Finally, meanings do not exist by themselves, they are always related to agents who use signs to perceive and transfer meanings" [17.27], p. 37. In the Sebeokian vein, we may add that meaning needs to have a form, and it is semiotic agents that create the form, giving meaning its shape, so to speak. What are transferred are meanings as forms.

The only compromising factor in Sebeok and Danesi's treatment of model, as I have discovered, is that both representational forms and concepts (as mental forms) seem to be oddly confused with signifiers and signifieds from what they call "traditional sign theory," which is precisely the Saussure's tradition that Sebeok himself critiqued as a minor trend in contemporary sign studies ([17.23], p. 182) and brushed aside as a "pars pro toto fallacy" (cf. [17.12], p, viii; [17.6], p. 4). The reason for the above confusion remains unclear, but it is still possible to rectify this confusion.

Here comes the "tweaking." All we need to do is concede, not without a considerable amount of courage, that models are not signs (or their Saussurean components), but any experienceable[2] forms of meaning that have a *significatory potential*, i.e., the potential of turning into signs. Models and signs are thus essentially and functionally different: ***models supersede***, whereas *signs signify*. To be specific, "models indicate an ever-present presence as the result of supersession, whereas signs always exist only inasmuch as they imply a significatory or semiosic relation" [17.30], p. 650. Therefore, when they are regarded in their own existence stripped of significatory relations, signs could become models. Conversely, models could turn into signs, that is, acquire the status of signs if the focus shifts to significatory relations. We can't reasonably have both models and signs at the same time working as interchangeable terms. This "tweaking" is very much in line with Sebeok's thought because

[2] While "perceptible" would be a more generally accepted term, I consider "experienceable" a better word in describing models—it includes but is not limited to "perceptible". The reason is that "perception" is too misleading and inadequate a term to capture the properties of Umwelt and Innenwelt, which defy the Cartesian dualism (cf. [17.30], p. 652).

the properties of ever-present presence and supersession accord fully with the ubiquity and operability of modeling, the points that Sebeok's works implied. Only in this way can model have its legitimate place in semiotics, and thus, serve as a uniquely powerful concept alongside the existing sign concepts, whether they are Saussurean or Peircean.

17.3 Modeling, Coupled with Semiosis

Now we can proceed to couple the concepts of modeling and semiosis. The relationship between modeling and semiosis is largely derived from the relationship between model and sign. See the following comment for details [17.30], p. 651:

> Modeling may function as the prerequisite and/or the outcome of semiosis, that is to say, models can be pre-semiosis or post-semiosis identities, so to speak, of the three elements in the tripartite semiosic relation. Instead of it being a derivative of semiosis ([17.26], p. 5), [...] I prefer to see modeling as being interlaid with semiosis, and vice versa, although they hardly coincide, in that they are two differing, albeit potentially relatable, dimensions of contemplation. In a manner of speaking, modeling is more pertinent to the issue of what may be called immediate reality, whereas semiosis, or sign activity, is more pertinent to the matter of significatory relations. Imagine them as the vase and the faces in "Rubin's Vase." At any given moment, only one dimension can be maintained, but both are valid ways of contemplation of semiotics. However, in terms of effects, semiosis can also be taken to be an instance of modeling, inasmuch as semiosis as a whole in itself is not in a semiosis relation. To put this in a nutshell, the term "semiosis," as well as what it stands for, is but a model that semioticians (not all of them) use.

The basic step in such coupling, therefore, is the simple recognition that in any act of modeling, models supersede and are brought to the front for salience, accessibility, and operability, whereas in the meantime the "modeled" (or the signified before the signs became models) recedes and "exists" in the background, inaccessible and inoperable [17.30], p. 647. Now let's apply this recognition to semiosis. See Figure 17.1.

This figure is mainly based on Peirce's definition of semiosis [17.18], CP 5, 484, and his elaborate explanations of the components (e.g., [17.18], CP 2, 228; [17.19], pp. 80–81). In all his writings that have been published so far, Peirce never seems to have provided any figure other than words to illustrate semiosis. Intuitively, it should be a triangle or another figure that approximates to a triangle, but all the figures of semiosis that subsequent generations

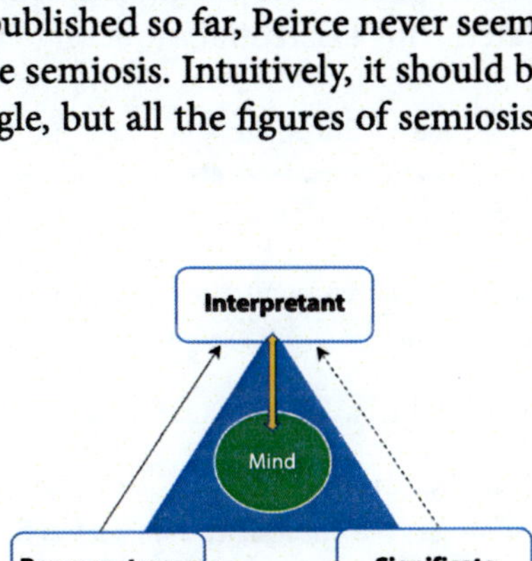

Figure 17.1 The revised complete Peircean triad of semiosis (from [17.31], p. 405).

of semiotically-minded scholars have intuited and developed over the past decades can only resemble (or model) what Peirce wished to convey, although they have served various purposes and they have served us well. Figure 17.1 is no exception because, like all figures before it and all those yet to come, it is and can only be a model of semiosis. I call it "the revised complete Peircean triad of semiosis" only in the sense that it includes all the three components (object,[3] representamen, and interpretant) and visualizes the relations of determination between them as Peirce described. It does not purport to exhaust everything that Peirce said about semiosis or its components.

According to Peirce, all these components are signs, including significate (object), which is "of the nature of a sign or thought" [17.18], CP 1, 539. Once stripped of their significatory relations and accentuated for their own sake, these components become models, because they now supersede and are brought to the front for salience, accessibility, and operability, whereas in the meantime, whatever was signified recedes and "exists" in the background, inaccessible and inoperable. Even at the level of semiosis, instead of its components, the same law could still apply: semiosis indicates a dynamic process of action, but the whole process in itself could become a model. And a peculiar thing about models is that they can be modeled. Hence, we have a semiotic web. Figure 17.2 provides a model of—a glimpse into—this web.

This figure is called "the growth of signs and semiosis," but it also reflects or models how the components of a semiosic process, as well as the process itself, can be further modeled and extended. Therefore, when we say that meanings grow, it is in fact the models, as the forms of meaning, that grow. Since meanings per se are models [17.26], p. 8, meaning-generation is simply an act of modeling. Given that there are two major types of modeling,[4] semiotic modeling and existential modeling, meaning-generation should not be a unidimensional phenomenon. On the one hand, in the case of semiotic modeling, there can be two dimensions for meaning-generation, one through representations and the other through mental forms, both in the Sebeokian sense. On the other hand, since Umwelten are worlds of meaningful objects [17.15], p. 651, and are also models [17.25], existential modeling may be considered a dimension for meaning-generation as well.

We can, therefore, identify three dimensions in which meaning-generation is actualized, namely the representational dimension, the interpretational dimension,[5] and the existential dimension[6]. The first two dimensions are both associated with semiotic modeling, whereas the third dimension is associated with existential modeling. These three dimensions form

[3] Following John Deely [17.11], p. 15, I have substituted "significate" for "object." Deely, perfectly in keeping with Peirce, was acutely aware that "not all things exist as significates, but only known things, things cognized, what has entered into awareness. But all objects exist as significates, whether they are also things or not" [17.11], p. 16.

[4] Semiotic modeling correlates with the use of representational and mental forms, at least in terms of psychical effects, whereas existential modeling can be thought of as another name for Umwelt building. For more details, see Yu [17.30].

[5] It is perhaps only in the interpretational dimension that René Descartes's "*cogito ergo sum*" would make any sense.

[6] The names of these dimensions could be adapted, depending on the subject matter under discussion. For instance, in a previous paper that addresses the semioticity of historical figures, Yu [17.31] uses slightly different names for these dimensions: the physical representation dimension (i.e., the representational dimension), the thinking dimension (i.e., the interpretational dimension), and the contact dimension (i.e., the existential dimension). The guiding principle is that they should roughly correspond to the Peircean triad.

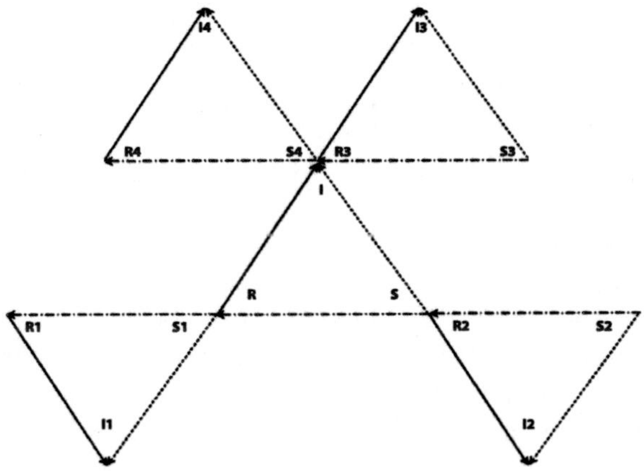

Figure 17.2 The growth of signs and semiosis (from [7.31], p. 407).

a system that corresponds roughly to the Peircean triad of semiosis. By "roughly" I mean existential modeling could only correspond to physical significates (or objects, in Peirce's term), whereas significates are not necessarily physical. Now, we have a comprehensive approach that considers meaning-generation as actualized in different dimensions. It is a framework derived from an updated integrative concept of modeling, coupled with Peirce's semiosis. To better know how integrative the concept of modeling can be, let's look at the following examples (cf. [17.30], pp. 646-647):

1. The Mantis shrimp has the ability to perceive its environment using as many as 16 types of photoreceptors, while humans possess only three. (Existential dimension)
2. In preparation for a space launch, rocket scientists use both mathematical models and computer-generated visual models. (Representational dimension)
3. A child plays with toy models of cats, dogs, and mice and knows that they are not "real." (Representational and interpretational dimensions)
4. A government think tank creates a report on weapons of mass destruction in another country. (Representational dimension)
5. An experienced military veteran recounts tales of their brave service in a military operation and is deeply moved by them. (Representational and interpretational dimensions)
6. A historian in the twenty-first century tells fascinating stories about Genghis Khan. (Representational dimension)
7. In front of an audience, a fashion model walks in the latest costume. (Representational dimension)
8. A husband feels paranoid and anxious, imagining what his wife may be doing while waiting for her. (Interpretational dimension)
9. A writer works hard on a thriller book and aspires to be the next Agatha Christie. (Representational dimension and interpretational dimension)
10. A new gymnastics team member observes their coach dancing to music. (Representational dimension)

The parentheses highlight the dimensions that are most relevant to these examples, but it would be impossible for human beings to experience meaning-generation without the existential dimension, which provides the ground for the other two dimensions.

This is particularly true of human danger aversion. There are generally two ways of danger modeling: learning through the effects of one's own actions and social/observational learning [17.1], p. 47. Both rely on existential modeling, because every one of us lives in the ever-presence of our own Umwelten and it is not possible to live in a complete vacuum of existence. Moreover, studies have shown that animals, including humans, don't simply modify any aspects of their phenotypes adaptively, but instead are predisposed to learn some things and not others. For instance, when rats try to figure out whether something new is safe to eat, there are certain items they never sample. Human babies also display similar predisposition that helps them avoid danger from unfamiliar plants (cf. [17.4], pp. 406-407). They already exist in their own immediate realities, created by existential modeling (Umwelten[7]), that are most meaningful to them. The items that rats don't touch might be food sources for other species; and plants poisonous to humans may appear not dangerous to other organisms. These are perfect examples of meaning-generation in the existential dimension. Physical significates, in terms of Peircean semiosis, may become different forms of meaning to different species of organisms.

Nevertheless, natural selection seems to favor danger modeling mechanisms that help prevent human children from making costly mistakes in avoiding danger as they grow up and learn about the world (their Umwelten). The reason is that children have a very slow process of maturation, and they obtain a significant portion of the necessary knowledge and skills for survival through semiotic modeling, especially in the representational dimension. But this is not limited to children. Adults rely heavily on semiotic modeling in danger aversion as well. A good example is the aversion to existential danger, explored by Marcel Danesi [17.7].

17.4 Danger Modeling

According to Danesi, existential danger is "any situation that imperils the existence of living things, if allowed to continue without refrain" [17.7], p. 3. It can be massive or small in scale. In his book, Danesi focuses on the global dangers, such as floods, volcanic eruptions, climate change, nuclear warfare, pandemics, unchecked human-generated pollution, and so on. But existential danger can be small in scale. In either case, Danesi's observation about semiotic modeling equally applies: the use of various representational systems helps in the avoidance of existential danger by enabling humans to conceptualize, envision, and even imagine hazardous scenarios. The representations are external models of danger, whereas the concepts, visions and imaginations elicited are internal models. All of them serve as the building blocks of meaning, each in their own dimensions, representational and/or interpretational.

For example, let's look at the fable of "cry wolf" to gain some insights into how humans would model (represent and interpret) threats to our lives and how meaning-generation unfolds in the three dimensions. The fable is a story of a shepherd who repeatedly cries out

[7] At the same time, species-specific Innenwelten also have a crucial role to play.

for help, even when there is no danger, causing people to ignore him when he is later facing actual danger and in need of help. Danesi analyzes the story with a framework of different angles, such as its use of metaphors, its place in folklore, and its narrative structure. In short, it is a framework of representational forms. Building on this framework, we can better understand at two major levels how humans represent danger and how different forms of representation can impact our knowledge of threats.

The first level is a macroscopic one, which pertains directly to Danesi's framework. At this level, the story is a cautionary tale that warns against the danger of losing others' trust as a result of raising false alarms. It exemplifies at least five of the seven semiotic principles of danger representation that Danesi outlines: representationality, relativity, metaphoricity, folkloricity, and narrativity [17.7], pp. 134-135. While the principles of metaphoricity, folkloricity, and narrativity are easy to grasp as the story is a folkloric narrative that serves as a metaphor for a particular form of social conduct, the principles of representationality and relativity are of particular importance to semiotics.

The principle of representationality involves the creation of representations by humans to encode and communicate danger in society [17.7], p. 135. In the case of "cry wolf," the danger is in relation to one's bearings in society and the moral teachings about the danger. This creates a kind model of behavior, embodying an otherwise abstract idea of comportment, that people are expected to follow. Therefore, it gives a shape to a meaning that is expected to be socially shared.

As for the principle of relativity, it is the idea that words, images, and other expressive structures shape meaning-generation. The "cry wolf" story has been retold in various forms and cultural communities over the past two millennia, and the general message of the moral warning has remained intact, but the verbal and non-verbal media in which the story has been told have had culturally marked impacts on how it has been received. This principle is closely related to the arrow pointing from Representamen to Interpretant in Figure 17.1. It attests to Peirce's idea that Representamen determine its effect (Interpretant) upon a person [17.19], pp. 80-81. In this case, meaning-generation maps from the representational dimension onto the interpretational dimension, creating or modifying a previous interpretational dimension (cf. [17.18], CP 1, p. 339). If the "cry wolf" story is told in a Peking opera, a musical, a cartoon movie, a poem, etc., the effects (meanings in the interpretational dimension) are likely to be different, although the danger at the macroscopic level remains the same.

The second level at which we can approach the "cry wolf" story is a microscopic one. At this level, we can also discover some insights. For example, the message of "wolf!" only has an impact when both the shepherd and the people he tricks share the same understanding—causal reasoning, which is a mental model—that wild wolves are dangerous animals that prey on livestock, fowl, and even humans. This understanding is acquired after birth through social learning (another instance of semiotic modeling) about the dangers of wolves from various sources of representational forms, such as folklore, myths, movies, and others. If a child who has never seen any depictions of wolves encounters one that is not displaying any aggression or violence, they might not react in the same way as someone who has learned about the dangers of wolves. This does not mean that one must have direct experience (through existential modeling) with wolves to fear them, but rather that representations can be just as effective in shaping our meaning-generation in relation to the danger of wolves.

In all these cases, there is the same potential for a semiotic web (see Fig. 17.2) to grow. When the shepherd in the "cry wolf" story uses the representational form "wolf!", both the shepherd and the people who receive the form will experience it as Representamen if it enters a significatory relation and becomes a sign, but they will experience different Interpretants. The people could be scared, alert, angry, or indifferent, depending on how many times they have heard the message. The shepherd could probably be amused to differing degrees or upset, etc. These interpretants and representamens, and even the "wolf" as a significate can turn into other identities. For instance, a representamen (R) can become a significate (S1) when people talk about the message; a receiver's interpretant (I) at any point can become a representamen (R3) if it leads to other thoughts or emotions at a later point. The list can go on. The point is that modeling and semiosis interlay and interplay, taking turns in creating and pushing forward meaning-generation in three dimensions.

But the interlaying and interplaying doesn't necessarily mean that we as semiotic agents could direct our attention to all three dimensions at the same time, even when we choose to analyze and reflect on them, as this chapter does, which is an instance of semiotic modeling that only ends up creating an external model, a representational form consisting of written words and figures, which are also models. The dimensions are the domains of meaning-generation, providing affordances for semiotic agents to create meanings. In this sense, we are merely limited beings going through and probing in the dimensions for meanings, which are in themselves merely models.

Another complicating factor is that, as studies have indicated, humans are superorganisms [17.21], more specifically known as holobionts. As much as half of all the biological matter in the human body is not human but made up of hundreds of trillions of viruses and bacteria. Some of them can cause illness, physical and psychical, whereas many others may simply coexist with the human body, forming part of our Innenwelten. Since Innenwelten work with Umwelten to form functional circles, to what degree these non-human organisms exert their influence on or participate in "our" meaning-generation in the three dimensions through existential modeling and semiotic modeling remains to be explored. Suffice it to say that we not only live to create semiotic webs but are also made up of semiotic webs within "ourselves" as superorganisms, jointly created by trillions of semiotic agents. This promises to be a fascinating topic for biosemiotics.

17.5 The Three Dimensions: Further Comments

Let's now return to the issue of meaning-generation in three dimensions. What matters in differentiating these dimensions and treating them under the lens of modeling, as understood in this chapter, is that they not only pertain to meaning-generation, but also introduce a counter-intuitive perspective that questions some old epistemological dogmas.

If we wish, the three dimensions could also be understood as the dimensions — or even "worlds" in Karl Poppers' fashion [17.20], p. 154 — of experience, existence, or being. But since meaning-generation is an act of modeling and any end result of modeling creates immediate reality, meaning-generation in three dimensions correlates closely with what I call "reality in three dimensions."

This requires some further elaboration, because "reality" has obviously become one of the most confusing terms over the years. In this chapter, by the word I mean the immediate effect of supersession created by modeling, existential or semiotic [17.30], p. 650. It designates the state of existence that every one of us directly experiences, constantly in the "now," as "real," even if it turns out to be "not real" compared with another distinguishable state of existence, if any, in a few seconds. After all, the "cry wolf" story, showcasing false alarm, does touch on an enduring theme in semiotics, the problem of reality, the boundary between real and unreal. This reminds us of a well-known point that Thomas Sebeok used to make about the main mission of semiotics [17.24], pp. 77-78:

> the central preoccupation of semiotics is an illimitable array of concordant illusions; its main mission to mediate between reality and illusion—to reveal the substratal illusion underlying reality and to search for the reality that may, after all, lurk behind that illusion.

But aren't illusions, lies (cf. [17.13]), and even hallucinations, directly experienced as "real" when they happen? Moreover, reality as we know it is just another word for Umwelt, or in mundane words, the world, and it could also be an illusion, especially considering quantum physics and embodied cognition, both of which accord closely with insights in Buddhism (cf. [17.10] [17.22] [17.28]).

Which reality is True or Real? Perhaps there is no definite answer to the question. Peirce alluded to the same kind of indeterminacy when he expressed to Lady Victoria Welby his "abhorrence of the doctrine that any proposition whatever is infallibly true" and said that "I cannot infallibly know that there is any Truth" [17.19], p. 73. In the Peircean vein, we might as well think of reality as whatever is experienced as "real" until another model, if any, is experienced to prove the first one wrong or unreal—even this thinking-of is in itself a model experienced and held to be true by the author, which is me. This, however, should not be confused with unlimited relativism, à la radical constructivism. On the contrary, it is grounded in the pragmatism doctrine (not dogma)—spearheaded by Peirce himself and applicable to natural selection—that stresses both validation and the history of failure/success. In this light, the commonly held "reality-vs-illusion" dichotomy, with which we are all too familiar, should perhaps be replaced with one: "reality-forming," or "infinite modeling" in the same spirit as Peirce's infinite semiosis. What is important is the function of the "reality-forming" or "infinite modeling": it is about *niche construction* in the broad biosemiotic sense, which entails not only the Umwelt niches, but also cultural and cognitive niches (cf. [17.3] [17.16]). Applying this perspective to the semiosis of human danger aversion, we could even identify a maladaptive (not necessarily in Darwinian sense) role of semiosis in creating dangers and not just in monitoring and knowing them. After all, many dangers come directly from cultural and cognitive practices (if we have to separate them) themselves when they serve as components of their semioses. The examples could be verbal violence and depressing thoughts, respectively. This perspective underlies and further validates the rationale for understanding (experientially) meaning-generation in the three dimensions as mentioned above.

In this chapter, however, the distinction between these three dimensions is merely an expedient one that makes most sense when it is intended for analysis and reflection, which are also instances of semiotic modeling. Even in such analysis and reflection, the three dimensions would unavoidably overlap and interlay. We may feel free to call the dimensions by any other name—even by such biosemiotically limited terms as "perceptions" and "sensations"—as long as the understanding remains that these dimensions are what we live in, serving as sources of affordances to create the only yet varied kinds of immediate realities we could possibly and directly experience. As John Deely once rightly pointed out, although Monism, the idea that all of reality is one single underlying whole, is not impossible, it is contrary to our experience [17.22], p. 18. Therefore, "reality," understood in biosemiotic terms, "can hardly be any more than one of the various models that we, as modelers, incessantly make by building or extracting out of pre-existing models that may eventually trace their origin to the modelers' enacted Umwelten." [17.30], pp. 650-651. Thus, the plurality of danger modeling is inevitable in the struggle for existence at various temporal and spatial scales.

Declaration

The author declares no conflict of interest.

References

[17.1] Bandura, A. (1986) *Social Foundations of Thought and Action: A Social Cognitive Theory*. Prentice-Hall, Inc., Upper Saddle River, USA.

[17.2] Barrett, H.C. and Broesch, J. (2012) Prepared social learning about dangerous animals in children. *Evolution and Human Behavior*, 33(5), 499–508.

[17.3] Bertolotti, T. and Magnani, L. (2013) Terminator niches. In: *VRIC '13: Proceedings of the Virtual Reality International Conference: Laval Virtual, March 20-22, 2013, Laval, France*. Simon Richir (ed.) Association for Computing Machinery, New York, NY, USA. Article No. 31, 1-10.

[17.4] Boyd, R. and Silk, J. B. (2021) *How Humans Evolved*. W. W. Norton & Company, New York, USA.

[17.5] Chao, Y. R. (1962) Models in linguistics and models in general. In: *Logic, Methodology, and the Philosophy of Science*. E. Nagel, P. Suppes and A. Tarski, (eds.). Stanford University Press, Stanford, USA: 558–566

[17.6] Cobley, P., Deely, J., Kull, K. and Petrilli, S. (2011) Introduction: Thomas A. Sebeok: Biography and 20th century role. In: *Semiotics Continues to Astonish: Thomas A. Sebeok and the Doctrine of Signs*. P. Cobley, J. Deely, K. Kull and S. Petrilli (eds.) De Gruyter Mouton, Berlin, Germany: 1–17.

[17.7] Danesi, M. (2022) *Warning Signs: The Semiotics of Danger*. Bloomsbury Academic, London, UK.

[17.8] Danesi, M. (2023) *Apocalyptic Visions: Pandemics in Literature, Art, and the Movies*. Nova Science Publishers, Inc., New York, USA.

[17.9] Dawkins, R. (2009) *The Greatest Show on Earth: The Evidence for Evolution*. Free Press, New York, USA.

[17.10] Dawkins, R. 2011. *The Magic of Reality*. Free Press, New York, USA.

[17.11] Deely, J. (2009) *Purely Objective Reality*. Mouton de Gruyter, Berlin, Germany.

[17.12] Deely, J., Williams, B., and Kruse, F. E. (1986) Editors' preface "Pars pro Toto". In: *Frontiers in Semiotics*. J. Deely, B. Williams, and E. K. Felicia, (eds.) Indiana University Press, Bloomington, USA: viii–xvii.

[17.13] Eco, U. (1979 [1976]) *A Theory of Semiotics*. Indiana University Press, Bloomington, USA.

[17.14] Lotman, J. (2011 [1967]) The place of art among other modeling systems. *Sign Systems Studies* 39(2–4), 249–270.

[17.15] Magnus, R. and Kull, K. (2012) Roots of culture in the Umwelt. In: *The Oxford Handbook of Culture and Psychology*. J. Valsiner (ed.) Oxford University Press, Oxford, UK.

[17.16] Odling-Smee, F. J., Laland, K. N., Feldman, M. W. (2003) *Niche Construction: The Neglected Process in Evolution*. Princeton University Press, Princeton, USA.

[17.17] Pearl, J. and Mackenzie, D. (2019) *The Book of Why: The New Science of Cause and Effect*. Penguin Books, London, UK.

[17.18] Peirce, C. S. (1931–1958 [1866–1913]) *The Collected Papers of Charles Sanders Peirce*, vols. 1–6. C. Hartshorne and P. Weiss (eds.); vols. 7–8, A. W. Burks (ed.) Harvard University Press, Cambridge, MA, USA. Cited as CP.

[17.19] Peirce, C. S. (1977) Letter from Peirce to Welby, dated Dec. 23, 1908. In: *Semiotic and significs: The correspondence between Charles S. Peirce and Victoria Lady Welby*. C. S. Hardwick (ed.) Indiana University Press, Bloomington, USA.

[17.20] Popper, K. R. (1979 [1972]) *Objective Knowledge: An Evolutionary Approach*. Clarendon Press, Oxford, UK.

[17.21] Pride, D. (2020) The viruses inside you. *Scientific American* 323(6), 46–53.

[17.22] Rovelli, C. (2017 [2014]) *Reality is Not What It Seems: The Journey to Quantum Gravity*. S. Carnell and E. Segre (trans.) Riverhead Books, New York, USA.

[17.23] Sebeok, T. A. (1977) Ecumenicalism in Semiotics. In: *A Perfusion of Signs*. T. A. Sebeok (ed.) Indiana University Press, Bloomington, USA: 180–206.

[17.24] Sebeok, T. A. (1986) *I Think I Am a Verb: More Contributions to the Doctrine of Signs*. Plenum Press. New York, USA.

[17.25] Sebeok, T. A. (1988) Language: How primary a modeling system? In: *Semiotics 1987*. John Deely (ed.). University Press of America, Lanham, USA: 15–27.

[17.26] Sebeok, T. A. and Danesi, M. (2000) *The Forms of Meaning: Modeling Systems Theory and Semiotic Analysis*. Mouton de Gruyter, Berlin, Germany.

[17.27] Sharov, S. and Tønnessen, M. (2021) *Semiotic Agency: Science beyond Mechanism*. Springer, Cham, Switzerland.

[17.28] Varela, F. J., Thompson, E. and Rosch, E. (2016) *The Embodied Mind: Cognitive Science and Human Experience*. MIT Press, Cambridge, MA, USA.

[17.29] Yu, H. (2019) *On Semiotic Modeling*. Soochow University Press, Soochow.

[17.30] Yu, H. (2021) Modeling in semiotics: An integrative update. *Chinese Semiotic Studies* 17(4), 639–659.

[17.31] Yu, H. (2022) Confucius the untouchable: on the semiotics of historization. *Chinese Semiotic Studies* 18(3), 391–412.

Part IV

MEANINGS IN HUMANS AND BEYOND

18

Anchors of Meaning: The Intertwining of Signs, Abduction, and Cognitive Niches

Lorenzo Magnani

Department of Humanities, Philosophy Section, and Computational Philosophy Laboratory, University of Pavia, Pavia, Italy

Online resources:
https://en.wikipedia.org/wiki/Lorenzo_Magnani
https://philpeople.org/profiles/lorenzo-magnani
https://unipv.academia.edu/LorenzoMagnani
https://scholar.google.at/citations?user=KbrqtfAAAAAJ&hl=de

Abstract

In this chapter, I describe how semiotic tools can be fruitfully integrated with studies on abduction, cognitive niches, and human creative cognition. It is extremely advantageous to see human brains as "semiotic" brains that generate a series of signs and that are involved in producing, showing, or responding to a succession of signs. Through this semiotic activity, we see, in a very naturalized way, that brains are engaged in "being minds," so in thinking intelligently and, occasionally, creatively. In this perspective, I explain that at the root of thinking abilities there is a process of externalization of the mind that presents a new cognitive perspective on the role of semiotic external models and representations as *anchors of meaning*. I consider this interplay critical in analyzing the relation between meaningful semiotic internal resources and devices and their dynamical interactions with the externalized semiotic materiality suitably stocked in those parts of the environment that recent evolutionary research has called *cognitive niches*. In sum, minds are material, semiotic, extended, and "artificial" in themselves, so to speak. The mechanism of externalization—that we can call *disembodiment of the mind*—also discloses a novel cognitive perspective on the mechanisms underlying the semiotic emergence of abductive processes—especially *manipulative abductions*—of meaning formation, capable of generating and at the same time anchoring, meanings. Indeed, a considerable part of human abductive thinking is occurring through an activity consisting in a kind of reification in the external environment and a subsequent re-projection and reinterpretation through new configurations of neural networks and chemical processes.

Email: lorenzo.magnani@unipv.it

Keywords: Evolution, ecological niches, cognitive niches, abduction, anchors of meaning, purposefulness in evolution, semiotic brains, disembodiment of the mind

18.1 Humans as Ecological Engineers and Chance Seekers: The Role of Abduction

As the father of the concept of abduction, which he revived many centuries after its Aristotelian introduction, Peirce's philosophical pragmatist framework informs my perspective in this chapter, which approaches "inference" *semiotically* (and not simply logically). Peirce distinctly says that all inference is a form of sign activity, where the word sign includes "feeling, image, conception, and other representation" [18.32], and adopts a wide interdisciplinary breath. When speaking of anchors of meanings, I am a strict follower of Peirce's conviction that our idea of something is our idea of its sensible effects: "Consider what effects, that might conceivably have practical bearings, we conceive the object of our conception to have. Then, our conception of these effects is the whole of our conception of the object" [18.33]. This rule founds the pragmatic procedure thanks to which it is possible to fix our ideas.

The recent investigation into abductive reasoning [18.20] [18.22] [18.23] [18.24] has clarified that humans make judgments and deal with issues in light of what Herbert Simon [18.38] [18.39] referred to as "incomplete information." When we create assumptions, we tend to influence the situation now available and the depictions we have of it to the end of acquiring fresh, valuable, and informative information sources. Hypotheses may originate from an abductive selection in a set of previously-stored assumptions (such as in the case of what I have called selective abduction, for example, in medical diagnoses) or can be the fruit of creative abduction, when they are new [18.20], such as, for example, in the case of scientific discovery.

When it comes to diagnostic situations, a shortage of information is inherent: occasionally, the facts at hand are insufficient to allow the doctor to make an accurate diagnosis. He or she must thus carry out further tests or perhaps try other manipulations to uncover symptoms and information that would otherwise be hidden. In this final instance, we may argue that the doctor is increasing the likelihood of coming up with a reliable diagnosis and the ensuing treatment. In my studies on abduction I have shown in great detail that, for instance, scientists manipulate intricate artifacts that are appropriately built in their study settings in order to gather additional data. This activity may be tacitly and pragmatically embodied rather than necessarily being achieved consciously and logically through exact sequences of relevant acts. Everyone is aware that manipulative skillfulness is a virtue that influences how people behave in everyday situations. In fact, they are always working to extract latent possibilities in order to get knowledge and information that was not before available: humans may be seen from this perspective as *chance seekers*.

In this chapter, I try to show how, in addition to being risk-takers and chance-seekers, people are also ecological engineers and constructors of the so-called cognitive niches. In fact, we should also emphasize a crucial conceptual implication for evolutionary epistemology: humans, like other species, actively alter and reconstruct their surroundings in order to find the best opportunities. In other words, organisms are not inert creatures influenced by their environments; rather, they actively contribute to the construction of the world by creating species-specific biorealities.

18.2 Semiotic Brains Construct Cognitive Niches

Recent biological research joins the biosemioticians in their belief that the idea of agenthood [18.37] is necessary to comprehend the meanings organisms give to the *signs* and the communicative structures they employ. By highlighting the fact that the majority of humans have what I have dubbed *semiotic brains* that generate a series of signs and that are involved in producing, showing, or responding to a succession of signs, we can further strengthen this viewpoint. Through semiotic processes, they are simultaneously engaged in "being cognitive agents."

In fact, some cognitive paleoanthropological studies tell us that the emergence of a modern language (speech) coupled with a high level of reflective consciousness in terms of thoughts about our own thoughts and about our feelings,[1] institutes a relevant condition of possibility for the emergence of the so-called "material culture," that is of cultural niches, so to speak.[2] According to this perspective, conscious activity is further defined by the fact that, as a result of those semiotic processes, we may claim that humans are, in reality, *ecological engineers* and *chance-seekers*, as I have already emphasized.

The human process of constructing *cognitive niches*—for example, we can immediately anticipate that, in the case of human animals, *natural language* is a fundamental example of a cognitive niche—is a useful further integration of the intellectual framework I have just described (cf. [18.18] [18.35] [18.36] [18.42]). The book written by Odling-Smee *et al.* [18.30] gives a thorough explanation of the doctrine of the cognitive niche. After natural selection, cognitive niche construction should be seen as a second fundamental player in evolution. Famous scholars including Schrödinger, Mayr, Lewontin, Dawkins, and Waddington, brought the concept of niche formation to the public's attention for the first time, first in the sphere of physics, and then in the actual subject of the theory of evolution—Waddington stressed the impact of organism growth in particular.

By altering their environments and partially influencing some of the energy and matter fluxes in their ecosystems, organisms (not just humans) have the potential to shift some of the natural selection pressures existing in their local settings. Natural selective pressure varies, particularly when the same environmental alterations occur often enough throughout generations and selective change. For example, in the case of spiders, webs may be a source of selection that leads to additional phenotypic changes in some species, such as marking their webs to enhance crypsis or the creation of dummy spiders probably to divert the attention of the birds that prey on them. Thus, as also clearly illustrated by the main tenets of the modern synthesis,[3] the dynamic act of altering the natural environment and producing diverse artifacts (including, in the case of human beings, technological tools) by people or animals, is referred to as niche construction. The environment's offerings in terms of cognitive chances are carefully chosen and/or produced to improve people's adaptation as chance-seekers thanks to the niche construction process. This ecological viewpoint,

---

[1] That is, consciousness not trivially seen as brute sensation.

[2] More details concerning these important studies can be usefully found in Mithen [18.26] [18.27] [18.28] [18.29].

[3] The reader interested in the connections between semiotics and the evolutionary extended synthesis (of which niche construction is a key component) and the further relationships with the evolutionary developmental biology (evo-devo) can refer to the Special Issue of the journal *Biosemiotics* "Semiotic Aspects of the Extended Synthesis" (Vol. 11(2), 2018): in particular, cf. [18.12] [18.34].

therefore, aims at understanding cognitive processes in terms of their *environmental situatedness* [18.23] [18.24].

18.2.1 Ecological Niche and Cognitive Niche Construction

According to the theoretical model described above, the environment provides limitless possibilities to living creatures, which only partially exploit them. Humans and other organisms are participating in a change of their environment to better benefit those outside factors that are to their advantage and eliminate or diminish the result of the undesirable ones.

It is crucial to define the theoretical and epistemological links between "cognitive" and "ecological" niche theories before moving on to discuss cognitive niches and their function in anchoring meanings. A setting that depicts how an organism interacts with its surroundings, as opposed to the habitat that represents its home area, is referred to as an ecological niche, in Gibson's opinion. According to Gibson, a niche may be seen as a collection of *affordances*, which are things that the environment offers. For instance, a chair "affords" the opportunity to sit, just as air, water, stairs, etc., "provide" the opportunity to breathe, and so on. The idea of affordance has two benefits. First, it shows the particular relationship between an actor and its environment as well as how interdependent they are. Second, this concept may also be utilized to describe people as chance-seekers more accurately.

Speaking in a generic way, human and non-human animals are able to fill fascinating, maybe unique *cognitive niches* that have significant practical and cognitive impacts; however, first of all organisms also regenerate and change their *ecological niche* to adjust the selective pressure of the local environment (by removing detrimental consequences), a process that, although not halting natural selection, does make it more challenging to explain.

Overall, by constructing ecological and cognitive niches, adaptability should account for both the environment's effect and the behavior of creatures that impacts it. We can now more readily understand the idea of animals acting as ecological engineers: organisms actively change and rebuild their environments in poietic ways in addition to just existing in them (cf. [18.8]).

Let us now explain the difference between ecoogical and cognitive niches in human and non-human animals. An unambiguous case of cognitive niche construction (that is not of a simple ecological niche construction) is illustrated by Clark by referring to the activities of a bartender: in order to better recall the cocktails she has to make without the use of a written list, the bartender at a busy bar organizes the glasses in front of her according to their shape after taking orders. She also adds embellishments like colorful straws and cocktail umbrellas. Of course, this description of a cognitive niche immediately brings to mind Odling-Smee, Laland, and Feldman's concept (however very different) of *ecological niche construction*, illustrated as follows:

> Niche construction occurs when an organism modifies the feature-factor relationship between itself and its environment, either by physically performing perturbing factors at its current location in space and time, or by relocating to a different space-time address, thereby exposing itself to different factors [18.30].

A clear example of an ecological niche construction is clearly one of beaver dams, where these species can thrive (Figure 18.1 shows various cases of ecological niches made by

Figure 18.1 Examples of ecological niche construction through which organisms shape their niche and realize a second inheritance system (ecological) which also favors the reduction of selective pressure by reducing uncertainty: bee colonies, bird nests, beaver dams, and human dwellings.

humans and animals). However, ecologically relevant environmental changes may not only be made to help humans survive better, they may also be made to help humans survive by improving their *cognitive* abilities. We are surrounded by structures that we engage with as *cognitive agents*, to the point where we are completely submerged in virtual niches as well [18.1] [18.2]. In order to make life simpler cognitively, we alter local settings, such as by using pencil and paper to construct a map, but we also create and post other kinds of *signs* that are as widely understood as possible.

According to Clark [18.5], one of the first and most spectacular instances of cognitive niche construction initiated is the deployment of *language* into our ecologies, which builds a powerful scaffolding for more sophisticated cognitive efforts. By using language, humans form their relevant cognitive niches, and this in turn generates a series of chances and constraints to cope with, reflecting the focus on selection pressure feedback cycles in ecological niche construction [18.30].

Odling-Smee, Laland, and Feldman assert that their "beaver example could mislead. Beaver dams are obvious artifacts; so much so, they can easily give the impression of being intentionally designed by intelligent animals for anticipated functions. However, most niche construction is entirely noncognitive in character, so intentionally cannot be ascribed to it, even when humans can ascribe function to it" [18.30], p. 335. This is not meant to imply that cognitive displays and ecological niche construction are mutually exclusive. Ecological niches can be "cognitive" inasmuch as they are conducive to—and occasionally even demand—the use of cognitive abilities, but they only become "cognitive niches" if changes made to the environment have a *cognitive character* insofar as they serve a cognitive goal.

In order to comprehend this separation and any potential overlap between the two types of niche formation at this early level, an example could be helpful. Consider that you are putting up a university library. You must arrange a large number of volumes. The weight and bulk of the books provide physical constraints for book storage. Constructing a size-appropriate bookshelf is the easiest method to do this. Even though it is obvious that it takes mental work, constructing bookshelves is equivalent to constructing an ecological niche. Never mind that books are cognitive artifacts; if that bothers you, consider stone specimens instead of books. After placing all of the books (or stones) on the shelves, you label them, create a catalog, and write the categories and contents on the shelves. This is the construction of a "cognitive" niche because it modifies the environment to enhance reasoning about a target domain, in this case, allowing people to retrieve the books they want without having to look through them all or relying solely on memory (Figure 18.2 illustrates various unambiguous cases of cognitive niches).

Odling-Smee and his colleagues observe that, when dealing with human animals' activities, we can speak of cognitive niche construction as *cultural niche construction*, that can in turn be described as a *three-way inheritance system*, namely genes, environment (ecological niche) and "culture." The example of healthcare is more than clear:

> Humans may respond to novel selection pressure (caused by a disease) either through cultural evolution, for instance, by constructing hospitals, medicine, and vaccines, or at the ontogenetic level, by developing antibodies that confer some immunity, or through biological evolution, with the selection of resistant genotypes. As cultural niche construction typically offers a more immediate solution to new challenges, we anticipate that cultural niche construction will usually favor further counteractive cultural niche construction, rather than genetic change [18.30], p. 261.

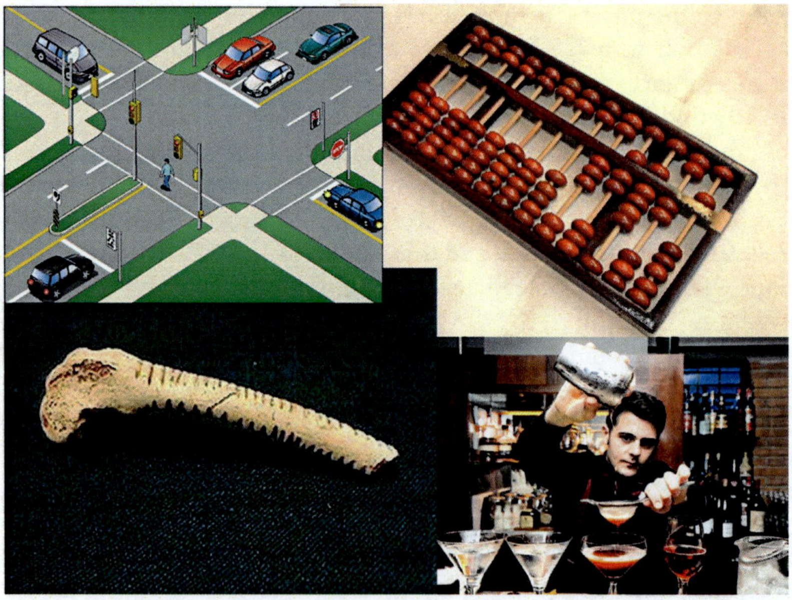

Figure 18.2 Cognitive niche construction is the process by which human animals (but also non-human animals) build physical structures that transform problem spaces in ways that aid (or sometimes impede) cognitive activities about some target domain or domains. Examples are street traffic regulation, abacus, ancient bone with apparent records of property items, and human bars as environments for communication and relaxation.

Hence, an upgraded definition of cognitive niche construction is the one I illustrated in a previous article I wrote with Tommaso Bertolotti, which reads as follows:

> Cognitive niche construction is the process by which organisms modify their environment to affect their evolutionary fitness by transforming their environment in ways that facilitate (or sometimes impede) the persistent individuation, the modeling, and the creation of cause-effect relationships within some target domain or domains. These mod ifications may combine with appropriate culturally transmitted practices to enhance problem-solving, and (in the most dramatic cases) they afford potential whole new forms of thought and reason [18.3], p. 4766-4767.

We have to note that, according to Odling-Smee *et al.* [18.30] the concept of fitness in this context must be considered loosely Darwinian because extragenetically informed behavior patterns might be generally both adaptive and maladaptive, and genetic evolution produces random variations while extra-genetic information does not.

At this point, the theoretical connection between ecological and cognitive niche construction has to be further and finally delineated. The two kinds of niches are surely not coextensive but we can easily say that a shift from one kind of niche to the other is possible. In the already quoted article [18.3] I have illustrated in detail a solution to this conundrum, that takes advantage of the exploitation of the concept of *enablement*. According to Longo and Montévil, the concept of "enablement," which accounts for the changes of "phase space" amply meant, and namely for the creation of a new, hitherto uncomputable observable, is the most rational approach to represent something resembling "causation" in biological systems: "in short, a niche *enables* the survival of an otherwise incompatible/impossible form of life, it *does not cause* it. More generally, niches enable what evolves, while evolving with it." For instance, in physics "gravitation, for example, causes a body to fall [...]. Gravitation instead, in embryogenesis, say, is a (fundamental) constraint, not a cause" [18.19], p. 205, as it "enables" a specific development of the embryo.

Ecological niche construction shifts into cognitive niche construction because enablements occasionally tend to *shift* into those cognitive "affordances" I have illustrated above, when dealing with Gibson's description of niches. The possible overlap is explained by the fact that there are many examples that appear to fall somewhere between the ecological and cognitive niches. A simple example could be of help: phenomena such as the New Caledonian crows' socially learnt toolmaking [18.44] could either be interpreted as complicated ecological niche construction or as early-stage cognitive niche construction. It depends on whether we emphasize the *ecologically enabling* character of manufacturing tools with bits of wire (which obviously adds a new observable), or the cognitive exploitation (*learning-affordance*) of the wires previously bent by other birds.

18.3 Cognitive Niches Construction Expresses an Evolutionary Sense of Purposefulness

To thoroughly understand cognitive niche theory and its link to extragenetic[4] information and the development of a sense of purposefulness[5] in evolution see [18.30], where a deep analysis in a biological and evolutionary perspective is proposed to treat the subject at hand: the authors present a fresh intellectual climate that goes beyond the conventional adaptationist framework. They clearly contend that "niche construction" should be seen as the second important player in evolution, immediately after natural selection. Because it adds "feedback" to the evolutionary processes, niche construction is a powerful evolutionary agent (ibid., p. 2).

In sum, we can conclude that organisms, including people, have the capacity to potentially change a few of the selective pressures present in the immediate surroundings thanks to some degree of control over the movement of matter and energy inside their ecological systems. This is especially true when environmental changes present enough repetition over many generations and selective change.

18.3.1 Two Inheritance Systems: General and Ecological

The so-called *general inheritance* (that asserts which individuals will persist among living beings to pass on their DNA to the following generation) is not the only process that affects evolution. One more process is the one brought about by the action of niche construction, which we have described above, which Odling-Smee and his collaborators refer to as *ecological inheritance*. It is common knowledge that the first process takes place, for instance, when sexual reproduction is occurring during the course of some organisms' lives. The second, however, is always realized by every living entity toward every other living entity, regardless of whether they are genetically related or simply "ecologically" related. According to this viewpoint, we can state that while organisms adapt to their environments, they also adapt to environments that they or other organisms create.[6]

Even acquired traits, like information that organisms have learned [18.30], p. 284, can contribute to evolution from the standpoint of niche construction, albeit in a non-Lamarckian manner. Because of the specific type of niche construction that the involved living creatures have generated, they can indeed have an impact on specific settings. Through what we have just referred to as "ecological inheritance," phenotypes construct niches that act as a source of selection that can correct genetic selection, allowing them to do more than just act as "carriers" of their genes. Even though we must add that non-human animals are not able

---

[4] It is important to remark that in the present chapter, the term extragenetic does not admit epigenetic factors.

[5] "Purpose has been, of course, forbidden intellectual territory for some time, presumably because it steers us dangerously close to the Platonic teleology that Darwinism has rightly put aside. Nevertheless, organisms are, if nothing else, purposeful creatures, and failure to acknowledge this frankly steers us in other dangerous directions. The important questions are: how are organisms purposeful, and how does the purposefulness work?" [18.43], pp. 348-349.

[6] Some controversies are related to the fact that it is sometimes unclear which alterations lead to cognitive niches that can affect evolutionary mechanisms. On this problem, cf. [18.40], as well as the criticisms highlighted by [18.7]. Important responses are given in [18.17] together with useful examples.

to accumulate information as human beings can, other organisms exhibit what we can refer to as protocultural learning behavior, which is essentially social. Additionally, experimental investigations have shown that food washing is a skill that monkeys can learn very rapidly and that spreads throughout a group of animals as a broad "social" habit.

We have stated that niche construction can have an indirect impact on genetic selection, but we must note that it also has other effects on culture and ontogenetic selection. As already noted in the previous section, when humans encounter a new selective pressure, they might react either by cultural evolution, for example, by building medical facilities, manufacturing drugs and vaccinations, by the creation of antibodies or via biological evolution with the selection of genotypes with resistance. Cognitive niche construction, as opposed to genetic modification, frequently will lead to a more immediate answer to new difficulties. In case of unsuccessful niche construction results, genotypes will grow and prosper.

Numerous cognitive *mediating structures* are used in human ecological inheritance, for example, diagrams, scientific tools used in laboratories, books, monuments, religious artifacts, works of art, flags and symbols, vital databases, tools used in medical therapy, financial institutions, written laws, etc. These are all examples of artifacts that make up cultural niches that act as brokers of cognitive opportunities that in turn can be adopted by people who live in them. These structures are the product of cognitive delegations to the environment that are the fundamental components of cognitive niche construction processes. These delegations are part of a semiotic process that is occurring either intentionally and consciously or unconsciously, in this last case as a kind of spontaneous cognitive process embedded in the so-called manipulative abduction, which I will describe in Subsection 18.4.1. I should note that docility, as Herbert Simon put it [18.39], explains the tendency of humans to accept and believe what is suggested by *ecological* resources, made available through cognitive niches.

18.3.2 Adaptive and Maladaptive Artifacts and Abductive Cognition

Artifacts, as *anchors of meaning*,[7] can serve as examples of cognitive niches that are considered to be more than abiota and less than biota. They are so well organized that they present a negative entropy, but lack the ability to defend themselves and to stay away from their "dissipation." As previously mentioned, some of the entities constructed by humans are cognitive niches that look *adaptive* but also *maladaptive*. Indeed, with respect to general perception, environmental innovations (produced by phenotypes) may be more likely to evolve than those caused by mutations. It is commonly recognized that managing multiple artifactual niches is a difficult task that contemporary humans are constantly faced with; the ethical studies on technology are somewhat directly tied to the problem.[8]

To better understand the relationship between cognitive niches and *abductive reasoning*, we must add a crucial observation. The information that in a significant way drives

[7] Adopting this expression I am taking advantage of Hutchins's idea of "material anchor" for conceptual blending. He says "that the physical objects themselves are input to the conceptual blending process. This is what I intend when I speak of 'material anchors' for conceptual blends. What is at stake here is the boundary of the conceptual blending process" [18.14], p. 1560. More details regarding conceptual blending will be illustrated in the last section of this chapter (Section 18.5).

[8] In [18.21], I have extensively described the main aspects of this problem.

niche construction is encoded at the genetic level and provided by the evolutionary mechanisms as a result of natural selection. There is little doubt that a large portion of the niche construction activity is tied to the Darwinian scheme. However, it is crucial to emphasize that niche construction is also *active* and not only *reactive*, *profitable* and not *goalless*, like natural selection, and, furthermore, always a deliberate activity of selection (governed by "plastic" cognitive processes such as memory and learning).[9]

Abduction is the process that first acknowledges the *opportunities* provided by the environment and that simultaneously generates the possible successive active "changes" in terms of sophisticated or unprecedented niche formation. As I have previously mentioned in the opening section of this chapter, abductive cognition refers to all the thinking processes (selective or creative) that result in interesting hypotheses (and consequently in behaviors that are possibly favored by to those hypotheses).

18.3.3 An Evolutionary Sense of Purposefulness

It is crucial to remark that owing to their inherited chosen genes, organisms are, in a way, "informed" about their prior natural selection history, which is about the sorts of niche construction activities considered legitimate by that heritage. According to this viewpoint, the most adaptive niche builders have a higher likelihood of being selected in front of active selective pressures. As I have previously indicated, there is another mechanism that Darwin did not consider, that goes beyond the blind selection processes. It is significant to highlight that in this second level, selection favors *purposive* organisms, that is good niche constructors capable of improving their "fitness," even if considered in a loosely Darwinian sense, as I have already illustrated above. A good niche constructor is—so to speak—a follower of Peircean pragmatism that sees meaning related to the practical production of habits: "Thus, we come down to what is tangible and conceivably practical, as the root of every real distinction of thought, no matter how subtle it may be; and there is no distinction of meaning so fine as to consist in anything but a possible difference of practice" [18.33]. In the course of their evolution and growth, organisms are essential because they play a crucial role in their development and evolution. This perspective contends that variation is both *blind* and *constructed* because the different "filtering" and "editing" processes that take place before and throughout the transmission determine which variations are inherited and what final form they take [18.15], p. 319. Obviously, this interaction has neither a teleological nor a Lamarckian character.

According to Turner, "a sense of purposefulness" can be added to evolution [18.43], pp. 348–349. The "grandmother theory," which claims that intellect and brain size would have coevolved in tandem with the nutritional change toward high-quality, calorie-dense, scarce food supplies, is an example of this. In general, it has been highlighted that even if developing the knowledge and skills necessary for specialized niche construction tasks requires time and a lengthy learning process, that compromises high productivity, later, compensation is provided by greater output throughout the adult years.

---

[9] Again, the readers interested in framing my discussion in the current perspective made in evolutionary theory thanks to the extended synthesis can refer to Winters' article [18.45] contained in the Special Issue of the journal *Biosemiotics* "Semiotic Aspects of the Extended Synthesis" (Vol. 11, 2, 2018).

One of the numerous elements that make up the specialty of the niche construction activity in both people and animals, beyond the impacts that are simply "behavioral" (including thought, observation, imitation, etc.), the "symbolic" one is crucial because it is directly connected—in the case of humans—with the value of *language* and other semiotic social instruments for the sharing of information and communication. I have already shown how abductive cognition in this circumstance serves as a key tenet of niche construction mainly because it allows for the development of novel hypotheses: they lead to the development of new niches and/or to the modification of the preceding ones. Abductive cognition, when related to an activity of a mere selection of hypotheses (diagnosis), such as in the case of diagnostic reasoning, even if not creative, is still crucial, especially in the case of niche maintenance and preservation.

18.4 Anchors of Meaning: External Semiotic Representations and the Disembodiment of the Mind

18.4.1 Construction and Manipulation of Meaning Using External Semiotic Anchors

In my most recent study on abduction [18.22] [18.23] [18.24], I have consistently emphasized my belief that the concept of niche construction is useful for understanding fundamental facets of human cognition and constitutes one of its key characteristics. We should also point out that current research on cognitive niches is unquestionably consistent with the results in the field of the so-called distributed cognition. The activities of problem-solving were initially perceived as processes in actual work circumstances according to the important achievements reached by Edwin Hutchins [18.13]: for the whole field of cognitive research, a novel and unexpected intellectual framework was presented. Agents make, construct, manipulate, alter, or sustain cognitive states using tools and equipment (and hence external semiotic delegations/representations endowed with cognitive worth). Additionally, the semiotic/cognitive representations hosted by the materiality at play have an *independent* impact on the resulting hybrid cognitive processes.

In sum, it is believed that cognition is held in common across individuals, objects, artifacts, and external representations. Hence, the interaction between internal (inside the brain) and external (in material entities) representations is crucial: this recent perspective has been strengthened by the mention of the widely-done research regarding the studies known as 4E cognition: embodied, embedded, extended, and enactive.

An interesting special issue of the journal *Pragmatics & Cognition*, which specifically focused on the subject of distributed cognition, illustrates many of the baffling theoretical issues that were later expanded in different ways in many debates. The study by Sutton [18.41], in particular, makes an important point about how distributed cognition is connected to Clark and Chalmers' [18.6] article on the "extended mind hypothesis" and other studies on embodied, embedded, situated, dynamical cognition, active and vehicle externalism, and numerous distinctions that I am unable to account for here. Sutton does a good job of outlining a taxonomy of the distributed resources that are investigated in various areas of study: 1) external cultural instruments and devices, artifacts, and symbol structures; 2) natural environment elements appropriately enriched with cognitive value (e.g., stars to help navigation); 3) intersubjective and social distribution or the so-called cognitive "scaffolding" (it is the case of language[10] that is capable of scaffolding thoughts); 4) embodied capacities and skills intricately entwined with our use of technological, natural, and social resources of the preceding cases; 5) internalized cognitive artifacts.

The last two cases for Sutton involve the illustration of the complicated systems created when embodied brains couple and blend with "cognition amplifiers" like entities, such as technologies, but also other individuals, through a process I have dubbed "re-embodiment of the mind" in [18.22], CH 3. This is a kind of "neural recapitulation" of cognitive aspects—for example, propositional, but also model-based, or both—that were appropriately distributed in the external environments. Re-embodiment of the mind is contrasted to the disembodiment of the mind that regards the various process of cognitive delegations to the external environment that I have sketched above.

An external representation can be accessed and modified in the appropriate cognitive niche, as a strong *anchor of meaning*, for example, in the case of the type of cognitive processing a human individual utilizes to think about a problem. The use of Roman numbers eliminates some of the most challenging additions through the use of external signs, whereas the Arabic system accomplishes the same for challenging multiplication. The result of the internalization of external forms of representation is the capacity for novel mental internal reasoning and cognition. We can have a variety of objectified knowledge and structures considering the external representations (such as physical symbols, written symbols, and other entities, forms, measurements, and models in three dimensions), as well as external rules, relations, and limitations integrated into physical symbols (for example spatial relations of written digits, physical limitations in mathematical diagrams and abacuses) [18.46]. The internal representations, which include knowledge and the organization of memory as well as propositions, schemas, models, prototypes, and pictures are, of course, contrasted with the exterior representations.

As I have repeatedly said in this chapter, humans can extend their "minds" into the surrounding physical world thanks to cognitive niche construction. Consequently, they make extensive use of resources outside of the human brain that are suitably exploitable because they have been "semioticized"; these externalities are used to support or facilitate reasoning and decision-making processes, to improve memory capacities, and when the need arises for the purpose of managing something. Humans create objects, instruments, depictions, novel physical artifacts, etc., for complex problem-solving processes, such as those that I have referred to as *manipulative abductions* [18.20] [18.22] [18.24] [18.31] (i.e., the development of hypotheses as a result of more or less covert exploitation and cognitive manipulation of the outside world). We may state the conclusion that there is a creative "thinking through doing" method for getting hypotheses, essential in the same way that internal thinking is. In this manner, the cognitive task on external representations is completed, exploiting them in place of internal representations. Here, action realizes a cognitive (or, for instance, an epistemic, as in science) role, related to abductive reasoning.

Manipulative abduction occurs when we think about something *while* doing rather than just thinking about doing, in the received sense of the so-called practical reasoning. Of course, understanding this type of reasoning is crucial for defining how abduction actually occurs in science as well as for the development of programs that are computationally

[10] Language as a prototypical case of cognitive niche is described in Subsection 18.2.1.

capable of rediscovering or discovering for the first time, for instance, scientific hypotheses and mathematical theorems or laws, but also for helping to advance the design of more effective and realistic robots as well as their hypothetical reasoning capabilities. We must keep in mind that the term "abduction" also refers to a variety of situations of hypothetical cognition like the ones that are taking place in diagnosis: in this case, we face cases of "selective" abduction, as I have already illustrated above.

When many external factors, which are typically inert from a semiotic perspective, can be transmuted into what I have called, in the case of scientific reasoning, "epistemic mediators" [18.20], they can play an important role in manipulative abduction. As *anchors of meaning* epistemic mediators produce fresh signs, and consequently new opportunities for interpretants, and new interpretations.

Manipulative abduction took place, for example, when ancient Greek geometricians increased elementary geometrical knowledge by manipulating an externally realized icon, such as a triangle, in search, through newly obtained signs, of new properties of the triangle itself. New signs as anchors of new meaningful properties, as in the geometrical example offered by Immanuel Kant [18.16] in the "Transcendental Doctrine of Method" of the *Critique of Pure Reason*. He refers to the extra-theoretical conduct that seeks to communicate fresh experiences in order to incorporate them into previously available and approved beliefs. Gooding [18.11] in turn provides another example of tangible manipulative reasoning by illustrating the function of so-called "construals" in science, which favor the reaching of conclusions thanks to apparatus and machine-based techniques.

We must acknowledge that physical manipulation of the environment is assigned great cognitive value given how important a role the environment structures play in affecting our semiotic representations and so cognitive processes. Actions alter the surroundings such that they develop a semiotic configuration that helps the actor in achieving a cognitive target. In this case, the cognitive processes are activities that usually an agent carries out in a semiotic setting merely to relieve the mind of a mental burden but especially to gather knowledge that is difficult to find through mere internal computational means (of course something similar, even if in a more passive way, is occurring in the case of a participant to the process): in sum, actions are the result of an obvious cognitive process of what we can call *conceptual blending*.[11]

I have to repeat that a manipulative action on the environment that aims at creating a specific *conformation of signs* carrying pertinent information may be seen as a cognitive semiotic process. *Cognitive manipulating*, according to our definition, is any alteration of the surrounding environment done in an effort to create exterior configurations that qualify as representations.

The diagrammatic celebrated illustration of a cognitive manipulation (usually called "demonstration") is shown in Figure 18.3, extracted from the area of basic geometry. Here, a simple triangle manipulation communicates crucial semiotic information about the internal angles of a triangle "anchoring," as a causal relationship, the possibility of obtaining new meanings.

The complete procedure in this case adopted by the agent expresses a physical activity that may be considered cognitive: the idea of manipulation may still be grasped by using the concept of manipulative abduction. This viewpoint indicates that manipulative abduction is

[11] More details are illustrated in Section 18.5 of this chapter.

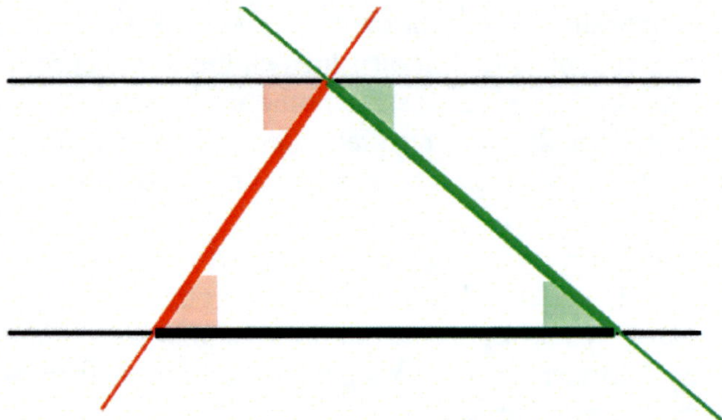

Figure 18.3 Diagrammatic demonstration (or discovery, in the case of the ancient Greek geometers) showing that the sum of the internal angles of any triangle is 180 degrees.

an extraordinary kind of cognitive manipulation, in which a person, when confronted with a challenging external circumstance to uncover fresh, useful, and meaningful characteristics of an entity, "selects" or "creates" an activity that alters the environment so that it gives updated information that would otherwise be inaccessible and that can be specifically used to obtain explanatory hypotheses and subsequent results.

The technique also demonstrates first of all a motor-originating constructive operation (placing the new segment end to end parallel to one side in the externally depicted provided triangle), followed by a sensory-originating procedure, "visual" (calculating the sizes of the now – clear – externally – "perceived" angles).

Similar to how they are used in logic programming, both abduction and induction play a part in this abductive example: 1) first of all abductive cognition increases—by geometrical construction—the intension (that is the cognitive structure) of well-known individuals—a given specific triangle in our case—(because abducible characteristics are made true for these individuals, for instance by presenting new offered "samples" as "anchors" that provide opportunities for more knowledge), but 2) without really having generalization impact on the observables in terms of generality (it does not increase their extension). In this scenario, it is essential to manipulate the external diagram to create new situated outcomes in order to enable an "induction" that can provide new "general knowledge" stating the new property found as true for all triangles, a truth that is not attainable through mere abduction.[12]

[12] The geometric example we have just reported might be interpreted as being an illustration of deductive reasoning. Actually, it presents an abductive case in which induction is also further implicated. Indeed, this cognitive process, if considered from the viewpoint of a textbook of Euclidean geometry (as an abstract and static system of knowledge) is a true "demonstration" and so an illustration of deductive reasoning, and obviously does not lead to a "discovery." Rather, we can say it is a kind of re-discovery. On the contrary, when ancient Greek pre-Euclidean geometers discovered the new property regarding the sum of the internal angles of a triangle, we dealt with a case of abductive cognition.)

18.4.2 The Semiotic Role of the Anchors of Meaning and the Disembodiment of the Mind: An Example

Thanks to another example I will describe in this subsection it is possible to understand how managing outside resources enables the human cognitive system to be "molded" by accidental environmental components. Again, this viewpoint leads us to the conclusion that some externalities, as semioticized cognitive niches, are crucial to human cognitive processes. Let me give a specific illustration of how the construction of externalities involves significant semiotic "effort."

As I mentioned above, *semiotic brains* generate a number of signs and are preoccupied with creating, manifesting, or responding to a collection of signs. Through these semiotic processes, they are also actively engaged in being "cognitive actors" (also known as "minds"), or at the very least, in thinking intelligently. As I have already anticipated, an essential component of this semiotic activity of the brain is the "disembodiment of the mind" process. The invention of handaxes and their innovative semiotic value are tied to the emergence of what Mithen [18.28] characterizes as the enhancement of technical intelligence of the early human mind. In fact, a big cortex can only offer an evolutionary benefit in the presence of a large amount of relevant information and knowledge about external mediators, which can only be possessed by a small, previously constituted collective of humans. Paleoanthropological evidence appears to support this viewpoint. In fact, some cognitive paleoanthropological studies tell us that the emergence of modern language (speech) coupled with a high level of reflective consciousness in terms of thoughts about our own thoughts and about our feelings (i.e., the consciousness not trivially seen as brute sensation) institutes a relevant condition of possibility for the emergence of the so-called "material culture."

Understanding the process of constructing cognitive niches surely takes advantage of the understanding of the semiotic function of material culture. By creating artifacts, the past hominids produced objects that might serve as *semiotic anchors* for a variety of "new" cognitive representations (for example, of supernatural entities). According to this point of view, first of all, material culture is most notable for its ability to *disembody thoughts through signs* that would otherwise quickly fade away if not shared with other people. Second, material culture realizes a structured *semiotic delegation* to the outside world, encouraging meaning creation precisely through the interaction of internal and external representations, a primitive kind of "thinking through doing" that was entangled with the use of signs as mediators. Just to use Mithen's example, the early human mind displayed two "different kinds of intelligence" for thinking about people and animals. Also, thanks to the semiotic mediation of material culture, it was possible for the modern human mind to arrive at the "new" notion of animals and people existing at the same time (that is a supernatural entity): a remarkable illustration of meaning invention thanks to the disembodiment of the mind is the carving of what is most likely the mythological person from the last ice age, 30.000 years ago, a half-human/half-lion figure carved from mammoth ivory unearthed in Hohlenstein-Stadel, Germany (cf. Figure 18.4). The new meaning emerged there, in the adequately artificialized outward material/physical environment, where the mind took it in.

Therefore, artifacts like the lion man were able to free those humans from the bonds and chains that were ingrained in their unorganized minds and made essential by their evolutionary past. Chains that, instead, inexorably shaped the neural pathways in the brains of

Figure 18.4 The lion man.

earlier humans like the Neanderthals. According to this perspective, the semiotic expansion of minds is a continuous process of their own disembodiment into the physical/material world around them. From the first handaxes to the computational domestication of physically ignorant entities (digital machines), which is also producing new novel cognitive embodiments (computation on non-digital substrates), this point of view contends that the development of the mind is inextricably linked to the development of extensive, interconnected, material and physical cognitive semiotic systems.[13]

In conclusion, I richly emphasized the role played by semiotic cognitive information processes in the disembodiment of mind from the perspective of the interaction between internal and external representations, as well as the inherent "semioticity" of human brains in the process of constructing cognitive niches. The relationship between significant internal semiotic resources and devices and their complex interrelationships with the pre-existing externalized semiotic materiality in the environment is also explained by this interplay. As a result, we may draw the conclusion that minds can be "extended" and artificial in nature; we have already shown that one significant result is that they actively contribute to the establishment of an "evolutionary sense of purposefulness."

[13] In [18.25] I have offered an entirely new dynamic perspective on the nature of computation, highlighting the role of unconventional computation as an incessant and terrific process of cognitive domestication of ignorant entities. The book also describes the kinds of cognitive creativity that differently characterize computational entities and human cognizers.

18.5 Material Anchors for Conceptual Blends

We have said that the human agent, through cognitive interaction and blending with external representations [18.14], is able to pick up and employ what is indicated by the constraints and characteristics intrinsic to their outward materiality and to their relative ascertained conventionality.

Let us go back to the case of the discovery of the sum of the internal angles of a triangle (cf. Subsection 18.4.1). The human agent—an "interpretant" in the Peircean sense—at first embodies a sign in the outside world, in this example an icon with deliberate assigned cognitive standard and public characters—that is "meanings." These meanings derive from several *already known and available properties* of pre-Euclidean geometry: a particular language and notation, the available definition of a triangle, the characteristics of parallel lines, etc. Applying the properties of the parallel lines thanks to the manipulations of the dynamical constructed[14]—externally depicted—diagram, the agent can reach the expected conclusion about the sum of the angles, seen out there, in the diagram, and picked up at it.

The general process of blending is related to the cognitive interplay of two cognitive aspects:

- the standard cognitive characteristics *delegated* to the external representation at stake and
- the characteristics that are *intrinsic to the materiality* of the model that favor fruitful manipulations.

The external model turns into a sort of "independent" cognitive *actor* that is made available to novel interpretations of the issue at hand. Because of the characteristics of the external materiality at stake, which predetermine movements along some pathways and not others, its existence compels the competent reasoner to trace sequences of interpretants in some directions and not others. Those characteristics present themselves to us as a collection of limitations as well as potentialities. At the end of the process of blending a newly acquired brain configuration, or "thought," can be fixed, after having been picked up: that is the new finding about the sum of internal angles, in the example I have illustrated.

Hutchins [18.14] fundamentally analyzed the role of constraints through the connection between the conceptual structure and the material structure in what he called "conceptual blending," as a crucial cognitive heuristic, using a wide range of remarkable and highly fascinating examples from real-world to scientific scenarios. First, it should be emphasized that some elements of the external representations embedded in material objects, which create a "blended" space, can be changed while other elements are constant, as also especially stressed by Turner and Fauconnier [18.9] [18.10]. Empirical findings have demonstrated the value of the stable character of representations and of their simple and portable features in aiding thinking: in sum, their capacity to function as both thought anchors and suppliers of fresh inferences and findings.

Hutchins outlines a variety of situations, such as the function of a "line" for patrons waiting in line at a theater in forming a spatial memory for the order in which customers

[14] The construction consists of "auxiliary" drawn features that extend the original icon of a triangle.

arrive: the blend, which gives rise to the queue, is made up of a combination of the line and directional ordering. "The world is its own best model," as Brooks [18.4] once famously stated. A mental (neural) conceptual structure and a mental representation (neural) of the material structure can be thought of as the two inputs to the blend from the standpoint of classical representationalism. Hutchins chooses to take a different stance and avoids providing a distinct mental picture of the material structure as an input field when he speaks about "material anchors for conceptual blends":

> Another alternative is to say that the physical objects themselves are input to the conceptual blending process. This is what I intend when I speak of "material anchors" for conceptual blends. [...] What are the advantages of these two different ways of thinking of the role of material structure in the formation of conceptual blends? One of the problems with the "real space" formulation is that it requires two stages of selectivity. First, there is the selectivity of perception that produces a filtered conceptual representation of the physical world. Second, there is selective projection in the process by which the prior conceptualization of the world (the "real space" representation) is blended with the other conceptual input. Is there any evidence that these are two separate processes? It seems preferable to assume that the selective attention to, and projection of, structure from the material world to the blended space is the perceptual process. That is, that selective perception is a conceptual process [18.14], p. 1559.

The stability of representations of the conceptual blend of the above example (sequential relationships among people in the queue) is the primary "emergent" quality of the blend itself. This opens up new opportunities for inferences and provides complete "cultural" models (habits, in Peirce's perspective).

In conclusion, the usage of the material anchor rather than its inherent qualities confers cognitive worth on it: conceptual aspects are mapped onto the materiality at hand and considered as proxies for finding relationships among them. It is in this sense that materiality, appropriately endowed with cognitive delegations, plays the role of a material anchor of meaning, as clearly explained by Hutchins [18.14], p. 1562.

18.6 Conclusion

In this chapter, I have discussed how studies on abduction, cognitive niches, and human creative cognition might be productively combined with a semiotic perspective. It has been argued that it is very beneficial to think of human brains as "semiotic brains" that are involved in creating "signs," but also manifesting or responding to a succession of them. I made an effort to demonstrate how brains are actively involved in "becoming minds"—that is, in thinking smartly and, on occasion, creatively—through this semiotic activity. My approach aimed at giving a novel cognitive perspective on the role of semiotic external models and representations as *anchors of meaning*, thanks to a process of externalization of the mind. In my analysis, I have stressed the crucial character of the relationship between significant internal semiotic resources and tools and their dynamic interactions with that externalized semiotic materiality that is appropriately provisioned in such spaces of the environment that recent evolutionary research has called *cognitive niches*. Thus, minds are inherently material, semiotic, extended, and also "artificial," so to speak. We may extend the mind through the externalization process, which also provides a unique cognitive viewpoint on the mechanics behind the semiotic formation of abductive processes—particularly *manipulative abductions*—of meaning formation capable of generating and simultaneously anchoring meanings. In fact, a significant portion of human abductive thinking involves a reification of some sort in the external world, followed by re-projection and reinterpretation using novel arrangements of brain networks and chemical reactions.

Lexicon on "Anchors of Meaning"

Affordances as a semiotic configuration of signs and abducibility: An affordance or chance, such as the availability of water or the possibility of locating recovery and hiding, is presented to the particular organism by the environment (a specific niche, for example). From an abductive perspective, affordances as environmental chances can be connected to the changeable (degree of) discoverability and diagnosticability of a semiotic configuration of "signs": the act of sitting results from a sign activity in which we observe certain physical characteristics (flatness, stiffness, etc.), thus we may ordinarily infer—abduce—that one possible behavior to deal with a chair is to sit on it.

Chance, discoverability, diagnosticability: Humans can be categorized as chance-seekers insofar as they are consistently building up and then extracting from a configuration of signs latent possibilities to configure discoverability and diagnosticability and thus potentially uncovering new valuable information and knowledge to generate abductions. That is, much like other species, human beings actively shape and modify their environment by constructing cognitive niches searching for advantageous chances. Chances are thus, from this perspective, that "information" is not already stored internally in memory or already present in an external reserve but instead must be "extracted" and then picked up from time to time.

Creative abduction: The process of generating a brand-new hypothesis is called creative abduction. Scientific discovery is an example of creative abduction. For instance, Kepler's discovery of the elliptic orbit of the planets or the discovery of a new illness and the effects it has in the area of medical knowledge.

Manipulative abduction: In the process of the so-called manipulative abduction, a hypothesis is developed and assessed by relying primarily on extra-theoretical and extra-sentential behavior. This behavior aims at producing communicable accounts of new experiences in order to incorporate them into pre-existing systems of experimental and linguistic (theoretical) practices. In order to handle things and pieces of knowledge that cannot be instantly represented or located internally, manipulative abduction is a type of redistribution of the epistemic and cognitive effort. The use of external diagrams in geometrical thinking as tools for observations and "experiments" to switch from one cognitive state to another, such as when learning new properties and theorems, is an example of manipulative abduction.

Model-based abduction and semiosis: Starting the novel semiotic perspective, Peirce stated that all thinking is in signs, and signs can be icons, indices, or symbols, and that all inference is a form of sign activity, where the word sign includes "feeling, image, conception, and other representation" [18.32] 5.283. This means that a sizeable portion of hypothesis creation and selection is model-based; it takes place in the interaction between brains and model-based features of external objects and instruments that have acquired cognitive and/or epistemological delegations.

Multimodal abduction: Multimodal abduction indicates the common hybrid aspects of abductive reasoning. We must recognize that abduction may be sentential, model-based, and manipulative. Abductive inference can be both verbal and visual; furthermore, the representation of data and hypotheses can both be done through different sensory modalities. Basic elements of this constitutive hybrid nature of multimodal abduction include words, sights, sounds, scents, images, and other sensory modalities, but also kinesthetic and motor sensations and other feelings like pain.

Niche construction and abduction: According to the hypothesis of cognitive niches, the environment functions as a kind of *global market* that offers all possible options to living things. In fact, not all of the opportunities provided by the environment can be taken advantage of by the human and non-human creatures that interact with it. The environment, for instance, gives creatures access to air to fly, water to swim in, flat surfaces to walk on, and so on. None of the species can completely utilize all of them, though. With the activity of niche construction, all organisms strive to adapt their environment in order to better exploit those components that fit them and remove or lessen the effects of the unfavorable ones. This action is highly connected to the hypothetical powers offered by abductive cognition. That is, the pertinent environmental factors are properly chosen and/or reconstructed in order to transform the local environment, which is cognitively inactive, into a cultural mediating structure that can provide effective chances for behavior regulation. By creating cognitive niches, animals not only shape their environment but also, to some extent, control the Darwinian selective pressure to which they and their offspring are exposed (as well as, of course, the selection pressures to which other species are exposed). When seen from this perspective, each abduced hypothesis may create a shift (and an opportunity) in the cognitive processes and develop fresh possibilities in the coevolution of the organism and its environment.

Selective abduction: When a hypothesis is chosen from a pre-stored encyclopedia of "abducibles" the inferential process is called selective abduction. Diagnostic reasoning in medicine is an illustration of selective abduction; it starts with patient data that is abstracted into clinical characteristics to be explained. Then, selective abduction produces plausible diagnostic explanatory hypotheses.

Declaration

The author declares no conflict of interest.

References

[18.1] Bertolotti T., Cinerari C. (2013) The gospel according to Google: The future of religious niches and technological spirituality. *European Journal of Science and Theology* 9(1), 41-53.

[18.2] Bertolotti T., Magnani L. (2015) Contemporary finance as a critical cognitive niche. *Mind & Society* 14(2), 273–293.

[18.3] Bertolotti T., Magnani L. (2017) Theoretical considerations on cognitive niche construction. *Synthese* 12(194), 4757–4779.

[18.4] Brooks R.A. (1991) Intelligence without representation. *Artificial Intelligence* 47, 139–159.

[18.5] Clark A. (2006) Language, embodiment, and the cognitive niche. *Trends in Cognitive Science* 10(8), 370–374.

[18.6] Clark A., Chalmers D.J. (1998) The extended mind. *Analysis* 58, 10–23.

[18.7] Dawkins R (2004) Extended phenotype - but not too extended. A reply to Laland, Turner and Jablonka. *Biology and Philosophy* 19, 377–397.

[18.8] Day R.L., Laland K., Odling-Smee F.J. (2003) Rethinking adaptation. The niche-construction perspective. *Perspectives in Biology and Medicine* 46(1), 80–95.

[18.9] Fauconnier G. (2005) Compression and emergent structure. *Language and Linguistics* 6(4), 523–538.

[18.10] Fauconnier G., Turner M. (2003) *The Way We Think*. Basic Books, New York.

[18.11] Gooding D. (1990) *Experiment and the Making of Meaning*. Kluwer, Dordrecht.

[18.12] Heras-Escribano M., de Jesus P. (2018) Biosemiotics, the extended synthesis, and ecological information: Making sense of the organism-environment relation at the cognitive level. *Biosemiotics* 11(2), 245–262.

[18.13] Hutchins E. (1995) *Cognition in the Wild*. The MIT Press, Cambridge, MA.

[18.14] Hutchins E. (2005) Material anchors for conceptual blends. *Journal of Pragmatics* 37, 1555–1577.

[18.15] Jablonka E., Lamb M.J. (2005) *Evolution in Four Dimensions. Genetic, Epigenetic, Behavioral, and Symbolic Variation in the History of Life*. The MIT Press, Cambridge, MA.

[18.16] Kant I. (1929) *Critique of Pure Reason*. MacMillan, London, translated by N. Kemp Smith, originally published 1787, reprint 1998.

[18.17] Laland K.N., Odling-Smee F.J., Feldman M.W. (2005) On the breath and significance of niche construction: A reply to Griffiths, Okasha and Sterelny. *Biology and Philosophy* 20, 37–55.

[18.18] Lewontin R. (1982) Organism and environment. In: *Learning, Development and Culture*, H. Plotkin (ed.). Wiley, New York, NY: 151–170.

[18.19] Longo G., Montévil M. (2014) *Perspectives on Organisms*. Springer, Berlin/Heidelberg.

[18.20] Magnani L. (2001) *Abduction, Reason, and Science. Processes of Discovery and Explanation*. Kluwer Academic/Plenum Publishers, New York.

[18.21] Magnani L. (2007) *Morality in a Technological World. Knowledge as Duty*. Cambridge University Press, Cambridge.

[18.22] Magnani L. (2009) *Abductive Cognition. The Epistemological and Eco-Cognitive Dimensions of Hypothetical Reasoning*. Springer, Heidelberg/Berlin.

[18.23] Magnani L. (2017) *The Abductive Structure of Scientific Creativity. An Essay on the Ecology of Cognition*. Springer, Cham, Switzerland.

[18.24] Magnani L. (2022a) *Discoverability. The Urgent Need of an Ecology of Human Creativity*. Springer, Cham, Switzerland.

[18.25] Magnani L. (2022b) *Eco-Cognitive Computationalism. Cognitive Domestication of Ignorant Entities*. Springer, Cham, Switzerland.

[18.26] Mithen S. (1989) To hunt or to paint? Animals and art in the upper paleolithic. *Man* 23, 671–695.

[18.27] Mithen S. (1996) *The Prehistory of the Mind. A Search for the Origins of Art, Religion, and Science*. Thames and Hudson, London.

[18.28] Mithen S. (1999) Handaxes and ice age carvings: Hard evidence for the evolution of consciousness. In: *Toward a Science of Consciousness III. The Third Tucson Discussions and Debates*, A.R. Hameroff, A.W. Kaszniak, D.J. Chalmers (eds.). MIT Press, Cambridge: 281–296.

[18.29] Mithen S. (2007) Creation of pre-modern human minds: Stone tool manufacture. In: *Creations of the Mind. Theories of Artifacts and Their Representation*. E. Margolis, S. Laurence (eds.). Oxford, Oxford University Press: 289–311.

[18.30] Odling-Smee F.J., Laland K.N., Feldman M.W. (2003) *Niche Construction. The Neglected Process in Evolution*. Princeton University Press, Princeton, NJ.

[18.31] Park W. (2017) On Lorenzo Magnani's manipulative abduction. In: *Handbook of Model-Based Science*, L. Magnani, T. Bertolotti (eds.). Springer, Switzerland: 197 –213.

[18.32] Peirce C.S. (1866–1913) *Collected Papers of Charles Sanders Peirce*. Harvard University Press, Cambridge, MA, vols. 1–6, Hartshorne, C. and Weiss, P., eds.; vols. 7–8, Burks, A. W., ed. (1931-1958).

[18.33] Peirce C.S. (1866-1913/1985) *Historical Perspectives on Peirce's Logic of Science: A History of Science*. Mouton, Berlin, vols. I-II, edited by C. Eisele, 1985.

[18.34] Peterson J.V., Thornburg A.M., Kissel M., Ball C., Fuentes A. (2018) Semiotic mechanisms underlying niche construction. *Biosemiotics* 11(2), 181–198.

[18.35] Pinker S. (1997) *How the Mind Works*. W. W. Norton, New York.

[18.36] Pinker S. (2003) Language as an adaptation to the cognitive niche. In: *Language Evolution*, M.H. Christiansen and S. Kirby (eds.). Oxford University Press, Oxford: 16–37. URL http://www.isrl.uiuc.edu/ amag/langev/paper/ pinker03LanguageAs.html

[18.37] Sharov A.A. (2016) Evolution of natural agents: Preservation, advance, and emergence of functional information. *Biosemiotics* 9(1), 103–120.

[18.38] Simon H.A. (1955) A behavioral model of rational choice. *The Quarterly Journal of Economics* 69, 99–118.

[18.39] Simon H.A. (1993) Altruism and economics. *American Economic Review* 83(2), 157–161

[18.40] Sterelny K. (2005) Made by each other: Organisms and their environment. *Biology and Philosophy* 20, 21–36.

[18.41] Sutton J. (2006) Distributed cognition: Domains and dimensions. *Pragmatics & Cognition* 14(2), 235–247.

[18.42] Tooby J., DeVore I. (1987) The reconstruction of hominid behavioral evolution through strategic modeling. In: *Primate Models of Hominid Behavior*, W.G. Kinzey (ed). Suny Press, Albany: 183–237.

[18.43] Turner J.S. (2004) Extended phenotypes and extended organisms. *Biology and Philosophy* 19, 327–352.

[18.44] Weir A.A.S., Kacelnik A. (2006) A New Caledonian crow (*Corvus moneduloides*) creatively re-designs tools by bending or unbending aluminium strips. *Animal Cognition* 9, 317–334.

[18.45] Winters A.M. (2018) Introduction: A structural and historical approach to understanding advancements in evolutionary theory. *Biosemiotics* 11(2), 167–180.

[18.46] Zhang J. (1997) The nature of external representations in problem solving. *Cognitive Science* 21(2), 179–217.

19

Levels of Translation, Levels of Freedom?

Kobus Marais

Department of Linguistics and Language Practice, University of the Free State, Bloemfontein, South Africa

Online resources:
https://philpeople.org/profiles/kobus-marais
https://www.researchgate.net/profile/Kobus-Marais
https://ufs.academia.edu/KobusMarais

Abstract
In this chapter, I investigate levels of semiosis, starting off with Pattee's work on the physics of symbols and linking that to Lotman's ideas that the emergence of culture implies a complexification of semiosis. If one takes the Peircean conceptualization of translation as the starting point, semiosis is a process of translating signs of different types and levels, into other signs of different types and levels. According Lotman's argument, a decrease in constraints over this process will increase the randomness, which will increase the potential for change and creativity. However, Lotman argues that culture is actually a complementary system of controlled and explosive dynamics. I am particularly interested in the way in which levels of translation could be said to relate to levels of freedom. The chapter therefore explores Deacon's paradoxical claim that constraints create freedom. Hoffmeyer and Stjernfelt's work shows how semiotic freedom emerged through the process of evolution. In the chapter, I argue that each new level of semiotic freedom entails a translational process through which new sets of constraints are imposed on the process. At lower levels of semiotic freedom, the translation process entails matter-energy only, but as semiotic freedom increases, translational processes increasingly also include ideation. The growth in semiotic freedom is, however, not the result of a removal of constraints but the result of the imposition of different sets of constraints.

Keywords: Translation, interpretation, constraint, thermodynamics, semiotic freedom

Email: jmarais@ufs.ac.za

19.1 Introduction

The relationship between materiality, i.e., matter-energy,[1] and ideationality, i.e., semiosis, has been of scholarly interest for a very long time, from a wide variety of angles and for a wide variety of reasons. For instance, Terrence Deacon recently wrote a 600-page book on how mind emerged from matter. Howard Pattee [19.43] [19.44] [19.45] [19.46] has similarly written many papers on the physics of symbols and the relationship between matter-energy and information. John Deely's [19.8] work is another example of a philosopher-semiotician who spent a lot of time thinking through the problem of turning things, i.e., matter-energy, into objects, i.e., ideas. Equally, Floyd Merrell [19.32] [19.33] [19.34] [19.35] has grappled with the levels of cognitive complexity as these relate to Peirce's theory of signs. To these, one could add the work of Stuart Kauffman [19.16] [19.17] [19.18] [19.19] [19.20] and Bruno Latour [19.21] [19.22] [19.23] [19.24], from two widely different angles, but both trying to conceptualize the ideational and the material in one framework. Then, I did not even begin to mention Marxism, poststructuralism and critical theory and their efforts to think about the relationship between matter-energy and semiosis. In biosemiotics, this relationship has found fresh interest and has led to interesting work on the upper and lower thresholds of semiosis, interspecies communication, and the complexification of semiosis as one moves into higher levels thereof.

Being from the humanities side of the scholarly neighborhood, I do not think that I have the knowledge to add anything to the physics, chemistry and biology of the problem under discussion. What I might be able to contribute are some thoughts about the role translation plays in these semiosic processes, which in my view simultaneously entail matter-energy and ideas. In particular, I am intrigued by the emergentist line of thought [19.3] [19.15] that posits levels of reality emerging from one another and the concomitant notion that higher levels of emergence imply higher levels of complexity—and eventually higher levels of semiotic freedom [19.12]. Even more specifically, I would like to explore this problem of levels of translation in terms of a seeming paradox, namely that levels of reality emerge through work performed by the constrained release of energy [19.7] [19.20]. While the universe tends towards entropy, life and society-culture works hard at negentropy. This leaves one with the fascinating paradox that it is the constraints on matter that enable semiotic freedom.

19.2 Conceptualizing Translation

The field of translation studies, which emerged during the 20th century, built its identity around conceptualizing translation as being "interlingual translation." This led to interlingual translation being dubbed "translation proper." The conceptualization was built mainly on Jakobson's [19.13] argument that, from a linguistic perspective, there are three types of translation, namely intralingual, interlingual and intersemiotic translation. Jakobson built this categorization on Peirce's view that the meaning of a sign is its translation into another

[1] I ascribe roughly to a performative new materialism [19.11] view of matter-energy, without necessarily endorsing every position held by its proponents.

system of signs. While Peirce did not specify the nature of any of the systems of signs that can be translated, Jakobson limited his conceptualization to language because of his interests as a linguist.

I argued elsewhere [19.28] that translation studies should free itself from the linguistic shackles imposed by its interpretation of Jakobson and return to the Peircean conceptualization.[2] This would mean that translation studies are primarily a semiotic endeavor, not primarily a linguistic one. Viewing translation as semiotic then also includes biosemiotics, which would allow for intra- and interspecies translation as well as translations between organisms using phytosemiosis, zoosemiosis, and anthroposemiosis. In this section, I try to clarify what I mean when I use the term "translation."

19.2.1 Thermodynamics and Translation

In recent research, I suggested an expansion of the concept of translation, namely to consider that a translation process entails the imposition of constraints on semiotic material, i.e., it is a thermodynamic process, more specifically teleodynamic, in which an agent performs semiotic or meaningful work [19.28] [19.30]. I base this argument in Peirce's conceptualization of translation. While Peirce also uses the term "translation" for interlingual translation and for mathematical translation, there are cases in which he uses the term translation to refer to the semiotic process more generally. A first example is his assertion that "... the conception of a meaning, which is, in its primary acceptation, the translation of a sign into another system of signs ..." [19.47], CP 4.127. I interpret translation in this formulation to refer to some action performed on signs. If one reads [19.47], CP 4.127, together with CP 2.89, it is clear that Peirce assumes some action that is performed on or that works on semiotic material: "Transuasion (suggesting translation, transaction, transfusion, transcendental, etc.) is mediation, or the modification of firstness and secondness by thirdness, taken apart from the secondness and firstness; or, is being in creating Obsistence." The words "mediation," "modification" and "creating" in this quote all refer to action performed on signs in my interpretation. One could also add Peirce's claim that "[t]he meaning of a proposition is itself a proposition. Indeed, it is no other than the proposition of which it is the meaning: It is a translation of it" [19.47], CP 5.427. While Peirce does not present details about the nature or agent of this action, I think there is enough evidence now to argue that living organisms are the agents that perform the action on and with signs [19.50]. What needs further clarification is the nature of this translational action.

Using Peirce's terminology, translation is the process through which a semiotic agent creates a new representamen from an existing representamen by relating the first representamen to an object through an interpretant [19.29]. To be more precise, the translation process entails relating a representamen and object so that an interpretant ensues, which interpretant is then the representamen for a next iteration of the process [19.32] [19.33] [19.34] [19.35], potentially ad infinitum. I suggest that one views this recursive process as a continuum of possibilities. On the one extreme, one could have photocopies of a text as a mechanical process that contributes very little if anything to meaning-making, and on the

[2] Readers who are unfamiliar with Peircean semiotics might find De Waal [19.5, 19.6] and Marais [19.28], pp. 83-114, to be valuable starting points.

other extreme, one could have literary writing or intense learning experiences that are rich in meaning-making. In between, one could conceive any number of possibilities.

The next question to answer is: What does this translational action entail? In my view, the action has to be conceptualized in terms of work because it entails the imposition of constraints, but it is a particular kind of work, namely teleological or functional work, i.e., the translation is aimed at "meaning" [19.47], CP 4.127, or "mediation" and "modification" (CP 2.89) in the future (CP 5.427). To be more specific, it is semiotic teleological work. As Deacon [19.7] makes clear, everything that exist does so because it was caused by either homeodynamic, morphodynamic or teleodynamic work. Work requires energy, which is created by the imposition of constraints on matter-energy. In the case of teleodynamic work, the constraints themselves are not only matter-energy but include, in Deacon's terms, absential systems [19.7] pp. 1-17, 107-142, 264-287. These absential systems co-evolved with the emergence of semiosis, i.e., the ability to translate and interpret, and it is this evolution that I trace in the next section of this chapter. It seems that a synthesis is emerging [19.30], which links thermodynamics, information and semiosis through translation as the semiotic work that is done to constrain semiotic material in order to perform some teleodynamic work. This line of argument usually raises the objection that thermodynamics is about matter-energy and not about ideas, and that ideas cannot constrain matter-energy, only matter-energy can. While this is a massively complex debate that would require a paper on its own, based on Kauffman's [19.20] latest work, I think we need to posit a circular relationship between matter-energy and ideas, and the basis of this relationship is the point that ideas are always materially instantiated, or objectified [19.30]. In this synthesis, ideas could indeed be viewed as contributing to a propensity [19.53] to constrain organisms to some extent. Because ideas are about differences in matter-energy and instantiated in matter-energy, though themselves not matter-energy, thermodynamics seems to be applicable to them, not as a metaphor but as part of the physics conceptualization of thermodynamics.

This raises another question, namely: What is semiotic work? Based on Deely's [19.8] [19.9] argument about the entangled nature of matter-energy and meaning,[3] I argue that semiotic action is always embedded in matter-energy systems such as living organisms. This does not mean that it can be reduced to matter-energy. Rather, if we view semiosis as the process of considering the relevance or significance of matter-energy systems for living organisms, it means that we need to consider the matter-energy as well as its relevance or significance because significance is the significance of matter-energy for a living organism or group of organisms, which itself is matter-energy. This would mean that semiotic work operates in the same way as other work, i.e., through constraints on the release of energy. This chapter is thus about considering the implications of growing semiotic freedom for semiotic work. I hence use "work" not as a metaphor but in the technical sense in which thermodynamics use it. As I indicated earlier, I do not have the knowledge or skill to contribute to quantitative debates about thermodynamics. I do foresee, however, that sufficiently skilled scientists would be able to do this. My own attempt entails finding ways to talk about thermodynamics in qualitative humanities studies in a way that, at the very least, is aligned with and does not contradict physics.

[3] I do not mean quantum entanglement here. I therefore use the term "entanglement" to refer to two things, matter-energy and meaning, that are in my view interdependent and mutually constitutive.

Concerning the entangled relationship between matter-energy and meaning, I also argue [19.30] that meaning is not a substance but a relationship between two matter-energy systems and its implications for a third (the interpretant). The third need not be matter-energy because it could be an idea, but even that idea is instantiated in a matter-energy system. Hence, meaning always entails matter-energy, and it flows logically from this that meaning is then always subject to the Second Law of Thermodynamics. Semiotic work, like any work, is aimed at countering the effects of the Second Law of Thermodynamics and is therefore negentropic at some level, though always entropic at a large scale.

To further clarify my conceptualization, I use the term "semiosis" to refer to the meaning-making and meaning-taking process in its broadest conceptualization. I then use translation to refer to the semiotic work performed by semiotic agents with teleodynamic intent, i.e., the imposition of constraints on semiotic material. This conceptualization usually also raises questions about interpretation [19.10] [19.51]. The word "interpretation" has at least two meanings, namely inferring the meaning of signs or the professional action of rendering one spoken language into another (also known as interpreting). I am exploring the first semantic domain here. Peirce himself seems hesitant to draw a distinction between translation and interpretation, and I found a number of cases in which he equated the two terms [19.48], p. 340, [19.49], p. 82. In earlier work where I discussed Peirce's views on these two terms, I concluded that one has a number of options. You could take the words "translation" and "interpretation" to be exact synonyms, and there is enough evidence from Peirce to do so. Alternatively, one could try to draw some theoretical distinction between the two. This could mean following suggestions by Eco [19.10] and Steiner [19.51] to conceptualize translation as the narrower concept, limited to interlingual translation, and interpretation as the broader concept, roughly synonymous to semiosis. In the earlier work to which I refer [19.28], pp. 107-114, I conceptualized translation as the semiotic process of relating representamen, object and interpretant in a never-ending process, and I conceptualized interpretation as the "… outcome or goal or function of the semiotic process" [19.28], pp. 111. I am no longer convinced of this conceptualization of interpretation. I think there may be a better reason for drawing a distinction between translation and interpretation.

I would like to suggest an argument that translation and interpretation could be seen as two aspects of the semiosic process that stand in a complementary and entangled relationship to one another. In other words, in the meaning-making and meaning-taking process, i.e., semiosis, two things happen simultaneously and in a complementary relationship to one another. Like the wave and particle functions of light that exist at the same time but cannot be observed at the same time, translation and interpretation are the complementarity of matter and idea (mind) as they are entangled in the web of experience [19.9]. In this conceptualization, translation refers to the process that turns representamens into interpretants, which interpretants become representamens for a further translational process. Here, one could observe and describe representamens, objects and interpretants regarding aspects such as their matter-energy, their relationships to one another and the particular logic that relates them. This whole translation process is based on and entails an interpretation process. Interpretation then refers to a different yet necessary aspect of the same process, namely the ways in which the translation is based on and flows from an interpretation by the organism[4] that performs the translation. It refers to the processes through which the organism interprets the relevance of the sign, and it results in an interpretant. The interpretation, which is empirically to a large extent hidden from observation, i.e., in a brain or in the metabolism of an organism, thus entails observing and describing absentials based on sentials. As a working conceptualization to try to clarify the relationship between translation and interpretation, I therefore suggest a complex, non-linear, entangled, complementary relationship between the two aspects of the semiosic process. Translation and interpretation are two aspects of the one semiotic process, and translation requires interpretation as much as interpretation requires translation. The term semiosis therefore seems to entail translation-interpretation, which is the semiotic work performed by semiotic agents based on teleodynamic intent, which teleodynamic intent entails interpretation of the relevance of the translation.

Should the argument above be too strong, ontologically speaking, one could make a softer epistemological argument that translation and interpretation refers to two distinguishable levels of observation and description. One could call it a double helix or one could refer to the wave and particle functions of light to indicate that translation and interpretation could be discernible as two different aspects of a semiosic process, but they are not different "things" or separate "things," or even different processes or separate processes. In this conceptualization, translation refers to a constraint-centered observation of semiosis while interpretation refers to an interpreter-centered observation. "Aspect" assumes some kind of abstraction, but in line with arguments in new materialism, the abstraction or the observation constitutes the aspect as much as the abstraction or observation is constituted by the aspect [19.2] [19.11]. The advantage of a more nuanced conceptualization of translation and interpretation is that it provides for a more fine-grained analysis of semiosic process.

I also consider, along the lines of arguments in new materialism, an onto-epistemological argument, suggesting that reality and its observation are entangled. Thus, translation describes a process of relating representamens and objects in order to create interpretants, through the imposition of constraints on semiotic material. Interpretation describes how the interest (contextual, internal) of the organism that interprets is entangled in and co-constrains the translation process. In this argument, Atã's formulation puts it elegantly:

> The words translation and interpretation pick out different levels, or properties, or maybe structural functions, of the semiotic process: one about transforming potentialities for interpretants, the other about realizing those potentialities in terms of actual interpretants [19.1].

The conceptualization above also raises a question about the relationship between semiotic and non-semiotic thermodynamics. As argued above, semiosis is a process that entails teleodynamic work, but what does that mean? First, as Deacon [19.7] p. 270, demonstrates, teleodynamics emerges from (or is hierarchically nested in) morphodynamics while

[4] The interpreting organism need not be singular, which means that communities could also be seen to interpret. In addition, tools created by interpreting organisms, e.g., computers, could also be said to interpret under certain conditions. Perhaps, then, one may want to talk about an 'interpretive situation' rather than an "interpreting organism." That said, I am worried that the more abstract conceptualization, i.e., interpretive situation, might detract from the biosemiotic ideas on the co-evolution of life and semiosis.

morphodynamics emerges from (or is hierarchically nested in) homeodynamics (see Figure 19.1). This means that all translational processes have morphodynamic and homeodynamic features that one could study if you were so inclined. Second, it is important to clarify the distinction between teleodynamics, on the one hand, and morpho- and homeodynamics, on the other hand. Teleodynamics is a matter-energy process that plays out under the auspices of non-matter-energy (absential) processes such as goals, intentions and significance. In other words, over and above the morpho- and homeodynamic constraints that operate on matter-energy systems, these systems are constrained by ideational or absential processes when they are teleodynamic systems. For instance, an idea such as "I'm in the mood for a movie" constrains my body to move to the movie theater or search Netflix on my computer screen. Elsewhere, I tried to demonstrate how absential or ideational processes themselves are objectified and are hence able to constrain matter-energy processes [19.30]. However, even if the intention has not been objectified, it is still able to drive various systems in the organism, e.g., the muscular system, the nervous system [19.50]. I do think that it might also be helpful to distinquish between teleodynamics, generally speaking, and semiotic teleodynamics. For instance, biologically one could argue that the function of the heart is to pump blood that provides oxygen to an organism. In this conceptualization, the function of the heart is teleodynamic, i.e., aimed at a goal, but not necessarily semiotic. Semiotic teleodynamics, or semiotic work as I referred to it above, could therefore be conceived of as a subset of teleodynamics.

When I use the term translation in the rest of this chapter, I am thus referring to semiotic work. Work is only possible through constraints on the release of energy, which means that we need to consider the constraints that operate in process of translation. What I would be interested in is the way in which the representamen and object constrain the sign process, what materialities are involved in the sign process and what the interpretant would be in each case. I would also be interested in how this interpretant is again translated into further interpretants, in other words in the semiotic process. Lastly, I would also be interested in the relation between the source and target texts, which has been explored in great depth in

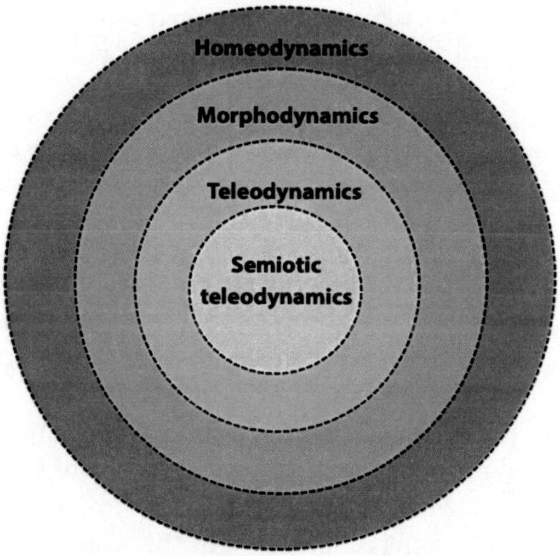

Figure 19.1 The nested nature of dynamic processes.

translation studies. This interest has resulted in some of the most famous binaries in translation studies such as formal and dynamic equivalence [19.38] [19.39] [19.40], documentary and instrumental translation [19.41] and word-for-word/faithful translation and idiomatic translation [19.37].

19.2.2 Lotman and Semiotic Emergence

As evolution and our knowledge thereof stands now, human culture entails the most evolved semiotic system on earth. Whether it is sustainable in the long run remains to be seen. My interest in this section is, however, limited to a broad understanding of the entanglement of matter-energy and ideas in human culture. This interest would include a consideration of the differences between physical, biological and social-cultural constraints and the ways in which the latter creates the propensity for more semiotic freedom.

Juri Lotman [19.27] is one example of a cultural theorist who made ample use of ideas on complex thermodynamics as espoused by the physicist Ilya Prigogine. He suggested that translation is the key driving force behind the emergence of culture. One of the things that puzzled me when I first read Lotman was his insistence that translation is always operating between at least two semiotic systems. These systems could be natural languages or systems of literature or other forms of culture, but the question I had was: Why at least two systems?

I would like to suggest that the answer lies in Lotman's use of thermodynamic theory. He [19.26], p. 36, suggests that any human experience and hence any aspect of human culture requires "internal semiotic diversity." In my view, this insistence on semiotic diversity is built on his understanding of energy, which is what thermodynamics is about, as not something or a substance but a relationship of difference between two physical systems. A little further in the same paper, he argues that culture emerges where any "stimulus calls forth at least two equally valid reactions" [19.26], p. 45, which seems to suggest that he is applying thermodynamic theory to culture. While Lotman does not write about thermodynamics all that openly, the following quotes make it unambiguously clear that he thinks in terms of thermodynamics:

> "Creative consciousness is impossible within an entirely isolated, monostructural (without the potential for internal exchange) and static system." [1.26], p.76.
>
> We will define meaning-making as the capacity of a culture as a whole or of individual parts to produce new, non-trivial texts. We will define new texts as those that arise as a result of irreversible (as Prigogine used the term) processes, that is, texts that are to a certain extent unpredictable. [19.26], p. 85.

In Lotman's [19.26], pp. 116-7, schema, culture emerges through explosions, which is a term he also seemingly borrowed from complexity thinkers in physics, and he views these explosions as translations. Lotman further posits that the crucial points at which creativity in culture occurs are "bifurcation points," which is another term from complexity and thermodynamics [19.26], p. 100. In complex systems theory, bifurcation points are points at which systems are poised to change their trajectory in one of two equally possible directions [19.4], pp. 150-180. This leads me to suggest that Lotman thought of culture in terms of the energy required for it to emerge, in terms of the work that needs to be performed for new forms of culture to emerge. The creative moments or bifurcation points in culture are points where "an event is released from the control of a strict chain of cause-and-effect relationships" [19.26], p. 117.

Conceptualizing the thermodynamic dimension of cultural emergence holds a number of implications. I think the first implication would be that semiosis and energy are entangled. In particular, semiosis is, apart from everything else that it may be, also about the constrained release of energy in order to perform work, i.e., to counter the negentropic tendency in reality. Like the famous quote that says that culture is the form nature take in humans, I would like to suggest that translation is the form thermodynamics takes in living organisms. From thermodynamics, we know that energy is not a substance but a relationship of difference between two matter-energy systems. Lotman seems to suggest that the difference in energy between semiotic systems drives the emergence of culture.

In order to make this argument, one has to demonstrate the entanglement between mind and matter-energy. Sharov and Tonneson [19.50] presented an eloquent argument to this effect, and I used their argument to argue that language and other semiotic systems are objectified ideas, which means that they are embodied but also en-mattered and en-energied [19.30]. Because ideas are always instantiated in the matter-energy, be that through channels of communication or in the internal neurology of the brain, ideas are always entangled with energy.

It is, however, important in my view to distinguish between the type of constraints that are operating on the release of energy in different spheres of reality. I suggest that we need to consider at least three spheres, or three nested types of constraints, namely a physiosphere (matter-energy), a biosphere (all living organisms) and an anthroposphere (human culture). In the physiosphere, it seems to me that constraints are of a physical-chemical nature, e.g., the laws of motion or the laws of chemical binding, and the effects of these constraints is the emergence of new forms of matter-energy that support the survival of the organism. The translation process starts and finishes with matter-energy, but it is clearly embedded in the organization of a living organism, which means that it now fulfills a function. In the biosphere, the physical constraints still apply, but it seems that the effect of constraints at this level is the opening up of possibilities that are not limited to matter-energy constraints. It seems to me that the tendency in the biosphere is for translation processes to move towards less immediate matter-energy experience and more ideational experience. This development clearly took millions of years and went through various stages as Hoffmeyer and Stjernfelt showed, but it seems to me that the translation process developed from a focus on immediacy to a focus on mediacy, which includes a growing role for habit, memory and pattern. In the anthroposphere, both physical and biological constraints apply, but the effect of the constraints seems to be directed towards alternatives to reality. Semiosis in the anthroposphere thus seems triply constrained seeing that the earlier constraints do not disappear but are subsumed in higher organization. As with other biological organisms, anthroposemiosis shows a tendency for semiosis to be less and less immediate, less and less constrained by matter-energy without the immediacy and physical ever disappearing. However, the nature of the interpretants created through translation in anthroposemiosis is different from those in biosemiosis. Whereas the latter focuses on memory and anticipation, the former includes alternatives, fantasies, dreams, ideals. It obviously also included reflexivity, as has been argued by many. It is not only searching for the possibilities that reality offers but for alternative possibilities, for unthought of possibilities.

The third implication is that culture seems to be an interplay between entropy and negentropy—as is all of reality. It seems that the emergence of culture is a negentropic process but that culture itself, as it plays out every day, has strong entropic features. What I mean is that the emergence of culture is a never-ending interplay between creativity and stagnation, or a never-ending process in which negentropy is followed by entropy is followed by negentropy, ad infinitum. Take for instance the writing of this paper. My typing on my keyboard is a negentropic process which constrains the energy in the computer and its computational abilities to create words on a screen, which might later be printed as words on a page. The words on a screen can be saved, which means that the negentropic work is juxtaposed to the entropy of preserving the text in an equilibrium. On printed pages, this seeming equilibrium is also not as stable as it looks because the ink can fade, or the pages can turn yellow or brittle over time. Even on a computer, information can get corrupted, which means that the larger-scale entropic process is continuing to play out. This would mean that both change and stagnation are constrained by one another. Entropy and negentropy are entangled in a dance that is neither absolutely chaotic nor in absolute equilibrium. In other words, our everyday living in which we exist in a culture means that we exist both in the creation and the preservation of patterns and habits. Thermodynamically speaking, culture cannot just be change. It must be an interplay between work and sedimentation of the work. Everyday life, even the art of everyday life, is based on memory of semiotic processes learned while at the same time looking for new or alternative possibilities. They might, to a larger or smaller degree, also be negentropic and hence creative, but they are not only negentropic. It seems then that what we call culture is the entropic remains of negentropic processes that have been committed to memory and from which we draw for living. From a larger perspective, entropy and negentropy should probably be at the edge of chaos in order for it not to be overly stable nor overly unstable. How this plays out in each case is, however, an empirical matter.

19.3 Levels of Translation and Complexity

The 19th century not only produced a significant body of work on thermodynamics, which introduced the notion of time into physics. It also gave us Lamarck and Darwin who, along different lines, argued that biological reality is "not only … a chain in space but also, or rather, a chain in time, where 'more perfect' forms had gradually emerged from 'less perfect' forms" [19.12], p. 7. Both inert matter-energy and living matter-energy need to be explained, among others, in terms of thermodynamic processes that follow the arrow of time. The implications of theories of thermodynamics and evolution seem to have found more philosophical implications from the late 19th century until the current time, leading to waves of complexity and emergence thinking during this period.[5] As Hoffmeyer and Stjernfelt's [19.12] reference to Lamarck indicates, this line of thought is informed by the idea that reality emerges in a process in which later phases build on earlier phases.

[5] See works like Coveney and Highfield [19.14], Lewin [19.25], Mitchell [19.36], Page [19.42] and Waldrop [19.52] for a historical perspective on complexity and Bedau and Humphreys [19.3]and Juarerro [19.14] [19.15] for a historical perspective on emergence.

A concomitant line of thought is that later phases in the process are more complex, as expressed by the following quote:

> "Thus we propose the great chain of semiosis with particular emphasis on how the cognitive and communicative potentials of species may be classified according to their plasticity and articulatory sophistication." [19.12], p. 7.

Based on these two assumptions, Hoffmeyer and Stjernfelt proceed to suggest a "theory of semiotic emergence" with the aim to explain how semiosis came to be as we know it but also to make the case that human semiosis did not arrive on earth de novo but is itself part of a long evolutionary process. Table 19.1 is a representation of Hoffmeyer and Stjernfelt's conceptualization. They discuss the semiosis of each stage in some detail, and I do not repeat their arguments here. Rather, I use their discussion to talk about the translation process at various levels. In addition, due to space limitations, I focus on only a number of their levels, which should suffice for the argument I am trying to make. For the same reason, I do not consider other conceptualizations of semiotic emergence. The point here is not whether Hoffmeyer and Stjernfelt's biology was correct but how the translation process plays out at different levels of emergence.

As all biosemioticians do, Hoffmeyer and Stjernfelt [19.12], p. 9, posit the co-emergence of life and semiosis, which means that one should start at the simplest forms of life to explain semiosis. For them, the first semiotic level is that of molecular recognition [19.12], pp. 10-11. There seems to be consensus among biosemioticians that the semiosis at this level is of a proto-semiotic nature, but the point for semiosis is that prokaryote cells needed to recognize molecules. As Hoffmeyer and Stjernfelt explain, this recognition entails more than just the physical attraction between molecules but requires a selection in order to bind with the correct molecule [19.12], p. 12. The translation at this level therefore also seems to be of a proto-translation nature. The representamen in such a semiotic process would

Table 19.1 Hoffmeyer and Stjernfelt's levels of semiosis [19.12].

11. Consciousness
10. Sentience
9. Learning and social intelligence
8. Collaboration, deception
7. Active information gathering
6. Behavioral choice
5. Sense perception
4. Phenotypic plasticity
3. Multicellularity of eukaryotes
2. Eukaryotic cell
1. Molecular recognition

be a molecule, i.e., a chemical sign-vehicle. The object would be the required property that the molecule should have in order to bind [19.12]. The interpretant would be a chemical action, namely binding to the molecule. Thus, both translation and interpretation at this level are chemical, which means that the process does not entail any cognitive or ideational activity but rather plays out at a chemical level. In other words, all three relata in the semiotic process operate at the level of matter (chemistry) and energy. Compared to later stages of translation, it also seems that a "faithful" translation would have been a matter of life or death for organisms at this level. In other words, if the wrong molecule was identified and bound with, the result could be death. Translation here is a matter of absolute equivalence between the source and the target representamens.

At another level (5 in Table 19.1), Hoffmeyer and Stjernfelt [19.12], p. 16, argue that the level of sense perception of itself does not ensure survival. Rather, the information gained through sense perception needs to be interpreted. Just like at the level of molecular recognition that I discussed in the previous paragraph, translation at this level of semiosis also assumes interpretation, and vice versa. At this level of semiosis, it is argued, habituation as a primitive form of learning can take place, and reference is made to polyp clones of sea anemones that change their behavior after a number of encounters with other clones [19.12]. In my interpretation, the representamen at this level of semiosis is no longer purely chemical. Rather, an encounter with another of the species suggests tactile or some other sensing, which may require pressure or tactile sensation of differences of pressure in the water, for instance. Here, the object is also not only at a chemical level anymore, i.e., it entails a biological organism, and the interpretant now clearly includes habitual behavior, a third in Peircean semiosis. This does not mean, and it is important, that the chemical semiosis at previous levels are no longer relevant. They are indeed, and they never stop being relevant, not even in the most fanciful imaginations of human semiosis, which is still instantiated in the chemistry of the brain. What it does mean is that translation, i.e., the work performed to create sign processes, is increasingly adding new forms of organization to the chemistry of the process. While always instantiated in chemistry, it seems that the gains of each level of semiotic development is that it adds translational processes that are less and less constrained by matter-energy. In this case, habituation means that the organism does not have to have an experience itself on which to respond directly. Rather, it responds to the representamen, which now stands for a habit (immediate object) and which now causes behavior and not only chemical binding. In other words, it seems that the goal of semiotic evolution is to limit the constraints of matter-energy and replace them by semiotic constraints. Habitual (ideational) patterns stored in the memory of organisms now create patterns of behavior that are new representamens for further translation.

The next level of semiotic evolution (Level 6 in Table 19.1) is called behavioral [19.12], p. 6. The major evolutionary development that made this choice possible is the emergence of some kind of central nervous system [19.12], p.17. At this level of evolution, one is able to talk about real behavioral choices for organisms. One also observes a sense of self or what Hoffmeyer and Stjernfelt call "egotism" at this level. This development in semiotic ability allows organisms "an increased range of potential response patterns" [19.12], p. 18. In my understanding, the representamen at this level of translation is no longer just at the level of chemistry. While it always involves chemistry, as indicated earlier, the translation task now is not to bind to a molecule but to respond with the body to any relevant information from the environment. The object here would be a variety of entities from the environment, of different forms, shapes, sizes and colors. The translation process at this level leads to multiple interpretants that emerge virtually simultaneously at various levels. Let us assume that an organism at this level has the ability of tactile recognition. It would be able to, for instance, feel vibrations in the earth or water around it and use this information as a sign of either prey or predator. The first instances of this semiosic process would be just the endosemiosis described in level three. These would enable the emergence of higher-level semiosis, without the earlier levels ever disappearing. The interpretant in this case is clearly not yet cognitive. Rather, it would be at the level of embodied response, muscular action to achieve some goal rather than another. The translation in this context need not be faithful anymore. While any semiotic process, including at this level, could go wrong and lead to death, semiotic mistakes or misinterpretations at this level do not automatically lead to death as it seems to do in earlier stages, although they might still have negative consequences. With choice, organisms have more than one way of attaining the same goal, which means that competing interpretants now have relative value, no longer the absolute life or death value. One of the implications of increased semiotic freedom then seems to be that interpretants themselves become the domain of competition and struggle.

As brains evolved, an organism became able to "displace itself according to the double determination of its inner needs and its outer sensations" [1.12], p. 18, i.e., move around to achieve its intentions. The differences that need to be interpreted now, crucially, include space [19.12], p. 19, but also sounds and smells and tastes. Because animals move, their environment, as it were, changes from moment to moment. Space becomes important because the relative position of both moving and non-moving signs are now relevant to the moving organism. Hoffmeyer and Stjernfelt [19.12], p. 19, offer a detailed description of the implications of the evolution of central nervous systems, which is worth quoting in full here:

1. the integration of subsequent perceptions of the same object or event from different viewpoints during the trajectory of the organism;
2. the integration of different sense-modality perceptions (vision, hearing, smell, taste, touch, electro communication, etc., of the same object or event);
3. the evolution of moveable sense organs to facilitate quick perception-action cycles permitting the focusing of attention on selected features of the environment, speeding up information gathering;
4. the integration of maps of the surroundings, more extended than the single situation, making possible the situating of the organism itself in the map;
5. recognition of conspecifics and individuals from biologically relevant species (predator, prey, competitors, etc.);
6. as indicated by Uexküll, this also makes possible the appearance of neutral objects, not immediately relevant for organismic needs, in the surroundings of the animal—objects whose affordances offer a new degree of semiotic freedom because the organism may now ontogenetically learn to deal with them, and in some cases, combine them;
7. the ability to autonomize the anticipation aspect of all cognition into imaginary scenarios of immediate future events, guiding action [...]

It seems clear to me that the translation at this level (Level 7 in Table 19.1) has changed significantly. The representamen can now be experienced through multiple modalities (2 in the list above). This would mean that the relevance of, for instance, color became part of the translation process. It would mean that distance (1 in the list above) also became a difference, the difference of which needs to be interpreted through the translation process. This development also means that objects increased exponentially. Color could now, for instance, stand for danger, but so could sounds and smells and tastes, which means that the amount of total information increased. The net effect of this development is a massive increase in interpretants. The key development would, however, not only be the number of interpretants but an increase in the type of interpretants. The emergence of "neutral objects," for instance, means the addition of an interpretant that did not really exist in previous levels of semiosis. At those levels, irrelevant things would probably not have registered. In addition, through the learning afforded by memory in organisms with central nervous systems, one also sees an increased role played by factors that are not in themselves matter-energy. Anticipation would be such a factor, as would maps of the environment. The ability to anticipate is built on patterns of experience, as are the maps of the environment. This means survival and thriving are now less and less influenced by matter-energy only because the role of cognition, though not yet thoroughly conceptual at this level, is playing an increasingly important role. Interpretants themselves now start to play a role in the translation process in the sense that a remembered pattern, rather than an immediate matter-energy experience, can drive an organism to act. The more the interpretants increase and the less they are influenced by matter-energy, the less necessary and, even less possible, are faithful translations. Rather than being faithful, translations are now seeking out new possibilities based on anticipation and maps. This would also mean that translations are now less about responses to an environment, though response never disappears fully, and more about the construction of engaging the environment. In other words, the semiotic agency of the organism increases and the semiotic role of non-matter-energy increases. Accordingly, Hoffmeyer and Stjernfelt [19.12], p. 18, calls this level "active information gathering." I think one might even consider that this increase in information itself is extending semiotic freedom. In addition, information itself is now seemingly becoming a biological cause as information is being translated into action.

Another level of semiotic evolution is called sentience (Level 10 in Table 19.1). Hoffmeyer and Stjernfelt [19.12], p. 22, defines sentience as follows:

> "We shall prefer here to use the term in a more-or-less sense, as a property that has something to do with mapping of an experience on feelings, where feelings are understood as mentally processed emotions."

Like with all of the levels, it seems that there are no clear boundaries between the levels and that sentience is not a permanent state in all organisms that sometimes experience sentience [19.12], p. 23. Thus, the amount of time spent in a sentient state is an indicator of the semiotic freedom of a species. In my interpretation, it seems that the evolutionary gain at this level is that, given the same kind of representamen and object, feeling is now added as a possible aspect of the interpretant. As Hoffmeyer and Stjernfelt [19.12], p. 22, suggest, the function of feeling is to optimize the connection between perception and action. It seems then that feeling emerges at an endosemiotic level. The relevance of this observation is that interpretants created in the organism seems to play a larger role as semiotic freedom grows, compared to earlier phases where the interpretant would mainly be a dynamic response. As the example of play among reptiles demonstrates, semiosis at this level requires even less of a faithful translation. Rather, some representamens will just be ignored while the same representamens might in some cases allow for play to emerge, and yet the same representamens might in other cases lead to other feelings and behavior, such as aggression.

The last level of semiotic evolution is consciousness (Level 11 in Table 19.1). They view consciousness as a complex emergent phenomenon that cannot be explained with simple or linear arguments [19.12], p. 24:

> In its human form, consciousness supports a long series of different processes and scaffolds perception, cognition, reasoning, categorization, action, communication, language, and a host of other abilities.

With consciousness, in particular self-reflexive human consciousness, one is in semiosis as Peirce described it. Representamens can be either matter-energy or idea, as can be objects. Equally, interpretants could be just ideas or matter-energy actions. It seems that we have come full circle from initial translations where all three relata of semiosis were matter-energy to translations in which all three relata of semiosis can be ideational. At this level, ideas as much if not much more than matter-energy create the propensity for organisms to create a wide variety of interpretants that may have any of an infinite number of relations to matter-energy. Whereas the ability to have feelings and ideas emerged historically for the sake of survival, these emergent properties now constrain organisms who subsequently act according to these absential feelings and ideas. The amount of semiotic freedom that human organisms have contributes to their phenomenological experience of agency that transcends the constraints that matter-energy offers.

It seems to be clear from the discussion in this section that semiotic freedom has increased through evolution. It seems that one of the implications of increased semiotic freedom is that semiotic processing is increasingly driven by interpretants or thirds. This is the symbolosphere to which Maran refers, and given the discussion above, one can understand why human beings think that they live in a world consisting of and solely determined by symbolic interpretants. However, the discussion also makes clear that the translation process that drives meaning-making is never free of the constraints of matter-energy [19.31]. Rather, translation seems to always be a process of entanglement between matter-energy and idea. In the next section, I turn to the implications of this view for culture.

19.4 The Paradox: Levels of Translation and Levels of Freedom

It seems a common assumption in biosemiotic thinking that reality emerged over time in an evolutionary process. Biosemioticians equally seem to assume that a hierarchical explanation of evolutionary history is the best option because, on the one hand, it fits the data and, on the other hand, it overcomes various philosophical problems such as dualism and reductionism. It also seems that the general assumption would be that as levels of translation become more complex, levels of freedom increase.

Thinking about the thermodynamics of culture, which I want to think about because I am interested in the relationship between matter-energy and ideas, brings us to the paradox that an increase in freedom is only to be achieved through an increase in constraint. In the first nanoseconds after the Big Bang, there were virtually no constraints. Anything was possible, and chaos reigned. Over time, constraints caused certain trajectories to develop, such as the formation of matter, later on galaxies with stars and planets, later on life and later on conscious life. All of these emergent processes were subject to constraints, which caused them to emerge [19.7]. When we then think about culture and its relation to thermodynamics, it seems to me that the same principle holds because culture, while not determined by matter-energy, is always embedded in the matter-energy from which it emerged. The development or evolution of a culture or of human culture as a whole seems to be aimed at an increase in semiotic freedom. This might mean freedom from the limitations in nature, i.e., rain and food, or freedom from social institutions like the church or a king. It might also mean freedom from patterns of thought like racism or sexism or liberal capitalism. However, becoming free from a semiotic trajectory or creating alternative semiotic trajectories seems to me to always involve the imposition of constraints. When we are "freeing" ourselves, it seems that we are substituting sets of constraints for other sets of constraints that, we think might offer us better alternatives. Freedom can never be the absence of constraints because absolute freedom of constraint would mean death. How far semiotic freedom can evolve is an open question, but at this point it seems to me that semiotic freedom is like the mythical juggernaut – it will only stop when it has destroyed itself. Whether and how this will happen is the stuff that post-apocalyptic fiction is about.

Acknowledgments

Apart from the anonymous reviewers who made invaluable contributions to this paper through their reviews, I would like to thank Alexei Sharov and Pedro Atã for engaging me with hard yet constructive questions.

Declaration

I have no conflict of interest to declare.

References

[19.1] Atã, P. (2023) Personal communication on February 6.

[19.2] Barad, K. (2007) *Meeting the universe halfway: Quantum physics and the entanglement of matter and meaning.* Duke University Press, London.

[19.3] Bedau, M.A. and Humphreys, P. (2008) *Emergence: Contemporary readings in philosophy and science.* MIT Press, Cambridge.

[19.4] Coveney, P. and Highfield, R. (1995) *Frontiers of complexity. The search for order in a chaotic world.* Fawcett Columbine, New York.

[19.5] De Waal, C. (2001) *On Peirce.* Wadsworth, Belmont.

[19.6] De Waal, C. (2013) *Peirce: A guide for the perplexed.* Bloomsbury, London.

[19.7] Deacon, T.W. (2013) *Incomplete nature: How mind emerged from matter.* WW Norman & Company, New York.

[19.8] Deely, J. (2001) *Four ages of understanding: The first postmodern survey of philosophy from ancient times to the turn of the twenty-first century.* University of Toronto Press, Toronto.

[19.9] Deely, J. (2009) *Purely objective reality.* De Gruyter/Mouton, Berlin.

[19.10] Eco, U. (2001) *Experiences in translation.* University of Toronto Press, Toronto.

[19.11] Gamble, C.N., Hanan, J.S. and Nail, T. (2019) What is new materialism? *Angelaki: Jounal of the Theoretical Humanities 24*(6), 111-134.

[19.12] Hoffmeyer, J. and Stjernfelt, F. (2016) The great chain of semiosis: Investigating the steps in the evolution of semiotic competence. *Biosemiotics 9*, 7-29.

[19.13] Jakobson, R. ([1959] 2012) On linguistic aspects of translation. In: *The translation studies reader*, 2nd, L. Venuti (ed.) Routledge, London: 138-143.

[19.14] Juarrero, A. (1999) *Dynamics in action: Intentional behaviour as a complex system.* Bradford Books, London.

[19.15] Juarrero, A. and Rubino, C. A. (2010) *Emergence, complexity and self-organization: Precursors and prototypes.* Emergent Publishers, Litchfield Park.

[19.16] Kauffman, S. (1995) *At home in the universe. The search for the laws of self-organisation and complexity.* Oxford University Press, New York.

[19.17] Kauffman, S. (2000) *Investigations.* Oxford University Press, Oxford.

[19.18] Kauffman, S. (2012) From physics to semiotics. In: *Biosemiotic gatherings.* S. Rattasepp and T. Bennet (eds.) University of Tartu Press, Tartu: 30-46.

[19.19] Kauffman, S. (2008) *Reinventing the sacred: A new view of science, reason and religion.* Basic Books, New York.

[19.20] Kauffman, S.A. (2019) *A world beyond physics: The emergence and evolution of life.* Oxford University Press, Oxford.

[19.21] Latour, B. (1993) *We have never been modern.* (H. Wheatsheaf, Trans.) Harvard University Press, Cambridge.

[19.22] Latour, B. (1996) On actor-network theory: A few clarifications. *Soziale Welt 47*(4), 369-381.

[19.23] Latour, B. (2007a) Can we get our materialism back, please? *Isis 98*(1), 138-142.

[19.24] Latour, B. (2007b) *Reassembling the social. An introduction to actor-network theory.* Oxford University Press, Oxford.

[19.25] Lewin, R. (1999) *Complexity: Life at the edge of chaos.* University of Chicago Press, Chicago.

[19.26] Lotman, J. (2019) *Culture, memory and history: Essays in cultural semiotics.* (M. Tamm, Ed., & B. J. Baer, Trans.) Palgrave Macmillan, Cham.

[19.27] Lotman, Y.M. (1990) *Universe of the mind: A semiotic theory of culture.* (A. Shukman, Trans.) Indiana University Press, Bloomington.

[19.28] Marais, K. (2019) *A (bio)semiotic theory of translation: The emergence of social-cultural reality.* Routledge, New York.

[19.29] Marais, K. (2022) Reconsidering the binaries in translation studies. In: *Exploring the implications of complexity thinking for translation studie.s* K. Marais and R. Meylaerts (eds.) Routledge, New York: 7-29.

[19.30] Marais, K. (Forthcoming) *Trajectories of translation: The thermodynamics of semiosis.* Routledge, New York.

[19.31] Maran, T. (2020) *Ecosemiotics.* Cambridge University Press, Cambridge.

[19.32] Merrell, F. (1997) *Peirce, signs and meaning.* University of Toronto Press, Toronto.

[19.33] Merrell, F. (1998) *Sensing semiosis: Toward the possibility of complementary cultural 'logics'.* St Martin's Press, New York.

[19.34] Merrell, F. (2000) *Change through signs of body, mind and language.* Waveland, Prospect Heights.

[19.35] Merrell, F. (2003) *Sensing corporeally.* Univerity of Toronto Press, Toronto.

[19.36] Mitchell, M. (2009) *Complexity. A guided tour.* Oxford University Press, Oxford.

[19.37] Newmark, P. (2001) *A textbook of translation.* Longman, Harlow.

[19.38] Nida, E. (2004) Principles of correspondence. In: *The translation studies reader* (2nd ed) L. Venuti (ed.), Routledge, London: 153-167.

[19.39] Nida, E. (1969a) The science of translation. *Language 45*(3), 483-498.

[19.40] Nida, E. (1969b) *The theory and practice of translation.* EJ Brill, Leiden.

[19.41] Nord, C. (2018) *Translating as a purposeful activity: Functionalist approaches explained.* (2 ed.). Routledge, New York.

[19.42] Page, S.E. (2011) *Diversity and complexity.* Princeton University Press, Princeton.

[19.43] Pattee, H. (2001) The physics of symbols: Bridging the epistemic cut. *Biosystems 60*, 5-21.

[19.44] Pattee, H. (2007) The necessity of biosemiotics: Matter-symbol complementarity. In: *Introduction to biosemiotics: The new biological synthesi.s* M. Barbieri (ed.) Springer, Dordrecht: 115-132.

[19.45] Pattee, H.H. and Kull, K. (2009) A biosemiotic conversation: Between physics and semiotics. *Sign System Studies 37*(1/2), 311-331.

[19.46] Pattee, H.H. and Raczaszek-Leonardi, J. (2012) *Laws, language and life: Howard Pattee's classic papers on the physics of symbols with contemporary commentary.* Springer, Dordrecht.

[19.47] Peirce, C.S. (1994) *The Collected Papers of Charles S. Peirce.* C. Harthone, P. Weis and A Burks (eds), Harvard University Press, Cambridge.

[19.48] Peirce, C.S. (1998) *Essential Peirce: Selected philosophical writings,* vol. 2, N. Houser (ed.), Indiana University Press, Bloomington.

[19.49] Peirce, C.S. (2016) *Prolegomena to a science of reasoning: Phaneroscpopy, semeiotic, logic.* E. Bisanz (ed.), Peter Lang, Frankfurt am Main.

[19.50] Sharov, A. and Tønneson, M. (2021) *Semiotic agency: Science beyond mechanism.* Springer, Dordrecht.

[19.51] Steiner, G. (1998) *After Babel: Aspects of language and translation* (3 ed.). Oxford University Press, Oxford.

[19.52] Waldrop, M. (1992) *Complexity: The emerging science at the edge of order and chaos.* Viking, London.

[19.53] Williams, M. (2021) *Realism and complexity in social science.* Routledge, London.

20

Towards a Biosemiotic Account of Memes as Units of Cultural Replication and Interpretation

Ivan Fomin

Institute of Scientific Information for Social Sciences of the Russian Academy of Sciences, Moscow, Russia

Online resources:
https://scholar.google.ru/citations?user=xNB7vysAAAAJ
https://orcid.org/0000-0003-4703-5262
https://www.researchgate.net/profile/Ivan-Fomin-4
https://scopus.academia.edu/IvanFomin

Abstract

The chapter seeks to apply semiotics to resolve some conceptual problems of the meme theory of cultural evolution. There are three aspects in which semiotics can be helpful in this respect. They correspond to three problems of memetics: the form/content problem, the interpretation problem, and the agency problem. The study shows how the semiotic concepts of vehicle-interpretant-object, secondary subagency, and tardo-sign can provide solutions to resolve these problems. Firstly, the form/content problem can be mapped onto the distinction of sign vehicle and interpretant, and thus the tension between memetic externalism and internalism can be transformed into a framework that would distinguish meme vehicles and meme interpretants as elements of memes. Secondly, the notion of tardo-sign can be used for a stricter definition of memes, conceptualizing them as degenerate signs that produce no final interpretants but provide inertia and stability in cultural development. Thirdly, the problem of whether memes are agents can be transformed into a question of what kind of agents they are. While not all cultural signs are agents, semiotics distinguishes a particular kind of secondary subagents that trigger the production of cultural signs, and memes can be considered a special case of these subagents. Overall, memes can be defined as secondary subagent-signs of cultural tardo-semiosis. In this sense, they are important for cultural evolution but are not the only kind of semiotic entities involved in it. Moreover, various other definitional options for the notion of meme can be distinguished based on whether memes are assumed to be necessarily agentive and post-semiotic, as well as whether the term "meme" has to refer to sign as a whole, its vehicle or interpretant.

Keywords: Semiotics, memetics, sign, meme, logonomic sign, tardo-sign, agency, cultural evolution

Email: fomin.i@gmail.com

20.1 Introduction

In 1976, Richard Dawkins introduced the term *meme*, trying to show parallels between bio-evolutionary processes and processes of cultural development. He coined this word in his seminal book *The Selfish Gene*, in which he suggested that the principle of evolution, understood as "differential survival of replicating entities" [20.11], p. 192, is a universal law that applies to all life in the universe and thus genes that are involved in the biological evolution of species are not necessarily unique evolutionary "replicators." According to Dawkins, there can be evolutionary processes involving other "replicators" while still adhering to the same fundamental principles. It is in this context that he suggested using the term *meme* to describe basic units of replication in the evolution of culture [20.11], p. 192:

> I think that a new replicator has recently emerged on this very planet ... It is still in its infancy, still drifting clumsily about in its primeval soup, but already it is achieving evolutionary change at a rate that leaves the old gene panting far behind. The new soup is the soup of human culture. We need a name for the new replicator, a noun that conveys the idea of a unit of cultural transmission, or a unit of *imitation*. 'Mimeme' comes from a suitable Greek root, but I want a monosyllable that sounds a bit like 'gene'. I hope my classicist friends will forgive me if I abbreviate mimeme to *meme*. If it is any consolation, it could alternatively be thought of as being related to 'memory', or to the French word *même*.

According to Dawkins, examples of memes are "tunes, ideas, catch-phrases and clothes fashions, ways of making pots or of building arches." He described memes as entities that can be "transmitted from one brain to another" and used different metaphors to represent them as agents with a certain behavior inherent in them that are similar to "parasites," "viruses" or other "selfish" living creatures that "leap from brain to brain" and actively compete against each other [20.11], p. 192.

Thanks to the writings of Dawkins as well as those of a few other scholars and enthusiasts [20.7] [20.19] [20.34] [20.43], a science of "memetics," which is the study of memes, gained some prominence in late 20th century. One might even say that memetics itself eventually turned out to be a "viral" meme. However, after a couple of decades, the excitement and interest surrounding this idea began to wane. In essence, as it turned out, this new evolutionary science of culture was unable to go beyond some basic metaphors and parallels between biological and cultural evolution. At the same time, in some sense, "the study of memes" turned out to be an extremely successful sphere of knowledge, but not in the way it was initially supposed to be. Namely, the term *meme* has been "hijacked" by the popular internet culture [20.65], so now it is mainly used in reference to various "viral" pieces of content that emerge, circulate, and evolve online (cf., e.g., [20.1] [20.8] [20.45] [20.69]).

During the time when memetics was particularly popular, some of the proponents of this emergent discipline launched a journal called *Journal of Memetics – Evolutionary Models of Information Transmission*, which was intended to serve as a "forum for academics and professionals to discuss their research on the spread of information from an evolutionary point of view" [20.32]. The journal appeared in 1997 but was shut down in less than 10 years, after 17 issues were published. In the last issue, Bruce Edmonds, one of the journal's key contributors, claimed that the concept of meme "has not provided any 'added value' in terms of developing a new understanding of phenomena" [20.23]. Still, Edmonds acknowledged that evolutionary theories of communication may be valid and effective in other contexts, without relying on memes or the "discredited label of memetics."

The crisis that memetics experienced was hardly caused by flaws in the very idea of cultural evolution or by any fundamental limitations that would have made this discipline unviable. It rather happened due to the fact that some of the supporters of memetics were often careless and sloppy in the use of biological and cultural metaphors. It happened that the project of memetics was developed mostly by people who were more interested in playing with words and metaphors than developing a consistent conceptual apparatus or conducting empirical studies of cultural evolution. As a result, according to Edmonds, those scientists who tried to seriously approach the domain of memetics "suffered in the respect that they are often confused with those on the penumbra for whom memetics is a fad" [20.23]. Another factor contributing to the failure of memetics was that its proponents were more interested in the parallel between memetics and genetics and paid little attention to the implications for the existing fields of cultural studies [20.39], p. 219. Moreover, even though Dawkins and his supporters were relatively successful in popularizing the evolutionary theory of culture, they were, in fact, neither the first (e.g., [20.37] [20.36], pp. 56-58) nor the last (e.g., [20.10] [20.30] [20.35] [20.55]) to investigate this domain.

Overall, today the task of developing a conceptual framework for transdisciplinary research of cultural and social evolution, regardless of whether we call it *memetics* or use a different term, is still largely unresolved. Can memetics become such a framework? Can this discipline be more than just a kit of bio-evolutionary metaphors applied to culture and more than just a study of internet memes? Maybe the very label of *memetics* is just too discredited. However, if there exists a way to move forward in the evolutionary approach of culture, under the label of memetics or without it, there is one essential theoretical framework that most certainly should not be ignored—the framework of *semiotics*, i.e., the "doctrine of signs" [20.50], CP 2.227.

In this chapter, I attempt to use semiotics to make sense of the concept of meme and show how the semiotic toolkit can be used to resolve some of the fundamental problems of memetics (form/content problem, agency problem, and interpretation problem). In particular, I am focusing on how a more nuanced understanding of relevant terminological and conceptual options involved in theorizing memes can be provided on the basis of semiotic accounts of representation, interpretation, and agency.[1]

20.2 What Is a Meme?

Before discussing how one can apply semiotics to make sense of memetics, we should first take a closer look at how the notion of *meme* emerged and developed. It all started with the original definition proposed in *The Selfish Gene,* according to which a meme is "a unit of cultural transmission, or a unit of imitation" [20.11], p. 192. This conceptualization was later updated by Dawkins himself in *The Extended Phenotype,* where *meme* is defined as "a unit of information residing in a brain" and having "a definite structure, realized in whatever

[1] Some fragments of this chapter reiterate, reformulate and further develop arguments that were outlined in [20.24] [20.25].

physical medium the brain uses for storing information" [20.12], p. 165. Similarly, Daniel Dennett describes a meme as "an information-packet with attitude—a recipe or instruction manual for doing something cultural" that is "made of information, which can be carried in any physical medium" [20.22], p. 176.

Another influential memeticist, Susan Blackmore [20.3], p. 51, describes memes in the following way:

> If we define memes as transmitted by imitation then whatever is passed on by this copying process is a meme. Memes fulfil the role of replicator because they exhibit all three of the necessary conditions; that is, heredity (the form and details of the behaviour are copied), variation (they are copied with errors, embellishments or other variations), and selection (only some behaviours are successfully copied).

In these definitions,[2] one can notice several conceptual problems. First, there is the *form/content problem*—an unresolved question of whether memes are "forms of behavior" or "information-packets." Second, there is the *problem of agency,* i.e., a question of to what extent memes can be regarded as *agents* that can "almost intentionally" [20.57], p. vi, "infect" human thinking and actively compete with each other. Third, there is the *interpretation problem* that emerges from meme theory and emphasizes the *copying* of cultural patterns and thus overlooks the aspects of *meaning-making* in cultural development.[3]

The *form/content problem* that concerns the ontological status of memes produced a schism in memetics. Namely, as Øyvind Vada [20.68] puts it, there is a divergence between memeticist *externalists* and *internalists*. The former argue that "memes are observable artifacts and behavior" that exist physically, "out here" in the environment. The latter "state that memes are manifested in the cognitive brain" as information patterns in an individual's memory. There is also a rather productive attempt by Michael Schlaile and Marcus Ehrenberger [20.57] [20.58] to resolve this issue by introducing a "three-dimensional" view of memes that is inspired by Karl Popper's evolutionary epistemology and his theory of "three worlds" [20.54]. This approach would distinguish p-memes as "world 3" objects ("primal memes"), i-memes as "world 2" objects (individual mental representations, for example, "elements of cultural schemata that help transform information into knowledge and serve as models of the world as well as models for the world"), and e-memes as "world 1" objects (environmental or artifactual representations of memes, "for example, as (codified in) tools, books, imitable behavior, practices, and the like") [20.57], p. 41.[4]

In a similar way, Werner Patzelt also maps memes in Popper's three worlds. He suggests that in "world 1" memes appear (1a) as *phemes* (cf. phenes in biology), i.e., physically existing artifacts and performed actions, and (1b) as vehicles that carry examples and instructions for how these artifacts and actions are produced. In "world 2" memes also can manifest themselves in two ways, (2a) as a state of consciousness in which a cultural pattern is recognized and (2b) as a state of consciousness from which a cultural pattern is transformed into "world 1" objects. Memes also exist as "world 3" objects, while at least one

---

[2] See also other examples of definitions of meme in [20.57].

[3] Some of more nuanced memetic approaches suggest that *replication* is not necessarily reducible to straightforward *copying,* but can involve reconstructive extraction of semantic content [20.5].

[4] A similar differentiation between i-memes and e-memes is proposed by Adam McNamara [20.46].

of four other ways of their existence (1a, 1b, 2a, or 2b) is maintained [20.49], pp. 192-193, (see also [20.48]).

As for the *agency problem*, it is dismissed sometimes as a mere result of misunderstanding of the memetic theory. For example, Blackmore emphasizes that representing memes as "selfish" intentional beings that "want" to be transmitted from one carrier to another is just a metaphorical shorthand [20.3], p. 5. However, as Schlaile points out, it is precisely this manner of agentive representation of memes that is attractive to some researchers [20.57], p. 18. Moreover, it is exactly the agency of memes that makes it possible to speak about people and memes as about hosts and parasites, respectively. Such theoretical—not merely metaphorical—representation of memes is far from being uncommon. In particular, as Marion Blute puts it, there are generally two basic views among memeticists about the fundamental metaphor of memetics [20.4]. In the first, memes are viewed as coding elements that effectively accumulate and transmit information in the processes of sociocultural evolution. This approach is known as a "gene-like view." Another approach is called a "virus-like view," which focuses exactly on how elements of human culture parasitize human biology.

Finally, the *interpretation problem* is also discussed in the literature on memetics. In particular, "quite unsurprisingly, given their views on human agency" [20.57], p. 186, many memeticists tend to emphasize the aspect of *blind* variation and advocate for a Darwinian approach to the human capacity for innovation. Thus, some versions of memetics end up neglecting purposeful meaning-making and radical deliberate inventions, essentially reducing cultural innovation to "design without a designer," similar to the bio-evolutionary process in its neo-Darwinian view [20.20], [20.57], p. 186, 188. At the same time, in other memetic approaches, it is assumed that even though people have an innate tendency to imitate, they are capable of choosing memes freely.[5]

20.3 Memes as Signs

One of the potential causes of the failure of memetics can arguably be found in the fact that this emergent discipline is detached from other fields and traditions that also examine the mechanisms of information transmission in culture [20.36] [20.39]. Semiotics is one of those fields. The lack of connections between the two disciplines is particularly striking because semiotics and memetics share some of the same goals. In particular, both disciplines aspire to serve as a meta-language for the study of human culture [20.47], p. 2, [20.21], p. 309. However, neither *The Selfish Gene* nor *The Extended Phenotype*, which both introduced the concept of meme, contain any references to the semiotic doctrine [20.11] [20.12]. It seems that the first joint discussion of semiotics and memetics appeared in Thomas Sebeok's 1979 book *The Sign and Its Masters*. In that book, Sebeok comments on Dawkins's concept of meme and overall speaks about this theory quite positively. In particular, Sebeok cited a couple of relevant pieces from Charles Peirce's writings [20.50], CP 2.222, 2.302, that resonate with Dawkins's agentive vision of memes. Sebeok says [20.60], p. xiii:

[5] See [20.29] for a more detailed discussion of these theoretical alternatives.

> The fundamental property of life belongs inalienably to signs, for "every symbol is a living thing, in a very strict sense that is no mere figure of speech" (2.222). To be a sign and to be a replicator—this is ultimately a statement of identity, for, while symbols propagate, "it is only out of symbols that a new symbol can grow" (2.302). Replicators—Dawkins's name for genes—are, in the last analysis, but signs which construct for themselves survival machines (containers, vehicles) to assure their continued existence.

However, in his later writings, Sebeok was more negative about memetics. In the 2000 book *The Forms of Meaning*, Sebeok and Danesi argued [20.61], pp. 163-164:

> Dawkins' case is, at its core, a deceptive one. *Genes* can be identified and separated from organisms, and then studied, altered, and even cloned physically. That is a scientific fact. A *meme*, on the other hand, is no more than Dawkins' own term for what we have called a *model* in this work. There is no empirical way to verify the reality of memes, as defined by Dawkins; they can only be talked about as if they existed. But it is possible to study the structure of mind in the structure of models…

According to Erkki Kilpinen [20.39], p. 231, Sebeok's attitude toward memetics supposedly changed not simply due to the dissatisfaction with the notion of the meme itself but rather because of the fuss that it had aroused. The fuss was probably especially irritating, as the idea of universal replication was, in fact, Sebeok's original concept, which he had formulated much earlier than Dawkins.

Sebeok, obviously, is not the only one who was skeptical about Dawkins's meme theory. Many other semioticians also have criticized it, suggesting semiotic reformulations, reconceptualizations, and adjustments [20.2] [20.6] [20.9] [20.13] [20.14] [20.38] [20.39] [20.41] [20.44] [20.56] [20.67]. In particular, one of the most common points of those criticisms consists in the idea that *meme* is just a new word for *sign*. This, in essence, was the position articulated by Sebeok and Danesi, who equated *meme* to *model*, which they defined as "a form that has been imagined or made externally (through some physical medium) to stand for an object, event, feeling, etc., known as a referent, or for a class of similar (or related) objects, events, feelings, etc., known as a referential domain" [20.61], p. 2.

Some scholars argue that the concept of *meme* is not just a new word for an old concept but, in fact, a concept that is "inferior" [20.39] to the concept of *sign*. In particular, Terrence Deacon emphasizes that Dawkins's theory of "replicators" wrongly focuses on the replicative properties of signs while ignoring their referential and functional properties [20.14] [20.15] [20.18]. Importantly, this criticism applies not only to Dawkins's ideas about memes but also to his account of genes. In a similar way, Timo Maran and Karel Kleisner [20.44], p. 198, point out that the idea of "signs as replicators" is "ignorant of semiosis as mediating and connecting meaning relation" and fails to account for semiosis as a targeted process that is "an independent factor in evolution." As an alternative, Maran and Kleisner suggest a framework of "evolutionary biosemiotics" that would allow accounting for both natural selection and semiosis.

Also commenting on how the concepts of *meme* and *sign* relate to each other, Kalevi Kull [20.41] suggests that these two terms belong to two different biological paradigms: *mechanistic biology* and *semiotic biology* correspondently. He argues that the main principle of cultural development cannot be reduced to copying, even though it might seem so from the point of view of mechanistic biology. In the semiotic biology paradigm, the main mechanism is not mere imitation but translation. As Kull formulates it, "the objects of copying are memes, whereas the objects of translation are signs" [20.41], as reprinted in [20.40]:

> [B]oth terms denote almost the same thing, and accordingly it would be easily possible to blend mutually the brilliant examples—still emphasizing the different sides of the coin, in one case the ability to propagate and compete, with all their consequences, in the other case the relatedness to creativity and symbiosis. However, if organisms would only copy information and not translate, i.e., not change and transform information like it always takes place in the process of translation, then they would never be able to predict, to expect, to intend, i.e., to live.

Another important point of criticism of memetics from the point of view of semiotics is formulated by Kilpinen. According to him, one more aspect in which the word *meme* is inferior to the term *sign* is that while the concept of *sign* opens a way to bridge the gap between the studies of nature and culture, the *gene/meme* dichotomy merely reproduces this cleavage [20.39], pp. 228-230.

However, even with all these blindspots and problems of memetics, *meme* as a term does not necessarily have to be discarded. Some semiotic theorists suggest that it can rather be reconceptualized in a more constrained and accurate manner. In particular, Deacon suggests that memes (as well as genes) can be understood as "the physical loci where the replicative and adaptational functions intersect" [20.15], p. 23. Importantly, the information for both functions is not fully contained in the meme but is constituted by the relationship between the physical pattern of the token and the system of processes in which it is embedded. Thus, as Deacon argues, memes are equivalent to the entities called *sign vehicles* (or *representamens*). Accordingly, if memes are sign vehicles, that means that they are "not some intangible essence that is passed from brain to brain" but rather concrete things or events that stand for some information: "It is something concrete, mostly outside of brains (though something like the instructions for making it may be in a brain), that can be easily and unambiguously copied, precisely because it is some tangible thing." [20.15], pp. 20, 23.

In contrast to Deacon's externalist reconceptualization, one might argue that it would be more accurate to consider *memes* not as *sign vehicles* but as *immediate interpretants* (i.e., as "determinations of the consciousness" that are produced as effects of signs [20.53], ILS 285). In fact, this is similar to the kind of conceptualization that is suggested by Alexei Sharov and Morten Tønnessen, who suggest that memes can be defined as a special kind of mental interpretants that can initiate chains of reproduction of the initial signs by which they were brought about [20.63]. Such an approach would be closer to the internalist understanding of meme as "a unit of information" [20.11], p. 192, or "an information-packet" [20.22], p. 176. Alternatively, they can be seen as equivalents of not *immediate* but *final* interpretants [20.53], ILS 285, i.e., as habits produced by signs. These parallels, however, seem quite problematic, as both initial and final interpretants are, in their very essence, products of interpretation, not copying.

Overall, even though it might be reasonable to equate the concept of *meme* to the concept of *sign vehicle* or *interpretant*,[6] it seems unlikely that such terminology will be consis-

[6] There is also a semiotically inspired version of meme theory that theorizes "neuromeme" as a self-replicating neurophysiological structure that appears as *both* a sign vehicle and an interpretant of two different semiotic triangles that are parts of a "semiotic circuit" [20.31].

tently used by many people. Probably, opting for either *meme as a vehicle* or *meme as an interpretant* approach will merely reproduce the existing schism between memetic externalists and internalists, respectively. A better way to use semiotics to resolve the conceptual issues of memetics would be to equate the *meme* to the *sign as a whole*. Then, one can still distinguish between *meme vehicle*, *meme interpretant*, and *meme object* as elements of memes (cf. *sign vehicle (representamen)*, *interpretant* and *object* as elements of sign in Peircean semiotics [20.50], CP 2.242), while the *form/content problem* of memetics will be thereby more or less relaxed if not resolved. In addition to that, one might also suggest distinguishing between *meme tokens (meme sinsigns)* and *meme types (meme legisigns)*, thus differentiating between individual existent artifacts and individual instances of behavior on the one hand and general cultural habits producing those artifacts and behaviors on the other.[7]

20.4 Memes as Logonomic Signs

Another important distinction that has to be accounted for when we discuss various modes of existence of memes is the distinction between two kinds of signs, of which the first is the signs that are concrete cultural artifacts and actions, whereas the second is the signs that convey instructions for how to produce cultural artifacts and actions [20.49], pp. 192-193. In order to grasp this distinction in semiotic terms, it might be useful to use the category of *logonomic signs* [20.26] [20.27] [20.28].

Logonomic signs are socially devised signs that constrain semiosis "by restricting who is able to produce and receive what signs under what circumstances" [20.28], p. 33. In other words, they convey "logonomic rules" [20.33], pp. 2-5, [20.24] [20.26] [20.27] [20.28], i.e., rules about when and how social semiosis has to be carried out. Some examples of logonomic rules can be found, for instance, in Dawkins's account of memes when he describes them as "clothes fashions, ways of making pots or of building arches" [20.11], p. 192.

Dawkins, however, narrowly focuses on the fact that these norms are transmitted through imitation, thus, effectively reducing memes to *iconic* logonomic signs. At the same time, one can notice that the same rules can be (at least partially) transmitted through other semiotic forms as well. In particular, they can be transmitted verbally, i.e., with *conventional* signs of language. Moreover, some rules can be conveyed *indexically*. For instance, the "ways of making pots" can be transmitted by the shape of a throwing table or other pottery tools.

Importantly, logonomic signs do not have to be something completely separated from other cultural signs. It is rather that in some signs the logonomic aspect can be a part of their functionality. Furthermore, iconic logonomic signs do stand out in some respect, because the iconic kind of logonomic semiosis is virtually omnipresent in culture. Any cultural artefact or action has iconic logonomic capacity, as it can be used as an example for similar cultural forms.

So, here we can notice three distinct definitional options for memetics. In particular, one can (1) loosely define memes as just *any cultural signs*, or (2) more strictly, as *logonomic signs*, or (3) ever more strictly, as *iconic logonomic signs*. Defining memes not as just any

[7] See (CP 2.243-246, 4.537) [20.50] [20.51] for Peirce's introduction of the notions of *sinsign (token)* and *legisign (type)*.

cultural signs but as *logonomic signs* might be useful, as it allows us to resolve the issue of distinguishing *memes* (as signs that carry instructions on how signs [artifacts and actions] should be produced) from *phemes* (as signs [artifacts and actions] that are produced). This way, the term *memes* can be used to refer to instructions, blueprints, recipes, etc., and contrasted to the term *pheme* that would indicate artifacts and actions that are produced based on those instructions, blueprints, and recipes.

Some additional definitional options that derive from the idea of conceptualizing memes as logonomic signs can also be considered. In particular, we can opt for defining memes either in an "externalist" manner, as *sign vehicles of logonomic signs* (i.e., forms that are used to convey logonomic rules), or in "internalist" manner, as mental *interpretants of logonomic signs* (mental effects produced by those forms). Memes can also be conceptualized as habits that enable instructions, blueprints, recipes, etc., to consistently produce particular kinds of cultural semiotic entities as their effects. In Peirce's terms, this would mean that memes are equated to *logonomic types* (*logonomic legisigns*). Alternately, one can equate memes exclusively to *logonomic tokens* (*logonomic sinsigns*), thus assuming that memes are not logonomic habits but rather individual things and events that convey logonomic rules.

20.5 Memes as Secondary Subagents

In the two previous sections, we considered how semiotics can help to resolve the *form/content problem* of memetics by introducing a well-developed conceptual apparatus that allows us to distinguish various kinds of semiotic entities and analyze them into elements. As to the *agency problem*, it can in fact also be at least partially tackled if we apply the toolkit of semiotics to it. In particular, in his biosemiotic critique of memetics, Terrence Deacon argues that Dawkins's account of memes is particularly problematic due to the "misplaced agency" that is implied in the concept of "replicator." This concept gives the impression that memes (as well as genes) "can be understood without considering their embeddedness in a dynamic system which imbues them with their function and informational content" [20.15], p. 20. Deacon points out [20.15], p. 17:

> The concept of meme misidentifies units of cultural information as active agents, which is the same "shorthand" that misleads our understanding of genes and obscures the dynamic logic of evolution. ... In treating memes as replicators, Dawkins fails to distinguish mere patterns from information (and replication from interpretation), which leads to the problem encountered in all realms of information processing: what counts as information is context dependent. Nothing is intrinsically meaningful, to be so it must be interpreted. In the evolution of both genes and words, replication has always been a multilevel affair in a dynamic system, from which what we conceive as "adapted" or "interpreted" units emerge.

When one considers memes (and genes) as signs, not merely as "replicators," it becomes clear that their functions are informed by a semiotic relationship that emerges between the physical form of the sign token and the system of processes in which the token is embedded [20.15], p. 23. In particular, in the process of cultural development, memes (as "units of pattern," i.e., sign vehicles) are "preserved over time because of the ways they contributed to the usefulness and replication of the larger systems in which they were embedded" [20.15],

p. 22. Taking this into account, one might notice that the concept of replicator "inverts the reality" [20.15], p. 20, as it attributes the function of the multi-level process of replication merely to a physical pattern that, in fact, is not a replicator but just a replica. It is because of this misconception that memetics ends up with the idea that memes are primordial "replicators" and "that the systems in which they are found are derived solely from their interactions" [20.15], p. 22.

At the same time, it is still possible to conceptualize memes as agents. In particular, such conceptualization can be derived from a theory of semiotic agency developed by Sharov and Tønnessen [20.62]. According to this theory, agents are defined as entities that are capable of acting purposefully and using signs to make informed choices. However, importantly, not all agents are self-reproducing, so they can be classified into *primary*, i.e., self-reproducing, agents, and *secondary subagents* that are "produced by primary agents and perform some of their functions."

Furthermore, some subagents belong to a category of so-called *cognits* ("cognitive subagents"), and a particular replicative kind of those cognits can be equated to *memes* [20.63]:

> The majority of mental signs are passive memory states that are not agents. However, some memory states are active agent-sign hybrids represented by activated cognitive subagents. In particular, mental interpretants are not only signs for cognition but also active cognitive subagents that initiate chains of downstream interpretation and activity. For example, seeing smoke in a house activates the interpretant of potential fire. Neurophysiologically, this interpretant is represented by an executive cognitive subagent (cognit) that starts planning emergency activity for triggering an alarm, escape, and rescue. Thus, a 'mental interpretant' is more than a sign, it is also a subagent that starts interpreting other signs. A special case of downstream activity of a mental interpretant is the reproduction of the initial sign that caused its activation. Such signs [...] are called 'memes'...

Even though Sharov and Tønnessen relate memes to mental interpretants, one might also suggest that a particular kind of sign vehicles that produce and are produced by those replicating cognits can also be called *memes* (in externalist sense). Moreover, in this context, we can also apply the principle that was already suggested in the previous section. Namely, we can use the word *meme* to refer not only to sign vehicles or interpretants but to the whole dynamic semiotic entity that consists of replicating cognits as interpretants and sign vehicles that produce them and are reproduced by them. The functioning of memes can thus be modeled based on Frederik Stjernfelt's theory of "natural propositions" [20.66], pp. 141-148, as a special case of functioning of perception-action cycles, in which performed dicisigns[8] are similar to perceived dicisigns, provided that perceived dicisigns are successful.

20.6 Memes as Tardo-Signs

The question remains whether we can really speak about signs and semiosis in cases when we deal with a mere reproduction of a habitual behavior. Above, I have already mentioned

[8] *Dicisign* is Peirce's term for a sign that is a proposition (CP 2.250) [20.50].

a fundamental disagreement between the memetic and semiotic approaches to cultural evolution: semiotics in its accounts of cultural development focuses on interpretation and purposefulness, whereas memetics emphasizes copying and blind variation. However, even though, from the perspective of semiotics, memetic mechanisms of copying are degenerate instances of semiosis, they can still be, in an insightful way, placed into a broader account of semiotic entities. Namely, from the semiotic point of view, the processes of "semiosis without interpretation" can be characterized as processes of functioning of post-semiotic signs. More specifically, they can be conceptualized as processes that involve *tardo-signs*.

A detailed theoretic account of tardo-signs, inspired by Sebeok's "six species of signs" [20.59], is proposed by Tyler James Bennett. According to Bennett [20.2], pp. 191-205, in addition to Peirce's famous distinction of fully-fledged triadic signs into icons, indices, and symbols, one can also introduce a typology of dyadic semiotic entities in which the triad of vehicle-interpretant-object has not yet emerged (proto-signs) or has degenerated (tardo-signs).

Overall, the doctrine of dyadic signs derives from the writings of Peirce, as he discusses some cases of semiosis in which the observation of an object or the production of an interpretant can be "diminished" [20.2], p. 193. Here is how Peirce introduces this issue [20.52], CP 5.473:

> Suppose, for example, an officer of a squad or company of infantry gives the word of command, 'Ground arms!' This order is, of course, a sign. That thing which causes a sign as such is called the *object* ... represented by the sign: the sign is determined to some species of correspondence with that object. In the present case, the object the command represents is the will of the officer that the butts of the muskets be brought down to the ground. Nevertheless, the action of his will upon the sign is not simply dyadic...

Peirce then compares the automated behavior of the soldiers who hear the military command to the rise in mercury of a thermometer. He emphasizes the aspect of automation in these processes and points out that interpretants are not necessarily "of a mental mode of being" [20.52], CP 5.473:

> For the acceleration of the pulse is a probable *symptom* of fever and the rise of the mercury in an ordinary thermometer [...] is [...] an *index*, of an increase of atmospheric temperature, which, nevertheless, acts upon it in a purely brute and dyadic way. In these cases, however, a mental representation of the index is produced, which mental representation is called the *immediate object* of the sign; and this object does triadically produce the intended, or proper, effect of the sign strictly by means of another mental sign; and that this triadic character of the action is regarded as essential is shown by the fact that if the thermometer is dynamically connected with the heating and cooling apparatus, so as to check either effect, we do not, in ordinary parlance speak of there being any *semeiosy*, or action of a sign, but, on the contrary, say that there is an 'automatic regulation,' an idea opposed, in our minds, to that of semeiosy. For the proper significate outcome of a sign, I propose the name, the *interpretant* of the sign. The example of the imperative command shows that it need not be of a mental mode of being.

As Bennett [20.2], pp. 193-194, points out, dyadic signs are generally characterized by a "lack of deliberation." More specifically, he suggests using the term *tardo-signs* to refer to a particular kind of dyadic signs in which the dimension of the interpretant is diminished (while the signs in which the diminished dimension is that of the object are called *proto-signs*) [20.64]. Tardo-signs are "post-semiotic" entities, "for which the symbolic threshold has already been crossed." In other words, these are entities that had once been full-fledged triadic signs, but then degenerated towards dyadicity, as their capacity to produce interpretants declined [20.2], pp. 193-194.

The concept of tardo-sign should be used with a certain terminological rigor, as it is important that tardo-signs are not signs with "no interpretants" [20.2], pp. 193, 197, but precisely signs with *reduced* capacity to engender interpretants. What seems to be lacking in them is just the *final* interpretant, i.e., the habit-change produced by the sign, while the *dynamical* interpretant, i.e., the "actual event" brought about by some sign (e.g., the action performed by the soldiers reacting to a military command) is fully present [20.53], ILS 285.

The example of a military command as a sign of which the effect is not a deliberation over the object, but an action, might be an important hint to how cultural signs can be theorized. In particular, the term *meme* can, in fact, be quite appropriate here as a way to refer to the functioning of tardo-signs of this kind, in which the effect produced by a sign is not a mental interpretant, but an action that is performed almost completely automatically, due to frequent repetition of both the command and the action that is triggered by it. In this respect, tardo-signs can also be considered as degenerate cases of iconic logonomic signs, in which copies of a particular cultural example are produced as dynamical interpretants, virtually without any final interpretants.

Thus, conceptualizing memes as tardo-signs does provide a promising opportunity to address the *interpretation problem* of memetics as well as relax the tension between semiotic (focused on meaning-making) and memetic (focused on replication) approaches to cultural evolution. By recognizing memes as a special kind of semiotic entities, we might, in fact, be able to theorize them as something distinct from other cultural signs. In particular, they can be understood as semiotic entities that enable cultural "inheritance" as an aspect of stabilizing inertia that is necessary for cultural evolution.

As Bennett [20.2], pp. 200-205, points out, approaching *memes* as the tardo-semiotic entities that enable the retention of cultural patterns resonates with Deacon's references to automatized skillful behavior that is performed "with minimum of conscious monitoring" and is inaccessible to introspection [20.17], p. 28, as well as to his accounts of "something akin to inertia associated with a change of mental content" [20.16].[9]

More generally, taking into account the cultural inertia that is enabled by memes can be important, as the studies of cultural evolution cannot exclusively focus on the production of new cultural meanings and solutions but also consider semiotic structures that enable their retention, stabilization, and reproduction. In fact, approaching cultural evolution this way would resonate with Kull's theory of evolution of semiotic systems, which suggests that such

[9] "Rapidly shutting down an ongoing dynamic in one area and just as rapidly generating another requires considerable work, both thermodynamic (metabolic) and morphodynamic. But consider the analogy to simple physical morphodynamic processes like whirlpools and Bénard convection cells. These cannot be generated in an instant nor can they be dissipated in an instant, once stable. The same must be true of the mental experiences in such cases of highly aroused shifts of attention. Prior dynamics will resist dissolution and may require structural interference with their attractor patterns (morphodynamic work imposed from other brain regions) to shut them down rapidly. They will in this sense resist an imposed change. This tension between dynamical influences at these two levels—the homeodynamics of metabolic processes and the morphodynamics of network dynamics—is inevitable in any forced change of morphodynamic activity" [20.16].

evolution requires not only the processes of finding new sign relations but also processes of turning those relations into inheritable habits. Thus, the theory of evolution of semiotic systems has to account for two fundamental tendencies, those being the tendency of *diversification* (i.e., semiotic systems being introduced with new distinctions) and the tendency of *standardization* (i.e., semiotic systems becoming more coherent) [20.42], p. 492. Memes, thus can be considered extreme degenerate cases of cultural semiosis, in which the aspect of inert reproduction overpowers the aspect of innovative meaning-making.[10]

20.7 Conclusion: Memes as Tardo-Semiotic Subagents and Sixty-Five Other Conceptualizations of Memes

Summing up, there are three aspects in which semiotics can turn out to be helpful for making sense of meme theory. These aspects correspond to the three problems of memetics that are the form/content problem, the agency problem, and the interpretation problem. Accordingly, the semiotic concepts of vehicle interpretant-object, secondary subagency, logonomic sign, and tardo-sign provide some tools to resolve or at least manage these problems.

In particular, the form/content problem can be insightfully mapped onto the semiotic distinction of sign vehicle and interpretant. Moreover, equating memes to signs makes it possible to relax the tension between memetic externalism and internalism by distinguishing meme vehicles, meme interpretants, and meme objects as elements of memes, not as mutually exclusive definitional options. Furthermore, the concept of *iconic logonomic signs* can be used for a stricter definition of memes, grasping that memes are signs that work as examples that iconically convey logonomic rules, prescribing the production of other similar signs.

The agency problem, seen from the point of view of semiotics, can be transformed from a problem focusing on *whether* memes are agents into a problem of *what kind* of agents they are. In particular, the semiotic approach to agency emphasizes that not all cultural signs are agents, but suggests that there does exist a particular kind of "executive" cognitive agents (cognits) that trigger the reproduction of signs that are similar to those that initiated the production of those cognits in the first place. Thus, such entities can be called memes in a stricter sense than just cultural signs in general. Importantly, even these entities cannot be considered primary (self-reproducing) agents, as they are necessarily *secondary subagents* that are produced by primary agents and perform some of their functions.

Finally, the semiotic category of *tardo-sign* helps to deal with the fact that memes are supposed to be not (or at least not only) meaning-making entities but, first of all, entities that enable cultural replication. This issue can be resolved if we conceptualize memes as tardo-signs, i.e., degenerate post-semiotic signs with diminished capacity to produce mental

[10] The overpowering aspect of inert reproduction does not necessarily imply that memes should be attributed with a pejorative status and imagined as entities that always should be resisted. However, I am inclined to partially agree with Bennett's approach to tardo-signs [20.2] and Hodge and Kress's theory of logonomic systems [20.33] in a sense that the lack of deliberation that is inherent in memetic processes does suggest that memes are appropriate objects for ideology critique and psychoanalysis through which, in some cases, they should be decathected.

interpretants. Such semiotic entities are essential for cultural development as they enable mental inertia and provide stability to cultural patterns.

Overall, we can conclude that—in the strictest sense—*memes are secondary subagent-signs of cultural tardo-semiosis*. Understanding them this way allows us to develop a more precise understanding of their functioning and characterize their stabilizing role in the processes of cultural evolution. Moreover, such a definition can be particularly insightful as it highlights that memes are by far not the only kind of units involved in cultural evolution. In essence, it shows that there is also a whole universe of other cultural semiotic phenomena, including fully-fledged triadic signs and non-agentive cultural semiotic entities, which also should be taken into account in theories of cultural evolution.

Even though defining memes as secondary subagent-signs of cultural tardo-semiosis seems the most accurate way to define this concept, it does not mean that other possible understandings of this term have to be abandoned. Instead, we can use semiotic categories to better navigate among the different terminological options. In particular, those options can be mapped onto the conceptual dimensions of *semiosis* and *agency* (Table 20.1). In the dimension of semiosis, different approaches to the conceptualization of memes can be distinguished based on whether we equate any cultural signs to memes or assume that only logonomic signs, only iconic logonomic signs, or only tardo-signs should be referred to as memes. Next, in the dimension of agency, we distinguish between the more relaxed approaches that do not require memes to be necessarily agentive and stricter ones that assume that memes are, in fact, agents.

A more nuanced map (Figure 20.1) can also account for the conceptualizations that differ in whether the term meme should apply to signs in general, only to tokens (sinsigns), or only to types (legisigns). In total, this gives us 12 distinct non-agentive definitions of memes (column 1 in Table 20.2) and 12 agentive ones (column 2 in Table 20.2). Then, another 42 definitions can be formulated to account for conceptualizations according to which the

Table 20.1 Definitional options distinguished based on different understandings of semiosis and agency of memes.

<table>
<tr><th></th><th></th><th colspan="3">Semiosis of memes</th></tr>
<tr><th></th><th></th><th>Any signs</th><th>(Iconic) logonomic signs</th><th>Tardo-signs</th></tr>
<tr><th rowspan="2">Agency of Memes</th><th>Not necessarily agentive</th><td>Memes are cultural signs</td><td>Memes are (iconic) logonomic signs</td><td>Memes are tardo-signs</td></tr>
<tr><th>Agentive</th><td>Memes are secondary subagent-signs</td><td>Memes are (iconic) logonomic secondary subagent-signs</td><td>Memes are tardo-semiotic secondary subagent-signs</td></tr>
</table>

word *meme* refers not to sign as a whole (columns 1-2 in Table 20.2) but to its vehicle (column 3 in Table 20.2) or interpretant (column 4 in Table 20.2).

Each of these numerous definitions suggests a different way of how the theory of memes fits into the semiotic approach to the study of cultural evolution. However, importantly, none of them implies that studying memes can be detached from semiotics. On the contrary, the broadest definition of memes (definition 1 in Table 20.2) would suggest that the terms *meme* and *sign* have the same meaning. In this case, it seems appropriate to suggest that no special discipline of memetics is required, as the domain of this discipline would

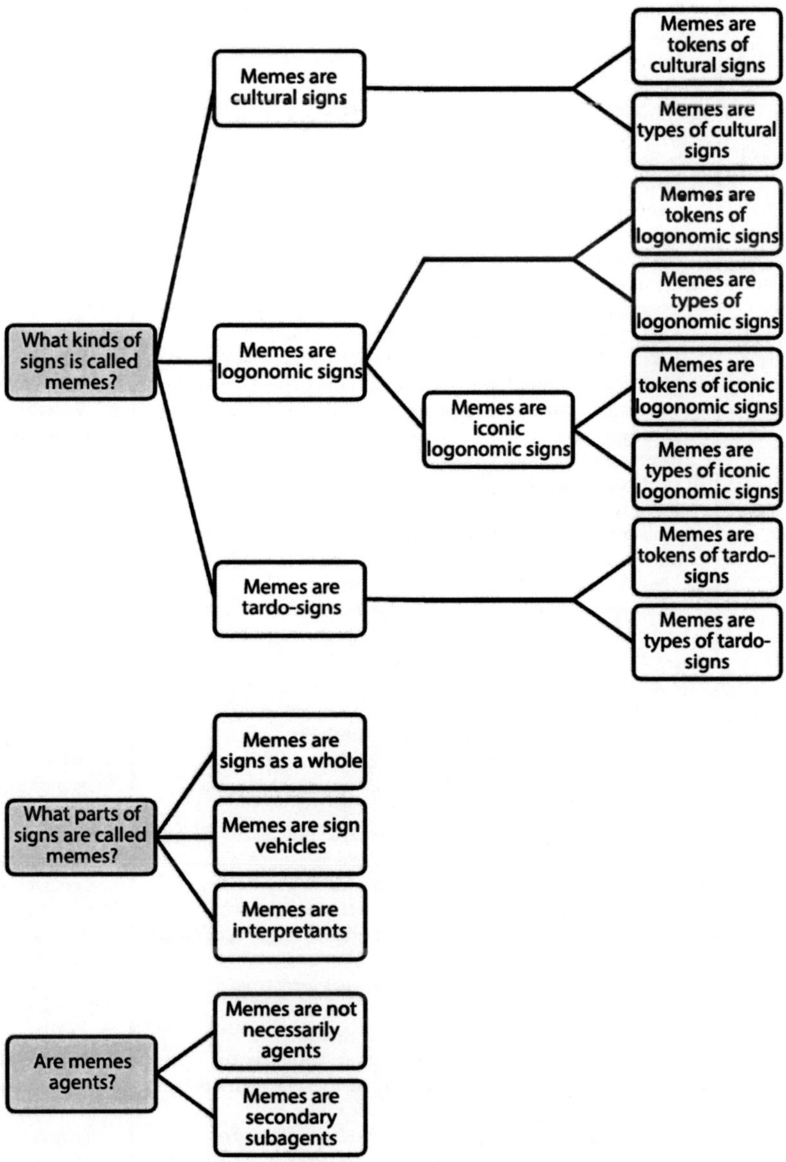

Figure 20.1 Definitional options that are available in semiotic conceptualizations of the notion of meme.

Table 2.2 Possible semiotic conceptualizations of the notion of meme.

Memes as signs	Memes as subagent-signs	Memes as sign vehicles	Memes as interpretants
1. Memes are cultural signs. 2. Memes are logonomic signs. 3. Memes are iconic logonomic signs. 4. Memes are cultural tardo-signs. 5. Memes are tokens of cultural signs. 6. Memes are tokens of logonomic signs. 7. Memes are tokens of iconic logonomic signs. 8. Memes are tokens of cultural tardo-signs. 9. Memes are types of cultural signs. 10. Memes are types of logonomic signs. 11. Memes are types of iconic logonomic signs. 12. Memes are types of cultural tardo-signs.	13. Memes are secondary subagent-signs in cultural semiosis. 14. Memes are secondary subagent-signs in logonomic semiosis. 15. Memes are secondary subagent-signs in iconic logonomic semiosis. 16. Memes are secondary subagent-signs in cultural tardo-semiosis. 17. Memes are tokens of secondary subagent-signs in cultural semiosis. 18. Memes are tokens of secondary subagent-signs in logonomic semiosis. 19. Memes are tokens of secondary subagent-signs in iconic logonomic semiosis. 20. Memes are tokens of secondary subagent-signs in cultural tardo-semiosis. 21. Memes are types of secondary subagent-signs in cultural semiosis. 22. Memes are types of secondary subagent-signs in logonomic semiosis. 23. Memes are types of secondary subagent-signs in iconic logonomic semiosis. 24. Memes are types of secondary subagent-signs in cultural tardo-semiosis.	25. Memes are vehicles of cultural signs. 26. Memes are vehicles of logonomic signs. 27. Memes are vehicles of iconic logonomic signs. 28. Memes are vehicles of cultural tardo-signs. 29. Memes are vehicles of tokens of cultural signs. 30. Memes are vehicles of tokens of logonomic signs. 31. Memes are vehicles of tokens of iconic logonomic signs. 32. Memes are vehicles of tokens of cultural tardo-signs. 33. Memes are vehicles of types of cultural signs. 34. Memes are vehicles of types of logonomic signs. 35. Memes are vehicles of types of iconic logonomic signs. 36. Memes are vehicles of types of cultural tardo-signs. 37. Memes are vehicles of secondary subagent-signs in cultural semiosis. 38. Memes are vehicles of secondary subagent-signs in logonomic semiosis. 39. Memes are vehicles of secondary subagent-signs in iconic logonomic semiosis. 40. Memes are vehicles of secondary subagent-signs in cultural tardo-semiosis. 41. Memes are vehicles of tokens of secondary subagent-signs in cultural semiosis. 42. Memes are vehicles of tokens of secondary subagent-signs in logonomic semiosis. 43. Memes are vehicles of tokens of secondary subagent-signs in iconic logonomic semiosis. 44. Memes are vehicles of tokens of secondary subagent-signs in cultural tardo-semiosis. 45. Memes are vehicles of types of secondary subagent-signs in cultural semiosis. 46. Memes are vehicles of types of secondary subagent-signs in logonomic semiosis. 47. Memes are vehicles of types of secondary subagent-signs in iconic logonomic semiosis. 48. Memes are vehicles of types of secondary subagent-signs in cultural tardo-semiosis.	49. Memes are interpretants of cultural signs. 50. Memes are interpretants of logonomic signs. 51. Memes are interpretants of iconic logonomic signs. 52. Memes are interpretants of tokens of cultural signs. 53. Memes are interpretants of tokens of logonomic signs. 54. Memes are interpretants of tokens of iconic logonomic signs. 55. Memes are interpretants of types of cultural signs. 56. Memes are interpretants of types of logonomic signs. 57. Memes are interpretants of types of iconic logonomic signs. 58. Memes are interpretants of secondary subagent-signs in cultural semiosis. 59. Memes are interpretants of secondary subagent-signs in logonomic semiosis. 60. Memes are interpretants of secondary subagent-signs in iconic logonomic semiosis. 61. Memes are interpretants of tokens of secondary subagent-signs in cultural semiosis. 62. Memes are interpretants of tokens of secondary subagent-signs in logonomic semiosis. 63. Memes are interpretants of tokens of secondary subagent-signs in iconic logonomic semiosis. 64. Memes are interpretants of types of secondary subagent-signs in cultural semiosis. 65. Memes are interpretants of types of secondary subagent-signs in logonomic semiosis. 66. Memes are interpretants of types of secondary subagent-signs in iconic logonomic semiosis.

just be the same as the domain of cultural semiotics. Alternatively, the stricter definitions imply that memetics can be considered as a sub-field of semiotics, focusing on particular kinds of dyadic semiotic entities. In any case, our analysis demonstrates that in order to provide memetics with an adequate conceptual apparatus, the term meme has to be more consistently positioned among other semiotic categories. A different solution would imply that memetics can try to develop further its own typology of memes, similar to the semiotic systematics of signs. However, it is unclear in what respect such a typology of memes can be more effective than the potent conceptual apparatus that is already provided by semiotics.

Speaking more generally, and even putting aside *both* the label of semiotics and the label of memetics, any theory of cultural evolution has to have a theory of meaning-making as its essential part. It has to be so even though some of the entities involved in the processes of cultural development might, in fact, have a diminished dimension of interpretation and be characterized by an overpowering inert functionality of stabilized patterns reproduction.

Funding

This work was supported by the Russian Science Foundation (Grant Number: 22-18-00383).

References

[20.1] Anapol, A. (2022) "Stay Alert, Control the Virus, Make Memes": A multimodal discourse analysis of UK internet memes during the COVID-19 pandemic. In: *Discourses, Modes, Media and Meaning in an Era of Pandemic*. Routledge: 227-244.

[20.2] Bennett, T.J. (2021) *Detotalization and retroactivity: black pyramid semiotics*. University of Tartu, Tartu.

[20.3] Blackmore, S. (2000) *Meme Machine*. Oxford University Press.

[20.4] Blute, M. (2005) Memetics and evolutionary social science. *Journal of Memetics-Evolutionary Models of Information Transmission* 9, http://cfpm.org/jom-emit/2005/vol2009/blute_m.html.

[20.5] Boudry, M. (2018) Replicate after reading: on the extraction and evocation of cultural information. *Biology & Philosophy* 33(3), 27.

[20.6] Bouissac, P. (1992) Why do memes die? *Semiotics*, 183-191.

[20.7] Brodie, R. (2011) *Virus of the Mind: The New Science of the Meme*. Hay House, Carlsbad, Calif., Reissue edition.

[20.8] Caldwell, D. Know Your Meme. https://knowyourmeme.com/

[20.9] Cannizzaro, S. (2016) Internet memes as internet signs: A semiotic view of digital culture. *Sign Systems Studies* 44(4), 562-586.

[20.10] Cousins, S.D. (2014) The semiotic coevolution of mind and culture. *Culture & Psychology* 20(2), 160-191.

[20.11] Dawkins, R. (2006) *The selfish gene*. Oxford University Press, Oxford ; New York, New Edition.

[20.12] Dawkins, R. (2016) *The extended phenotype: The long reach of the gene*. Oxford University Press.

[20.13] Deacon, T.W. (1997) *The symbolic species: The co-evolution of language and the brain*. WW Norton & Company, New York.

[20.14] Deacon, T.W. (1999) Memes as Signs. *The Semiotic Review of Books* 10(3), 1-3.

[20.15] Deacon, T.W. (2004) Memes as signs in the dynamic logic of semiosis: beyond molecular science and computation theory. In: *Conceptual Structures at Work*. K.E. Wolff, H.D. Pfeiffer and H.S. Delugach, (eds.) Springer, Berlin, Heidelberg. 3127: 17-30.

[20.16] Deacon, T.W. (2011) *Incomplete Nature: How Mind Emerged from Matter*. W. W. Norton & Company, New York; London.

[20.17] Deacon, T.W. (2012) Beyond the Symbolic Species. In: *The Symbolic Species Evolved*. T. Schilhab, F. Stjernfelt and T. Deacon, (eds.) Springer Netherlands, Dordrecht: 9-38.

[20.18] Deacon, T.W. (2021) How Molecules Became Signs. *Biosemiotics* 14(3), 537-559.

[20.19] Dennett, D.C. (1993) *Consciousness Explained*. Penguin UK.

[20.20] Dennett, D.C. (1995) *Darwin's dangerous idea: evolution and the meanings of life*. Simon & Schuster, New York.

[20.21] Dennett, D.C. (2001) The Evolution of Culture. *Monist* 84(3), 305-324.

[20.22] Dennett, D.C. (2003) *Freedom evolves*. Viking, New York.

[20.23] Edmonds, B. (2005) The revealed poverty of the gene-meme analogy – why memetics per se has failed to produce substantive results. *Journal of Memetics - Evolutionary Models of Information Transmission* 9, http://cfpm.org/jom-emit/2005/vol2009/edmonds_b.html.

[20.24] Fomin, I. (2019) Memes, genes, and signs: Semiotics in the conceptual interface of evolutionary biology and memetics. *Semiotica* 2019(230), 327-340.

[20.25] Fomin, I. (2021) How can economists study memes? What does (econo)memetics lack? (Reflections on Michael Schlaile's «Memetics and Evolutionary Economics»). *METHOD: Moscow Yearbook of Social Studies* 11, 422-442.

[20.26] Fomin, I. (2022) Constraining, standardization, and habituation in interpretive systems. How to imagine the logonomic aspects of social semiosis? *METHOD: Moscow Quarterly of Social Studies* 2, forthcoming.

[20.27] Fomin, I. (2022) Constraining, standardization, and habituation in interpretive systems. How to imagine the logonomic aspects of social semiosis? *METHOD: Moscow Yearbook of Social Studies* 12, forthcoming.

[20.28] Fomin, I. (2022) Logonomic signs as three-phase constraints of multimodal social semiosis. *Semiotica* 2022(247), 33-54.

[20.29] Gill, J. and Price, I. (2022) A meme-based research programme for management and organization studies. *International Journal of Management Reviews* 24(3), 415-432.

[20.30] Henrich, J., Boyd, R. and Richerson, P.J. (2008) Five Misunderstandings About Cultural Evolution. *Human Nature* 19(2), 119-137.

[20.31] Herrmann-Pillath, C. (2021) Biosemiotic Foundations of a Darwinian Approach to Cultural Evolution. *Cultural Science Journal* 13(1), 16-33.

[20.32] Heylighen, F. Journal of Memetics - Aims and Scope. http://pcp.vub.ac.be/jom-emit/index.html

[20.33] Hodge, R. and Kress, G.R. (1988) *Social semiotics*. Cornell University Press, Ithaca, N.Y.

[20.34] Hofstadter, D.R. (1986) *Metamagical themas: questing for the essence of mind and pattern*. Bantam Books, New York.

[20.35] Jablonka, E., Lamb, M.J. and Avital, E. (1998) 'Lamarckian' mechanisms in Darwinian evolution. *Trends in Ecology & Evolution* 13(5), 206-210.

[20.36] Jahoda, G. (2002) The ghosts in the meme machine. *History of the Human Sciences* 15(2), 55-68.

[20.37] James, W. (1880) Great men, great thoughts, and the environment (Lecture delivered before the Harvard Natural History Society). *Atlantic Monthly*.

[20.38] Jiazu, G. (2009) The theoretical foundation of dynamic semiotics. *Chinese Semiotic Studies* 1(1), 223-227.

[20.39] Kilpinen, E. (2008) Memes versus signs: On the use of meaning concepts about nature and culture. *Semiotica* 2008(171), 215-237.

[20.40] Kull, K. Copy versus translate, meme versus sign: development of biological textuality. *Zooloogia- ja Botaanika Instituut*, http://www.zbi.ee/~kalevi/copytr.htm.
[20.41] Kull, K. (2000) Copy versus translate, meme versus sign: development of biological textuality. *European Journal for Semiotic Studies* 12(1), 101-120.
[20.42] Kull, K. (2014) Towards a Theory of Evolution of Semiotic Systems. *Chinese Semiotic Studies* 10(3), 485-495.
[20.43] Lynch, A. (1998) *Thought Contagion*. Basic Books, New York, N.Y. Great Britain.
[20.44] Maran, T. and Kleisner, K. (2010) Towards an Evolutionary Biosemiotics: Semiotic Selection and Semiotic Co-option. *Biosemiotics* 3(2), 189-200.
[20.45] Marino, G. (2022) Semiotics of virality. *Signata. Annales des sémiotiques / Annals of Semiotics*(13).
[20.46] McNamara, A. (2011) Can We Measure Memes? *Frontiers in Evolutionary Neuroscience* 3, #1.
[20.47] Morris, C. (1938) *Foundations of the theory of signs*. University of Chicago Press, Chicago, Ill.
[20.48] Patzelt, W.J. (2007) *Evolutorischer Institutionalismus: Theorie und exemplarische Studien zu Evolution, Institutionalität und Geschichtlichkeit*. Ergon, Würzburg.
[20.49] Patzelt, W.J. (2018) Genes, memes and signs. *METHOD: Moscow Yearbook of Social Studies* 8, 185-211.
[20.50] Peirce, C.S. (1932) *The Collected Papers of Charles Sanders Peirce, Vol. II: Elements of Logic*. Harvard University Press, Cambridge.
[20.51] Peirce, C.S. (1933) *The Collected Papers of Charles Sanders Peirce, Vol. IV: The Simplest Mathematics*. Harvard University Press, Cambridge.
[20.52] Peirce, C.S. (1934) *The Collected Papers of Charles Sanders Peirce, Vol. V: Pragmatism and Pragmaticism*. Harvard University Press, Cambridge.
[20.53] Peirce, C.S. (2014) *Illustrations of the Logic of Science*. Open Court, Chicago.
[20.54] Popper, K.R. (1979) *Objective knowledge: an evolutionary approach*. Clarendon Press ; Oxford University Press, Oxford [Eng.] : New York, Revised Edition.
[20.55] Sasaki, J.Y. (2013) Promise and Challenges Surrounding Culture–Gene Coevolution and Gene–Culture Interactions. *Psychological Inquiry* 24(1), 64-70.
[20.56] Schaden, G. and Patin, C. (2018) Semiotic systems with duality of patterning and the issue of cultural replicators. *History and Philosophy of the Life Sciences* 40(1), 4.
[20.57] Schlaile, M.P. (ed.) (2021) *Memetics and Evolutionary Economics: To Boldly Go Where no Meme has Gone Before*. Springer International Publishing, Cham.
[20.58] Schlaile, M.P. and Ehrenberger, M. (2016) Complexity, Cultural Evolution, and the Discovery and Creation of (Social) Entrepreneurial Opportunities: Exploring a Memetic Approach. In: *Complexity in Entrepreneurship, Innovation and Technology Research: Applications of Emergent and Neglected Methods*. E.S.C. Berger and A. Kuckertz, (eds.) Springer International Publishing, Cham: 63-92.
[20.59] Sebeok, T.A. (1975) Six species of signs: some propositions and strictures. *Semiotica* 13(3), 233-260.
[20.60] Sebeok, T.A. (1979) *The Sign and Its Masters*. University of Texas Press.
[20.61] Sebeok, T.A. and Danesi, M. (2000) *The Forms of Meaning: Modeling Systems Theory and Semiotic Analysis*. De Gruyter Mouton, Berlin ; New York.
[20.62] Sharov, A. and Tønnessen, M. (2021) Conceptualizing Agency. In: *Semiotic Agency: Science beyond Mechanism*. A. Sharov and M. Tønnessen, (eds.) Springer International Publishing, Cham: 153-188.
[20.63] Sharov, A. and Tønnessen, M. (2021) Semiosis. In: *Semiotic Agency: Science beyond Mechanism*. A. Sharov and M. Tønnessen, (eds.) Springer International Publishing, Cham: 189-223.

[20.64] Sharov, A.A. and Vehkavaara, T. (2015) Protosemiosis: Agency with Reduced Representation Capacity. *Biosemiotics* 8(1), 103-123.

[20.65] Solon, O. and Dawkins, R. (2013) Richard Dawkins on the internet's hijacking of the word 'meme'. *WIRED UK*.

[20.66] Stjernfelt, F. (2014) *Natural Propositions: The Actuality of Peirce's Doctrine of Dicisigns*. Docent Press, Boston (Mass.).

[20.67] Tønnessen, M. (2012) Introducing semetics. In: *Semiotics in the Wild*. Tartu University Press, Tartu: 47-54.

[20.68] Vada, Ø. (2015) What happened to memetics? *Emergence: Complexity & Organization* 17(3).

[20.69] Valensise, C.M., Serra, A., Galeazzi, A., Etta, G., Cinelli, M. and Quattrociocchi, W. (2021) Entropy and complexity unveil the landscape of memes evolution. *Scientific Reports* 11(1), #20022.

21

Astrobiosemiotics and Its Frontier with Astrobiology

Julian Chela-Flores

[1]*The Abdus Salam International Centre for Theoretical Physics, an IAEA (International Atomic Energy Agency)/UNESCO Institute, Trieste, Italy*

[2]*IDEA, Fundación Instituto de Estudios Avanzados, Caracas, Bolivarian Republic of Venezuela*

Online resources:
https://users.ictp.it/~chelaf/
https://en.wikipedia.org/wiki/Julian_Chela-Flores

Abstract

Success in finding reliable biosignatures of life in solar system exploration would force upon us a new problem in biosemiotics, consisting, not only of receiving a communication from elsewhere in the universe, even though such an event alone would be a major scientific achievement. Rather, the success would be to learn that meanings exist in living beings other than those confined to the Earth. The interpretation of a first contact defines a new study, for which we introduce the name "astrobiosemiotics." Meanings need not be constrained to humans alone, according to the ethicist Peter Singer and others. They propose the following statement, in a different context than has been incorporated in this work: "In certain cases, basic aspects of humanity should be extended to genera beyond our own species.". In Singer's ethical context, the extension refers to non-human animals based on the assumption of a principle of equality, which is essentially a principle of equal consideration of interests. In our context, also based on a principle of equality, we step beyond the boundaries of astrobiology. We discuss some implications for astrobiology and especially for astrobiosemiotics. For completeness' sake, we begin this chapter with a general background of the origin of life in the cosmos, especially of species beyond humans, not only animals, but beings with intelligence comparable, or superior to ours.

Keywords: Astrobiology, origin of life, biosemiotics, astrobiosemiotics, ethics, natural theology, astrotheology, evolutionary convergence

Email: chelaf@ictp.it

21.1 Introduction: Astrobiology and Its Frontier with the Humanities, Especially Biosemiotics

Astrobiology is a science covering four distinct aspects of life in the universe. For simplicity we refer to them as "chapters": the origin, evolution, distribution and destiny of life in the universe [21.6]. Astrobiology is a wide-ranging scientific topic, whose frontiers extend as far as the humanities [21.7] [21.8]. But to understand astrobiology and its implications on semiotics and its extensions (when meanings and signals might go beyond our own species), it is convenient to begin our discussion with a general overview of the above-mentioned four chapters, to make more evident its relevance with respect to semiotics.

21.1.1 The Origin of Life in the Universe

We begin with the first chapter of astrobiology. This inevitably takes us back to the work of Stanley Miller [21.35]. From the 1950s a group of researchers attempted to rationalize how the macromolecules of life emerged in conditions analogous to those that ruled Archean terrestrial geophysics. Miller's seminal paper has given us deep insights on the origin of life. (Biographical notes on Miller, as well as some of the other scientists mentioned in this chapter, have been gathered in Table 21.1.) In his doctoral thesis, Miller attempted to rationalize the emergence of the amino acids that conform to the proteins of living organisms. To simulate the early terrestrial conditions the primitive atmosphere was assumed to contain methane (CH_4), ammonia (NH_3), and hydrogen (H_2). By exposing such a gas mixture to electrical discharges, the gas was tested for the presence of amino acids. Miller's success consisted in having identified glycine and α- and β-alanine, which are fundamental monomers of proteins. Later on, other scientists, including Cyril Ponnamperuma and John Oro obtained monomers of DNA and RNA in analogous experiments in prebiotic conditions; the work of these pioneers, as well as the contributions of Alexander Ivanovich Oparin and Sidney Fox have been reviewed in several of the Trieste Conferences [21.39].

These results formed the basis of what is known as chemical evolution towards the origin of life on Earth [21.45]. A hypothetical step in this evolution is RNA world, where self-replicating RNA molecules are assumed to function as catalysts [21.23]. Another scenario called lipid world is based on the idea of self-aggregation of hydrophobic molecules (e.g., lipids) in water which collectively regulate the chemical composition of self-reproducing lipid vesicles [21.47]. Hence, the stage is set to concern ourselves with biosemiotics: the preliminary interpretation of signs and codes in a living world.

21.1.2 The Evolution of Life in the Universe

The basis of all biology is based on the work of Charles Darwin, the great naturalist from Shrewsbury, England. The consideration of evolution, as we currently interpret it in astrobiology, is in terms of a universal Darwinism. This is an old idea, going back to the commemoration at the University of Cambridge of the Centenary of Darwin's Death [21.2]. It was evident, even then, that if life were to be considered as a cosmic phenomenon, alien life should be ruled by certain principles that are basic to all life, prominent among them ought to be Darwinism [21.15]. Astrobiology, which began its robust progress in the present

Table 21.1 Short biographies of scientists mentioned in the text.

Arrhenius, Savante (1859–1927): The Swedish 1903 Chemist Nobel Laureate was among the founders of physical chemistry and climate science. He became director of the Nobel Institute.
Calvin, Melvin (1911–1997): The American chemist was one of the discoverers of the Calvin cycle. He received the 1961 Nobel Prize in Chemistry. He held a professorship at the University of California, Berkeley.
Conway Morris, Simon (1951-): An English paleontologist, evolutionary biologist, and astrobiologist known for his study of the fossils of the Burgess Shale and the Cambrian explosion. He has held the Chair of Evolutionary Paleobiology in the Department of Earth Sciences, University of Cambridge, since 1995.
Darwin, Charles Robert (1809–1882): An English naturalist, author of the theory of evolution through the mechanism of natural selection published in *On the Origin of Species by Means of Natural Selection, or the Preservation of Favored Races in the Struggle for Life*. Darwin avoided discussing the origin of life, a very reasonable attitude for the late 19th century. Since then, the question of the origin of life began being addressed with experimental science by, amongst others, Alexander Oparin, Stanley Miller, Sidney Fox, John Oro and Cyril Ponnamperuma.
Drake, Frank (1929–2022): In 1960 he conducted the first radio search for extraterrestrial intelligence with a radio telescope 26 meters in diameter at the U.S. National Radio Astronomy Observatory in Green Bank, West Virginia. The telescope was aimed at Sun-like stars that are about 11 light-years from the Earth.
Gould, Stephen Jay (1941–2002): An American paleontologist and evolutionary biologist. He was a Harvard Professor and a remarkable science popularizer as well. A major contribution to the theory of evolution was due to him in collaboration with Niels Eldredge, which they called "punctuated equilibrium."
Haught, John. F. (1942-): an American theologian who is a Distinguished Research Professor at Georgetown University. His field of expertise is Roman Catholic systematic theology. He has a particular interest in physical cosmology, evolutionary biology, geology, and Christianity.
McKay, Christopher P. (1954-): An American planetary scientist at NASA Ames Research Center, studying planetary atmospheres (especially Titan), astrobiology (the origin and evolution of life), and terraforming. He is a co-investigator in the Huygens probe, the Mars Phoenix lander, and the Mars Science Laboratory. He has also done research on extremophiles in Death Valley and ice-covered lakes in Antarctica, and in the Atacama Desert, Chile.
Miller, Stanley Lloyd (1930–2007): Under the guidance of Harold Urey Miller, he published a paper in 1953 on the origin of terrestrial amino acids in conditions that were thought to be similar to the Archean Earth. The 50th anniversary of the publication of his paper was celebrated with a scientific event in Trieste at The Abdus Salam International Centre for Theoretical Physics [21.27].

(*Continued*)

Table 21.1 Short biographies of scientists mentioned in the text. (*Continued*)

Oró, John/Joan/Juan (1923–2004): A Spanish biochemist whose research in the USA focused mainly on the experimental study of the chemical reactions that occurred in the primitive Earth. In 1960 he accomplished the pre-biotic synthesis of adenine from HCN, a key component of nucleic acids. In 1961 he published a theory on the role of comets on the formation of biochemical compounds that led to the appearance of life on Earth. He was involved with the detection of amino acids in the Murchison meteorite. Oro was responsible for the correct interpretation of the experiments on the Viking lander on Mars. In the year 2003 he was named Marquess of Oró by Royal Decree.
Ponnamperuma, Cyril (1923–1994): He came in contact with the Nobel Laureate Melvin Calvin in 1962. This led to a series of papers on the synthesis of DNA components, extending in a significant way the pioneering work initiated by Stanley Miller. He became principal organic analysis investigator for Project Apollo and also worked on the Viking and Voyager programs. In the analysis of the meteorite that fell in Murchison, Australia in 1969, he and his co-workers provided evidence for extraterrestrial amino acids. His contact with the Nobel Laureate Abdus Salam led to the Trieste series of conferences in astrobiology.
Rawls, John Bordley (1921–2002): An American philosopher, he was the author of the influential book *A Theory of Justice* (1971). He was a faculty member at several universities in the USA and Britain, principally Harvard. We interpret Rawls' contribution in a wider scenario of justice applied to space exploration.
Sagan, Carl (1934–1996): An American astronomer and distinguished science popularizer, he searched for intelligent life in the cosmos. Sagan began researching the origins of life in the 1950s and went on to play a leading role in every major U.S. spacecraft expedition to the planets in the last century.
Sebeok, Thomas Albert (1920–2001): An Hungarian-born American polymath, semiotician, and linguist, he was one of the founders of biosemiotics. He was influenced in his work by his predecessor, the Estonian biologist Jakob von Uexküll.
Singer, Peter Albert (1946-): An Australian moral philosopher who is Ira W. DeCamp Professor of Bioethics in the University Center for Human Values at Princeton University. Applied ethics is the field where he has made some major contributions. His most influential book was the 1975 *Animal Liberation: A New Ethics for Our Treatment of Animals.* He is also the author of the 1993 book *Practical Ethics* [21.28], which influenced the present work. He is a co-editor of the journal *Bioethics.*
Von Uexküll, Jakob (1864–1944): He was an Estonian biologist who worked in the field of animal behavior studies and the cybernetics of life, amongst other topics, from which biosemiotics emerged as a topic of research [21.49]. Later on, the distinguished semiotician Thomas Sebeok (cf., the above entry) was influenced by his work.

century, incorporates this old view of universal Darwinism. On this basis, we are able to discuss biosemiotics in the early terrestrial environment, essentially with genera beyond our own species, non-human animals, prior to the ascent of human beings. In spite of overwhelming support and universal consensus on the biology based on Darwinism, there is some additional recent research based on biology [21.38].

21.1.3 The Distribution of Life in the Universe

The third chapter of astrobiology is relevant for the general consideration of biosemiotics to genera beyond humans that may be of non-terrestrial origin, having arisen from a Second Genesis (cf. Glossary, Table 21.2). We have the advantage of space exploration, both within the Solar System, as well as in the observation of worlds that are now known to exist around stars other than our Sun (referred to as exoworlds).

In this context, current missions are: firstly, the ESA (European Space Agency) JUICE (Jupiter Icy Moons Explorer) that was launched in April 2023 to the Jovian System [21.26].

Table 21.2 Glossary.

Astrobioethics: This is the subsection within astrobiology that concerns itself with studying the moral implications of implementing practical applications of the search for extraterrestrial life.
Astrobiosemiotics: An area of biosemiotics, and also astrobiology, proposed in this chapter, which studies the possibility of production and interpretation of signs, codes and their communication in the universe.
Astrotheology: A branch of theology providing a critical analysis of space sciences for the purpose of constructing an understanding of our human condition within a cosmic context.
Bioethics: This is an area of moral philosophy that refers to ethical matters that are concerned with health-related issues of medical practice. Compare this definition with *astrobioethics*.
Biosemiotics: An area of semiotics and also biology. It studies production and interpretation of signs and codes and their communication in the living world.
New Atheism: proposes the view that religious belief should be challenged by "rational" arguments. We have strongly emphasized that "rational" should categorically exclude "scientific." If we were to include this term, the Galilean tradition in science based on experiments would be abandoned. Galileo Galilei's belief on experimental verification of all theoretical statements, which has been adopted by all scientists, has now been persevering for a respectable four centuries. Abandoning the pathway first advocated by the Florentine scientist would make statements unacceptable whenever they are made beyond the frontiers of science subject to experimental verification. Forbidden non-Galilean extrapolations are often made outside the scope of serious scientific discussions. This is the case in the questions related to the work of the New Atheists.
Principle of Equality: This is a principle on which the equality of all human beings rests. It is based on the principle of equal consideration of interests. In Singer's *Practical Ethics*, Chapter 2 [21.48], understandably equal consideration of interest is confined to the interests of terrestrial life alone. However, when our canvas is that of an astrobiologist, the interests in question must be extended to all life in the cosmos. Indeed, we should "give equal weight in our moral deliberations to the like interests of all those affected by our actions."
Second Genesis: This is a suggestive term that was introduced by Christopher P. McKay to denote the emergence of life in places other than Earth [21.34].
Semiotics: A philosophical theory of signs and symbols and the role they play in language. A sign is something that communicates a meaning to whoever is attempting to interpret it.

Secondly, search for reliable biosignatures in the icy Galilean moons will also be performed by the forthcoming NASA (National Aeronautics and Space Administration) Europa Clipper mission [21.36]. Besides, there is an ongoing effort to verify whether in Earth-like exoworlds there are conditions conducive to habitability [21.31]. Finally, habitability in the universe can be approached from several viewpoints. For instance, on geologic terms, the early onset of biogenesis on Earth points towards life being a frequent phenomenon in the Universe. According to Lineweaver and Davis [21.32], the observational constraints on the rapidity of biogenesis on Earth can be used for estimating the probability of the origin of life on terrestrial planets with the same unknown probability as the origin of life on Earth. In addition, our current knowledge of extremophilicity, the capability of life to emerge in extreme conditions [21.42], is a basis for discussing the possible emergence of life in environments other than the terrestrial ones, as explained in the work of Joseph Seckbach [21.46]. The question of panspermia—the suggestion of how life could be transported in the universe, rather than the origin of life itself —first raised by the Swedish Chemist Nobel Laureate Savante Arrhenius [21.1], has also some bearing on the distribution of life in the universe.

21.1.4 The Destiny of Life in the Universe

The fourth chapter of astrobiology is bound to be a significant frontier with biosemiotics. With Europe's Euclid space telescope launched on a mission, we will enjoy an improved understanding of dark energy (responsible for acceleration of the expansion of the Universe) and dark matter (responsible for the growth of cosmic structures). It will measure the acceleration of the universe by observing how it evolved over the past 10 billion years [21.22]. This new vision of the cosmos will not only add depth to our understanding of the past in the universe, but it will also give new insights into the (future) destiny of life in the universe. Thus, biosemiotics will be appreciated on a considerably wider canvas. An additional question was pointed out by Frank Drake and Dava Sobel [21.19]: "If the capability for total destruction lay within the reach of every technologically advanced civilization, how likely were we to find anybody out there?"

Finally, habitability of the universe has been discussed by Sagan [21.43] and Chyba [21.11]: the main point made by both authors is that only the study of a Second Genesis can help us to side-step the difficulties of arriving at definite conclusions based on our understanding of terrestrial biology.

21.2 Philosophical Overlapping of Humanities and Astrobiology

21.2.1 Epistemology

Most philosophers would agree that this branch of philosophy concerns the study of the nature and the frontiers of human knowledge. Its relevance for astrobiology is evident, as soon as we raise the question: How certain are we of the existence of life in other worlds, inside and outside the solar system? Neither efforts with space travel to reach habitable environments, nor direct visual contact are possible at the present time.

21.2.2 Cognitive Science

Some of the questions that are raised within the scope of cognitive science include the following three: What is intelligence and how could its diverse evolutionary stages be compared with what we know in our own world? Are there universal laws for the evolution of intelligence? If the projects for establishing bases on the Moon and Mars are successful, would there be changes in the human mind?

21.2.3 Semiotics

Astrobiology should also consider another significant frontier, namely semiotics, especially as far as the life sciences are concerned. Communicating information amongst living beings is a subdiscipline called biosemiotics [21.9], which includes zoosemiotics and phytosemiotics. These areas have been considered in the past, strictly in the context of the evolving Earth, and the evolution of life. However, astrobiology research forces upon us the consideration of information exchange, either intentionally or unintentionally, between terrestrial receivers and emitters from other worlds [21.21]. We suggest that this extension of biosemiotics could receive the name of astrobiosemiotics, taking into consideration our search for other intelligence in the cosmos. In this work we intend to discuss different aspects of astrobiosemiotics that are suggested by the science of astrobiology.

21.3 Three Remarkable Frontiers of Astrobiology: Semiotics, Biosemiotics and Astrobiosemiotics

21.3.1 New Frontiers of Astrobiology

Returning to the second and third chapters of astrobiology—discovering whether evolution of life has taken place elsewhere—would open a new frontier of astrobiology [21.6]. This achievement would be beyond previous philosophical aspects, closer to the main topic of the present book. Codes are understood as symbols for meaningful communication. Indeed, we could be confronted with a system of standardized signals for conveying information from non-human animals, or from beings from an exoplanet, or even an exomoon, sometimes also referred to as an exosatellite. In any case, these worlds would be separated from us by distances of at least several light years.

These possibilities would put us in touch with a new field of study. First, we should recall what we mean in this new frontier of astrobiology: Semiotics is a philosophical doctrine that covers signs and symbols in relation to languages. But signs and meanings exist, not only in human minds and texts, but also in living organisms [21.49]. The study of the production and interpretation of signs and codes and their communication in the living world is referred to as "biosemiotics" [21.9]. In this work, we include the possibility of the existence of other living beings from an extra solar system (astrobiosemiotics).

21.3.2 What Does Astrobiology Share with Astrobiosemiotics?

Astrobiology is an interdisciplinary science that not only concerns itself with the origin, evolution, and distribution of life in the universe, but it is much richer in its contents. This is evident when we recall the frontiers of astrobiology with the humanities, especially ethics, semiotics, and theology.

Following the Estonian biologist Jakob von Uexküll [21.49], in biosemiotics, we assign a different significance to the term "meaning," as used in either bioethics or astrotheology. This is clearly stated in the Preface to this book:

> "In biosemiotics, signs and meanings are considered in a much wider sense than in linguistics, and include non-mental signification. A biosemiotic sign is a tool that thanks to its form rather than matter or energy prompts an agent (e.g., organism, cell, or protein) to perform some meaningful activity or change, which tends to be beneficial to the agent itself or associated agents [...]."

Consequently, astrobiosemiotics is a hybrid science that ranges from philosophy (i.e., philosophy of signs, or semiotics) to science (astrobiology). In other words, success of a first contact would render biosemiotics as an additional chapter in the already established new movement known as "experimental philosophy," according to Knobe and Nichols [21.30].

21.3.3 Astrobiosemiotics and Other Humanistic Studies

Unlike astrobiology, or astrobioethics, the significance of meaning in astrotheology is based on revelation, or tradition [21.5] [21.10]. Hence, in astrobiology there is an intrinsic novelty regarding the question of meanings: When the phenomenon of life is viewed from a cosmic perspective, the basis of theology or any of its extensions is irreducible to an experimental approach, unlike astrobiosemiotics, biosemiotics, or bioethics. Therefore, from the point of view of theology meanings are intrinsically related to faith alone, since meanings would be interpreted in terms of revelation [21.37]. Understanding the above distinction between a philosophical concept—semiotics—and any epistemological discussion of theology, serves as a basis for avoiding misinterpretations in the fundamental dialogue of science and religion, which is often misdirected by not only religious writers, but especially by scientists abusing their own discipline, when they go beyond the boundaries of their own fields. (But we underline that the New Atheists [21.16] [21.17] [21.27] [21.29] are entitled to their own opinions, provided they discard their scientific cloak, which unfortunately is not always the case.)

21.4 Is It Reasonable to Suppose a First Contact Will Influence Astrobiosemiotics?

21.4.1 Should Aspects of Humanity be Extended Beyond Our Own Species?

We adopt the Australian ethicist Peter Singer's proposal (paraphrased in a different context): In certain cases, basic aspects of humanity should be extended to genera beyond our own species [21.48]. In his ethical context, the extension is confined to non-human animals. We take a step beyond, now including Singer's limited extension to forms of evolution elsewhere in the cosmos, beyond non-human animals.

In addressing all of life, as we do in astrobiology, we could move beyond the ideas of the Australian ethicist, which were intended to apply specifically to animals. Indeed, this is a topic that has attracted much attention among astrobiologists and philosophers. Those efforts are attempting to develop ethics to apply to all life [21.12] [21.13] [21.40].

We find ourselves in this line of thinking, strictly confined to the possibility of life in the solar system and base our work on the Australian moral philosopher, especially since he has carefully discussed the basis for a principle of equality (a principle of equal consideration of interests, cf. Glossary, Table 21.2): It is an all-embracing principle on which we can base a defense of all humans. In the context of astrobiology, we even bypass all the implications that are fundamental for the welfare of humans, and relegate them to the background in our present attempt to develop an ethical framework that would be acceptable and applicable in our future relations with potential extraterrestrial life forms.

Singer's discussion is based on the earlier work of the moral philosopher John Rawls, who developed his ideas in his major work: *A Theory of Justice* [21.41]. With the principle of equality discussed by Singer (cf. Table 21.2), we extend it into the context of astrobiology *beyond* non-human animals. Indeed, we also assume that [21.48]:

> Having accepted the principle of equality as a sound moral basis for relations with others of our own species, we are also committed to accepting it as a sound moral basis for relations with those outside our own species—the non-human animals.

21.4.2 Are We Able to Receive Signals from Other Worlds?

This topic is part of the wider issue of biosignatures for astrobiology discussed by Cavalazzi and Westall [21.3], but in the context of this chapter we refer to the work of Duner [21.20]. Keeping in mind the question of semiotics of biosignatures, we should restrict ourselves to the Search for Extraterrestrial Intelligence (SETI). This is a time-honored research project that is fundamental in our discussion of the future of semiotics. A possible contact is expected to be in terms of signals received by means of specialized radio-telescopes. In the words of Frank Drake and Dava Sobel [21.19]:

> So much has happened...since SETI began—all of it good. We have seen clear evidence of the unrelating metamorphosis of cold and barren interstellar clouds into rich planetary systems. Here the glimmer of the faint beacons of intelligent life may overcome the darkness and still be detectable amid the cacophony of unintelligent signals from the objects of the mechanical universe.

Drake's conviction has always been shared by a significant part of bioastronomers, as testified in the large volume *A Roadmap for the Search for Extraterrestrial Intelligence,* SETI 2020 [21.21].

21.4.3 Evolutionary Convergence

Another major idea coming from the life sciences, may be relevant to biosemiotics. Natural selection can be sufficiently efficient to give selective advantage to organisms that develop nervous systems and brains, independent of the details of their morphology. This type of evolutionary convergence has been strongly opposed by the distinguished American paleontologist and evolutionary biologist Stephen Jay Gould [21.25]:

> Arguments of this form [a metaphor of rewinding the clock of evolution] lead me to the conclusion that biology's most profound insight into human nature, status and potential lies in the simple phrase, the embodiment of contingency: Homo sapiens is an entity, not a tendency.

In contrast with this view that emphasizes contingency, the English paleontologist, evolutionary biologist and astrobiologist Simon Conway Morris strongly supports the effect that evolutionary convergence might force upon us [21.14]:

> What we are interested in is not the origin, destiny, or fate of a particular lineage, but the likelihood of the emergence of a particular property, say consciousness. Here the reality of the tape of life, to use Gould's metaphor, can be run as many times as we like and in principle intelligence will surely emerge.

What Conway Morris is mainly interested in is not the possible emergence of alternative forms of intelligence on Earth, but rather, the possibility of other intelligences in a context relevant for astrobiology. Two examples of alternative forms of non-human intelligence are: the dolphin, pointed out by Lori Marino [21.33], and the octopus discussed by Peter Godfrey-Smith [21.24].

In passing, it should be recalled that evolutionary convergence can be tested in the forthcoming decade during the exploration of the Jovian moon Europa [21.4]. Evolutionary convergence is an aspect of the evolutionary phenomenon that constrains randomness. Christian de Duve, the Belgian Nobel Laureate, further supports the inevitability of life's emergence [21.18]:

> Life is an integral part of the universe; it is even the most complex and significant part of the known universe. The manifestations of life should dominate our world picture, not be excluded from it.

21.5 Astrobiosemiotics after Contact: A Tentative Framework

21.5.1 Speculations on the Content of a Message from a First Contact

Regarding the question of meaning, from the point of view of philosophers (existentialists and radical atheists), the emergence of life is the result of random events in an otherwise lifeless cosmos. It has been argued that a first contact probably would not alter the present skepticism [21.28].

Against this background, in order to build a framework, we consider two topics:

Firstly, research on semiotics is also relevant to one of the central problems of astrobiology. For human communication various systems of signs are used. A question is forced upon us as to the system of communication that would be needed for establishing contact with an extra-terrestrial civilization. The term semiotics is used for the general science of signs. The SETI Project raises the question of the interpretation of an interstellar message in the future by a first contact.

Secondly, at a deeper level, Sagan [21.44] pointed out possible difficulties that we could encounter in the future by a contact with another intelligence. Sagan presents us an intriguing imaginary scenario, if we were to receive a message within the context of a SETI program, which besides the remarkable implications for astrobiosemiotics, could expose implicit signals of another intelligence, which could unequivocally be from intelligent beings:

> "This is not a beacon. It's an announcement signal ...We get strange patterns of pulses from quasars and pulsars...But prime numbers are very specific, very artificial...It is hard to imagine some radiating plasma or exploding galaxy sending out a regular set of mathematical signals like this. The prime numbers are to attract our attention."

21.5.2 Implications of a First Contact for Astrobiosemiotics

For astrobiology, there is the possibility of probing for the presence of other life in the universe. Such a discovery would have an important implication: We would have to extend the existence of living organisms to evolution of life elsewhere. The success of the SETI Project will not be to receive a communication from elsewhere in the universe, but rather such contact would imply that meanings do exist for living beings, other than those confined to the planet Earth. Meanings need not be constrained to humans alone. We have taken one further step and have proposed to extend meanings to all forms of evolution at the level of any complex organism anywhere in the cosmos.

21.6 Conclusions

With the few examples we have discussed, it is evident that the time-honored topic of biosemiotics acquires a special relevance in the context of astrobiology, a relatively new science that is concerned with the search for life in the universe. The new term astrobiosemiotics intends to cover an important area that lies beyond the more restricted area out of which semiotics arose. The prefix "astrobio-" in the new term we have introduced is justified, since in the so-called marriage of natural sciences with semiotic work, astrobiosemiotics is appropriate, We do know, in principle, how to make contact with other intelligences. This potential contact could eventually be achieved following the pioneering work of the late Frank Drake.

Indeed, there is an observational technique—a radio telescope—bringing astrobiosemiotics to a position of being able to be tested. Other borderline subjects, namely, astrotheology, by its very nature relying on tradition and revelation, differs from the other two approaches discussed in this chapter (astrobioethics and moral philosophy) regarding the question of meanings. Astrotheology does not aim at experimental, or even observational testing, being unnecessary for a branch of theology.

Acknowledgments

The author would like to thank Dr. Atish Dabholkar, the International Atomic Agency and UNESCO for their hospitality at The Abdus Salam International Centre for Theoretical Physics, Trieste, where this chapter was completed.

References

[21.1] Arrhenius, S. (1908) *Worlds in the Making: The Evolution of the Universe*. Harper & Row, New York.

[21.2] Bendall, G.S., ed. (1983) *Evolution from molecules to men,* Cambridge University Press, London.

[21.3] Cavalazzi, B. and Westall, F. (eds.) (2019*) Biosignatures for Astrobiology*. Springer, Dordrecht. ISBN: 978-3-319-96175-0.

[21.4] Chela-Flores, J. (2003) Testing Evolutionary Convergence on Europa. *International Journal of Astrobiology* 2, 307-312.

[21.5] Chela-Flores, J. (2005) Fitness of the universe for a second genesis: Is it Compatible with Science and Christianity? *Science and Christian Belief*, 17 (2), 187-197.

[21.6] Chela-Flores, J. (2011) *The Science of Astrobiology A Personal Point of View on Learning to Read the Book of Life*. Springer, Dordrecht, The Netherlands, 2nd.

[21.7] Chela-Flores, J. (2019) *Astrobiology and Humanism: Conversations on the frontiers of science, philosophy and theology.* Cambridge Scholars Publishing, Newcastle upon Tyne, UK.

[21.8] Chela-Flores, J. (2023) Frontiers of astrobiology and the humanities. *International Journal of Astrobiology* 22 (3), 229-237. DOI: https://doi.org/10.1017/S1473550422000453.

[21.9] Chon-Torres, O. A. (2021) Disciplinary nature of astrobiology and astrobioethic's epistemic foundations. *International Journal of Astrobiology* 20, 186–193. https://DOI.org/10.1017/S147355041800023X

[21.10] Chon-Torres, O. A. and Szocik, K. (2021) A brief epistemological discussion of astrotheology in the light of astrobiology. *International Journal of Astrobiology* 21, 1–8. DOI:10.1017/S1473550421000367.

[21.11] Chyba, C. F. (1997) The Origin of Life in a Cosmic Context. In: *Carl Sagan's Universe.* Y. Terzian and E. Bilson (eds.). Cambridge University Press, Cambridge, UK: 64-74. ISBN 0-521-57286-X

[21.12] Cockell, C. S. (2005) Duties to Extraterrestrial Microscopic Organisms. *Journal of the British Interplanetary Society* 58, 367–373.

[21.13] Cockell, C. (2011) Ethics and extraterrestrial life. In: *Humans in Outer Space—Interdisciplinary Perspectives.* U. Landfester, N.L. Remuss, K.U. Schrogl, J.C. Worms (eds.). Studies in Space Policy, 5. Springer, Vienna: 80-101. DOI: https://doi.org/10.1007/978-3-7091-0280-0_7

[21.14] Conway-Morris, S. (1998) *The crucible of creation,* Oxford University Press, #14

[21.15] Dawkins, R. (1983) Universal Darwinism. In: *Evolution form molecules to men,* D.S. Bandell (ed.), Cambridge University Press, London: 403-425.

[21.16] Dawkins, (2007) *The God Delusion*. Black Swan, London, UK. ISBN 978-0-552-77429-1.

[21.17] Dennett, D. (2007) *Breaking the Spell: Religion as a Natural Phenomenon*. Penguin, London, p. 464. ISBN 978-0141017778.

[21.18] de Duve, C. (1995) *Vital dust: Life as a cosmic imperative*. Basic Books, 291.

[21.19] Drake, F. and Sobel, D. (1992) *Is Anyone Out There? The Scientific Search for Extraterrestrial Intelligence*. Dalacorte Press, NY, pp. 62 and 235.

[21.20] Duner, D. (2018). Semiotics of Biosignatures. *Southern Semiotic Review*, Issue 9, 47- 63.

[21.21] Ekers, R. D., Kent Cullers, D., Billingham, J. and Scheffer, L. (eds.), (2002) *SETI 2020: A Roadmap for the Search for Extraterrestrial Intelligence*, SETI Press, The SETI Institute, California, USA.

[21.22] Euclid Mission (2021) Euclid telescope ready for extreme space environment. https://www.esa.int/Science_Exploration/Space_Science/Euclid_telescope_ready_for_extreme_space_environment.

[21.23] Gilbert, W. (1986) Origin of life: The RNA world. *Nature* 319 (6055), 618.

[21.24] Godfrey-Smith, P. (2018) *Other Minds: The Octopus and the Evolution of Intelligent Life*. William Collins.

[21.25] Gould, S. J. (1989) *Wonderful Life: The Burgess Shale and the Nature of History*, London UK, Penguin Books, #320.

[21.26] Grasset, O., Dougherty, M. K., Coustenis, A., Bunce, E. J., Erd, C, Titov, D., Blanc, M., Coates, A., Drossart, P., Fletcher, L. N., Hussmann, H., Jaumann, R, . Krupp, N., Lebreton, J. P., Prieto- Ballesteros, O., Tortora, P., Tosi, F., Van Hoolst, T. (2013) JUpiter ICy moons Explorer (JUICE): An ESA mission to orbit Ganymede and to characterise the Jupiter system. *Planetary and Space Science* 78, 1-21.

[21.27] Harris, S. (2004) *The End of Faith: Religion, Terror, and the Future of Reason*.W.W. Norton & Company, New York, USA. ISBN 978-0-7432-6809-7.

[21.28] Haught, J. F. (2001) Theology after contact: Religion and Extraterrestrial Intelligent Life. *Ann. NY Acad. Sci.* 950, 296-308.

[21.29] Hitchens, C. (2019) *The Four Horsemen: The Conversation That Sparked an Atheist Revolution*. New York: Random House. ISBN 978-0-525-51195-3.

[21.30] Knobe, J. and Nichols, S., eds. (2009) *Experimental Philosophy*, Oxford University Press, United Kingdom, vols. 1 and 2.

[21.31] Lammer, H., Bredehöft, J. H., Coustenis, A., · Khodachenko, M. L., Kaltenegger, L., Grasset, O., Prieur, D., Raulin, F., ·Ehrenfreund, P, Yamauchi, M., Wahlund, J.-E., Grießmeier, J.-M., Stangl, G., Cockell, C. S., Kulikov, Yu. N., Grenfell, ·J. L. and Rauer, H. (2009) What makes a planet habitable? *Astron Astrophys Rev.* 17, 181–249.

[21.32] Lineweaver C. H., Davis TM. (2002) Does the rapid appearance of life on Earth suggest that life is common in the universe? *Astrobiology* 2(3), 293-304. DOI: 10.1089/153110702762027871.

[21.33] Marino, L. (1998) A comparison of encephalization between odontocete cetaceans and anthropoid primates. *Brain, Behavior and Evolution* 51 (4), 230-238.

[21.34] McKay, C. P. (2001) The Search for a Second Genesis in the Solar System. In: *The First Steps of Life in the Universe*, J. Chela-Flores, T. Owen and F. Raulin (eds.). Kluwer Academic Publishers, Dordrecht, The Netherlands: 269-277.

[21.35] Miller, S. L. (1953) Production of amino acids under possible primitive earth conditions. *Science* 117, 528–529.

[21.36] Pappalardo, R.T., S. Vance, F. Bagenal, B.G. Bills, D.L. Blaney, D.D. Blankenship, W.B. Brinckerhoff, Connerney, J. E. P., Hand, K. P., Hoehler, Tori M., Leisner, J. S., Kurth, W. S., McGrath, M. A., Mellon, M. T., Moore, J. M., Patterson, G. W., Prockter, L. M., Senske, D. A., Schmidt, B. E., Shock, E. L., Soderlund, K. M., Smith, David, E. (2013) Science Potential from a Europa Lander. *Astrobiology* 13, 740-773.

[21.37] Peters, T. (ed.) (2018) *Astrotheology: Science and Theology Meet Extraterrestrial Life*. Cascade Books, Eugene, Oregon, USA.
[21.38] Pigliucci, M. and Müller, G.B. (2010) Elements of an extended evolutionary synthesis. In: *Evolution - The Extended Synthesis*. M. Pigliucci and G.B. Müller, (eds.) MIT Press, Cambridge MA, London: 3-17.
[21.39] Ponnamperuma, C. and Chela-Flores, J. (eds.) (1995) *Chemical Evolution: The Structure and Model of the First Cell*, Kluwer Academic Publishers, Dordrecht, The Netherlands.
[21.40] Randolph, R., and McKay, C. (2014) Protecting and expanding the richness and diversity of life, an ethic for astrobiology research and space exploration. *International Journal of Astrobiology* 13, 28–34.
[21.41] Rawls, J. (1971) *A Theory of Justice*. Harvard University Press, Cambridge Mass.
[21.42] Rothschild L. J. and Mancinelli R. L. (2001) Life in extreme environments. *Nature* 409 (6823), 1092-101. DOI: 10.1038/35059215.
[21.43] Sagan, C. (1974) The Origin of Life in a Cosmic Context. *Orig. Life Evol. Biosph.* 5, 497-505.
[21.44] Sagan, C. (1985) *Contact*, Pocket Books, a division of Simon & Schuster Inc., New York, p. 76.
[21.45] Seckbach, J., Chela-Flores, J., Owen, T. and Raulin, F. (eds.) (2003) *Life in the Universe: From the Miller Experiment to the Search for Life on other Worlds*. Kluwer Academic Publishers, London, 387 pp. ISBN: 978-1-4020-2372-5.
[21.46] Seckbach, J and Chela-Flores, J (2011). Biosignatures for detecting unicelluar and multicellular organisms in extreme environments Invited Paper: *Instruments, Methods, and Missions for Astrobiology XIV Conference OP409*, 21 - 25 August, San Diego Convention Center, San Diego, California United States.
[21.47] Segré, D., Ben-Eli, D., Deamer, D. W. and Lancet, D. (2001) The lipid world. *Orig. Life Evol. Biosph.* 31 (1-2), 119-145.
[21.48] Singer, P. (1993) *Practical Ethics*, Cambridge University Press, United Kingdom, 2nd Edition. 55-62.
[21.49] von Uexküll, J. (1982) The Theory of Meaning, *Semiotica* 42(1), 25-82.

22

Time Horizons and Biosemiotic Adaptation: Taking Seriously Variable Temporalities in the Evolution of Possible Life Forms

Yogi H. Hendlin[1,2,3*] and Constantijn-Alexander Kusters[1]

[1]*Erasmus School of Philosophy, Erasmus University, Rotterdam, Netherlands*
[2]*Dynamics of Inclusive Prosperity Initiative, Erasmus University, Rotterdam, Netherlands*
[3]*Environmental Health Initiative, University of California, San Francisco, California, USA*

Online resources:
www.yogihendlin.com
https://nl.linkedin.com/in/constantijn-kusters-41a16572

Abstract

Semiotics occurs beyond our planet; that we can be sure of. Astrobiology requires astrosemiotics; and as life on this planet may well have arrived from an extraterrestrial source, rather than (only) endogenously, we must consider the basis for extraterrestrial semiosis. This extraterrestrial semiosis suggests that the adaptability of life is far greater than conventionally considered. Similarly, the role variable temporalities have in adapting to divergent astroecological conditions also requires reassessing Einstein's taking for granted the universality of time in the ontological extension of relativity theory. Following Bergson and Whitehead's rejection of temporal universality we focus attention on the importance of variable temporal *Umwelten* to understanding astrobiological but also terrestrial contexts. Especially in terms of the massive die-offs in the current anthropogenic Sixth Great Extinction, now involving aided adaptation moving plants and animals around to viable habitats to perpetuate their species, a better conception of time horizons in semiosis, adaptation, and fitting deserves investigation. If lifeforms in theory are infinitely adaptable (even across different planets), but in practice require different timespans to adapt to changing environmental conditions, then *time-appropriate semiosis* offers a route towards ethical space exploration, decelerating the unreflective current craze of quick space exploration and colonization. Slow space exploration—starting with stabilizing the biotic conditions on our own planet—presents a biosemiotically-informed perspective on how organisms of various complexities can fit with (drastically) new environments with different temporal needs. By looking at the variances that time embodies in our own terrestrial settings we argue that astrobiology has much to gain from

**Corresponding author*: hendlin@esphil.eur.nl

considering radicalizing temporal differentiations found at home. At the very least, comparing possible temporal distinctions can assist the practice of locating blind spots resulting from our anthropocentric outlook. The implications for astrobiology of the different time horizons of various complex organisms requires rethinking and coming to terms with the more conservative rates of change possible for complex organisms, suggesting that any future space exploration might have to contend with unintuitive dimensional variations.

Keywords: Evolution, temporality, organisms, astrobiology, astrobiosemiotics, thought experiments dimensionality

> *"What is perceived is not the nature of things, but their action upon us, and what is known or believed of this nature is beyond the range of intuition."*
>
> – Friedrich Schleiermacher [22.45]

22.1 Introduction

It would be a mistake to believe that the same varieties of bacteria and other microbiota that accompany us on this planet can be found on other planets. Assuming the extra-planetary origin of life hypothesis, known as panspermia [22.53], which supposes that microorganisms arrived on asteroids and somehow were able to make home on this planet [22.47], they still would have had to adapt quickly to the very particular mix of gravity, solar exposure, chemical composition, and elemental atmosphere here on Earth [22.17]. Not quite yet an ecosystem, such an environment bereft of life still maintained strong attractors.[2]

Whether organisms encountered on other planets are motile like animals or sessile like plants will likely determine their ability to manipulate matter in ways similar to how organisms on earth have. These attributes may in turn be based on how strong gravitational pulls of a given planet are, amongst other factors. A dense, high-gravity environment will likely favor organisms that stay close to the planet's surface (whatever their equivalent of terrestrially or aquatically might be); for example, rather than expending the tremendous metabolic energy to grow skywards or move quickly. What such thought experiments can teach us, is that the many aspects of our *Umwelt* that we take for granted may not be obtained, or exist in unrecognizably different form, for life on other planets.

Exploration of biosemiotic analysis of astrobiology is not new. David Dunér [22.19] already notes that in evaluating biosignatures we must take note of both the variability of possible signs and our limited capacity for interpreting biosignatures as biosignatures. This chapter seeks, in part, to expand on this problematization by breaching the topic of time in biology as a variable which brings about both biological and cultural complexities in recognizing and meaningfully engaging with astrobiological semiosis.

Taking up an ecosemiotic motif, Baird Callicott [22.10] mentions how "[r]eality is mediated by our species-specific sensory receptors and by both species-specific and culturally variable conceptual frameworks." Applied to the purposes of this chapter, this suggests that

---

[2] Such a pre-Daisyworld Earth would have required some sort of semiotic fitting for the arriving organisms and the barren and turbulent environment they encountered [22.54]. Of course, once sufficiently adapted to survive, the organisms promptly began going about their metabolic business, which in turn geoengineered the planet to their benefit, sowing an environment conducive for further evolution.

our understanding of the bounds of space and time is relative to how we perceive both as a particular type of organism in a particular environment. Thus, if we posit life on another planet, not only are there varieties of physical, chemical, biological, and biophysical constraints and affordances at play in determining time perception and time scales, but cultural variation as well.

The importance of this latter variable can be seen in the ways that cyclically versus linearly oriented time structures are assumed in cultures of different peoples on Earth. The cyclical time of the Mayan calendar, or the Yugas of Hindu religion in which death and rebirth are understood as reciprocal nodes in a feedback framework, contrast with the linear time of Christianity beginning with the creation of the world and ending in Apocalypse. These contrasting temporalities are but a glimpse into the possible schemata which would entirely change alien organisms' experience of time.

On Earth, we have microbes that reproduce in 20 minutes and Redwood trees that enjoy millennia-long lifespans. It is possible to imagine that on a planet with a much longer orbital period such changes would alter growing seasons, create animate and inanimate variation of state, such as tardigrades in outer space able to induce cryptobiosis [22.23], stopping all metabolic processes, and then springing back into metabolizing when returned to hospitable ecological conditions.

We can also see extreme variation occurring in the rate of genetic changes, with bacteria reproducing every 4 to 20 minutes [22.33], their rates of adaptive evolution can be much faster than those of humans with much slower generational cycles [22.4]. If an intelligent complex organism such as humans were able to reproduce many orders of magnitude faster, adaptation would produce incredible creatures. If trees could evolve their genetics at the rate of bacteria, they would likely outrun the evolutionary arms race with their blights, wilts, mildews, and rots, such as anthracnose fungi. If human genetic evolution occurred at bacterial speed, we cannot imagine the result, perhaps completely obviating our dependency on external technologies.[3]

The debate in animal philosophy surrounding whether some animals feel more pain from the same injuries that we can suffer [22.41] [22.57] [22.58], raises the possibility that our ability to rationalize and adapt our preferences cognitively minimizes or mitigates suffering, helping us deal with degraded conditions. In contrast, other animals may suffer much more intensely without the same supervening cognitive abilities our species commands to rationalize, justify, or consciously minimize abject pain [22.30] [22.46]. Such considerations are relevant for biosemiotic approaches to astrobiology, because we cannot know *prima facie* what sort of interactions may be disruptive or harmful to another alien species. Calls to investigate the ethics of *a priori* ignorance of the effects of astrobiological research on extraterrestrial life are not new. Worries about the effects a fission surface powered (a type of nuclear reactor) cryobot might have on unknown life on Mars (via radiation poisoning)

[3] Extended minds take a lot of effort to build, maintain, and use (such as computers and smartphones). If an organism could remember and recall this information *sui generis*, or be connected to another form of internet (via rhizomes, for example) without external devices, capacities, judgment, and decision-making could be greatly improved. (This doesn't even take into account the inefficiencies of external technologies as not actually answering the question we're asking specifically, or giving us a bunch of noise along with the signal.)

has, for instance, been suggested as a component of astrobiological research ethics dealing with this *a priori* ignorance [22.12].

As long as our experience and knowledge of harm to extraterrestrial life is limited (though one could speculate without much effort that generating space trash might have biosemiotically unethical consequences [22.51]), our precautions will have to be based initially on the unintended harm we find out about on our own planet. The goodwill to be nonviolent (*ahimsa*) can ground our action in the precautionary principle, inspired by an analysis of unintended terrestrial harm while attempting to lessen the impact of our anthropocentrism. As it stands, we have much to learn from both the intended and unintended harm we are causing right here on Earth to terrestrial organisms. We might, for instance, consider that seemingly innocuous forms of communication could actually be experienced as harmful death rays, in the same way that certain sound frequencies at sufficient wattage can put out fires [22.25], or how whales can go deaf from the sonic warfare training and testing of belligerent navies or oil exploration sonar [22.55]. Recognizing harmful externalities of our instrumental actions at home can therefore already gear us towards the types of considerations necessary for exploring the ethical concerns of astrobiological research and its possible harm on other planetary systems.

Similarly, we need to face up to the fact that we may be outmaneuvered. It is quite possible that we are continuously interacting with living beings we are almost totally unaware of, who are having a decisive impact on our experience of reality. It might in this sense not only be us who have an effect on other life, but also other life knowingly or unknowingly having an effect on us. The dimensions of biosemiosis are far wider than a single stream.[4]

This thought experiment can be further explored through the claims that there are ten, eleven, or even twenty-six dimensions reported by string theorists. Whatever the exact number, if indeed there exist more spatial-temporal dimensions than we normally take into account, this could have far-reaching consequences. Except for certain parts of physics, our chemical and biological, let alone our psychological and social (not to mention our political, international, and economic) sciences, operate as if only four dimensions exist, and could possibly have any bearing on the outcomes we observe. It is unlikely that those other dimensions are simply unused, extra, or optional, like we used to erroneously think was the case with "junk" DNA [22.42]. Instead, what if those other dimensions are not only teeming with activity, but are in many ways supervening on our activity, just as our activity (superseding the four dimensions we normally consider their impact domain) may supervene on theirs?

While at first glance an excess of spatiotemporal dimensions might seem like a far stretch, it is no more unlikely than it would be for a medieval farmer to believe that the human body, which appears to be an integrous whole, is actually made up of tens of thousands of different species of organisms, in addition to our endlessly differentiated human cells.

---

[4] "The interactivity of semiotic phenomena produces a sensorially *overdetermined experience for the organism* (i.e., more is going on than can be perceived, consciously or otherwise), and an *underdetermined individuated sign* accessible to reified analysis (i.e., analysis of any sign cannot possibly take into account the total influencing/interceding diffractions resulting from sign interactions, both in parallel and across channel levels)" [22.60]. Dimensionality has several overlapping meanings as we use it here. We mean quite plainly different dimensions of space and time, in a purely physicalist sense. But we also mean that there are dimensions (or contributing factors) to biosemiosis which we are often completely unaware of – such as the impact of biophysics (and light) on biosemiosis, including on our own sense of mind or self.

The difference is that we currently have no reliable way to measure the existence or causal interactions of the phenomena that occur between the dimensions we account for, and those we do not (so far).

The other dimensions that string theorists postulate leave plenty to be explored without leaving home. Without positing that we reorient biology and other disciplines to get overly distracted with dimensions that we have not yet found how they interact with what we already know, instead, we highlight these discrepancies between disciplines (i.e., physics and biology), clarifying that the epistemic models we are using are always going to be behind a fuller picture we might sketch in the future. For astrobiological research, which specifically aims at knowledge beyond our current ken, this might then require an uneasy principle of cognitive openness balanced with what can be defended as reasonable avenues of investigation.

We are still discovering new processes and pathways of health and disease in our own endosemiotic microbiota. For example, the existence of the neuronal microbiome itself is very recent, yet with weighty implications [22.26] [22.35]. If we are inhabiting (and inhabited by) differential consortiums of bacteria and other microorganisms in our brains, then how do varying consortia influence thought and action? How do different configurations of microbiota consortia in our heart microbiome influence our sense of feeling (and thinking)? And how do all these constellations of various "omics" (the study of different aspects of the totality of cellular biological processing) [22.40] work together to produce the end result of health or disease, capacities or disablements?

These are not only pressing questions for better understanding the interspecies origins of all thought, feeling, and action, but necessary for understanding astrobiological theorizing as well. What is the probability that extra-planetary organisms will be analogs to eukaryotes, especially those that are composed of different species of organisms? Can we expect intelligibility if life on earth is relatively individuated and we encounter much more ecologically integrated exo-organisms evolving according to a type of extreme Gaia or yoked Daisyworld model in which the entire ecosystem is integrated into what we would consider a single "mega organism" [22.27]? What does it mean if these exo-organisms have more democratic and egalitarian awareness of their interspeciality? Or where these organisms have solved the trick of parasitism, on the one hand, and on the other have a lower threshold of self- and other-distinctions, which does not encounter other organisms as separate? What would the result be if extra-planetary organisms are aware of the different metabolisms and reckonings of their fellow interdependent co-organisms? Would accounting for their differential life cycles, goals, frustrations, and perspectives give rise to some sort of emergent intelligence like our own, only much more sensitive, vulnerable, and gentle? There's every reason why an alien society that has conquered fear and separation would be thoroughgoingly nonviolent, and that the vast amount of science fiction speculations of life-desacralizing alien colonizers is little more than a projection of European colonialism's own bloody history.

With these questions in mind, in this contribution we take up the question of the role time plays as a variable in astrobiology. Time will be considered as irreducible to mathematics, or a single universalizable standard, by touching upon the difference between time in a mathematical and meaningful (semiotic) context. Following this, time will be considered in biological life as something which can take very different forms than we might expect from an anthropocentric point of view, a difference which could be further accentuated in astroecological contexts foreign to those we are familiar with in our solar system. Similarly, cultural influences, such as biological time differences, might have to be considered, showing that time as variable in cultural issues is already a highly relevant topic even in a purely anthropocentric context, let alone in astrobiological contexts. The final part of the chapter will consider dimensional uncertainties in astrobiology which go beyond time, problematizing the limitations of our expectations by reflecting on the possibility of unknown dimensionality. But first time will be considered as something more than a mathematical model.

22.2 Time as Model and Life

The famous Einstein-Bergson debates are illuminating to contextualize the role of time in astrobiosemiotics. Situated in the first quarter of the 20th century, their debate, while initially propelled by intentional international scientific cooperation to reduce the chance of war in Europe, these debates would come to be exemplar of the divide between the natural and humanistic sciences. Where at first both interlocutors didn't expect their differences to be so significant—after all, both were in agreement about the move from Newton's absolute time towards a new relative time—their eventual disagreement seemed to uncover a disagreement which had more to do with the status of time in relation to meaningful experience. Contemporaries would even come to interpret this renowned debate reductively as a debate between Vitalism (Bergson) and Mechanization (Einstein) [22.11]. Yet, neither Bergson nor Einstein ought to be dismissed as such, as Bergson's critique is closer to an epistemological issue with the reduction of the meaningfully experienced to mathematical models, while Einstein's relativity theory—groundbreaking as it is—does not require the metaphysical baggage thrust on it by Einstein in a scientific milieu which increasingly sought to distance itself from the humanities. As a change of paradigm, however, especially in the context of time, it does illuminate why time in terms of meaning-making is easily ignored, illuminating pitfalls opened up by this divide. When Albert Einstein and Henri Bergson focused on their respective theories of the relativity of time, Bergson did not in fact disagree with the predictive power of Einstein's mathematics but was fundamentally in disagreement with what he saw as the ontological extension unnecessarily added to this mathematical model. While Einstein distinguished only between physical and psychological time, Bergson argued that while yes, there is a distinction in time between clock and lived time, we cannot take a fixed notion of time as the objective time of science and have it transplant an authoritative ontology onto the time which slows down and speeds up in our everyday lived experience. Time experienced, for Bergson, clearly requires some additional consideration which cannot be subsumed under a model for relativity.

Bergson averred that contrary to the "twin" thought experiment, which posits time as a separate or separatable entity from our own existence, one cannot refer to a fixed third object—a clock, in this case—as the mediator of time. Pretending this third object (the clock) could be the grand arbiter of time *simpliciter* is a grave error in Bergson's view. He would later be proven wrong on the specific issue that the twin thought experiment is erroneous, as experience with atomic clocks and the necessity to consider time dilation with regards to satellites are unmistakable. Regardless, this does not necessarily disprove

Bergson's argument against the metaphysical speculation which he saw as unnecessary for the predictive power of Einstein's theory.

In *Duration and Simultaneity* [22.6], Bergson argues that clocks only serve to illustrate our own already embedded, species-specific form of apprehending duration. When we use a clock, it is in service to *our* understanding of time, which we have imbued with a generalizability and universality beyond our *Umwelt* (a so-called Archimedean point). Clocks, even atomic clocks, on their own, serve no other purpose; they are neither sacrosanct arbiters of some cosmic truth, nor is their reference point beyond the human. Humans make them, calibrate them, and use them as instruments for our needs. They have no interpreter without us. Even cosmic clocks, such as pulsars, insofar as they are picked up on our human-made sensors, become legible according to the categories we have already devised in order to make sense of the world. We translate their workings according to our own cataphatic categories. When they don't conform to our predictions and concepts of regularity, our tendency is to interpret them as random, or gloss over them as noise. Noise-to-signal ratios of signal clarity are always species-specific. What one species or resonant body cannot detect as a pattern gets written off as noise, while the exact same pattern may contain rich, invaluable actionable stimuli for another.

In distinguishing between Discrete (actual) and Continuous (virtual) Multiplicities, Bergson discusses how these different ways of attempting to co-track space and time according to quantity and measurement are recursive to a higher-order arbiter of countering or measuring that one must make reference to. But change in quantity, according to Bergson, does not and cannot occur without a concomitant change in *quality*, that is, changing the required instrument of measurement at each moment of change [22.16]. But time is not (only) a magnitude. It is perceived and experienced differently according to context. This is precisely the ancient Greek distinction between *kronos* (equally tempered cyclical time) and *kairos* (moment-quality temporality allowing for compression and stretching based on interpretant/observer state).

For Whitehead, the unwavering correspondence between space and time, and the existence of some god-like Archimedean point of clock-time commits both the fallacy of misplaced concreteness (reification) and the fallacy of simple coordinates (space-time simplification). Instead, "We must not slip into the fallacy of assuming that we are comparing a given world with given perceptions of it. The physical world is in some general sense of the term a deduced concept" [22.56]. To therefore assume from mathematical deduction the illusionary character of psychological time, or posit a final generalizable ontological status of space-time, is for Whitehead misunderstanding the relation between mathematics and physics. For Whitehead this ultimately comes down to a kind of mistaken "model-centrism." To combat this, he takes up an understanding of Natural time which begins from human time experience and only from there moves onto a Natural understanding of time. In doing so, time is given a "Epochal" thickness, which takes seriously the reality of the past and present in the now, but also teleological aspects of Nature. Such a description, even more so in the context of our ignorance of non-terrestrial matters of life, opens up to modes of temporal ordering not generally considered in an anthropocentric context. This is especially the case if, as Whitehead argues, our understanding of time is best understood from human experience, which, once again, makes clear the possibility of anthropocentric blind spots with regards to our understanding of time itself.

22.3 Semiotic Understandings of Time as Adaptation

From a biological perspective, possible variations between modes of temporal experience and biological organization are to be taken seriously as a sufficiently divergent set might reduce our ability to recognize non-earthlike temporal relations. The fact that the weight of the past and the promise of the future influence not just the psyche but also biological and ecological relations, shows that temporality as a biological concept plays an important role in structuring meaningful/semiotic relations. (This is related to the concept of supervenience.) Taking both Bergson's and Whitehead's critique to heart, we must try and move away from the idea that any one model can span the variations which space-time might play in all possible biological relations. Moving towards a less anthropocentric and more pluralistic understanding of time experience might thus be illuminating in situations where a general theory of relativity might come up short explanatorily, such as in cases concerning lifeworlds.

A strong early proponent for moving towards a less anthropocentric understanding of experience can be found in the Baltic German theoretical biologist Jakob von Uexküll (1864–1944), who is considered, among other things, the father of biosemiotics. In the first half of the 20th century, von Uexküll sought to establish an approach for the qualitative research of the subjective life world of animals which he called *Umwelt-Forschung*. Researching the *Umwelt*, or lifeworld, of animals was for von Uexküll, who saw himself as Kantian, not an investigation into some lifeworld which could be investigated in-itself. The Kantian tendency to not engage the *ding-an-sich* (thing-in-itself) led von Uexküll to consider that the research into the *Umwelt* could itself only be done through the lifeworld of the investigator themselves; in this sense, the *Umwelt* as something observed is always mediated through the eyes of the human observer, never reaching the asymptote of the *Umgebung*, or actual reality. Nor does *Umgebung*, according to von Uexküll, come to an animal as some kind of raw sense data which is immediately converted into an *Umwelt* and reacted to as some kind of automaton. Instead, von Uexküll showed through a variety of experiments that sense data itself was modified by the inner world (*Innenwelt*) of the animal, which subsequently had an effect on this *Umwelt* generation, creating a recursive loop between the *Umwelt* and *Innenwelt* [22.52].

Taking this framework vis-à-vis the relation between, on the one hand, space-time as model, and on the other hand, time as aspect of experience or biological organization, reminds us of the fundamental limitation of using *Umgebung* exclusively to get to species-specific *Umwelten* and *Innenwelten*. We might say that according to how our personal *Umwelt* rules out the possibility of coming to this mythically generalizable *ding-an-sich* at all, it also limits our capacity to reach a full understanding of the *Umwelten* of others including a generalizable temporality to which this *Umwelt* relates. This insight might then guide us as both a precautionary and exploratory principle.

What the biosemiotic approach can therefore add to this analysis of space-time is interpreting it interdisciplinarity as playing a role in the meaning-making process of biological systems and their *Umwelten*. Understanding that even if there exists an objective measure of time (*Umgebung*), we will not be able to apprehend it, is useful for expectation management. (Assuming that the *Umgebung* has a unitary sense of time itself may not be warranted.) Drawing on these considerations, biosemiotic analyses of time can investigate the possibility of non-human-centric models of meaningful temporality, without uncritically losing sight of our own human blind spots.

The fact that time plays a significant role in the *Umwelt* of animals was already established by Karl Ernst von Baer (1792–1876) whose research on the influence of time on the construction of the animal subjective lifeworld was later also used by von Uexküll to show time's role in the creation of an *Umwelt*. A snail who, for instance, is shown a slowly vibrating stick, will refrain from attempting to climb on top of it. If, however, the stick is made to vibrate quickly, the supernormal speed overrides the snail's ability to discern movement versus stasis, and it is thus interpreted by the snail as a non-vibrating stick, prompting the snail to once again try and climb on top of it as if it were stationary (predictably for us, but not the snail, this attempt proves unsuccessful) [22.52].

In an astrobiological context, such insights underwrite potentially radically different temporal relations emerging in astroecological contexts very different from the regularity of the earth-sun-moon cycles. While on Earth we are "lucky" that the regularity of circadian clocks can to a certain extent be "hard-coded" into the biological structures, this does not mean that our terrestrial circadian rhythms are not variably responsive to both fluctuating light/dark patterns and intracellular information exchange, and thus not completely set in stone [22.38]. Regardless, the fact that our evolutionary ancestors could for the most part rely on regularities, is more energetically efficient and significant in all terrestrial life at least as complex as a cell. This general rhythm is often even capable of slowing down the body significantly on seasonal or daily bases in more complex creatures, a highly advantageous adaptation to survive more extreme conditions. To deal with non-ideal circumstances, for instance, fish going into winter conditions will enter torpor to lower energy expenditure. The induced torpor state has been reported to provide not just advantages in energy economy but, for Zebrafish, has also been shown to reduce the effects of radiation exposure [22.9]. Astrobiologically, this opens torpor up as a possible adaptation for intermittent increased radiation exposure due to astronomical events. Similarly, fishing bats are known to habitually go into and out of torpor condition during winter months, although for these bats this relatively frequent switching in and out of torpor still tends to follow regular daily cycles [22.3]. State change based on predictable but never perfect periodicity is an incredible adaptation that might very well apply to many xeno-organisms.

From a non-terrestrial-centric point of view we might then consider that a semi-chaotic return of extreme conditions—thinkable in a multi-star system—might introduce extreme forms of temporal ebb and flow as an adaptive defensive mechanism against the semi-irregular passing orbit of a secondary or tertiary star. While we all understand that clock times will be different in variable astroecological contexts, this multi-periodicity can also be seen as a driver for different and irregular biological time adaptations. Some biological time cycles which might form due to unforeseen differences in terrestrial and non-terrestrial astroecological rhythms become almost incomprehensible in terms of our own biological makeup. While these differences might be unthinkable in our milieu, on our own planet this oblong use of multiple times, such as the irregular use of torpor, is a good example of the necessity to understand temporal relations in our own biosphere. It has already been proposed that the capacity for irregular torpor—as opposed to more strict seasonal torpor/hibernation—might show significant advantages for species exposed to the extremities of the mounting anthropogenic climate changes [22.39]. We suggest such time must not be seen only as an aspect of biological evolution, but even within the life of a single organism, as something possibly open to variation from moment to moment, or season to season. Therefore, such state changes must be critically investigated. Biology influenced by its relations to our star system's rhythms plays a significant role in evolutionary adaptation, making its possible variations in plausible forms of life an important and highly variable subject for astrobiology.[5]

22.4 Temporal Horizons

If we take some form of interstellar universality of physiochemical relations for granted, possible longevity-related outcomes of torpor analogs—or generally slower/faster biological structures—could create significant hurdles in our capacity to recognize extraterrestrial life. Additionally, it might also form serious barriers for interspecies intelligibility, even in the unlikely case that the biological consciousness and experiencing of time is comparable enough to expect intelligibility. While Star Trek's version of humanoid panspermia gave us the idea of a universe full of humanoids, ready and able to talk and think in our tempos, the chances of this being realistic are slim. Possible cultural implications of biological adaptations in a species are likely to create problems of reference we are ill-equipped to handle compared to the differences we are challenged with terrestrially. But, while nonhuman time horizons might throw up serious issues, even culturally driven interhuman temporal differences are capable of creating intelligibility issues even where biological difference between groups are relatively irrelevant. For example, one of the reasons why European conquest was so swift, was because of a notion of time and history based on linear progress, manifest destiny, and entitlement. We will touch upon some of these specific anthropocultural aspects later on. Clock time—as a time with which the experienced time is culturally enmeshed, yet is a human construction—contextualizes the temporal experience much further than the just happened and nearly there, influencing the experience of the "now" in relation to history. For instance, we contextualize our lives in relation to the monuments and thoroughfares scattered throughout a city. We reflect upon our place in relation to what was before through the artifacts of the past and our understanding of them. This is a function of memory and cultural symbolic behavior found as an expression of, among other things, our clock, calendar, but also our biological time, as life and death influence culture and vice versa.

This seeming detour into the historical concept of time holds relevance for time as biosemiotic adaptation because the capacity for culturally historical time mediation is heavily influenced by the societal capacity for intergenerational continuity. This continuity relies on generational temporalities such as life and death, parental roles, family and group. Prior to the 20th century, for example, most human generations reproduced on the scale of 15–20 years, while the contemporary average generation interval is more in the realm of 25–35 years for many developed countries. Thus, the concept of a generation itself has shifted, with concomitant implications for intergenerational caretaking and knowledge transfer.

---

[5] These conclusions do not consider the manifold cultural repercussions with which time is enmeshed, which would further affect extraterrestrial *Umwelten* [22.21] [22.31] [22.48].

In this sense, we cannot take for granted a lifeform's temporal structure, let alone on an interstellar scale, as the speed and regularity of generational change might be wildly different. If life expectancy is considerably different, relations between the present and the past could also become divergent to such an extent that interspecies contracts become almost impossible to adhere to. Geopolitical alliances hold up maybe a few hundred years if we are lucky; this is no surprise, as memory of historical events, treatises, and the consequences of circumventing peaceful relations weakens over generations [22.34], while the manner by which this societal memory is transformed over generations is a culturally variable process ranging from religious inundation to simple stories or forgetting altogether. Any biological variation in this generational shift—for example, a generational change every 230 terrestrial years—would require significant changes to our own dealing with our relationship to the past, even in dealing with a very humanoid species. Nobody takes kindly to having their grandchildren renege on a contract, but it happens; however, reneging on a contract you made yourself *not so long ago* is much worse. Agreeing on what is not so long ago, in the context of this thought experiment, is not so universal.

Returning to the critique of Einstein's temporality, from a semiotic perspective the metaphysical extension of relativity theory misses the functional aspect of time in biological, psychological, cultural, and ecological contexts. If we therefore take seriously time as variable in astrobiological investigation, the first step will be to explore and deconstruct the variable role time plays in our own terrestrial complexities.

Eduard Kohn applies biosemiotic insights to establish the counter-intuitive concept of a non-anthropocentric anthropology. Taking seriously the historical enmeshing of ecology and culture, Kohn sets out to detail how the experiences of the Runa people, indigenous to the rainforests of Ecuador, are irreducibly enmeshed with the rhythms and history of the forest [22.29]. By referencing not merely the socio-cultural aspects of the Runa worldview, but encapsulating it as part of a much wider historic-ecosemiotic web of relations, time is shown to be interwoven into the complexity of these nature-cultural interrelations.

By moving away from a purely symbolic anthropology, the rhythms of ecological relations begin to blur distinctions taken for granted by traditional humanistic approaches to dividing scientific disciplines. Similarly, we ought not to take time for granted as a model which shows a generalizable conception of temporality. Instead, we must also take up the variability of a time which plays an active role in evolutionary ecological relations. The recentering of human blind spots and the attempt at a less homogeneous but more relational approach to time might thus lead us towards softening the prejudices of our received and constructed *Umwelt*. This approach will also be fruitful for keeping open the possibility that non-terrestrial life may experience completely heterodox formations of relating past, present, and future. Or that in encountering whatever forms this might take in a biological and ecological context, we make sure that the temporal divergences from our norm are not mistakenly taken as non-life.

How ecology, history, and time come together to highlight the importance of temporality underlies quite a few of the most significant challenges humans are facing. In his book *Hyperobjects* [22.37], Timothy Morton argues that some objects are so spatio-temporally distributed that we lack the ability to localize them—calling them "hyperobjects." A prominent example of this class of objects Morton discusses is anthropogenic climate change. Because the effects of climate change are so varied, non-constant, spread out, and often gradual, climate change diffuses itself beyond the horizons of our evolutionarily instinctual temporalizing of our *Umwelt*. Cultural strategies of narrativizing, however, can pull together these spatio-temporal heterogeneous aspects of climate change and braid them together. Still, the span of this spatio-temporal spread of the narrative tends to create room for counter-narratives and psychological self-defense mechanisms to insert distrust and disassociation from the issue at hand. This capacity—we might even say evolutionary adaptation—to easily disassociate from these kinds of negative thoughts is what *The Worm at the Core* [22.49] describes as the negative side of our capacity not to let thoughts about death continuously obsess our mind. This is generally considered a positive adaptation, as it prevents among other things, a debilitating fear of death. It can, however, also keep out of our mind the long-term negative effects of habits like smoking and drinking, or the consequences of climate change. In a sense this is the "Vital Lie" which underlies much of the human character, the incapacity to deal with the world in its full intensity [22.5]. The further these consequences are diffused over space-time, the easier it is to disassociate from them. Dealing with temporal issues is, therefore, hardly a field which astrobiology must tackle alone. It will be on Earth where many of our issues in dealing with temporal differentiations will be uncovered. It will then also be on Earth where the effects of failing to deal with these issues will manifest themselves before we are likely to bother extraterrestrials.

It is important to note that the issues aired here are not anthropomorphic per se. It is a fundamental and non-circumscribable limit that *Umwelten* are always necessarily less all-encompassing than the set of all possible subjective lifeworlds. Any creature trying to make sense of experience vis-à-vis their *Umwelt* will run into these issues (consciously or not). Understanding the world is thus accomplished through the limitations and prejudices (per the hermeneutic denotation) of the localized biosemiosis which mediates the capacity to characterize *Umwelten* in others. This throws up issues for the fusion of horizons necessary to agree on contingent political relevancies. These issues will only be aggravated in an interstellar context. Strategies of horizon fusion and openness to other temporal rhythms than those favored by our direct environment are thus not only at issue in astrobiological context, but they take center stage in dealing with terrestrial challenges.

22.5 Dimensionality in Biosemiotics

In his novel *Contact*, later turned into a Hollywood movie starring Jodie Foster, Carl Sagan suggests that perhaps space travel has less to do with *moving* in space, than tapping into other adjacent possibilities here and now (e.g., via wormholes, as the movie and astrophysics describes them). Space travel, in this sense, has less to do with external movement (from here to there), than it does, in this instance, with internal movement (figure-ground shifts in perceiving what is already there). This shift from obsessing over constantly new stimuli introduced into our *Umwelt* from a static perspective to exploring other aspects of our *Umgebung* through expanding our *Umwelt* via changing our perspective and internal state, is precisely the work needed for ecological restitution on Earth, and with each other [22.36].

Many ancient organic technologies, from the Incas in Peru to the Vedas in India and the Tibetan Buddhists, understood that connection with other realms beyond those clustered around the Default Mode Network of the brain can be attained through a change in *internal* rather than external state [22.7] [22.50]. Commonly glossed over as meditation, these varieties of experience confront the limits of one's own *Umwelt* via the *Innenwelt*. In other words, niche construction occurs not in the physical *Umwelt*, but via skin-in physiology and psychology, a sort of emotional harmonization of all of the disparate and contradictory tensions. Such traditional practices that enable deliberate curation of the garden of our filters permit certain individuals blessed with the good fortune of natural ability and dedicated, progressive, disciplined commitment, to work towards interfacing regularly with entities and energies (sometimes definable in our preexisting meaning-making apparatuses, sometimes not) which they may exchange with. Such interactions in these traditions are warned to be accompanied by potential dangers, but expanding perceptions are nonetheless deemed essential for keeping the collective culture on a good path to adapt to an evolutionary arms race which may or may not only be confined to this planet alone.

One of the implicit assumptions of deliberately fostering certain interdimensional relations as opposed to others, is similar (analogically) to cultivating a mutualistic gut microbiome (or through education to learn not to fear "eye floaters," the tiny sporadic floating bacteria-like images in vision which are entropic cell debris or protein deposits within the eye's vitreous humor between the vitreous and the retina). Unless we deliberately eat foods which benefit the symbionts in our gut (in a milieu of synthetic ultra-processed food), we may inadvertently be led to being the vessel for the growth of parasites that can overrun our native microflora and fauna, and make us and those around us quite sick. In other words, there are all sorts of pathogens out there that may not ask for consent nor care if we are conscious of the fact that they are engaging in interchange with humans. The hope, or premise, then, is that by consciously cultivating deliberate habits and relations with co-symbionts (mutualists), we can prevent unconsciously harming them and ourselves.

The limits of our temporal sensemaking suggests that just like the snail in the earlier example, semiosis is affecting us in ways both far faster and far slower that we can consciously perceive. This does not mean, however, that we are not being influenced by, and responding to, these phenomena. Instead, the thought experiment of attending to the other dimensions that other organisms might intersect or wholly inhabit, can bring our attention to the ways in which we remain ignorant of the "aboutness" of bacterial experience [22.28], which itself, temporally and otherwise, certainly is quite alien to our phenomenology.

Taking seriously the limitations of the human or any *Umwelt* vis-à-vis our *Umgebung*, when we extend this analysis interdimensionally towards that which we cannot see with the senses we have been trained to constrict, interspecies interstellar interactions do not necessarily operate in a wholly disembodied realm. To the contrary, these other possible dimensions (e.g., 26-4 = 22) we are normally not taught to be aware of in globalized homogenized Westernized culture (*qua* the culture industry [22.24] [22.32]), do not just cease to exist for us or cease to supervene because we actively ignore them in our daily lives and sciences (with the exception of theoretical physics). Instead, they may interact with us in ways we are simply not aware of, perhaps that we have been *trained* to become not aware of (perhaps in order to protect and shroud their status as agents in anonymity).

Such diversionary tactics are no stranger to the natural world. For example, the cuckoo bird interpolates its own eggs in another's nest, or parasitic blood suckers like leeches release chemicals to condition the blood not to clot on feeding. This is neither a rallying cry to look out for imposters among us, nor a warning to watch out for interdimensional vampires. Rather, exploring these temporal and dimensional aspects of interspecies experience is a call for taking seriously the limits of both the known and its effect on our expectations, both terrestrially, and beyond.

22.6 Conclusion

Approaching the legitimate concerns of Fermi's Paradox, perhaps we are already long in contact with other intelligent extra-planetary organisms, but because they do not show up as our popular imagination represents them, and announce themselves to us in the way we expect from conquistadors, we overlook their presence. Fermi's Paradox also begs the question of intelligence; are we talking about just "intelligent like us"? (Which may be a rather low bar, if we end up blowing ourselves up for having, in the words of E.O. Wilson, "god-like technology, medieval institutions, and stone age instincts.") Or, do forms of distributed intelligence involve ethical stopgaps we cannot even imagine? While an ethical extraterrestrial might only interact with us based on our awareness and consent—that is, having self-reflective consciousness that we are in fact in the presence of an alien intelligence and that we agree to tentatively engage with that organism in an exploratory interpenetration—this may not necessarily occur only or even primarily on those levels of consciousness or existence we are most habituated to pay attention to (especially if doing so would trigger our resistance to an ideological phantom image we might be conditioned to conjure up—such as the perception of a loss of control—to interpret the experience we are actually having of semiosis we have not learned how to integrate on a conscious level). Indeed, the primary channels of communication of an extraterrestrial intelligent being may be (a) nonlocal – able to transmit over long distances or times, and/or operating on a (b) *kairos* rather than *kronos* temporality. (*Kronos* being a relatively recent adaptation on this planet, and only among one species.)

In Daniel Quinn's *Ishmael* [22.44], he suggests that our goal as a species is to help other life forms on earth gain sufficient agency to become sentient.[6] Just as we evolved from our primate ancestors, and are fervently if misguidedly trying to evolve ourselves out of our own self-destruction [22.43], we could design our habitat to be conducive to multispecies selection, in a nonzero-sum game [22.8]. Consilience Institute director Daniel Schmachtenberger [22.59] has phrased our current predicament as "institutional choice architecture which is perverse to the sum totality of what the world needs." Going a step further, in addition to engineering our world to meet the various polycrises that colonization and domination have wrought [22.20], if we take the Extended Evolutionary Synthesis of biology seriously, we must also for selfish reasons, but also could for altruistic ones, de-pollute and inoculate the earth to serve and support succession ecologies to replenish and evolve further.

Could we do this? It suggests that the creator role, historically limited to gods, lies not exclusively in geoengineering or genetical modification, but through engaging in semiosis with all life to the point where other lifeforms show up as sentient to us and this collective sentience provides us all with yet higher harmonics in the echelons of relatedness. The greater-than-the-sum-of-its-parts attainment of multimodal nonintentional pan-species semiosis may be the potentiality of evolving the natural order for us to become more in sync with an ecosystem of semiosis. Nonconscious semiosis could even be positioned as a higher caliber of semiosis, because the burden of trying to account for semiosis is offloaded efficiently and immaculately, so that organisms can live their purpose without interference

[6] For an array of possible mechanisms of how this could work, see Baluška and Mancuso [22.1], Barbieri [22.2], Cheney and Seyfarth [22.13], Collins [22.14], Corning and Szathmáry [22.15], and Geus and Höxtermann [22.22].

from the divided mind and trauma-filled body.[7] When semiosis simply has become a reflex, part of the infinite reflex arc of harmonic response that Dewey spoke of [22.18], attention to nonconscious stimuli while obviating the need or burden for thought, may signal more advanced evolutionary action in particular milieus. The key to flourishing in such contexts would be attending to multispecies ecological succession which allow life and death to proceed uncomplicated, and without accumulated traces or remainders. Such semiotic exactness that comes from accurate nonintentional action creates an unmistakable clarity of semiosis, manifesting ethical demands for those organisms and those present within but also adjacent to their *Umwelt*.

The possible discrepancies between our own temporal—or dimensional—attention and that of all possible extraterrestrial ordering will likely require more than a search for terrestrial modes of life in an extraterrestrial context. Yet, if we are not even ready to deal with those variations and relations found on our own evolutionary turf, how do we expect to be capable of dealing with someone else's?

Declaration

The authors declare no conflicts of interest.

References

[22.1] Baluška, F., Mancuso, S. (2009) Deep evolutionary origins of neurobiology: Turning the essence of "neural" upside-down. *Communicative & Integrative Biology* 2, 60–65. doi: 10.4161/cib.2.1.7620

[22.2] Barbieri, M. (1987) *The semantic theory of evolution*, 2. print. Harwood Acad. Publ, Chur.

[22.3] Barratt, A.E., Gonsalves, L., Turbill, C. (2023) Winter torpor and activity patterns of a fishing bat (Myotis macropus) in a mild climate. *Journal of Mammalogy* 104, 76–85. doi: 10.1093/jmammal/gyac061

[22.4] Barreiro, L.B., Quintana-Murci, L. (2010) From evolutionary genetics to human immunology: how selection shapes host defence genes. *Nature Reviews Genetics* 11, 17–30. doi: 10.1038/nrg2698

[22.5] Becker, E. (1997) *The Denial of Death*, 1. paperback ed. Free Press Paperbacks, New York.

[22.6] Bergson, H. (1999) *Duration and Simultaneity: Bergson and the Einsteinian Universe*, Clinamen Press Ltd., Manchester, 2nd.

[22.7] Brewer, J.A., Worhunsky, P.D., Gray, J.R., Tang, Y.-Y., Weber, J., Kober, H. (2011) Meditation experience is associated with differences in default mode network activity and connectivity. *Proc Natl Acad Sci* 108, 20254–20259. doi: 10.1073/pnas.1112029108

[7] This is commonly, if reductively, known as "cognitive offloading." The fact that when I tie my shoe, I do not have to do this consciously, but my fingers fumble to find the laces on their own, does not constrain me, but rather offloads to my basal ganglia what otherwise would occupy my much more metabolically expensive prefrontal cortex. In other words, turning our philosophy of mind on its head, it could very well turn out to be the case that the more actions we engage in unconsciously, the freer we are to contemplate other things. Of course, such an ontological flip would also challenge the lionization of control of one's environment and others as the *sine qua non* of intelligence, agency, and (individuated) evolution.

[22.8] Bull, J.W., Maron, M. (2016) How humans drive speciation as well as extinction. *Proceedings of the Royal Society B: Biological Sciences* 283, #20160600. doi: 10.1098/rspb.2016.0600

[22.9] Cahill, T., da Silveira, W.A., Renaud, L., Williamson, T., Wang, H., Chung, D., Overton, I., Chan, S.S.L., Hardiman, G. (2021) Induced Torpor as a Countermeasure for Low Dose Radiation Exposure in a Zebrafish Model. *Cells* 10, #906. doi: 10.3390/cells10040906

[22.10] Callicott, J.B. (2015) Science as Myth (Whether Sacred or Not), Science as Prism. *Journal for the Study of Religion, Nature and Culture* 9, 154–168. doi: 10.1558/jsrnc.v9i2.27264

[22.11] Canales, J. (2015) *The physicist & the philosopher: Einstein, Bergson, and the debate that changed our understanding of time.* Princeton University Press, Princeton, NJ Oxford.

[22.12] Chela-Flores, J. (2023) Frontiers of astrobiology and the humanities. *International Journal of Astrobiology*, 22(3), 1–9. doi: 10.1017/S1473550422000453

[22.13] Cheney, D.L., Seyfarth, R.M. (2008) *Baboon metaphysics: The evolution of a social mind, Paperback ed.* University of Chicago Press, Chicago, Ill.

[22.14] Collins, C. (2013) *Paleopoetics: The Evolution of the Preliterate Imagination*. Columbia University Press.

[22.15] Corning, P.A., Szathmáry E. (2015) Synergistic selection: A Darwinian frame for the evolution of complexity. *Journal of Theoretical Biology* 371, 45–58. doi: 10.1016/j.jtbi.2015.02.002

[22.16] Deleuze, G. (1990) *Bergsonism, Reissue edition*. Zone Books, New York.

[22.17] Demets, R. (2012) Darwin's Contribution to the Development of the Panspermia Theory. *Astrobiology* 12, 946–950. doi: 10.1089/ast.2011.0790

[22.18] Dewey, J. (1896) The reflex arc concept in psychology. *Psychol Rev* 3, 357–370. doi: 10.1037/h0070405

[22.19] Dunér, D. (2018) Semiotics of biosignatures. *Southern Semiotic Review* 47–63. doi: 10.33234/ssr.9.4

[22.20] Eisler, R., Fry D.P. (2019) *Nurturing Our Humanity: How Domination and Partnership Shape Our Brains, Lives, and Future*. Oxford University Press, New York.

[22.21] Fox-Keller, E. (1995) *Refiguring life: Changing metaphors in twentieth-century biology*. Columbia University Press, New York.

[22.22] Geus, A., Höxtermann, E. (2007) *Evolution durch Kooperation und Integration: zur Entstehung der Endosymbiosetheorie in der Zellbiologie; Faksimiles, Kommentare und Essays*. Basilisken-Presse, Marburg.

[22.23] Guidetti, R., Altiero, T., Rebecchi, L. (2011) On dormancy strategies in tardigrades. *Journal of Insect Physiology* 57, 567–576. doi: 10.1016/j.jinsphys.2011.03.003

[22.60] Hendlin, Yogi Hale (2016) Multiplicity and Welt. *Sign Systems Studies*, 44 (1/2), 94-110.

[22.24] Horkheimer, M., Adorno, T.W. (2002) *Dialectic of enlightenment: Philosophical fragments*. Stanford University Press, Stanford.

[22.25] Jain, S., Luthra, M., Sharma, S., Fatima, M. (2020) Sonic Fire Fighting System. In: *International Conference on Computer Communication and Informatics (ICCCI)*. 1–5.

[22.26] Jameson, K.G., Olson, C.A., Kazmi, S.A., Hsiao, E.Y. (2020) Toward Understanding Microbiome-Neuronal Signaling. *Mol Cell* 78, 577–583. doi: 10.1016/j.molcel.2020.03.006

[22.27] Janković, S., Katić, A., Ćirković, M.M. (2022) Gaia as Solaris: An Alternative Default Evolutionary Trajectory. *Origins of Life and Evolution of Biospheres* 52, 129–147. doi: 10.1007/s11084-022-09619-6

[22.28] Kauffman, S. (2003) Molecular autonomous agents. *Philosophical Transactions of the Royal Society A*, 361, 1089–1099. doi: 10.1098/rsta.2003.1186

[22.29] Kohn, E. (2013) *How forests think: toward an anthropology beyond the human*. University of California Press, Berkeley.

[22.30] Kuperus, G. (2007) Attunement, Deprivation, and Drive. In: *Phenomenology and the Non-Human Animal*. C. Painter, C. Lotz, (eds.) Springer, Dordrecht, the Netherlands, 13–27.

[22.31] Lakoff, G., Johnson, M. (2003) *Metaphors We Live By.* University of Chicago Press, Chicago. 1st.

[22.32] Latour, B. (2018) *Down to Earth: Politics in the New Climatic Regime, 1st edition.* Polity, Cambridge, UK ; Medford, MA.

[22.33] Lewis, N.E., Hixson, K.K., Conrad, T.M., Lerman, J.A., Charusanti P., Polpitiya, A.D., Adkins, J.N., Schramm, G., Purvine, S.O., Lopez-Ferrer, D., Weitz, K.K., Eils, R., König, R., Smith, R.D., Palsson, B.Ø. (2010) Omic data from evolved *E. coli* are consistent with computed optimal growth from genome-scale models. *Molecular Systems Biology*, 6, #390. doi: 10.1038/msb.2010.47

[22.34] Margalit, A. (2004) *The ethics of memory*, Harvard University Press, Cambridge, Mass. London.

[22.35] Marietta, E., Horwath, I., Taneja, V. (2018) Microbiome, Immunomodulation, and the Neuronal System. *Neurotherapeutics* 15, 23–30. doi: 10.1007/s13311-017-0601-4

[22.36] McGilchrist, I. (2009) *The Master and his Emissary*. Yale University Press, New Haven.

[22.37] Morton, T. (2013) *Hyperobjects: philosophy and ecology after the end of the world.* University of Minnesota Press, Minneapolis.

[22.38] Nomura, N., Muranaka, T., Tomita J., Matsuno, K. (2018) Time from Semiosis: E-series Time for Living Systems. *Biosemiotics* 11, 65–83. doi: 10.1007/s12304-018-9316-0

[22.39] Nowack, J., Stawski, C., Geiser, F. (2017) More functions of torpor and their roles in a changing world. *Journal of Comparative Physiology B*, 187, 889–897. doi: 10.1007/s00360-017-1100-y

[22.40] Obara, T., Hendlin, Y.H., Matsumoto, H. (2021) *Omics and the Biosemiotic Interaction of Food with Body, Mind, and Health.* In: Hendlin Y.H., Hope J. (eds) Food and Medicine. Springer International Publishing, Cham: 183–196.

[22.41] Pain, S.P. (2008) Inner representations and signs in animals. In: *Introduction to biosemiotics.* M. Barbieri, (ed.) Springer, Dordrecht, the Netherlands: 409–455.

[22.42] Pennisi, E. (2012) ENCODE Project Writes Eulogy for Junk DNA. *Science* 337:1159–1161. doi: 10.1126/science.337.6099.1159

[22.43] Persson, I., Savulescu, J. (2012) *Unfit for the Future: The Need for Moral Enhancement.* Oxford University Press, Oxford.

[22.44] Quinn, D. (1995) *Ishmael, Reissue edition*. Bantam, New York, NY.

[22.45] Schleiermacher, F. (1799) *On religion: Speeches to its cultured despisers.* Cambridge University Press, Cambridge UK; New York, NY, USA.

[22.46] Segner, H. (2012) *Fish: Nociception and pain; a biological perspective.* Federal Office for Buildings and Logistics, Bern.

[22.47] Sharov, A.A., Gordon, R. (2018) Life Before Earth. In: *Habitability of the Universe Before Earth.* Elsevier: 265–296.

[22.48] Sheets-Johnstone, M. (2016) *Insides and Outsides: Interdisciplinary Perspectives on Animate Nature.* Imprint Academic, 1st.

[22.49] Solomon, S., Greenberg, J., Pyszczynski, T. (2015) *The Worm at the Core: On the Role of Death in Life.* Random House, New York.

[22.50] Tomasino, B., Chiesa, A., Fabbro, F. (2014) Disentangling the neural mechanisms involved in Hinduism- and Buddhism-related meditations. *Brain and Cognition* 90, 32–40. doi: 10.1016/j.bandc.2014.03.013

[22.51] Tønnessen, M., Beever, J., Hendlin, Y. (2015) Introducing Biosemiotic Ethics. *Zeitschrift fur Semiotik* 37, 3–12. doi: 10.14464/zsem.v37i3-4.362

[22.52] von Uexküll, J. (2010) *A Foray into the Worlds of Animals and Humans: With a Theory of Meaning.* Minneapolis London.

[22.53] Vukotić, B., Gordon, R. (2021) Life in the Milky Way: The Panspermia Prospects. In: *Planet Formation and Panspermia, 1st ed.* B. Vukotić, R. Gordon and J. Seckbach, (eds.) Wiley: 41–51.

[22.54] Watson, A.J., Lovelock, J.E. (1983) Biological homeostasis of the global environment: the parable of Daisyworld. *Tellus B,* 35B, 284–289. doi: 10.1111/j.1600-0889.1983.tb00031.x

[22.55] Weilgart, L.S. (2007) The impacts of anthropogenic ocean noise on cetaceans and implications for management. *Canadian Journal of Zoology* 85, 1091–1116. doi: 10.1139/Z07-101

[22.56] Whitehead, A.N. (1915) Space, Time, and Relativity. *Proceedings of the Aristotelian Society* 16, 104–129.

[22.57] Wiech, K., Ploner, M., Tracey, I. (2008) Neurocognitive aspects of pain perception. *Trends in Cognitive Sciences* 12, 306–313. doi: 10.1016/j.tics.2008.05.005

[22.58] Wild, M. (2012) *Fische: Kognition, Bewusstsein und Schmerz ; eine philosophische Perspektive.* Bundesamt für Bauten und Logistik BBL, Bern.

[22.59] Schmachtenberger, D. (2022) *Bend Not Break Part 5. The Great Simplification* #50. https://www.youtube.com/watch?v=Kep8Fi_rUUI

Index